pre-calculus
mathematics

pre-calculus mathematics

hal g. moore
BRIGHAM YOUNG UNIVERSITY

john wiley & sons

NEW YORK   LONDON   SYDNEY   TORONTO

Library of Congress Cataloging in Publication Data:

Moore, Hal G., 1929–
   Pre-calculus mathematics.

   1.   Mathematics—1961–   I. Title.
QA39.2.M66   515   72–11635

ISBN 0-471-61455-6

Printed in the United States of America

10 9 8 7 6 5 4 3 2 1

preface

The primary purpose of this text is to prepare students to study calculus. A closely allied secondary purpose is to teach mathematics. The book has been written with the recommendations of the CUPM in mind; its approach is the study of elementary functions and coordinate geometry. Yet the material here includes *more* than would be required for the intensive Mathematics 0 course recommended by the committee. Most of the material covered in a standard college algebra and trigonometry course is here. (The chief omission is probability.) There is even more trigonometry than is usual in such standard courses. Therefore, courses ranging from the short 45-class-day Math 0 to a year's experience for well prepared high school students are possible using this text.

While topics are covered from a modern point of view, the plan is to build each concept intuitively before making any attempt to be rigorous. In fact, rigor for its own sake has been avoided. The traditional ("new mathematics") treatment of the real number axioms and basic theorems derived from them, found in the beginning of most standard texts, has been placed in the appendix. I have done this so that the student can get to the "meat" of the important mathematical ideas early rather than consuming time on side trips into the foundations of algebra. A brief introduction to complex numbers in terms of ordered pairs of real numbers, does, however, appear in Chapter 1.

On the other hand, people take mathematics courses in mathematics departments from mathematicians, presumably because mathematicians think somewhat differently from normal people. Mathematical thought does have a contribution to make to a person's education. I like to think of myself as a mathematician as well as a teacher. I also believe that some students will not end their mathematical study with this course, or with calculus, but some might even find their way into a linear or an abstract algebra course in the future. This book ought to prepare them, too. Hence, while I have tried not to impose a well-polished logical structure, arrived at after centuries of refinement, on the mathematically unpolished and unrefined student, I see no harm in alluding to this structure from time to time.

In relegating the axiomatic treatment of the real number system to the appendix and in omitting most of the detail in the construction of the complex number system, I frankly assumed that these details were, or will be, covered in some other course for those who are interested.

It is also frankly expected that the student who uses this book will have completed a year and one-half of high school algebra and the usual high school geometry before beginning this course. The first part of Chapter 1 may, therefore, repeat old material. The teacher should decide how much of this chapter is essential. It has been my experience, in over twenty years of teaching mathematics (Does that indicate that I am not to be trusted?), that in both

preface

calculus and in pre-calculus course of this nature and particularly in any further mathematical study, the topics in Chapter 1 and those of Chapter 2 are precisely those which the student needs to learn or review. Most of the terms used in the rest of the book are explained in Chapter 1. Chapter 2 is, admittedly, a bit of a digression. It could be postponed, used as an appendix, assigned as reading, or omitted. I like to do this material in the order in which it appears in the text.

Throughout the book there are a great many worked examples and more than an ample supply of exercises. These exercises may be "drill" for the purpose of fixing concepts, a review of the manipulative aspects of basic algebra, or problems of a theoretical nature which challenge the student to think about the mathematical ideas involved. Answers to many of the exercises appear at the end of the book.

I have used this material both in large classes, taught by the lecture-recitation-section method, and in traditionally organized classes. There is enough material here for a course which meets daily for one semester. A more relaxed pace is possible by omitting some of the material in Chapters 2, 6, or 7, depending on the goals of the particular class, or by meeting three times a week for two semesters or two quarters. Shorter courses can also be constructed using this material. Some possible uses for one-quarter or one-semester courses in *pre-calculus mathematics* or in *elementary functions and coordinate geometry* are suggested in the table on page vii, the selection of optional sections or chapters to be determined by the needs of the individual class.

I wish to express my deep appreciation to the following people who reviewed the manuscript and made many suggestions for improvement. Most of their suggestions were gratefully accepted; any remaining defects are clearly not their responsibility. Professor Gerald M. Armstrong of Brigham Young University, Provo, Utah; Professor Robert W. Negus, Rio Hondo Junior College, Whittier, Calif.; Professor Constantine Georgakis, DePaul University, Chicago, Ill.; Professor Charles H. Neumann, Alpena Community College, Alpena, Mich.; Professor John Morton, Baylor University, Waco, Tex.; Professor Don E. Edmondson, University of Texas, Austin, Tex.; and Professor Harold S. Engelsohn, Kingsborough Community College, Brooklyn, N.Y. My thanks also go to Mary Heywood who typed most of the manuscript, and to Frederick C. Corey and the staff of John Wiley & Sons, for their considerable help and valuable suggestions.

Provo, Utah   H. G. M.

POSSIBLE COURSE OUTLINES

	pre-calculus mathematics		elementary functions and coordinate geometry	
	Daily for 1 Quarter	Daily for 1 Semester	3 Days/Week For 1 Quarter	3 Days/Week For 1 Semester
	CHAPTER 1: Section 1.5 is optional.	CHAPTER 1:	CHAPTER 1: Concentrate on Sections 1.6–1.9.	CHAPTER 1: Cover Sections 1.1–1.5 lightly. Concentrate on Sections 1.6–1.9.
	CHAPTER 2: is optional	CHAPTER 2: is optional.	CHAPTER 2: omit.	CHAPTER 2: omit.
	CHAPTER 3:	CHAPTER 3:	CHAPTER 3:	CHAPTER 3:
	CHAPTER 4:	CHAPTER 4:	CHAPTER 4: Combine Sections 4.4 & 4.5.	CHAPTER 4:
	CHAPTER 5: Omit Sections 5.1 & 5.12. Combine Sections 5.2 & 5.3 with Section 5.11.	CHAPTER 5: Sections 5.1 & 5.12 are optional.	CHAPTER 5: Omit Sections 5.1 & 5.12. Combine Sections 5.2 & 5.3 with Section 5.11.	CHAPTER 5: Sections 5.1 & 5.12 are optional.
	CHAPTER 6: is optional.	CHAPTER 6:	CHAPTER 6: omit.	CHAPTER 6: is optional.
	CHAPTER 7: is optional.	CHAPTER 7: Sections 7.5, 7.7, & 7.8 are optional.	CHAPTER 7: Sections 7.5, 7.7, & 7.8 are optional. Combine Sections 7.1 & 7.6 and Sections 7.3 & 7.4.	CHAPTER 7: Sections 7.5, 7.7, & 7.8 are optional.

contents

contents

contents

fundamentals

In this chapter we shall consider some fundamental ideas which form the basis for the vocabulary and development of the rest of the book and for further study of mathematics. While you are not expected to completely master all of the details of the concepts presented here, you should be familiar enough with them to be able to apply them to your subsequent mathematical work.

1.1. sets: a brief look

The notation and terminology of set theory forms the basic vocabulary of all present-day mathematics. Developed in the nineteenth century principally by George Boole (1815–1864) and Georg F. L. P. Cantor (1845–1918), it has had profound effect upon both the method and content of twentieth century mathematics. The inclusion of certain aspects of this topic in elementary and secondary school mathematics courses and texts seems, at times, the sole criterion for judging the class to be "the new math."

It is fortunately not necessary for our purposes to travel the long and perilous road of a thorough treatment of set theory. We adopt an unsophisticated approach which will enable us to make use of its helpful language in the rest of this book.

A **set** in mathematics means about what it does in ordinary usage in English; i.e., a collection of things under some unifying property: a set of china, a chess set, a set of golf clubs, a set of carving knives, "the jet set," etc. Besides *collection*, other synonyms for set are *herd* (of cattle), *gaggle* (of geese), *flock* (of sheep), *clique*, and so on. The individual objects which make up the set are called its **elements** or **members**.

fundamentals

In this book, sets generally will be denoted by capital letters, $A, B, C, \ldots$ etc., while the elements are designated with lower case letters, $a, b, x, y, z, \ldots$. The sentence

The object x belongs to (is an element or member of) the set S,

will be abbreviated as

$$x \in S$$

The negation of this will be written $x \notin S$; that is, "x does not belong to S."

We have not given a mathematical definition of a set; indeed, we take *set* to be a "*primitive term.*" You will gain an intuitive conception of sets by looking at the following examples.

examples of sets

(1) The set of all students at this school.
(2) The set of all planets in our solar system.
(3) The set of all red-headed coeds taking this course.
(4) The set of all letters in the English alphabet.
(5) The set of all citizens of the United States.
(6) The set of all months in the year.
(7) The set N of all natural numbers.
(8) The set Z of all integers: positive, negative, or zero.
(9) The set Q of all rational numbers.
(10) The set R of all real numbers.
(11) The set C of all complex numbers.
(12) The set of all real numbers x satisfying $0 \leq x \leq 1$. This set is called the **unit interval**.
(13) The set of all former presidents of the United States.
(14) The set of all possible bridge hands.
(15) The set of all the letters in the word "Mississippi." ●

We shall discuss the sets listed in (7)–(11) above in greater detail in the next section.

In describing certain sets, some additional notation is frequently handy. One generally uses braces around either a list of the elements of the set, the *roster notation*, as

$$A = \{a, b, c\}$$

for the set consisting of the first three letters of the (English) alphabet; or, if all the elements x of the set have some distinguishing property $P(x)$ which they satisfy, we use the *set builder notation*

$$\{x \mid x \text{ has } P(x)\}$$

For example, the set of all the former presidents of the United States could be written as

{Washington, Adams, Jefferson, . . . , Eisenhower, Kennedy, Johnson, . . .}

or

$$\{p \mid p \text{ is a former president of the United States}\}$$

sets: a brief look

Two sets A and B are called equal,

$$A = B$$

if and only if they contain precisely the same elements. Thus, $\{2,4,6,8\} = \{8,8,6,2,4,\}$, even though the symbol 8 appears twice in the second set. The set (15) of all letters in the word "Mississippi" is most economically written $\{M,i,s,p\}$.

A set A is called a **subset** *of a set B,*

$$A \subseteq B$$

provided that each member of A also belongs to B; i.e., $x \in A$ implies $x \in B$. Note that $A = B$ if, and only if, $A \subseteq B$ and $B \subseteq A$. Further note that for every set S, $S \subseteq S$. If $A \subseteq B$, *but $A \neq B$, we often write $A \subset B$ and call A a* **proper subset** *of B.*

It is also convenient to consider a special set, **the empty set** $\varnothing$, which has no elements; it is empty. From the definition of subset, we see that $\varnothing \subseteq S$ for every set S.

One should be very careful to avoid one of the perils of set theory: to distinguish between an object x and the set $\{x\}$ whose only element is x. This is somewhat akin to differentiating between a student and a special reading course for which he is the only registrant, or between a can containing one peanut and the peanut. All the sets considered in this text will be subsets of some large universal set—usually the set of real numbers—called the **universe,** U.

Use of set theoretic ideas is frequently aided by means of pictorial representations called *Venn diagrams.* For example, if A and B are sets (subset of U) such that $A \subset B$, we can picture this in the Venn diagram of Figure 1.1.1.

That part of the universe U which is not in the set B is called the **complement** of B and is denoted by $\tilde{B}$. In Figure 1.1.1, $\tilde{B}$ is everything outside of the large circle. The annular region inside the circle B (in Fig. 1.1.1) but outside of the circle A is the complement of A in B and is denoted by $B \backslash A$ (some authors use $B - A$).

Given two sets S and T we can consider new sets formed by them. We give the formal definitions first and then sketch Venn diagrams which illustrate them.

(1) The **complement** of T in S, $S \backslash T$ is defined by

$$S \backslash T = \{x \,|\, x \in S \text{ and } x \notin T\}$$

(2) The **union** of S and T, $S \cup T$ is defined by

$$S \cup T = \{x \,|\, x \in S \text{ or } x \in T\}$$

Here "or" means "either $x \in S$ or $x \in T$ or x is in *both.*"

figure 1.1.1

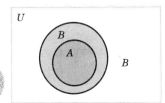

fundamentals

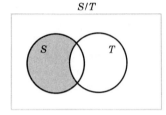
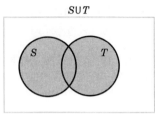
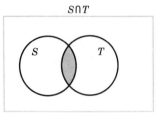

figure 1.1.2

(3) The **intersection** of S and T, $S \cap T$ is defined by

$$S \cap T = \{x \mid x \in S \text{ and } x \in T\}$$

examples

(16) Let $S = \{a,b,c,d\}$, $T = \{b,d,f,h\}$, $R = \{a,e,i\}$. Then we have the following.

Union:
$$S \cup T = \{a,b,c,d,f,h\}$$
$$S \cup R = \{a,b,c,d,e,i\}$$
$$T \cup R = \{a,b,d,e,f,h,i\}$$

Intersection:
$$S \cap T = \{b,d\}$$
$$S \cap R = \{a\}$$
$$T \cap R = \varnothing$$

Complement:
$$T \backslash S = \{f,h\}$$
$$S \backslash T = \{a,c\}$$
$$S \backslash R = \{b,c,d\}$$
$$R \backslash S = \{e,i\}$$

(17) Let

$$P = \{p \mid p \text{ is a former president of the United States}\}$$
$$A = \{l \mid l \text{ is a world politician (leader) who was assassinated}\}$$
$$V = \{v \mid v \text{ is a former vice president of the United States}\}$$

Then we have

$P \cap V = \{$Adams, Jefferson, Van Buren, Tyler, Fillmore, Andrew Johnson, Arthur, T. Roosevelt, Coolidge, Truman, Lyndon Johnson$\}$

$P \cap A = \{$Lincoln, Garfield, McKinley, Kennedy$\}$

(18) Let T be the set of positive integral multiple of 2,

$$T = \{2n \mid n \in N\}^*$$

4

*Remember that N denotes the set of Example 7 consisting of the natural numbers

$$N = \{1, 2, 3, 4, \ldots\}$$

sets: a brief look

and S be the set of positive integral multiples of 3, $S = \{3k | k \in N\}$.
We have $S \cup T = \{x | x = 2n \text{ or } x = 3n, \text{ some } n \in N\}$; i.e.,

$$\{2,3,6,8,9,10,12, \ldots\}$$

$S \cap T = \{x | x = 6k, k \in N\}$; i.e.

$$\{6,12,18, \ldots\} \quad \bullet$$

Consider an arbitrary set X. The set of all subsets of X is denoted by 2^X and is called the **power set** of X. For example, let $X = \{\text{Tom, Dick, Harry}\}$. The the subsets of X are X, $\emptyset$, $\{\text{Tom}\} = T$, $\{\text{Dick}\} = D$, $\{\text{Harry}\} = H$, $\{\text{Tom, Dick}\} = R$, $\{\text{Tom, Harry}\} = S$, and $\{\text{Dick, Harry}\} = W$. Thus $2^X = \{X, \emptyset, T, D, H, R, S, W\}$.

If $S \cap T = \emptyset$, then S and T are called **disjoint**.

Another method by which new sets are obtained from old ones involves the **Cartesian* product** of two sets. If A and B are two sets, then we define

$$A \times B = \{(x, y) | x \in A \text{ and } y \in B\}$$

This set will be exploited in the particular case where A and B are each the set R of real numbers.

example

(19) Let $A = \{a,p,m,e\}$, $B = \{p,e,a,r\}$.
Then $A \cup B = \{a,p,m,e,r\}$
$A \cap B = \{a,p,e\}$
$B \setminus A = \{r\}$
$A \setminus B = \{m\}$
$A \times B = \{(a,p), (a,e), (a,a), (a,r), (p,p), (p,e), (p,a), (p,r),$
$(m,p), (m,e), (m,a), (m,r), (e,p), (e,e), (e,a), (e,r)\}$

Note that none of (m,m), (r,r), (p,m) belongs to $A \times B$. $\bullet$

These operations on sets have many formal consequences, some of which are included in the exercises. We forego, however, a complete study of the "algebra of sets."

exercises

1. Use the roster method to describe each of the following sets.

(a) the set of all vowels in the English alphabet
(b) $\{x | x \text{ is an integer between } -3 \text{ and } 3\}$
(c) $\{x | x^2 - 3x + 2 = 0, x \text{ a real number}\}$
(d) $\{x | x + 1 = 2x, x \text{ an integer}\}$
(e) $\{z | z^4 = 16, z \text{ a real number}\}$

2. Given the following four sets,

$$A = \{1, 2\} \ B = \{1, 2, 3\} \ C = \{2, 3\} \ D = \{3\}$$

5

*After the French mathematician René Descartes (1596–1650).

fundamentals

describe:

(a) $A \cup B$

(b) $A \cap B$

(c) $B \cap C$

(d) $B \backslash A$

(e) $B \backslash D$

(f) $A \times B$

(g) $A \times C$

(h) $A \times (B \times C)$

(i) $A \cup (B \cup C)^{\circ}$

(j) $A \cap (B \cup C)$

(k) $A \cup (B \cap C)$

(l) $A \cap (B \cap C)$

(m) 2^A

(n) 2^B

(o) 2^C

(p) 2^D

3 Determine each of the following subsets of the set P of all former presidents of the United States (see any almanac).

(a) $\{p \mid p$ is still living$\}$

(b) $\{p \mid p$ served more than two terms$\}$

(c) $\{p \mid p$ served more than one term$\}$

(d) $\{p \mid p$ served two nonconsecutive terms$\}$

(e) $\{p \mid p$ is a woman$\}$

4. Given the sets of Exercise 2, which of the following are correct?

(a) $A \subseteq B$

(b) $B \subseteq C$

(c) $D \subseteq C$

(d) $A \cup C = B$

(e) $A \cap D = \varnothing$

(f) $A \subseteq A \times B$

(g) $A \times D \subseteq A \times B$

(h) $A \cup B = B \cup A$

5. Use Venn diagrams and shade the region $A \cap B$ when each of the following is true.

(a) $A \subset B$

(b) $B \subset A$

(c) $A \cap B = \varnothing$

(d) $A = B$

(e) $A = \tilde{B}$

(f) $A \backslash B = \varnothing$

6. Use Venn diagrams and shade the region $A \cup B$ for each of the situations in Exercise 5.

7. (a) Write down all the subsets of the set $\{x, y, z\}$. (There are eight of them.)

(b) How many elements are there in 2^X if X has n elements, n a positive integer?

8. Using Venn diagrams verify the following concerning any subsets A, B, C of the universe U.

(a) $A \cup B = B \cup A$ $\left.\right\}$ *commutative laws*

(b) $A \cap B = B \cap A$

(c) $A \cup (B \cup C) = (A \cup B) \cup C$ $\left.\right\}$ *associative laws*

(d) $A \cap (B \cap C) = (A \cap B) \cap C$

(e) $A \cap (B \cup C) = (A \cap B) \cup (A \cap C)$ $\left.\right\}$ *distributive laws*

(f) $A \cup (B \cap C) = (A \cup B) \cap (A \cup C)$

(g) $\varnothing \subseteq A \cap B \subseteq A \subseteq A \cup B$

(h) $\varnothing \subseteq A \cap B \subseteq B \subseteq A \cup B$

9. Use the formal definitions given to prove each of the following statements concerning sets A, B, C.

(a) $A \subseteq B$ and $B \subseteq C$ implies $A \subseteq C$.

(b) $A \subseteq \varnothing$ if, and only if, $A = \varnothing$.

(c) $B \backslash A = \tilde{A} \cap B$.

(d) $A \subseteq B$ if, and only if, $\tilde{B} \subseteq \tilde{A}$.

10. Use Venn diagrams to indicate that the following *De Morgan laws* hold for arbitrary subsets A, B, and X of U.

(a) $X \backslash (A \cup B) = (X \backslash A) \cap (X \backslash B)$

(b) $X \backslash (A \cap B) = (X \backslash A) \cup (X \backslash B)$

sets of numbers

11. Answer each of the following questions.

    (a) If A is any set, is $A \subseteq A$?
    (b) If A is any set, is $A \in A$?
    (c) If A is any set, does $A \subset \varnothing$?
    (d) How many elements are there in $\varnothing$?
    (e) How many elements are there in $\{\varnothing\}$?
    (f) If A is any set, is $\varnothing \in A$?
    (g) If A is any set, is $\varnothing \subset A$?
    (h) For any set A, $A \cap \tilde{A} = ?$
    (i) What is $\tilde{\varnothing}$?
    (j) For any set A, what is $(\tilde{\tilde{A}})$?
    (k) For any set A, $A \cup A = ?$
    (l) For any set A, $A \cap A = ?$

12. Prove each of the following formally.

    (a) $(\widetilde{A \cap B}) \subset \tilde{A} \cup \tilde{B}$.
    (b) If $A \subset B$, then $A \cup B = B$.
    (c) If $A \cap B = \varnothing$, then $A \cup \tilde{B} = \tilde{B}$.
    (d) If $A \subset B$, then $A \cap B = A$.

1.2. sets of numbers

In this text we shall, for the most part, be concerned with sets of numbers rather than sets in general. In particular we shall frequently use the set of real numbers for our universe. As was the case with our study of sets, we shall approach the real numbers on a largely intuitive basis. A rigorous development of the real number system more properly belongs to advanced courses. One can refer to such texts as E. Landau, *Foundations of Analysis*, Chelsea, 1951, or J. M. H. Olmsted, *The Real Number System*, Appleton-Century-Crofts, 1962, for such a development. In Appendix I we list the field axioms and basic theorems which one will meet in the usual Intermediate Algebra course development of the real number system.

We take the concept of a **real number** to be a *primitive term*—we don't try to define it. We describe some of the properties of real numbers. To begin with, consider the important subsets of the set of real numbers.

Natural Numbers. The set N of natural numbers consists of those numbers which arise naturally from counting, namely, 1,2,3,4, ...

$$N = \{1,2,3, \ldots\}$$

Integers. The set Z of integers consists of all the natural numbers, their negatives and the number zero. Thus,

$$Z = \{\ldots, -4, -3, -2, -1, 0, 1, 2, 3, 4, \ldots\}$$

The *natural numbers* are often referred to as *positive integers*, and the natural numbers with zero included as *non-negative integers*.

Rational Numbers. The set Q of rational numbers consists of those real numbers which can be expressed as the quotient of the two integers. Hence,

$$Q = \{x \mid x = p/q, p \in Z, q \in Z, q \neq 0\}$$

7

fundamentals

Therefore, $1/2 \in Q$, $5/16 \in Q$, $-27/38 \in Q$, $1.141 \in Q$, $3.14159 \in Q$. But $\sqrt{2} \notin Q$, and $\pi \notin Q$. Since $p = p/1 \in Q$ for every integer p, we have $Z \subset Q$.

Irrational Numbers. The irrational numbers are those real numbers which are not rational: the set $R \backslash Q$. Thus $\pi \in R \backslash Q$, $\sqrt{2} \in R \backslash Q$, $1 - 3\sqrt{2}$ is also irrational, as is $3^{1/3}$, $5^{1/7}$, etc.

Interestingly enough, the ancient Greeks, particularly the Pythagoreans (500 to 400 B.C.) had as an article of faith that the essence of all things in geometry and music, as well as in the practical and theoretical affairs of man, could be explained in terms of the intrinsic properties of the whole numbers and their ratios; i.e., in terms of rational numbers. The dialogues of Plato indicate that the Greek mathematical community was stunned by the discovery, probably some time before 410 B.C., that even within geometry itself, the rationals were not adequate. But even after the character of irrational numbers was recognized, the classical Greeks refused to accept them as actual numbers. Fractions and whole numbers had an obvious physical meaning, whereas irrationals only represented certain lengths. The classical Greeks accepted them only as lengths, rejecting them as numbers. This attitude severely retarded the development of algebra among the Greeks. They converted concepts which could have better been represented numerically as irrationals into lengths, areas and volume, and even solved quadratic equations geometrically. It was the Hindus and Arabs who decided that one could calculate with irrationals just like one does with other real numbers, i.e., $\sqrt{2}\,\sqrt{3} = \sqrt{6}$, etc.

In fact, the negative numbers had a similarly difficult history, being introduced by the Hindus about 600 A.D., but not gaining acceptance for about 1,000 years.

The history of the complex numbers is similar. They first appeared about 1540, but only 200 years were required for these to be used freely.

Complex Numbers. The set C of complex numbers consists of all those numbers which can be written in the form $a + bi$, where a and b are real numbers and $i^2 = -1$. Thus, $1 - 3i \in C$ and $\sqrt{2} - \pi i \in C$. Since $a + 0i = a \in C$, for any real number a, we have $R \subset C$. The numbers bi in the set $C \backslash R$ are called **pure imaginary**. Complex numbers are considered in greater detail in Section 1.5.

The following Venn diagram illustrates the various sets of numbers which have been described.

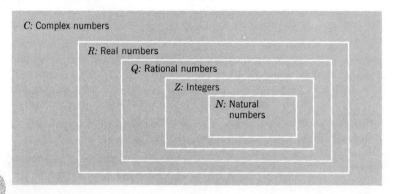

8

figure 1.2.1

sets of numbers

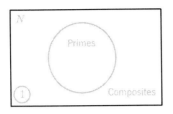

figure 1.2.2

The set N' of those natural numbers larger than 1 is frequently subdivided into two subsets. The **prime** numbers are those natural numbers p other than 1 whose only divisors are 1 and p itself. Thus, 2, 3, 5, 7, 11, 13, 17, 19, . . . are primes. If a natural number other than 1 is not prime, it is called **composite** (see Figure 1.2.2). Thus, 4, 6, 8, 10, . . . are composite.

Each real number x has a decimal expansion of the form $k . d_1 d_2 d_3 \ldots$, where $k \in Z$ and each d_t is one of the digits 0, 1, . . . , 9. For example, $3 = 3.0000 \ldots, 4/3 = 1.333 \ldots 3 \ldots, \pi = 3.14159 \ldots, 22/7 = 3.14285714\text{-}2857 \ldots$, etc. The decimal expansion of a rational numbers is "repeating"; i.e., a certain pattern of the digits recurs in succession as the "142857" in the expansion for 22/7 or the "3" in that for 4/3. The decimal expansion of an irrational number is nonrepeating.

The **number line** or **real axis** provides us with a convenient geometric representation of the real numbers. One generally chooses a horizontal line L and selects two points. The left-hand point is made to correspond with the number 0 and is called the **origin** and the right-hand point with 1. The set Z is then placed in one-to-one correspondence with certain points of L by copying the unit segment the appropriate number of times, using right of 0 for positive integers and left of 0 for the negatives. The elements of Q are then made to correspond with points on L by the classical Greek construction method of dividing a segment in any given ratio. (See Figure 1.2.3)

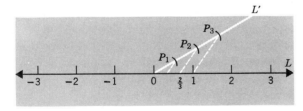

figure 1.2.3

For example, to construct the rational number 2/3 one constructs any other line L' which intersects L at the point 0. Then, using any convenient length, mark three equally spaced points P_1, P_2, and P_3 on L' beginning at 0. Connect P_3 with the point 1 on L. Then the line through P_2 parallel to this line cuts L at the number 2/3 while the line through P_1 parallel to it cuts L at 1/3. The formal proof of this is contained in most high school geometry books.

That this method does not exhaust all the points of L is easily seen by the simple construction of a segment of length $\sqrt{2}$ as the hypotenuse of a right triangle whose other two legs have length 1. However, we do have a one-to-one correspondence between each point of L and the set of all real numbers R. This correspondence is apparent when one locates a point on L first in the segment between two integers n and $n + 1$, then in a segment of length 1/10, and then in one of length 1/100, etc. In this way, the decimal expansion of the

fundamentals

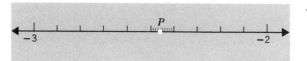

figure 1.2.4

number associated with that point is determined. For example, we note that in Figure 1.2.4 the point p is associated with the real number $-2.46 \ldots$. Conversely, in this manner each unending decimal determines a unique point of L.

For convenience, we shall abuse the language and call the points of L by the real numbers with which they are associated. There is, of course, a logical difference between the point and its **coordinate**, the real number associated with it, but we shall not often bother to make that distinction.

That part of the real line which lies to the right of 0 corresponds to the distinguished subset of **positive reals** R^+. The set R^+ has the following properties:

(1.2.1)

   (i)   For each $x, y \in R^+$ the sum $x + y \in R^+$.

   (ii) For each $x, y \in R^+$ the product $x \cdot y \in R^+$.

   (iii) For every $x \in R$ *exactly* one of the following is true:

$$x \in R^+, \qquad x = 0, \qquad \text{or} \qquad -x \in R^+$$

Let $a, b \in R$. *We say that* a *is* **less than** b $(a < b)$ *or* b *is* **greater than** a $(b > a)$ *provided* $b - a \in R^+$ (*i.e.,* $b - a$ *is a positive real number*).

Thus, it follows that $a > 0$, and $a \in R^+$ say the same thing. Equations (1.2.1) become:

(1.2.2)

   (i)   If $x, y \in R$ and $x > 0$ and $y > 0$, then $x + y > 0$.

   (ii) If $x, y \in R$ and $x > 0$ and $y > 0$, then $xy > 0$.

   (iii) For each real number x exactly one of the following is true:

$$x > 0, \qquad x = 0, \qquad \text{or} \qquad x < 0$$

We say, as a consequence of this, that the real numbers are ordered by $>$ (also by $<$). The concept of an ordered field is discussed in Appendix I. We shall discuss the question of inequalities in more detail in the next section. We also write $r \le t$ to mean that either $r < t$ or $r = t$. Similarly, $a \ge b$ means either $a > b$ or $a = b$; i.e., a is not less than b.

example 1

We have $2 < 5$, $-100 < 2$, $18 > -47$, $\pi > 3$, $-12 \ge -48$, etc.   ●

We are frequently interested in the *distance* between two points on the number scale. In the next section we shall discuss this concept in more detail. However, while -100 is less than 2 we *feel* that there is something about -100 that is *more* than 2. Of course, -100 is farther from 0 than 2, its magnitude is somehow larger. It is its direction that is different. The magnitude of a real number x or its undirected distance from 0 on the number scale is denoted by $|x|$, and is called its **absolute value**. Thus $|-100| = 100$ is larger than $|2| = 2$.

Formally, absolute value is defined for real numbers by the following. *For each* $x \in R$:

$$|x| = \begin{cases} x, & \text{if } x \ge 0 \\ -x, & \text{if } x < 0 \end{cases}$$

10

sets of numbers

That is, $|x|$ is the non-negative number x when x is itself non-negative; but when x is negative, $|x|$ is the *positive* number $-x$.

As an immediate consequence of this definition we have the desired result that $|x|$ is always non-negative; i.e.,

$$|x| \geq 0$$

example 2

$|3| = 3;\ |-3| = -(-3) = 3;\ |5/2| = 5/2;\ |-3/4| = -(-3/4) = 3/4.$ ●

Suppose $|x - 3| = 4$. Then if $x - 3$ is non-negative $|x - 3| = x - 3 = 4$ and $x = 7$. On the other hand, if $x - 3$ is negative $|x - 3| = -(x - 3) = 4$. Thus $-x + 3 = 4$, and $x = -1$. The set $\{-1, 7\}$ is called the **solution set** for the equation $|x - 3| = 4$, while $x = -1$ and $x = 7$ are called the **solutions** of the equation.

example 3

Solve the equation $|2x - 3| = 5$. This means find those numbers x, the solutions, which make the given equation a true statement. Since $|2x - 3|$ equals either $2x - 3$, or $-(2x - 3) = 3 - 2x$, according to whether the number $2x - 3$ is non-negative or negative, respectively, the two equations to be solved are as follows.

$$2x - 3 = 5 \qquad\qquad 3 - 2x = 5$$
$$2x = 8 \qquad \text{and} \qquad -2x = 2$$
$$x = 4 \qquad\qquad x = -1$$

Therefore, the solution set for the equation $|2x - 3| = 5$ is the set $\{-1, 4\}$; the solutions are $x = -1$ and $x = 4$. That these numbers actually do make the given equation into a true statement is seen by direct substitution. Thus

$$|2(-1) - 3| = |-2 - 3| = |-5| = 5$$

and

$$|2(4) - 3| = |8 - 3| = |5| = 5 \quad ●$$

Most often we shall not bother with the more accurate term "solution set." We shall merely name its members as *solutions*.

exercises

1. Categorize each of the following numbers in the smallest of the sets $N \subset Z \subset Q \subset R \subset C$ of which it is a member.

(a)	-7	(g)	-6	(m)	$1 - \sqrt{\pi}$
(b)	2	(h)	$\sqrt{5}$	(n)	$1 + \sqrt{2}$
(c)	4	(i)	$\sqrt{-5}$	(o)	$3 - 7i$
(d)	π	(j)	$\sqrt[3]{27}$	(p)	$\sqrt{-16}$
(e)	$22/7$	(k)	$\sqrt[3]{-8}$	(q)	1
(f)	3.1416	(l)	π^2	(r)	0

fundamentals

2. Complete the "organization chart" below by adding the missing names for the number sets described in the text.

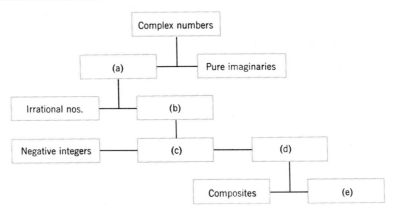

3. Show that $\pi \neq 22/7$ by comparing the first five digits in their decimal expansions. (A proof that π is irrational is a bit more difficult.)

4. Give straight-edge and compass constructions for the following rational numbers.

(a) $\dfrac{1}{3}$ (f) $\dfrac{5}{3}$

(b) $\dfrac{5}{8}$ (g) $3\dfrac{1}{2}$

(c) $\dfrac{3}{4}$ (h) $-2\dfrac{3}{4}$

(d) $-\dfrac{1}{3}$ (i) $-\dfrac{6}{5}$

(e) $-\dfrac{2}{5}$ (j) $-\dfrac{17}{3}$

5. List all the prime integers between 2 and 50.

6. Every integer can be factored as the product of its prime factors. For example $6 = (2)(3)$; $8 = (2)(2)(2) = 2^3$; $-24 = -(2^3)(3)$. Factor each of the following into its prime factors.

(a) 12   (f) 3
(b) 27   (g) -14
(c) 36   (h) -25
(d) 49   (i) -33
(e) 50   (j) 37

7. What is true about the number x if:

(a) $|x| = 3$ (g) $|x + 3| = 2$
(b) $\sqrt{x^2} = x$ (h) $|1 - x| = 4$
(c) $\sqrt{x^2} = -x$ (i) $|2x - 2| = 2$
(d) $|x| \leq 0$
(e) $|2x - 2| = 2 - 2x$ (j) $\left|\dfrac{x}{2} - 1\right| = 2x + 3$
(f) $|x + 3| = x + 3$

12

inequalities

8. Solve the following equations if possible.

(a) $|x - 1| = \dfrac{x}{2}$

(b) $|3x + 5| = 5x - 3$

(c) $|7x - 1| = \dfrac{x + 1}{4}$

(d) $\left|\dfrac{x}{3} - 1\right| = \dfrac{5 - x}{4}$

(e) $|(x - 1)^2| = 0$

(f) $\left|\dfrac{1}{x}\right| = 2 - x$

(g) $|x^2| = 2x - 1$

(h) $|x^2| = 6 + x$

(i) $|2x - 1| = x^2$

(j) $|x^2| = x$

9. Order the following real numbers on the number line using $<$. That is, write $-1 < 2 < 5 < 75$, etc.

$$-\pi, \ \frac{1}{2}, \ \pi, \ \frac{\pi}{2}, \ \sqrt{2}, \ 1, \ 3, \ -16, \ -\sqrt{2}, \ -\sqrt{3}, \ \frac{7}{3}, \ 0, \ -2, \ \frac{\pi}{3}, \ 1 - \pi$$

10. Show that if k is an odd positive integer, then k^2 is odd also. Show that if k is an even positive integer, then k^2 is even also.

11. Let $a \in R$ have the repeating decimal expansion $0.212121 \ldots$. Note that $100a = 21.21212121 \ldots$ so $99a = 100a - a = 21$. Thus $99a = 21$ and $a = 21/99 = 7/11$. Make use of this same method to write the following rational numbers as the quotient of two integers (reduced to lowest terms).

(a) $a = 1.191919 \ldots$

(b) $a = 3.03030303 \ldots$

(c) $a = 0.202202202 \ldots$

(d) $a = 0.1111111 \ldots$

12. Prove that there is no rational number a such that $a^2 = 2$.
Hint: Suppose there is $a \in Q$ such that $a^2 = 2$. Set $a = p/q$ where p and q are reduced to lowest terms; hence, have no factors in common. Use the results of Exercise 10 to arrive at a contradiction.

13. Show that if $a, b, c \in R$, with $a < b$, then $a + c < b + c$.
Hint: Remember that $a < b$ means $b - a \in R^+$ ($b - a$ is positive). Then use (1.2.1)(i).

14. Show that if $a, b, c \in R$ with $a < b$ and $c > 0$, then $ac < bc$.

15. Show that if $a, b, c \in R$ with $a < b$ and $c < 0$, then $ac > bc$.

16. Show that if $a, b, c \in R$ with $a < b$ and $b < c$, then $a < c$.

17. Show that if $a, b, c, d \in R$ with $a < b$ and $c < d$, then $a + c < b + d$.

18. Show that if $a, b, c,$ and d are as in Exercise 17, it may be false that $ac < bd$.

1.3. inequalities

Much of what is done in this text and a large part of basic calculus involves sets of real numbers and their ordering on the real line. In Section 1.4, we shall describe many of these sets in terms of inequalities. In this section, we shall expand the discussion of this ordering which we began in Section 1.2.

The following properties of "$<$" are immediate consequences of statements (1.2.1) and equations (1.2.2) of the previous section.

fundamentals

theorem 1.3.1

For a, b, $c \in R$ we have the following:

(i)   For every $a \in R$, $a \le a$.
(ii) If $a < b$, $b < c$, then $a < c$.
(iii) if $a < b$ and $c \in R$, then $a + c < b + c$.
(iv) If $a < b$ and $c > 0$, then $ac < bc$.
(v)   If $a < b$ and $c < 0$, then $ac > bc$.
(vi) If $a < c$ and $b < d$, then $a + b < c + d$.

proof of (ii). Since $a < b$, $b - a > 0$, while $b < c$ implies $c - b > 0$. There-fore, $(b - a) + (c - b) > 0$ (1.2.2 (i)). So $c - a > 0$, and $a < c$ as desired.

The proofs of the rest of the theorem are exercises. (See the previous section, Appendix I, and Exercise 7 of this section.)

Use is made, because of the Theorem (ii), of the notation $a < b < c$ to indicate that b is **between** a and c; i.e., $a < b$ and $b < c$. Recall the symbols $\ge$ and $\le$, where $a \le b$, read as a **less than or equal to** b, means that a is *not greater* than b. Similarly $b \ge a$ means that $b > a$ or $b = a$, but $b \not< a$. Then the statement $a \le b \le c$ would mean that b is *between a and c, or perhaps equal to one or the other* (or both if $a = c$). It is easy to see that Theorem 1.3.1 is true for $\le$ and $\ge$ as well as for $<$ and $>$.

A statement of the form $A < B$ is called an **inequality**. If either A or B involves a variable, the truth of the statement depends upon which real numbers may be substituted for the variable. To *solve an inequality* one determines which real numbers can be substituted to make the statement true.

example 1

The truth of the statement $x + 3 < 4$ depends upon what value(s) x has. Thus $0 + 3 < 4$ is a true statement, but $2 + 3 < 4$ is false. Theorem 1.3.1 is used to solve $x + 3 < 4$ as follows.

$$x + 3 < 4$$

so adding (-3) to both members gives

$$x + 3 - 3 < 4 - 3$$

or

$$x < 1$$

Therefore, any real number which is smaller than 1 may be substituted for x, and the original statement will be true. The set $\{r \mid r \in R \text{ and } r < 1\}$ is called the *solution set* for this inequality. ●

This example was a particularly easy one. We turn to several examples to illustrate the general procedure. In each case one proceeds with operations based on Theorem 1.3.1 to obtain simpler inequalities which have the same set of solutions.

14

inequalities

example 2

Solve $5x - 3 > 2 + 3x$. Again using the theorem we have the following sequence of steps

$$5x - 3 > 2 + 3x$$

$$5x - 3x - 3 + 3 > 2 + 3 + 3x - 3x$$

$$2x > 5$$

$$x > \frac{5}{2}$$

Therefore, the solution set will be $\{r \mid r \in R, \ r > 5/2\}$. This is, of course, an infinite set and corresponds to a ray on the real line. The graph of this set appears in Figure 1.3.1, where the open circle indicates that the end-point 5/2 *does not* belong to the set. ●

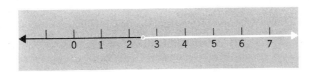

figure 1.3.1

example 3

Solve the inequality $3x^2 - 2x < 8$. The first step is to write this in the form

$$3x^2 - 2x - 8 < 0$$

The left member can be factored as

$$(3x + 4)(x - 2)$$

Therefore, we have the statement that

$$(3x + 4)(x - 2) < 0$$

that is, the product of two real numbers is negative. This can only be true when one of them is negative and the other is positive. Consider each. If $(3x + 4)$ is negative, we must have $(x - 2)$ positive. Write

$$3x + 4 < 0 \qquad \text{and} \qquad x - 2 > 0$$

Then

$$3x < -4 \qquad \text{and} \qquad x > 2$$

or

$$x < -\frac{4}{3} \qquad \text{and} \qquad x > 2$$

This cannot be true, since a real number cannot simultaneously be smaller than $-4/3$ and larger than 2. It must, therefore, be true that $(3x + 4)$ is positive and $(x - 2)$ is negative or

$$3x + 4 > 0 \qquad \text{and} \qquad x - 2 < 0$$

Thus,

$$x > -\frac{4}{3} \qquad \text{and} \qquad x < 2$$

fundamentals

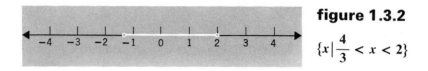

figure 1.3.2

$$\{x \mid \tfrac{4}{3} < x < 2\}$$

This is true for all real numbers lying between $-4/3$ and 2 on the real line. This is symbolized by

$$-\frac{4}{3} < x < 2$$

The above is the solution to the given inequality. The solution set for the given inequality is $\{x \mid x \in R, \ -4/3 < x < 2\}$ the graph of which is sketched in Figure 1.3.2.

The discussion given in the solution of this inequality can be represented graphically. In Figure 1.3.3 we mark with $+$ and $-$ signs those subsets of the line where each factor is positive and negative. Then the appropriate interval which is the solution becomes apparent as the intersection of the set where $3x + 4 > 0$ with the set where $x - 2 < 0$. ●

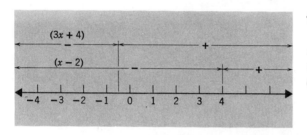

figure 1.3.3

Positive and negative values of the factors $(3x + 4)$ and $(x - 2)$.

In Section 1.2 we introduced the absolute value $|x|$ of a real number x and interpreted this as the distance of the point x from 0. Let x and y be any two real numbers. Then the distance from the point x to the point y on the real line is the real number $|x - y|$. These definitions of distance on the real line are compatible with each other and with the discussion in Section 1.2, as you can readily show.

It is of frequent concern to solve inequalities which involve absolute values. Consider a few examples.

example 4

Solve $|x - 2| \leq 4$. This is really asking us to locate all those points on the real line whose distance from 2 is no larger than 4. It is fairly obvious, when stated in these terms, that this will be a segment of the real line with the point 2 at its center stretching from -2 to $+6$. To see that this actually follows from the definitions and theorems given we proceed as follows.

$$|x - 2| \leq 4 \text{ is true when}$$

$$x - 2 \leq 4 \qquad \text{and when} \qquad -(x - 2) \leq 4$$

depending upon whether $x - 2$ is a positive or a negative number. That is, we have

$$0 \leq x - 2 \leq 4, \qquad \text{or} \qquad 0 \leq -(x - 2) \leq 4$$

16

inequalities

The second of these statements can be rewritten using Theorem 1.2.1 (v) as

$$0 \geq x - 2 \geq -4$$

and the two statements can be combined into the expression

$$-4 \leq x - 2 \leq 4$$

Therefore, this time using Theorem 1.3.1 (iii),

$$-4 + 2 \leq x - 2 + 2 \leq 4 + 2$$

or

$$-2 \leq x \leq 6$$

As we predicted the solution set is the set $\{x \mid -2 \leq x \leq 6\}$. ●

In the middle of this example we illustrated a convenient alternative form for an inequality of the type

$$|x - a| < b$$

You may prove the folowing theorem as Exercise 8. The proof follows the argument given in the example.

theorem 1.3.2

For real numbers x, a, and b, $|x - a| < b$ if, and only if, $-b < x - a < b$.

example 5

Solve $|x^2 - 3x + 2| > 2$. This will be true, on the one hand, when $x^2 - 3x + 2 > 2$, since then, *a fortiori**, $x^2 - 3x + 2 > 0$. On the other hand, if $x^2 - 3x + 2 < 0$ the given statement is true when $-(x^2 - 3x + 2) > 2$, or $x^2 - 3x + 2 < -2$. Consider each possibility:

Case 1		Case 2
$x^2 - 3x + 2 > 2,$	and	$x^2 - 3x + 2 < -2$

These become

$x^2 - 3x > 0,$	and	$x^2 - 3x + 4 < 0$

Factoring the left side of the first case yields

$$x(x - 3) > 0$$

The left side of case 2 is not factorable. It will be demonstrated in Chapter 3 that, in fact, $x^2 - 3x + 4$ is always positive for each real number x. Thus, $x^2 - 3x + 4 < 0$ is not possible. For the first, and now only, case

$$x(x - 3) > 0$$

implies both x and $(x - 3)$ must have the same algebraic sign. Thus,

$$x < 0 \qquad \text{and} \qquad x - 3 < 0$$

which is true for $x < 0$; or

$$x > 0 \qquad \text{and} \qquad x - 3 > 0$$

*With all the more force.

fundamentals

which is true for $x > 3$. Therefore, the original statement is true for all real numbers x except those which lie between 0 and 3. ●

exercises

1. Solve the following inequalities.

   (a) $x + 5 < 3$
   (b) $4x > 12$
   (c) $3 - x < 5$
   (d) $4 - 3x \geq 6$
   (e) $5x - 1 \leq 3x + 7$

   (f) $3x - 1 > 7 - x$
   (g) $3 - 6x < 9 - x$
   (h) $x + 6 \leq 2 - 3x$
   (i) $x(x - 2) < 0$
   (j) $x(x - 2) > 0$

2. Solve the following inequalities.

   (a) $x^2 - 25 < 0$
   (b) $x^2 - 9 \geq 0$
   (c) $16 - x^2 \leq 0$
   (d) $(x - 2)^2 < 0$
   (e) $x^2 - 4x + 3 \leq 0$

   (f) $2x^2 - x - 8 \leq 2$
   (g) $3x^2 - 4x + 1 \leq 5$
   (h) $2x^2 + 7 < 9x$
   (i) $5x^2 - 3 \geq 14x$
   (j) $2x^2 \geq 3x + 5$

3. Solve the following inequalities.

   (a) $\dfrac{x - 1}{x} < 0$

   (b) $\dfrac{x}{x - 1} > 0$

   (c) $\dfrac{x - 2}{x + 2} < 0$

   (d) $\dfrac{x - 2}{x + 2} > 0$

   (e) $\dfrac{x + 1}{x} > 1$

   (f) $\dfrac{x - 1}{x} < 1$

4. What is true about x if:

   (a) $|x| = 2$

   (b) $\left| \dfrac{x}{2} - 1 \right| = 5$

   (c) $\left| \dfrac{3x}{2} - 2 \right| = 2 - x$

   (d) $|x^2 - 4| = 2x - 4$

   (e) $|x| \geq 5$
   (f) $|x - 1| < 3$
   (g) $|x^2 - 4| \leq 5$
   (h) $|x^2 - 1| < 3$
   (i) $|x| \leq 0$
   (j) $|x^2 + 1| < 0$

5. Solve the following inequalities.

   (a) $|2x - 1| \leq 7$
   (b) $|3 - 2x| \leq 7$
   (c) $|4 - 3x| > 10$
   (d) $|2x - 5| < 3$
   (e) $|4 - x| \leq 8$
   (f) $|5 - 2x| \leq 15$
   (g) $|x^2 - 25| \geq 0$

   (h) $\dfrac{|x - 1|}{|x|} < 0$

   (i) $\dfrac{|x|}{|x - 1|} < 1$

   (j) $|2x^2 - 3x| > 5$

6. Given that $0 < r < s$ are real numbers. Justify or give a counter example for each of the following.

   (a) $-r > -s$

   (b) $r < \dfrac{r + s}{2} < s$

   (c) $r^2 < s^2$
   (d) $|r| < |s|$

(e) $\dfrac{1}{r} < \dfrac{1}{s}$

(f) $\dfrac{r+s}{4s} < \dfrac{s}{r+s}$

(g) $r^2 + s^2 < 2rs$

(h) $\dfrac{r^2}{s} + \dfrac{s^2}{r} > r + s$

7. Prove that for every real number a, $a \le a$.

Hint: Consider what $\le$ means.

8. Prove Theorem 1.3.2.

9. Prove that, for every number x and y and positive real numbers a and b with $a \ne b$,

(a) $|x| = |-x|$

(b) $|x|^2 = x^2$

(c) $|x - y|^2 = (x - y)^2$

(d) $\sqrt{x^2} = |x|$

(e) $(a + b)^2 > a^2 + b^2$

(f) $\dfrac{a}{b} + \dfrac{b}{a} > 2$

(g) $\dfrac{a + b}{2} > \sqrt{ab}$

(h) $|xy| = |x|\,|y|$

(i) $|x + y| \le |x| + |y|$

(j) $|x - y| \ge |x| - |y|$

10. Show that if x and y are real numbers, then the distance on the real line between the points associated with x and y is $|x - y|$. That is, show that the definitions of absolute value and of distance on the real line are compatible. See Section 1.2.

11. Show that the mid-point of the segment between x and y in Exercise 10 is the point $(x + y)/2$.

12. Find two rational numbers r and s such that $s - r < 0.001$ and $r < \pi < s$.

13. Find two rational numbers r and s whose difference is less than 0.001 and

(a) $r < \dfrac{\pi}{2} < s$

(b) $r < \dfrac{\pi}{3} < s$

(c) $r < \dfrac{\pi}{4} < s$

(d) $r < \dfrac{\pi}{6} < s$

(e) $r < \dfrac{2\pi}{3} < s$

(f) $r < \sqrt{2} < s$

(g) $r < \sqrt{3} < s$

(h) $r < 1 - \sqrt{2} < s$

(i) $r < 1 + \sqrt{2} < s$

(j) $r < \dfrac{1 + \sqrt{3}}{2} < s$

14. Prove that if a is any real number, then $a^2 > 0$.

Hint: Use Theorem 1.3.1 (iv) and (v), consider also the tricotomy property.

15. Prove that there is no real number x such that $x^2 + 1 = 0$.

16. Show that if $|x - 3| < 0.001$ then $|2x - 6| < 0.002$.

Hint: Use Exercise 9 (h).

1.4. the real line & the plane

As we pointed out in the last section, most of the sets which are encountered in this book are subsets of the real numbers and the corresponding subsets of the real line. Others can be considered as subsets of the plane. We first single out those sets of real numbers which correspond to segments or to rays of the real line for special consideration. Such sets are called **intervals** and are classified according to whether or not their end-points are included as members of the sets.

fundamentals

For $a, b \in R$, $a \leq b$ we have the following types of intervals.

The **open interval** $(a,b) = \{x | a < x < b\}$ (here $a < b$).
The **closed interval** $[a,b] = \{x | a \leq x \leq b\}$.
The **half-open intervals** $[a,b) = \{x | a \leq x < b\}$ and
$$(a,b] = \{x | a < x \leq b\}.$$

Notice that brackets denote the inclusion and parentheses the exclusion of the end points from the interval. For example, $3 \in [3,4)$ but $4 \notin [3,4)$. Figure 1.4.1 indicates some examples of graphical representation of intervals.

figure 1.4.1

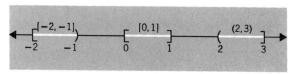

Particularly when dealing with real numbers which have large absolute value, one is concerned about sets of real numbers which correspond to rays on the real line. To facilitate our discussion, we introduce the symbols ∞ (infinity) and $-\infty$. Then the rays beginning at a real number a can be considered as intervals as follows:

$$(a, \infty) = \{x | x \in R \text{ and } x > a\}$$

$$(-\infty, a) = \{x | x \in R \text{ and } x < a\}$$

Each of the above rays is an open interval. We can include a in the interval and denote the resulting intervals by $[a, \infty)$ or $(-\infty, a]$ in the obvious way. *Note:* One should not mistake ∞ and $-\infty$ for real numbers. Neither is a member of R. While an arithmetic can be developed for an extended real number system $R \cup \{\infty, -\infty\}$, we do not do it here. One should be cautious of writing such things as $1/\infty$, $3 + \infty$, etc., because logical difficulties may arise.

Let a be an arbitrary real number. In the development of the theory of limits in calculus, one frequently is concerned with *neighborhoods* of a. *By a* **neighborhood** $N(a)$ *of* a, *one means simply an open interval with a as mid-point.* Let h be a positive number. Then $N(a) = (a - h, a + h)$ is a neighborhood of a of length $2h$. When the length is important, we will write $N(a)$ as $N_h(a)$, which we call an h *neighborhood of* a.

A real number x belongs to an h neighborhood of a provided its distance from a is less than the distance from a to the end-point of the interval; i.e., provided $|x - a| < h$.

It is also convenient to adopt the terminology "a neighborhood of infinity (or of minus infinity)." An open interval with ∞ (or $-\infty$) as mid-point doesn't exist, so we say that a set $N(\infty)$, $(N(-\infty))$, is a "neighborhood of infinity" (of $-\infty$) if it is an open interval (ray) of the form (s, ∞) (or $(-\infty, s)$), for some real number s. That is: A real number x is in the neighborhood (s, ∞) of ∞ provided $x > s$.

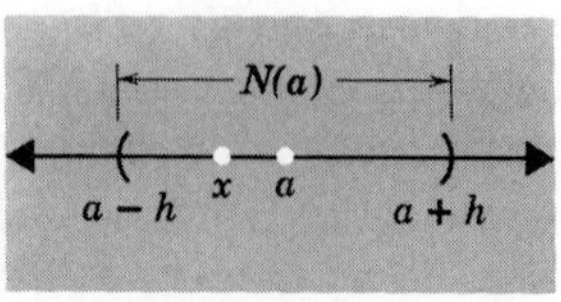

figure 1.4.2

the real line & the plane

example 1

Find the interval which is the solution set for the inequality $|x - 1| < 3$.

We have $|x - 1| < 3$ if and only if $-3 < x - 1 < 3$. Why? Then $-2 < x < 4$. Thus, the solution set is the open interval $(-2, 4)$. It is an open interval because x cannot *equal* -2 or 4 and still satisfy the *strict* inequality as stated. ●

The closed interval $[0, 1]$ is used so often as an example that we call this set **the closed unit interval**.

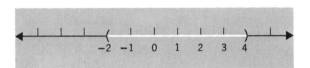

figure 1.4.3

We have already seen how a coordinate system can be introduced on a line by assigning a real number to each point and a point to each real number. We proceed now to discuss a coordinate system for the plane. The system which we use is called a **Cartesian coordinate system** after the French mathematician and philosopher René Descartes (1506–1650), who did not invent the idea, but who was the first to exploit it in the invention of analytic geometry.

Intuitively, then, we consider two copies of the real line. We place these lines in the plane Γ in such a way that they intersect at their corresponding 0 points with one line horizontal and the other vertical. The horizontal line is called the **x axis** and the vertical line the **y axis**. By X and Y, respectively, we designate the sets of real numbers corresponding to these lines.

Now if P is any point in the plane Γ, we associate with P a unique member of the Cartesian product $X \times Y$ as follows: Drop a perpendicular from P to the x axis. This determines a point P_1 called the **projection** of P onto the x axis. The point P_1 has a unique **coordinate** (real number) x in X. We also project P to the point P_2 on the y axis and determine a unique real number $y \in Y$. Then the element (x, y) of $X \times Y$ corresponds to the point P. The numbers x and y are called the **coordinates** of P. On the other hand, if (x, y) is any pair of real numbers, there is a unique point P in Γ with these coordinates. To locate P, locate x on the X axis and y on the Y axis. Then erect perpendiculars to the axes at these points. These two perpendiculars intersect at P.

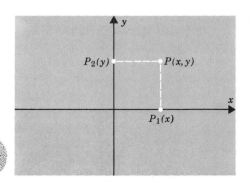

figure 1.4.4

fundamentals

In this way, a one-to-one correspondence is obtained between points in the plane and *ordered* pairs of real numbers. We emphasize the word *ordered* here because, of course, (2,1) and (1,2) are different elements of $X \times Y$ and determine different points in the plane.

Since the axes themselves are also part of the plane, they too should have elements of $X \times Y$ associated with each of their points. The method described clearly assigns $(x,0)$ to the point P_1 and $(0,y)$ to the point P_2. The intersection point of the axes, $0(0,0)$ is called the **origin**.

example 2

Plot the points whose coordinates are $(0, 1)$, $(1, 3)$, $(-1, 4)$, $(\pi, 0)$, $(-\pi/2, 1)$. These points appear in Figure 1.4.5. ●

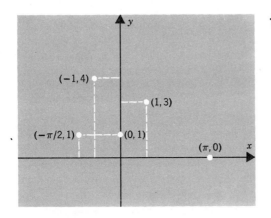

figure 1.4.5

The coordinate axes divide the plane into four parts called **quadrants**. These quadrants are numbered counterclockwise as in Figure 1.4.6.

Since we have a complete identification of points in the plane with elements of the set $X \times Y$, both x and y belong to R, the set of real numbers, we shall be sloppy, as is the custom, and refer to *the point* (a, b) rather than the more accurate "the point P whose coordinates are (a, b)." We shall also frequently designate the plane by $X \times Y$ or by $R^2 = R \times R$.

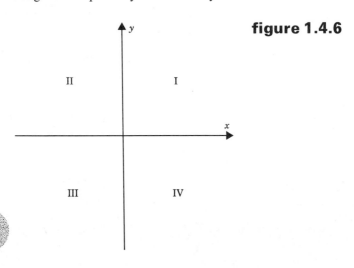

figure 1.4.6

the real line & the plane

Recall that the notation (a, b) is also used to designate the open interval $\{x | a < x < b\}$ of the real line. This confusion, while unfortunate, is not really troublesome since the context always makes clear what usage is meant.

In Exercise 14 of this section, you will be asked to demonstrate that the mid-point of the line segment represented by the interval $[a, b]$ (or (a, b)) has coordinate $(1/2)(a + b)$ (see also Section 1.3, Exercise 11). We now proceed to determine the mid-point of the segment joining *any* two points in the plane.

theorem 1.4.1

If $P_1(x_1, y_1)$ and $P_2(x_2, y_2)$ are points of the plane, then the point $M[(x_1 + x_2)/2, (y_1 + y_2)/2]$ is the mid-point of the segment $\overline{P_1 P_2}$.

proof. The projections of P_1 and P_2 on the x axis are the points $A(x_1, 0)$ and $B(x_2, 0)$, respectively (see Figure 1.4.7), and the mid-point of the segment $\overline{AB}$ has coordinates $(1/2(x_1 + x_2), 0)$ (see Exercise 11 of Section 1.3). Similarly, the projections C and D of P_1 and P_2 have coordinates $(0, y_1)$ and $(0, y_2)$, respectively. The mid-point of $\overline{CD}$ has coordinates $(0, 1/2(y_1 + y_2))$. Then the point $M(1/2(x_1 + x_2), 1/2(y + y_2))$ is the mid-point of $\overline{P_1 P_2}$, by the properties of similar triangles.

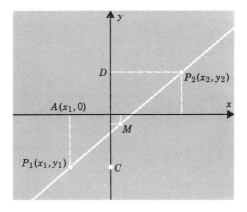

figure 1.4.7

In Section 1.3 we indicated that the distance between two points x_1 and x_2 on the line was given by the real number $|x_2 - x_1|$. The concept of distance in the plane is compatible with this definition. We have the following theorem which serves as a basis for the definition of distance.

theorem 1.4.2

If $P_1(x_1, y_1)$ and $P_2(x_2, y_2)$ are points in the plane, then the distance between P_1 and P_2, i.e., the length of the segment $\overline{P_1 P_2}$, is given by

(1.4.1) $$d = |P_1 P_2| = \sqrt{(x_2 - x_1)^2 + (y_2 - y_1)^2}$$

Outline of proof. (See Figure 1.4.8.) Determine the point Q so that the triangle $P_1 Q P_2$ is a right triangle with the right angle at Q. (We assume here that the line through P_1 and P_2 is neither vertical nor horizontal.) Then Q has coordinates (x_1, y_2). Why? The length of $\overline{P_1 Q}$ is $|y_2 - y_1|$ and the length of $\overline{QP_2}$ is $|x_2 - x_1|$.

23

fundamentals

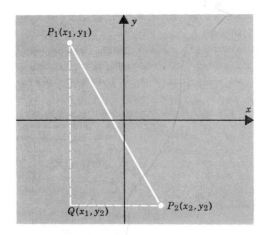

figure 1.4.8

Thus, by the Pythagorean theorem

$$|P_1 P_2|^2 = |P_1 Q|^2 + |Q P_2|^2$$

and the result follows. You should complete the details of this proof.

Formula (1.4.1) is called the **distance formula**.

Let $R(x, y) = c$ be an *equation* involving a pair of real numbers x and y. Then the set

$$S = \{(x, y) \mid R(x, y) = c, \ x, y \in R\}$$

is called the *solution set* for the equation. But S is a subset of $R^2 = X \times Y$, and, hence, one can identify which subset of the plane corresponds to S. This subset of the plane will also be denoted by S and is called the *graph of the equation* $R(x, y) = c$. This equation is called the *equation of S*.

example 3

Suppose we have the equation $x + 2y = 2$. Then the points $(0, 1)$, $(2, 0)$, $(1, 1/2)$ are on the graph of this curve since they satisfy the equation, while the points $(1, 1)$, $(2, 2)$ are not. A sketch of the solution set (graph of the curve) determined by this equation is the line of Figure 1.4.9. ●

example 4

Sketch the graph of the circle whose equation is $x^2 + y^2 = 1$.

The points $(\pm 1, 0)$ and $(0, \pm 1)$ satisfy this equation while the points $(0, 0)$ and $(1, 1)$ do not.

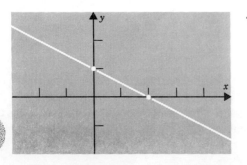

figure 1.4.9

the real line & the plane

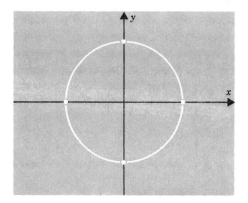

figure 1.4.10

A look at the distance formula tells us that the distance of any point (x, y), which satisfies this equation, from the origin is 1 since

$$1 = \sqrt{(x - 0)^2 + (y - 0)^2}$$

So the points do in fact lie on a circle with center $(0, 0)$ and radius 1. ●

We shall return to these ideas in later chapters.

exercises

1. List all the integers in the interval $(-2, 4)$. What about all the integers in $[-2, 4]$? What about a list of all the rational numbers in $(-2, 4)$?

2. What is the largest real number in the interval $[0, 6]$? What is the smallest real number in the interval $[0, 6]$? What about the largest or smallest number in $(0, 6)$?

3. Describe a *neighborhood of length* 2 for the real number $1/2$.

4. Describe a *neighborhood of infinity* which contains π.

5. Which of the following real numbers belong to the closed unit interval? $1, 0, 1/2, -1, -1/2, 3, \pi, \pi/2, \pi/4, \sqrt{2}, \sqrt{2}/2, \sqrt{3}, 1 - \sqrt{2}$.

6. Plot the points $(0, 1)$, $(-1, -2)$, $(\pi, -2)$, $(3/2, -2/3)$, $(-\sqrt{2}, -1)$ on a Cartesian coordinate system.

7. State the interval which is the solution set for $|2x + 1| < 5$. Sketch a graph of this interval.

8. Give as intervals the solution sets and sketch the corresponding segment or ray of the real line for each of the following inequalities.

(a) $x^2 < 9$

(b) $|3 - x| < .01$

(c) $|x + 2| < .001$

(d) $\dfrac{1}{x} > 10$

(e) $|x + 4| \geq .01$

(f) $\dfrac{1}{x} < 10$

(g) $x^2 \leq 4$

(h) $x^2 > 4$

(i) $|x - 1| < \dfrac{x}{2}$

(j) $|x + 1| < 2x + 1$

fundamentals

9. Let A be the open interval $(0, 1)$ and B be the open interval $(0, 3)$; describe and sketch a graph of:

(a) $A \cup B$

(b) $A \cap B$

(c) $B \backslash A$

(d) $A \backslash B$

(e) $A \times A$

(f) $B \times B$

(g) $A \times B$

(h) $(B \backslash A) \times A$

10. Find the smallest closed interval which contains the set of all (positive) prime integers less than 100.

11. Let $A_n = (0, 1/n)$. Consider the collection of all such intervals as n takes on all positive integral values.

(a) What is $\bigcap_{n \in N} A_n$?

(b) What is $\bigcup_{n \in N} A_n$?

12. What is the smallest closed interval which contains the non-negative integers? What is the smallest closed interval which contains all the non-negative rationals whose square is less than 2?

13. Let I denote the *unit interval* $[0, 1]$. Describe the set $I \times I$. Sketch it.

14. Prove that the *mid-point* of the interval $[a, b]$ is the point $c = (1/2)(a + b)$. What is the *mid-point* of (a, b)? (See Exercise 11 of Section 1.3.)

15. Let $X = \{0, 1, 2, 3, 4, 5\}$ and $Y = \{-2, -1, 0, 1, 2\}$. Sketch (a) $X \times X$, (b) $Y \times Y$, (c) $X \times Y$, (d) $Y \times X$ in the plane.

16. Sketch $Z \times Z$, the set of *lattice points* in the plane.

17. Let $A = [0, 1]$ and $B = [0, 3]$. Sketch a graph of the following sets on a Cartesian coordinate system: (a) $A^2 = A \times A$, (b) $B^2 = B \times B$, (c) $A \times B$, (d) $B \times A$.

18. Find the mid-points of the segments determined by the following points.

(a) $(0, 0)$ and $(2, 2)$

(b) $(1, 3)$ and $(-2, 1)$

(c) $(-1, 1/2)$ and $(-1, 1)$

(d) $(-4, -2)$ and $(-1, -1)$

(e) $(\pi, 3)$ and $(\pi/2, -1)$

(f) $(\sqrt{2}, 1)$ and $(-1, \sqrt{2})$

19. Find the length of each of the segments defined by the points in Exercise 18.

20. Complete the details of the proof of Theorem 1.4.2 including the case where $P_1 P_2$ is vertical and the case where it is horizontal.

1.5. complex numbers

In discussing sets of numbers in Section 1.2, brief mention was made of complex numbers. In that section it was pointed out that complex numbers are those numbers which can be expressed in the form $a + bi$, where a and b are real numbers and $i^2 = -1$. Since you have already shown (Exercise 14, Section 1.3) that the square of every real number is non-negative, the number i is clearly not a real number.

Complex numbers were invented to fill several needs. For example, in Chapter 3 you will see that real numbers are not adequate to handle situations involving solutions to polynomial equations. The equation $x^2 + 1 = 0$ is a case in point. It became apparent, during the 200 year period beginning about the middle of the sixteenth century that one needed to expand the system of numbers to include numbers whose squares could be negative. This expanded set of numbers should contain the real numbers and should retain, as far as possible, the usual properties of real numbers; i.e., the usual properties of

complex numbers

arithmetic and algebra (see Appendix I for these). Thus, we would, as a consequence, want a complex number to satisfy the following.

$$(a + bi) + (c + di) = (a + c) + (b + d)i$$

$$(a + bi)(c + di) = (ac - bd) + (ad + bc)i$$

To actually construct the set of complex numbers, beginning with the set of real numbers and the building blocks of set theory, requires a slightly less intuitive, perhaps, but logically more sound approach that we have given. A brief sketch of this approach is given here, but the details are left as exercises.

definition of the complex numbers

The system C of complex numbers consists of a set C, which is the set $R \times R$ of all ordered pairs (a,b) of real numbers, with operations $+$ and $\cdot$ defined on C so that the following are true.

(i) $(a,b) = (c,d)$ if, and only if, $a = c$ and $b = d$, as real numbers

(ii) $(a,b) + (c,d) = (a + c, b + d)$

(iii) $(a,b) \cdot (c,d) = (ac - bd, ad + bc)$

As you can see by identifying the symbol $a + bi$, which contains the mysterious number i, with the not so mysterious ordered pair (a,b), these definitions are motivated by the expected properties of complex number addition and multiplication previously listed. The justification for this identification is straightforward and will be provided shortly.

You will have the opportunity in the exercises to verify that, with the above definitions, the system of complex numbers does, indeed, satisfy the postulates for a field as given in the Appendix. We verify one of these properties as an example.

example 1

Show that for each complex number $(a,b) \neq (0,0)$ there exists a complex number (a', b') such that

$$(a,b)(a',b') = (1,0)$$

Since $(a,b) \neq (0,0)$, $a \neq 0$ and $b \neq 0$. Therefore, $a^2 + b^2 > 0$. Then, using (iii), we see that $aa' - bb'$ must equal 1 and $ab' + a'b = 0$. The following calculation shows that $a' = a/(a^2 + b^2)$ and $b' = -b/(a^2 + b^2)$:

$$(a,b)\left(\frac{a}{a^2 + b^2}, \frac{-b}{a^2 + b^2}\right) = \left(\frac{a^2}{a^2 + b^2} - \frac{-b^2}{a^2 + b^2}, \frac{-ab}{a^2 + b^2} + \frac{ba}{a^2 + b^2}\right)$$

$$= \left(\frac{a^2 + b^2}{a^2 + b^2}, \frac{-ab + ab}{a^2 + b^2}\right)$$

$$= (1,0)$$

Thus,

$$(a',b') = \left(\frac{a}{a^2 + b^2}, -\frac{b}{a^2 + b^2}\right)$$

and every nonzero complex number has a multiplicative inverse. ●

fundamentals

While the development of the complex number system is an important and mathematically interesting question, we shall not pursue it in great detail here. We proceed to show that the symbol $a + bi$ for a complex number is not completely devoid of meaning by identifying it directly with the pair (a,b). To do this first notice that

$$(0,1)(0,1) = (-1,0)$$

Also notice that there is an obvious one-to-one correspondence between pairs of the form $(a,0)$ and the set of real numbers; i.e., $a \leftrightarrow (a,0)$, for each $a \in R$. Thus, $(-1,0)$ corresponds with -1.

Consider any complex number (a,b). Then write

$$(a,b) = (a,0) + (0,b)$$

Now notice that $(0,b) = (b,0)(0,1)$. Thus,

$$(a,b) = (a,0) + (b,0)(0,1)$$

Call the pair $(0,1)$, i and make the identification of $(a,0)$ with a, and of $(b,0)$ with b. Thus $i^2 = -1$, and we have the desired identification of (a,b) with the symbol $a + bi$. In making this identification we have also shown that it is meaningful to consider the real numbers to be a subset of the complex numbers, $R \subset C$.

The symbol $a + bi$ is called the **standard form** of the complex number (a,b). The form $a + ib$ is also standard and sometimes easier to read. For example, $i\sqrt{3}$ is clearer than $\sqrt{3}i$, since sometimes the i tends to slip under the radical sign by mistake.

example 2

$1 - i, 2 + i\sqrt{5}, 7i, i\sqrt{2}/3$ are all in standard form. However, $(3 - 2i)(6i - 5)$, and $(3 + i\sqrt{2})/(1 - i\sqrt{2})$ are not. To put these two in standard form, we carry out the indicated operations:

$$(3 - 2i)(-5 + 6i) = -15 + 12 + i(18 + 10) = -3 + 28i$$

For the second one, write

$$\frac{3 + i\sqrt{2}}{1 - i\sqrt{2}} = \frac{(3 + i\sqrt{2})(1 + i\sqrt{2})}{(1 - i\sqrt{2})(1 + i\sqrt{2})} = \frac{1 + 4i\sqrt{2}}{1 + 2}$$

$$= \frac{1}{3} + \frac{4i\sqrt{2}}{3}$$

In this last case we "realized" the denominator by making use of the "good-ole" algebraic identity $(x + y)(x - y) = x^2 - y^2$. ●

You have, no doubt, already noticed that since complex numbers are ordered pairs of real numbers, they can be identified with points in the plane as described in the previous section. It was, in fact, probably this identification with something geometrical, given by the Norwegian surveyor Wessel and the French bookkeeper, Robert Argand, that eventually led to the acceptance of complex numbers *as numbers*. Using the symbols $a + bi$, we graph some complex numbers in Figure 1.5.1 on what has come to be known as an **Argand diagram.**

Complex numbers of the form bi, that is $0 + bi$, are called **pure imaginaries** (indicating their original sin). Since they are the pairs $(0,b)$, they will correspond

complex numbers

to points on the y axis. At the same time, real numbers are complex numbers of the form $a + 0i$ and correspond to points on the x axis. In an Argand diagram the y axis is called the **imaginary axis** and the x axis is the **real axis**.

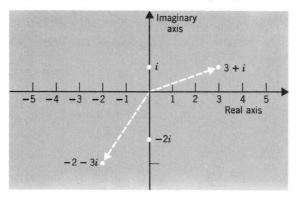

figure 1.5.1

Argand diagram.

example 3

Sketch the complex numbers i, $-2i$, -5, $3 + i$, and $-2 - 3i$ on an Argand diagram. This amounts to plotting the five points $(0,1)$, $(0,2)$, $(-5,0)$, $(3,1)$ and $(-2,-3)$. This is done in Figure 1.5.1 with the appropriate complex labels. ●

In Sections 1.3 and 1.4, we discussed the concepts of distance on the real line and in the plane. In the first case we associated distance with absolute value. This association is also made in the plane using complex numbers. Let $z = x + yi$ be a complex number. The **absolute value** or **modulus** $|z|$ of z is defined by

$$|z| = \sqrt{x^2 + y^2}$$

Notice that $|z|$ is a non-negative real number. Further, notice that $|z|$ is the distance, using the distance formula, that the point $x + yi = z$ is from the origin in an Argand diagram.

example 4

(a) $|2 + 5i| = \sqrt{2^2 + (5)^2} = \sqrt{29}$

(b) $\left| \dfrac{1}{2} - \dfrac{i\sqrt{3}}{2} \right| = \sqrt{(1/2)^2 + (\sqrt{3}/2)^2} = \sqrt{1/4 + 3/4} = 1$

(c) $|-3i| = |0 - 3i| = \sqrt{0^2 + (-3)^2} = \sqrt{9} = 3$

(d) $|-4| = |-4 + 0i| = \sqrt{16} = 4$ ●

Associated with each complex number $z = x + yi$ is another complex number called its **conjugate** $\bar{z} = x - yi$. The conjugate of a complex number is interesting for several reasons, some of which will turn up in later chapters. Here, we notice the following important facts about z and its conjugate $\bar{z}$. If $z \in C$, $z = x + yi$, then

(i) $\bar{z} = x - yi$

(ii) $(\bar{\bar{z}}) = z$

(iii) $z \cdot \bar{z} = |z|^2$

(iv) $z + \bar{z} = 2x$ is real

(v) $z - \bar{z} = 2yi$ is a pure imaginary if $y \neq 0$

fundamentals

Each of these is immediately verified by direct computation. For example, (iii) says $(x + yi)(x - yi) = x^2 + y^2 + i(yx - xy) = x^2 + y^2 = |z|^2$. The following properties are so important we state them formally as a theorem.

theorem 1.5.1

Let z, $w \in C$ be complex numbers and let $\bar{z}$, $\bar{w} \in C$ denote their conjugates. The following are true:

(i) $\overline{z + w} = \bar{z} + \bar{w}$
(ii) $\overline{zw} = (\bar{z})(\bar{w})$
(iii) $(\bar{\bar{z}}) = z$, *if, and only if, z is a real number*
(iv) $|z| = |\bar{z}|$

proof. We prove (ii) and leave parts (i), (iii), and (iv) for exercises. Let $z = x + yi$ and $w = r + si$. Then $zw = (xr - ys) + (xs + yr)i$.

$$(\bar{z})(\bar{w}) = (x - yi)(r - si) = [xr - (-y)(-s)] + [x(-s) + (-y)r]i$$

$$= (xr + ys) - (xs + yr)i$$

$$= \overline{zw}$$

example 5

Let $z = 2 - i$ and $w = (1/2) + (1/3)i$. Then

$$|z| = \sqrt{5} \text{ and } |w| = \frac{\sqrt{13}}{6}$$

$$z + w = \frac{5}{2} - \frac{2}{3}i$$

$$zw = \frac{4}{3} + \frac{1}{6}i$$

Taking conjugates gives the following.

$$\bar{z} = 2 + i, \ \bar{w} = \frac{1}{2} - \frac{1}{3}i$$

$$\overline{z + w} = \frac{5}{2} + \frac{2}{3}i = \bar{z} + \bar{w}$$

$$\overline{zw} = \frac{4}{3} - \frac{1}{6}i = (\bar{z}\bar{w})$$

Notice also that

$$\frac{\bar{z}}{\bar{w}} = -\frac{2 - i}{(1/2) + (1/3)i}$$

To write this number in standard form, multiply numerator and denominator by $\bar{w}$.

$$\frac{z}{w} = \frac{z\bar{w}}{w\bar{w}} = \frac{z\bar{w}}{|w|^2} = \frac{(2 - i)(\frac{1}{2} - \frac{1}{3}i)}{(\frac{1}{2} + \frac{1}{3}i)(\frac{1}{2} - \frac{1}{3}i)} = \frac{\frac{2}{3} - \frac{7}{6}i}{\frac{13}{36}}$$

$$= (\tfrac{2}{3} - \tfrac{7}{6}i)(\tfrac{36}{13}) = \tfrac{24}{13} - \tfrac{42}{13}i$$

complex numbers

We shall use these facts and others about complex numbers throughout the rest of the text. In the last section of Chapter 5 we shall introduce the polar form of a complex number. This will make it easier to determine the answers to questions concerning the complex numbers; for example, what is $\sqrt{i}$?

exercises

1. Plot each of the following complex numbers on an Argand diagram.

(a) $-i$

(b) 5

(c) $1 - i$

(d) $1 + i$

(e) 0

(f) $6i$

(g) $\dfrac{3}{2} - \dfrac{2}{3}i$

(h) $\dfrac{1}{2} - \dfrac{3}{5}i$

(i) 15

(j) $5 - 7i$

(k) π

(l) $\pi - i$

(m) $\dfrac{1}{2} - \dfrac{\sqrt{3}}{2}$

(n) $1 - i\sqrt{2}$

(o) $\sqrt{3} - i\sqrt{2}$

2. Write the conjugate for each of the complex numbers in Exercise 1.

3. Plot the complex numbers of Exercise 2 on an Argand diagram.

4. Find the absolute value of each of the complex numbers in Exercise 1.

5. Perform the indicated operations and put the answer in standard form.

(a) $(2 - i) + (3 + 7i)$

(b) $(4 - i) - (6 - 2i)$

(c) $(1/2 - i)(2/3 - (1/6)i)$

(d) $(\pi i) + \left(\dfrac{\pi}{2}i\right)$

(e) $(6/5 - i\sqrt{3}/2) + (1/5 + i\sqrt{3}/2)$

(f) $(1 + i)(1 - i)$

(g) $(2 - 3i)(1 + 2i)$

(h) $(1/2 - i\sqrt{2}/2)(-1/2 - i\sqrt{2}/2)$

(i) $(\sqrt{3} - i)(i + \sqrt{2})$

(j) $(\pi + i)(1 - \pi i)$

6. Perform the indicated divisions by multiplying both numerator and denominator by the conjugate of the denominator. Put the answer in standard form.

(a) $\dfrac{-11i}{2 - 7i}$

(b) $\dfrac{3 + i}{6 + i}$

(c) $\dfrac{-2 + 3i}{-2 - 3i}$

(d) $\dfrac{1 + i}{1 - i}$

(e) $\dfrac{5 - 2i}{7 + 3i}$

(f) $\dfrac{-5 + 2i}{1 + 3i}$

(g) $\dfrac{-2 - i}{1 - 3i}$

(h) $\dfrac{1}{1 + i}$

(i) $\dfrac{13 + 10i}{-7i}$

(j) $\dfrac{\sqrt{3} - i\sqrt{2}}{\sqrt{2} - i\sqrt{3}}$

7. Find each of the following. Put the answer in standard form.

(a) i^2

(b) $(-i)^2$

(c) $1/i$

(d) i^3

(e) $1/i^3$

(f) $|i|$

(g) $|i^3|$

(h) i^{10}

(i) i^4

(j) i^{27}

(k) i^{4n}, $n \in Z$

(l) i^{4n+2}, $n \in Z$

8. Prove that for all complex numbers $(x,y) \in C$ addition satisfies the following field postulates.

(a) closure, $(a,b) + (c,d) \in C$

(b) commutativity, $(a,b) + (c,d) = (c,d) + (a,b)$

(c) associativity, $(a,b) + [(c,d) + (e,f)] = [(a,b) + (c,d)] + (e,f)$

fundamentals

(d) additive identity, $(a,b) + (0,0) = (a,b)$

(e) additive inverse, $(a,b) + [-(a,b)] = (0,0)$. What is $-(a,b)$?

(f) Show how $(0,0)$ can be construed to be the real number 0.

9. Prove that for all complex numbers $(x,y) \in C$ multiplication satisfies the following field postulates.

(a) closure, $(a,b)(c,d) \in C$

(b) commutativity, $(a,b)(c,d) = (c,d)(a,b)$

(c) associativity, $(a,b) [(c,d)(e,f)] = [(a,b)(c,d)] (e,f)$

(d) multiplicative identity: Show that $(1,0)$ is the multiplicative identity for C. Example 1 shows the existence of multiplicative inverses. That is,

$$(a,b)^{-1} = \left(\frac{a}{a^2 + b^2}, \frac{-b}{a^2 + b^2} \right)$$

10. Show that complex multiplication distributes over complex addition;

$$(a,b) [(c,d) + (e,f)] = (a,b)(c,d) + (a,b)(e,f)$$

11. Prove Theorem 1.5.1 (i) that the conjugate of a sum is the sum of the conjugate.

12. Prove Theorem 1.5.1 (iii) that the conjugate of a complex number z equals z if and only if z is a real number.

13. Prove Theorem 1.5.1 (iv) that a complex number has the same absolute value as does its conjugate.

14. Given $z = x + yi$, a complex number, show that

(a) $\bar{\bar{z}} = z$

(b) $|-z| = |z|$

(c) $|z| = 0$ if and only if $z = 0$

(d) $|z| = \sqrt{z\bar{z}}$

(e) $\bar{z} + z$ is real

(f) $\bar{z} - z$ is pure imaginary unless z is real

15. Solve the following equations for x and y. Remember $a + bi = c + di$ if and only if $a = c$ and $b = d$ as real numbers.

(a) $5 - yi = x + 3i$

(b) $x/2 + 3i = 1 - yi$

(c) $-4 + 6yi = x/2 - i\sqrt{2}$

(d) $2x - i = i(x - yi)$

(e) $x + 5 - 2iy = i(x + yi)$

(f) $(x + 2y) + i(4x - 3y) = (2x - 1) + i(y - 6)$

16. Suppose z and w are any two complex numbers. Prove that

(a) $|zw| = |z| \, |w|$

(b) $|z/w| = \dfrac{|z|}{|w|}$ if $w \neq 0$

(c) $|z + w| \leq |z| + |w|$

(d) $|z - w| \leq |z| + |w|$

17. Use an Argand diagram to show that if r is a fixed positive real number then the set

$$\{z \in C \mid |z| = r\}$$

is a circle of radius r.

18. Show that if $z \in C$ is in fact a real number then the definitions of absolute value $|z|$, given here and in Section 1.2, are the same.

19. Show that if a point P in the plane is located by the complex number $z = a + bi$, then multiplying z by i rotates the point in a counterclockwise direction through $90°$.

32

20. Show that multiplication by $-i$ rotates points clockwise $90°$.

1.6. functions

The concept of a function is all pervasive in mathematics. It is, no doubt, on the level of the concepts of *number* and *shape* as a fundamental building block of mathematical ideas. As with both of these, functions are abundant in nature. One encounters them in such statements as "the cooking time depends upon the altitude," or "the area of a circle is a function of its radius," or "the administration of justice these days seems to be a function of how the judge liked his breakfast."

Like many other notions in mathematics, the concept of a function has undergone considerable evolution, beginning with Descartes who, in 1637, used it to mean some positive integral power x^n of a variable x. The influence of the present-day French school of mathematicians, who have dubbed themselves Nicholas Bourbaki, has sharpened and refined the concept considerably in the past few years. Because of this, the student may detect some slight variance among mathematics texts regarding some of the terminology involved. Once the concept is mastered, however, the difficulties caused by these differences are easily surmounted.

Before we attempt to give the mathematically formal definition of a function, we shall investigate several examples and try to develop the intuitive idea involved. The function concept involves three things: (1) a set D of objects, called the **domain** of the function, (2) a second set R called the **range** of the function, and (3) a rule which assigns to each member of D a unique member of R. In the previously mentioned examples, numbers which measure altitude (above sea level) make up the domain of the "cooking time" function, while numbers which measure time make up the range; the rule of correspondence between the two is not formally stated in this case. For the "area of a circle function," since any positive real number can measure the length of the radius of a circle and since any such number could also be an area for some circle, both the domain and the range of this function are the set of positive real numbers. The rule of correspondence is given by the formula $A = \pi r^2$, A denoting the area, r the radius.

Another example of a function is found in the fact that every worker in the United States (more or less) is assigned a Social Security number. In this case the domain is a set of people (workers in the U.S.) and the range is a collection of nine-digit positive integers; the rule of correspondence being on file somewhere in the Federal Social Security Administration.

A familiar function to which we shall often refer is the *postage stamp function*. The rule of correspondence for this function is found in the postal regulations. To send a letter or package by first class mail, one must pay postage in the amount of $p*$ cents per ounce or fraction thereof. The domain of the postage stamp function is the set of all positive real numbers (weights of letters), while the range is the set of all positive integral multiples of p.

There are a number of synonyms for *function* in use in mathematics. Among these are the words *mapping*, *transformation*, and *correspondence*.

The geographic theory of map making further illustrates the function

*Since postage rates tend to change more rapidly than the publication time of texts, the student is invited to substitute the *current* postal rate for p.

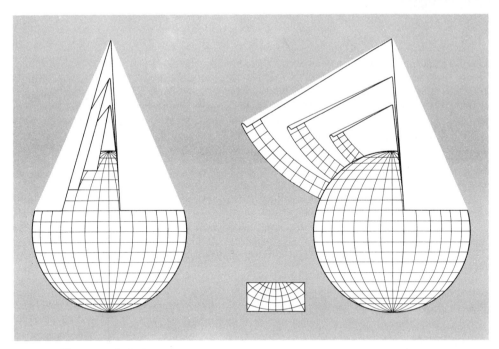

figure 1.6.1

Conical projection.

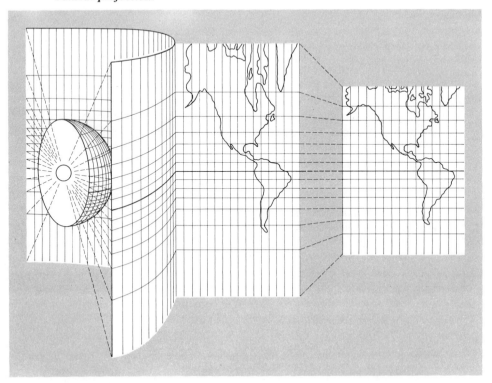

figure 1.6.2

Central projection of the globe upon a cylinder.

functions

concept. One common method for making a map of a portion of the earth's surface is to place a cone tangent to the surface of the earth at a central point of the area to be mapped. Then the points on the earth are "projected" onto the surface of the cone. The portion of the cone is then slit open and unfolded to give the flat map.

In the well-known Mercator projection, a cylinder rather than a cone is used. This accounts for the large amount of distortion near the poles.

Either of these techniques is an example of a function. The points on the surface of the earth make up the domain and the points on the cone (the eventual map) the range.

A common method by which numerical functions are described is by the use of an algebraic expression or formula which gives the rule for finding the member of the range when a typical element from the domain is given. Thus, if $x \in D$ (the domain) and $y \in R$ (the range), the "square" function can be expressed by

$$y = x^2$$

where the domain is "understood" to be the set of all real numbers and the range, of course, is the collection of all positive real numbers.

We turn now to a first definition of *function* suggested by the preceding examples.

first definition

A **function** f from a set D into a set R is a rule which assigns to each member x of the set D a unique element of R. The set D is called the domain of the function, the set R is called the range (codomain, target), and the member of R assigned by f to $x \in D$ is denoted by $f(x)$.

The symbol $f(x)$ is read "f of x" and is called the *image* of x under f, or the *value* of the function f at x. For instance, if f denotes the square function from the real numbers to the real numbers, $f(1) = 1$, $f(2) = 4$, $f(-2) = 4$ and in general, $f(x) = x^2$ for every real number x.

In symbols we denote the fact that f is a function from D into R by

$$f: D \rightarrow R$$

and also the diagram

$$D \xrightarrow{f} R$$

We may also depict this with Venn diagrams as is shown in Figure 1.6.3. This figure points out that it is not necessary that every element of the range R be assigned to some member x of D. The set of points which are assigned is denoted by $f(D)$.

figure 1.6.3

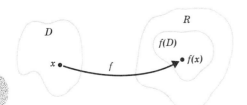

fundamentals

example 1

Consider the following "final grade sheet" for a small class. This illustrates a function with domain a set of students and range the set of possible letter grades, {A,B,C,D,E,W}.

table 1.6.1

	Student	Grade
1.	Anderson, Robert	B
2.	Andrews, George	C
3.	Baker, Robert B.	W
4.	Bills, Alice	A
5.	Cole, Charles	C
6.	Christensen, Marsha	B
7.	Davis, Samuel	B
8.	Newman, A. Paul	E
9.	Welch, Rachel T.	A

Note that we could represent this function with the diagram in Figure 1.6.4. Every member of the domain (student) is assigned a grade, yet the grade "D" is not used. Also we note that more than one student can receive the same grade: three were assigned a "B." This is typical of functions in general. ●

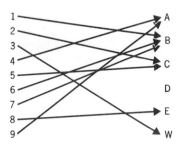

figure 1.6.4

Consider now several examples of functions. Each of them occurs often enough that it has its own name.

example 2. the identity function

Consider a given set X. The function

$$i: X \to X$$

which associates each element of X with itself is called the *identity function.* Thus, $i(x) = x$. If X is the set R of real numbers, a partial table which describes some of the behavior of the identity function on R is given in Table 1.6.2.

36

functions

table 1.6.2

x	0	1	-1	π	0.003	2	$\sqrt{2}$
$i(x)$	0	1	-1	π	0.003	2	$\sqrt{2}$

example 3. constant functions

Let X and Y be any two sets, and let c be a fixed element of Y. The function

$$f_c: X \rightarrow Y$$

defined by

$$f_c(x) = c, \text{ for each } x \in X$$

assigns the same member (constant) of Y to each member of X. This is the *constant function* f_c. In particular, if X and Y are both the set R and $c = \pi$, we have

$$f_\pi(x) = \pi$$

A function table for f_π would contain Table 1.6.3.

table 1.6.3

x	0	1	-1	2	$\sqrt{2}$	π	1/2
$f_\pi(x)$	π	π	π	π	π	π	π

The special constant function $f_o: X \rightarrow R$ is called the *zero function*.

example 4. the absolute value function

This function from R to R was defined in Section 2. We have

$$\mathrm{abs}(x) = |x| = \begin{cases} x, & \text{if } x \geq 0 \\ -x, & \text{if } x < 0 \end{cases}$$

example 5. the greatest integer function

Like the absolute value function, we employ a special symbol for this function from the set of real numbers into itself.

$$g: R \rightarrow R$$

The function g is defined as follows: $g(x) = [x]$ where, for each real number x, we let $[x]$ denote the largest integer not greater than x. Thus,

$$[0] = 0, \quad [2] = 2, \quad [\pi] = 3, \quad [-1.3] = -2$$

fundamentals

If x is a positive real number, and $n.d_1 d_2 d_3 \ldots$ is its decimal expansion, then $[x] = n$. If x is negative (hence, n is negative), then $[x] = n - 1$. You will have the opportunity, in Exercise 8f, to construct a partial function table for this function. ●

example 6.   the conjugation function

Consider the function conj: $C \to C$ from the set C of complex numbers into itself given by conj$(z) = \bar{z}$. Recall, if $z = a + bi$, then $\bar{z} = a - bi$. The properties of the conjugate of a complex number were discussed in detail in Section 1.5. Therefore, conj$(\pi) = \pi$, and conj$(\pi i) = -\pi i$, while conj $(\sqrt{2} - i) = \sqrt{2} + i$. ●

example 7.   the negation function

This function also has domain and range the same set; viz., the set R,

$$n: R \to R$$

The function is given by the rule $n(x) = -x$. For instance, $n(2) = -2, n(-3) = 3,$ $n(\pi) = -\pi, n(-\sqrt{3}/2) = \sqrt{3}/2$, etc. ●

Let f and g be functions on some common domain X. Let S be a subset of X. We say that f and g are **equal on** *S provided*

$$f(x) = g(x)$$

for each $x \in S$. Of course, if S is all of X, then f and g are **equal as functions**.

example 8

Consider the correspondence $f: R \to R$ given by the statement(s)

$$f(x) = \begin{cases} \sqrt{x}, & \text{for } x \geq 1 \\ |x|, & \text{for } x \leq 0 \end{cases}$$

This statement tells us that f equals the square root function on the set of all positive real numbers 1 or larger, and that f equals the absolute value function on the set of all non-negative real numbers. It also tells us that f is not a function on all of R. Its domain is the set $S = R \setminus (0, 1)$ consisting of the real line with the "chunk" between 0 and 1 removed. The value of $f(1)$ is $= \sqrt{1} = 1$, and the value of $f(0)$ is $|0| = 0$. Additional sample values of f, for members of S, are the following: $f(4) = \sqrt{4} = 2$; $f(-4) = |-4| = 4$; $f(3/4) = \sqrt{3}/2$; $f(-3/4) = 3/4$, etc. ●

example 9

Consider the functions f and g given by the formulas

$$f(x) = \frac{x^2 - 4}{x + 2}, \qquad g(x) = x - 2$$

Notice first that the domain of f is not all of R since the formula gives no value for $f(-2)$. The domain of f is the set $D = R \setminus \{2\}$. The domain of g is all of R. Since $D \subset R$, $R \cap D = D$. On the set D, note that $x + 2 \neq 0$.

functions

Therefore

$$f(x) = \frac{x^2 - 4}{x + 2} = \frac{(x - 2)(x + 2)}{(x + 2)} = x - 2 = g(x)$$

Thus, while f and g are not equal functions, they are equal on D. ●

In a subsequent section we shall sketch graphs of these functions.

exercises

1. Consider each of the following diagrams. State whether or not the diagram defines a function from the set $A = \{1, 2, 3\}$ to the set $B = \{x, y, z\}$.

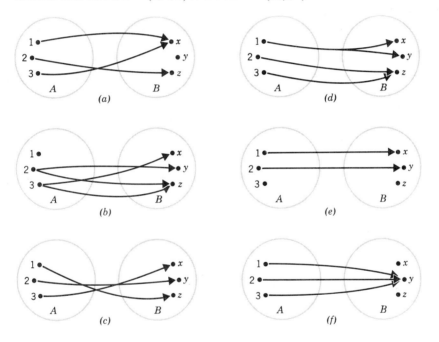

2. Let S denote the set of all states in the United States, and let C denote the set of all cities in the United States. For each state $s \in S$, let $k(s)$ denote the capital city. Is the correspondence $k : S \rightarrow C$ a function? Construct a partial function table showing the capitals of about ten states.

3. Let S be, as in Exercise 2, the set of all states. Let U be the set of members of the United States Senate. For each $s \in S$ let $f(s)$ denote both the United States senators from the state. Is f a function? What about the assignment $s \rightarrow t(s)$ where $t(s)$ is the senior United States senator from that state?

4. Let P be the set of all former presidents of the United States, and let V denote the set of all former vice-presidents of the United States. For each $p \in P$, let $v(p)$ denote the vice-president with whom he served at his first inauguration as president. Is the correspondence $p \rightarrow v(p)$ a function? Explain.

5. Consider the function $f : R \rightarrow R$ defined by $f(x) = 3x^2 - 6x - 9$, for each $x \in R$. Find $f(0), f(1), f(\sqrt{2}), f(3), f(-3), f(h), f(x + h), f(x + h) - f(x)$.

39

fundamentals

6. Let X be a set, and let E be any subset of X, and let $B = \{0, 1\}$. Consider the correspondence $\chi_E \colon X \to B$ given by

$$\chi_E(x) = \begin{cases} 1, & \text{for } x \in E \\ 0, & \text{for } x \in X \backslash E \end{cases}$$

This is a function called the **characteristic function of E**. Construct a function table for χ_E in each of the examples given below.

    (a) $X = \{2, 4, 6, 8, 10, 12\}$, $E = \{6, 12\}$.
    (b) X the set of letters in the English alphabet, E the set of vowels.
    (c) X the set of letters in the English alphabet, E the letters in the word "Mississippi."

7. Consider the set S of states in the United States and the set U of senators (Exercise 3) and the correspondence $s \to f(s)$ which assigns to each state its senators. Is this a function with domain S and range 2^U? If so, make a small partial function table.

8. Make partial function tables (different from those already given in the text) for each of the following functions.

    (a)   the constant function f_2 (e)   the postage stamp function
    (b)   the constant function $f_{1/2}$ (f)   the greatest integer function
    (c)   the zero function   (g)   the conjugation function
    (d)   the absolute value function   (h)   the negation function

9. Consider the following *successor functions*. Make a partial function table for each.

    (a) $s \colon Z \to Z$, where $s(n) = n + 1$, $n \in Z$.
    (b) $s \colon 2Z \to 2Z$, where $s(k) = k + 2$, $2Z$ is the set of even integers.
    (c)   Let A be the set of letters of the English alphabet ordered lexocographically (the usual way), and let $s(\alpha)$ denote the next letter for $\alpha \in A$, $\alpha \neq z$, and $s(z) = a$.

10. Describe the postage stamp function with a formula using the greatest integer function.

11. For each of the following numerical functions find $f(0)$, $f(1)$, $f(-1)$, $f(2)$, $f(-2)$, $f(\pi)$, $f(3/2)$, $f(\sqrt{2})$, and $f(1.6)$.

    (a) $f(x) = x^2$ (f) $f(x) = [x] + 1$
    (b) $f(x) = \sqrt{x}$, the *positive* square root of x (g) $f(x) = [x + 1]$
    (c) $f(x) = 2x + 1$ (h) $f(x) = 2[x]$
    (d) $f(x) = |x| - 1$ (i) $f(x) = [2x]$
    (e) $f(x) = |x - 1|$

12. Consider the following formulas. In each case decide whether the correspondence $x \to f(x)$ defines a function $f \colon R \to R$. If it does, find $f(0)$, $f(1)$, $f(-1)$, $f(2)$, $f(\pi)$, $f(1/2)$, and $f(\sqrt{2})$.

    (a) $f(x) = x^2 + x + 1$

    (b) $f(x) = \begin{cases} 1, & \text{for } x \leq 0 \\ x - 1, & \text{for } x > 0 \end{cases}$

    (c) $f(x) = \begin{cases} x^3, & \text{for } x \leq 0 \\ x^2, & \text{for } x > 0 \end{cases}$

    (d) $f(x) = \begin{cases} 1 + x, & \text{for } x \in [-1/2, 1/2] \\ 1 - x^2, & \text{for } x \notin [-1/2, 1/2] \end{cases}$

    (e) $f(x) = \dfrac{1}{1 - x}$

    (f) $f(x) = \dfrac{x^2 + 4}{x^2 - 4}$

    (g) $f(x) = \dfrac{1}{|x + 1|}$

13. Suppose a farmer has 100 feet of fencing material with which to make a rectangular pen. Write a formula which gives the area of the pen as a function of its width.

14. Write a formula which expresses the Centigrade temperature as a function of the Fahrenheit temperature.

1.7. properties of functions

Let us consider two sets X and Y and a function f with X as its domain and Y as its range.

$$f: X \to Y$$

We depict this in the Venn diagram (see Figure 1.7.1).

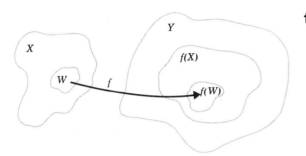

figure 1.7.1

The subset $f(x)$ of Y, which consists of the images in Y of all the members of the domain X and is defined by

$$f(X) = \{y \in Y \mid y = f(x) \text{ for some } x \in X\}$$

is called the **image** *of the function f. If $W \subset X$, is any subset of the domain of f, then $f(W) \subset Y$, consists of all the images of elements of W. Thus,*

$$f(W) = \{y \in Y \mid y = f(x) \text{ for some } x \in W\}$$

and is called the **image** *of W.*

In general, the image $f(X)$ can be a proper subset of the range of f. As examples we notice first the postage stamp function, whose range could be considered to be all of R, but whose image is only the positive integral multiples of the current first class rate p. In the case of the constant function

$$f_\pi: R \to R$$

while the function is into R, the image $f(R)$ is the singleton set $\{\pi\} \subset R$. The range of the absolute value function is again R, but its image is only R^+, the set of non-negative real numbers. This is also true for the square function $y = x^2$.

On the other hand, for some functions (notably the identity function, the conjugation function and the negation function to name a few), the image $f(X)$ coincides with the range: $f(X) = Y$. Consider Figure 1.7.2.

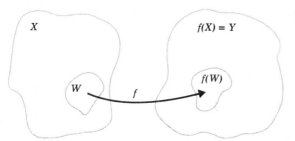

figure 1.7.2

fundamentals

When $f(X) = Y$, the function f is said to be **onto** Y or **surjective.** (Notice that *sur*-jective has a French prefix: $sur = $ on.) A surjective function is a **surjection**. The identity function, the negation function, the conjugation function, and the successor function are all surjections. While the distinction between an *onto function* and an arbitrary function is important in some contexts, it is unimportant in others for the simple reason that one can always take the range set for a given function to be exactly its image. Thus, one can consider the range of the postage stamp function to be just the positive integral multiples of p, the range of the absolute value function, and the square function to be just the non-negative reals, etc. In this case, each of these becomes an onto function (surjection). Hence, whether a function is onto its range or not depends upon what we want its range to be.

example 1

The greatest integer function

$$g: R \to R$$

where $g(x) = [x]$ is *not onto* R. However, the function

$$\hat{g}: R \to Z$$

where $\hat{g}(x) = [x]$ is clearly *onto* Z (hence, a surjection). ●

You will have noticed that there are many ways to describe functions. A function may be defined by:

a formula
a regulation or a descriptive rule
a table of values
a diagram or graph

Some functions readily accept definition by a combination of these methods while others do not. The larger the variety of possible descriptions a given function has, the greater is our understanding of it.

Given a function

$$f: X \to Y$$

where $X = Y = R$. We consider a technique for sketching a graph of f. *By the* **graph of f** *we mean the collection of all the points* $(x, f(x))$ *in* $R \times R = X \times Y$. Thus, the graph of the function f is the same as the graph of the equation $y = f(x)$. We have already seen that subsets of $R \times R$ can be represented as subsets of the plane. Thus, if we plot the elements $(x, f(x))$ as points in the plane, we have a sketch of the graph of f.

example 2.   the negation function

We have

$$n(x) = -x,$$

so we plot points $(x, n(x)) = (x, -x)$ as in Figure 1.7.3. ●

properties of functions

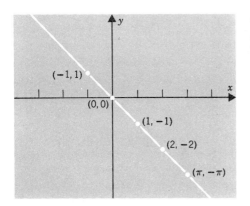

figure 1.7.3

The negation function.

example 3.   the greatest integer function
To sketch this graph we plot the points $(x, [x])$.   ●

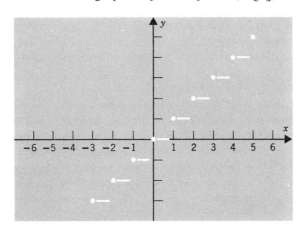

figure 1.7.4

The greatest integer function.

example 4.   the absolute value function
To sketch the graph of the absolute value function, we sketch the graph of $y = x$ for $x \geq 0$ and $y = -x$ for $x < 0$. Thus, we have the following graph.   ●

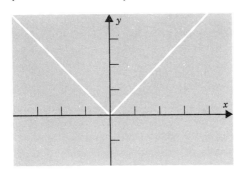

figure 1.7.5

The absolute value function.

This graphing can be done even for functions where X and Y are not sets of numbers. Consider the grading function in Table 1.6.1 of the previous section. If X is the set of students in the class and Y is the set of possible grades, we sketch $X \times Y$ and the function in Figure 1.7.6. Compare this with Figure 1.6.4.

43

fundamentals

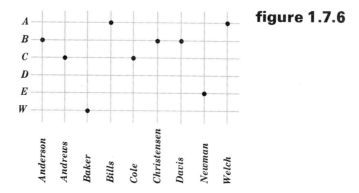

figure 1.7.6

Because our first definition of function, although highly intuitive, uses the words "*rule of correspondence*" which are no more primitive terms than the word "*function*" itself, a more "rigorous" definition based on set theory is given by most mathematicians. This essentially involves identifying a function with its graph.

second definition

A function f from a set X into a set Y is a collection of ordered pairs (x,y) such that each element of X appears as the first member of *exactly* one pair.

Thus, a function f can be considered to be a special subset of $X \times Y$. This definition uses only the primitive terms and previously defined concepts of set theory.

example 5

The identity function $i: R \to R$ is the set

$$i = \{(x,x) \mid x \in R\}$$

The negation function $n: R \to R$ is the set

$$n = \{(x, -x) \mid x \in R\}$$

The square function is the set

$$\{(x, x^2) \mid x \in R\} \quad \bullet$$

An *onto* function (surjection) can also be defined in these set theoretic terms as follows: *The function*

$$f: X \to Y$$

is onto (surjective) if each element of Y appears as the second element of at least one ordered pair $(x,y) \in f$.

The definition of a function, requiring that each element of X appear in as the first coordinate *exactly one* pair (x,y) in f, has a geometric interpretation which is useful. If a vertical line is drawn through any point in the domain X (on the X axis) of f, it must intersect the graph of f in *precisely* one point.

44

properties of functions

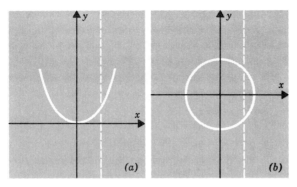

figure 1.7.7

(a) $X = Y = R$;
(b) $X = Y = [0,1]$

Consider the two graphs in Figure 1.7.7. The first one (a) is a graph of the square function $y = x^2$ and satisfies this vertical line criterion. The graph of the second, the unit circle $x^2 + y^2 = 1$, is not a function.

The square function, considered to have range R, is into R, but not onto R. The geometric interpretation for the definition of an onto function (surjection) is that a *horizontal* line through any point of the range (Y axis) must intersect the graph of f in *at least* one point. Figure 1.7.8 shows that the cube function $y = x^3$ is surjective.

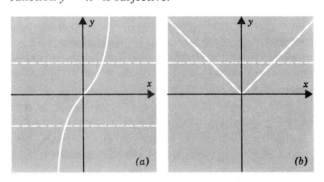

figure 1.7.8

(a) $f(x) = x^3$
one-to-one
(injective)
(b) $f(x) = |x|$
not one-to-one
(not injective)

Notice that if f is the square function $f(x) = x^2$, we have $f(2) = f(-2) = 4$. But if f is the cube function, $f(x) = x^3$, then $f(a) = f(b)$ if and only if $a = b$. Similarly, if f is the absolute value function $f(x) = |x|$, then $f(a) = f(-a) = |a|$ for any real number a; and if $f(x) = [x]$, then $f(a) = f(b)$ as long as a and b lie in the same unit interval $[n, n + 1)$, $n \in Z$. On the other hand, if f is the negation function $f(a) = f(b)$ if and only if $a = b$.

The negation function and the cube function have the property that *any element in the range is the image of at most one element of the domain.* Such functions are called **one-to-one functions** or **injections**.

*A function $f: X \to Y$ is **one-to-one (injective)** if, and only if, $f(a) = f(b)$ implies $a = b$.*

This can be geometrically pictured by noting that any *horizontal* line can intersect the graph of f in *at most* one point (note Figure 1.7.8).

*A function which is both one-to-one (injective) and onto its range (surjective) is called a **one-to-one correspondence**, or a **bijection**.* We say that X and Y are *in one-to-one correspondence* if there exists a function

$$f: X \to Y$$

from X onto Y which is also *one-to-one.*

fundamentals

The cube function $y = x^3$ is both one-to-one and onto all of R, hence it is a one-to-one correspondence (bijection) and, of course, the absolute value function is neither.

example 6

The function $f: R \to R$ given by $f(x) = 2x + 1$ is a bijection. To see this, first note that it is one-to-one (injective) since, if $f(a) = f(b)$, then

$$2a + 1 = 2b + 1$$

Therefore, $a = b$. Similarly, if y is any real number in the range $f(x) = y$ for $x = y - 1/2$. This x is obtained by solving the equation $y = 2x + 1$ for x. Hence, f is also onto R (a surjection). ●

The following Venn diagrams summarize the definitions of this section.

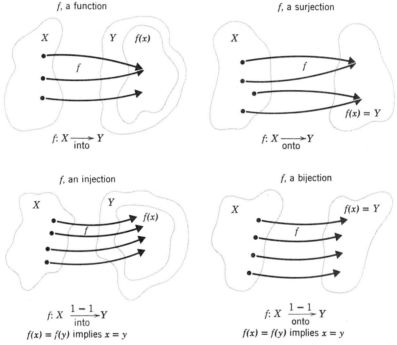

f, a function   f, a surjection

f, an injection   f, a bijection

figure 1.7.9

example 7

Sketch a graph of the function f of Example 8 in the previous section. This function had domain $S = R \setminus (0, 1)$, and was described by the formulas

$$f(x) = \begin{cases} \sqrt{x}, & \text{if } x \geq 1 \\ |x|, & \text{if } x \leq 0 \end{cases}$$

Thus, the graph does not exist over the open interval $(0, 1)$. Over the ray $[1, \infty)$ we plot points $(x, \sqrt{x})$, and on the negative half of the plane we plot points $(x, |x|) = (x, -x)$. The resulting graph is Figure 1.7.10. ●

46

properties of functions

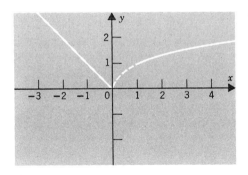

figure 1.7.10

$$f(x) = \begin{cases} \sqrt{x}, & \text{if } x \geq 1 \\ |x|, & \text{if } x \leq 0 \end{cases}$$

exercises

1. Determine which of the functions given below are *onto* their range, which are *one-to-one*, and which are both (*bijective*). The range and domain are as given.

    (a) the constant function $f_{1/2}: R \to R$ $\quad f(x) = \frac{1}{2}$ not onto here
    (b) the zero function $f_0: R \to Z$
    (c) the absolute value function $a: R \to R$
    (d) the identity function $i: R \to R$
    (e) the fifth power function $p_5: R \to R$ where $p_5(x) = x^5$ bijective
    (f) the conjugation function conj: $C \to C$
    (g) the postage stamp function $p: R \to Z$
    (h) the greatest integer function $g: R \to Z$
    (i) the successor function $s: Z \to Z$ where $s(n) = n + 1$ bijective

2. Do the same as in Exercise 1 for the following functions $f: Z \to Z$.

    (a) $f(n) = n + 3$ (d) $f(n) = 2n - 1$
    (b) $f(n) = n^2 + n$ (e) $f(n) = 5 - n$
    (c) $f(n) = n^3$

    (f) $f(n) = \dfrac{n + 1}{n - 1}$

3. Which of the following could be functions? Indicate a possible domain and range if they are. Explain.

    (a) $f = \{(0, 1), (1, 2), (2, 3)\}$
    (b) $g = \{(1, 0), (0, 1), (1, 2), (0, 3)\}$
    (c) $h = \{(x, x) \mid x \in R\}$
    (d) $j = \{(x, 2) \mid x \in R\}$
    (e) $k = \{(x, y) \mid x^2 + y^2 = 2, x, y \in R\}$
    (f) $1 = \{(x, y) \mid x = 1/y, x, y \in R\}$
    (g) $m = \{(x, y) \mid x^3 = y^2, x, y \in R\}$
    (h) The set Z^2 in the plane

4. Sketch graphs for each of the following functions. Determine their domains and images.

    (a) $f(x) = x/2$ (g) $f(x) = |x + 1|$
    (b) $f(x) = x - 3$
    (c) $f(x) = -[x]$ (h) $f(x) = \dfrac{|x|}{x}$
    (d) $f(x) = [x - 1]$
    (e) $f(x) = \sqrt{4 - x^2}$ (i) $f(x) = x^4$
    (f) $f(x) = 2[x] - 1$ (j) $f(x) = x^3 - 1$

47

fundamentals

5. Which of the following statements is *always true* for any function f?

   (a) If $a = b$, then $f(a) = f(b)$.
   (b) If $a \neq b$, then $f(a) \neq f(b)$.
   (c) If $f(a) = f(b)$, then $a = b$.
   (d) If $f(a) \neq f(b)$, then $a \neq b$.

6. Which of the statements in Exercise 5 is true only if f is one-to-one (injective)?

7. Sketch the graphs of each of the following functions

$$g: R \to R.$$

   (a) $g(x) = \begin{cases} 1, & \text{if } x \leq 0 \\ 0, & \text{if } x > 0 \end{cases}$

   (b) $g(x) = \begin{cases} 0, & \text{if } x \leq 0 \\ x, & \text{if } x > 0 \end{cases}$

   (c) $g(x) = \begin{cases} -x^2, & \text{if } x < 0 \\ x^2, & \text{if } x \geq 0 \end{cases}$

   (d) $g(x) = \begin{cases} x + 1, & \text{if } x < 0 \\ 0, & \text{if } x = 0 \\ x - 1, & \text{if } x > 0 \end{cases}$

   (e) $g(x) = \begin{cases} x^2 + 1, & \text{if } x \leq 0 \\ x - 3, & \text{if } x > 0 \end{cases}$

   (f) $g(x) = \begin{cases} |x| - 1, & \text{if } x \leq 0 \\ 1 + |x|, & \text{if } x > 0 \end{cases}$

   (g) $g(x) = \begin{cases} x - [x], & \text{if } x \leq 0 \\ [x] + 1, & \text{if } x > 0 \end{cases}$

8. Can a constant function ever be one-to-one (injective)?

9. Can a constant function ever be onto its range (surjective)?

10. Which of the functions in Exercise 7 are bijections?

11. Consider each of the following graphs of subsets of the plane. Determine which could be graphs of functions from X' to Y, where the domain X' is a subset of X. Of these, which are injective (one-to-one)?

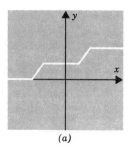

(a)

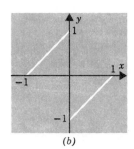

(b)

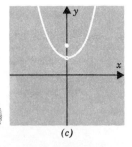

(c)

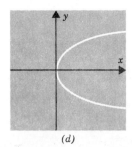

(d)

properties of functions

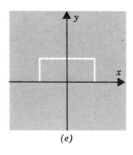

(e)

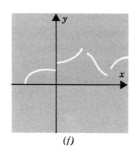

(f)

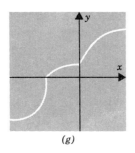

(g)

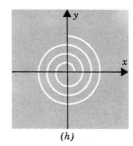

(h)

12. Let $f: R \to R$ be a function with the properties that

$$f(a - b) = f(a) - f(b)$$

Is f always injective? Prove!

13. Let $abs: C \to R$ be a function defined by the following: If $z \in C$, $z = x + iy$, $x, y \in R$, then

$$abs(z) = \sqrt{x^2 + y^2}$$

(a) Is $abs: C \to R$ one-to-one (injective)?
(b) Is $abs: C \to R$ onto R (surjective)?
(c) Make a portion of a function table for abs.

14. *A set X is said to have cardinal number n (often written $o(X) = n$) if, and only if, there exists a one-to-one correspondence (bijection) o between X and the set $\{1, 2, \ldots, n\}$ of the first n positive integers.* Find $o(X)$ for each of the following sets.

(a) $X = $ the set of letters in the alphabet.
(b) $X = $ the set of all presidents of the United States.
(c) $X = $ the set of all current members of the U.S. Senate.
(d) $X = $ the set of all current members of the U.S. House of Representatives.
(e) $X = $ the set of members of the U.S. Supreme Court.
(f) $X = $ the set of students registered for this course.
(g) $X = $ the set of full-time students registered at this school.

15. A set X is called **countable**, and its cardinal number is $\aleph_0$ (aleph-zero), if there exists a bijection between it and the set of *all* the positive integers. Which of the following sets is countable?

(a) the set $2N$ of even positive integers
(b) the set of odd positive integers
(c) the set of letters in the English alphabet

(d)   the set Z of integers
(e)   the set of prime integers
(f)   the set of rational numbers
(g)   the set of real numbers
(h)   the set of complex numbers
(i)   the set of current members of the U.S. Senate

1.8.   an algebra of functions

Let f and g be two functions with the same domain X and range R, the real numbers. *Functions, such as f and g, whose range is R are called* **real functions**. Since R has an algebraic structure with operations (of addition and multiplication) defined, we can construct corresponding operations on real functions as follows.

Function Addition.   *The function $f + g$ is defined by*

$$(f + g)(t) = f(t) + g(t)$$

for each $t \in X$.
    For example, if $f(t_0) = 3$ and $g(t_0) = 2$, for some $t_0 \in X$, then $(f + g)(t_0) = 3 + 2 = 5$. As a second example, suppose

$$f(t) = t^2 + 1 \qquad \text{and} \qquad g(t) = 2t \qquad \text{for each} \qquad t \in X$$

Then,

$$(f + g)(t) = t^2 + 2t + 1 = (t + 1)^2$$

Function Multiplication.   The function (fg) is defined by

$$(fg)(t) = [f(t)][g(t)]$$

Hence, if $f(t) = 3$ and $g(t) = 2$, for some $t \in X$, then $(fg)(t) = 3 \cdot 2 = 6$. As another example, if $f(t) = t^2 + 1$, and $g(t) = 2t$ for each $t \in X$, then

$$(fg)(t) = (t^2 + 1)(2t) = 2t^3 + 2t$$

for each $t \in X$.
    These definitions tell us that the image of an element $x \in X$ under the **sum** $f + g$ of two functions f and g is the *sum* of the images under f and g separately, and the image under the **product** fg, of f and g is the *product* of the separate images.
    Since these definitions depend only on the algebraic structure of the range, they could also be given when the functions f and g have the complex numbers as range. In this case we call them **complex functions**.

example 1

Let $f\colon C \to C$ be the *conjugation* function, $f(z) = \bar{z}$. Let $g\colon C \to C$ be given by the correspondence $x + iy \to y + ix$. Then for $z = x + iy$, $(f + g)(z) = x - iy + y + ix = (x + y) + (x - y)i$, and $(fg)(z) = (x - iy)(y + ix) = 2xy + (x^2 - y^2)i$. ●

50

an algebra of functions

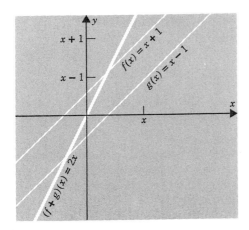

figure 1.8.1

$f + g$

example 2

Let $f(x) = x + 1$ and $g(x) = x - 1$ be two real functions. We sketch the graphs of f, g, and $f + g$ in Figure 1.8.1, and of fg in Figure 1.8.2. ●

We note the following properties of function addition and multiplication which are inherited from the corresponding properties of addition and multiplication in their range, R (or C). For real (complex) function f, g, h, etc., we have the following properties (compare them with Appendix I):

Commutativity	$f + g = g + f$
	$fg = gf$
Associativity	$(f + g) + h = f + (g + h)$
	$(fg)h = f(gh)$
Zero	$f + f_0 = f$, where $f_0(x) = 0$ for all $x \in X$
Identity	$(ff_1) = f$, where $f_1(x) = 1$ for all $x \in X$
Distributivity	$f(g + h) = fg + fh$

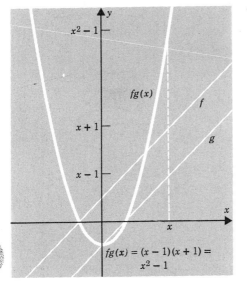

figure 1.8.2

fg

fundamentals

Function subtraction and division can also be defined. Thus,

Subtraction $\quad (f - g)(t) = f(t) - g(t)$ for all $t \in X$

and

Division $\quad \dfrac{f}{g}(t) = \dfrac{f(t)}{g(t)},$ for all $t \in X$ for which $g(t) \neq 0$

example 3

Let $\varphi(x) = x^2 - 4$ and $\psi(x) = x - 2$. We sketch these functions, the function $\varphi - \psi$, and the function φ/ψ in Figure 1.8.3. Notice that φ/ψ has a hole in its graph at $x = 2$ since $\psi(2) = 0$. Thus, 2 is not in the domain of φ/ψ. We have $(\varphi - \psi)(x) = x^2 - x - 2$, and $\varphi/\psi(x) = (x^2 - 4)/(x - 2)$. ●

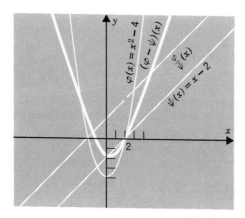

figure 1.8.3

A particular type of function "multiplication" is important enough to single out for special consideration. Let $f\colon X \to R$ be any real function, and let $c \in R$ be an arbitrary, but fixed, real number. We recall that the constant function $f_c\colon X \to R$ maps every $x \in R$ onto the real number c. Thus, the product function $f_c f$ is given by $(f_c f)(t) = cf(t)$. We shall henceforth identify the function cf, where c is any real number, with the product function $f_c f$. This introduces an operation on the class of real functions which is called **scalar multiplication**.

theorem 1.8.1

Let f and g be a real function and let c,d be real numbers. Then,

$$c(f + g) = cf + dg$$
$$(c + d)f = cf + df$$
$$c(fg) = (cf)g$$
$$(cd)f = c(df)$$
$$1f = f$$

This theorem follows immediately from the commutative, associative and distributive properties for function addition and multiplication and the identification $c \to f_c$ described preceding the theorem.

an algebra of functions

example 4

Let $f: X \to R$ be defined by $f(x) = x^2 - 1$. Then $3f: X \to R$ is given by $3f(x) = 3(x^2 - 1)$.

If $g(x) = x + 1$, then $-2g(x) = -2(x + 1)$. Thus,

$$(3f - 2g)(x) = 3x^2 - 2x - 5$$

Note too that $(1/2)f + (5/8)g\,(x) = (x^2/2) + (5/8)x + (1/8)$. ●

If f is a real function and n is a positive integer, we define the function

$$f^n$$

by $f^n(x) = [f(x)]^n$. We generally avoid the notation f^n, when n is negative, because f^{-1} has a special meaning, different from $1/f$. We shall see this in the next section. Thus, if we wish to use negative exponents, to avoid confusion we will only write

$$[f(x)]^{-2}$$

when we mean $1/[f(x)]^2 = (1/f^2)(x)$. Of course, $1/f^2$ is only defined for those $x \in X$ such that $f(x) \neq 0$.

A final, and highly significant, operation on functions which we consider in this section is termed **function composition**: one function followed by another. Let us introduce it with an example.

example 5

At many universities, students are assigned an identification number for use in computerized registration, etc. Consider the students in the class of the example of the grading function given in Section 1.6. (See Table 1.6.1.) The instructor receives a form on which to report the final grades to the records office. The grading sheet looks something like Table 1.8.1.

We see in this the table of two functions: the *student number assigning function*, φ, and a *grading function*, θ. The computerized record system files the grades

table 1.8.1

Student Name	I.D. Number	Final Grade
Anderson, Robert	107141	B
Andrews, George	108211	C
Baker, Robert B.	113110	W
Bills, Alice	117018	A
Cole, Charles	119141	C
Christensen, Marsha	041411	B
Davis, Samuel	200011	B
Newman, A. Paul	204151	E
Welch, Rachel T.	259171	A

fundamentals

by I.D. number rather than by name; i.e., $\theta\,(200011) = $ B. Thus, if one desires to know the grade Alice Bills received, he must first find her I.D. number, φ (Alice Bills) $= 117018$; and then the grade that I.D. number is associated with; i.e., $\theta\,(117018) = $ A. ●

Let $f: X \to Y$ be a function with domain X and range Y; let $g: Y \to W$ have domain Y. The **composition,** *$g \circ f$, of g and f is the function*

$$h: X \to W$$

defined by $h(x) = (g \circ f)(x) = g(f(x))$, for each $x \in X$.

In our example Alice Bills' grade is $(\theta \circ \varphi)$(Alice Bills). We see that the grading function of Table 1.6.1 is, in fact, $\theta \circ \varphi$.

Consider the Venn diagram in Figure 1.8.4, which illustrates the general idea of function composition.

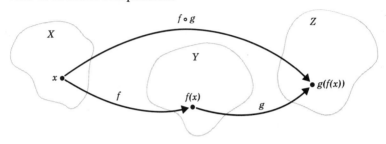

figure 1.8.4

We also diagram this as follows:

$$h: X \overset{f}{\to} Y \overset{g}{\to} W$$

example 6

Suppose $X = Y = W = R$, the real numbers, and f and g are two real functions

$$R \overset{f}{\to} R \overset{g}{\to} R$$

defined by

$$f(x) = 1 - x$$
$$g(x) = 2x + 3$$

Then the composition $g \circ f$ is defined by

$$(g \circ f)(x) = g(f(x)) = 2(f(x)) + 3$$
$$= 2(1 - x) + 3$$
$$= 5 - x$$

We can, in this case, because the domains and ranges match properly, also consider $f \circ g$.

$$R \overset{g}{\to} R \overset{f}{\to} R$$

$$(f \circ g)(x) = f(g(x)) = 1 - g(x)$$
$$= 1 - (2x + 3)$$
$$= -2x - 2 \quad ●$$

an algebra of functions

We notice that, unlike function multiplication, function composition is *not commutative*. The above example also shows us that, in general,

$$f \circ g \neq fg$$

You can readily verify this. However, we do have the following properties of function composition.

theorem 1.8.2

Let f, g, h be three functions with

$$X \xrightarrow{f} Y \xrightarrow{g} W \xrightarrow{h} T$$

Then

$$h \circ (g \circ f) = (h \circ g) \circ f$$

proof. Let $t \in X$ be arbitrary. Then

$$[h \circ (g \circ f)](t) = [h(g \circ f)](t) = h(g[f(t)]) = (h \circ g)(f(t)) = ((h \circ g) \circ f)(t)$$

example 7

Let three real functions f, g, and h be given by the following formulas.

$$f(x) = x + 1$$
$$g(y) = (1/2)y^2$$
$$h(z) = 2z - 1$$

Then

$$(h \circ g \circ f)(x) = h(g(f(x))) = h(g(x + 1)) = h((x + 1)^2/2) = (x + 1)^2 - 1$$
$$= x^2 + 2x$$

We illustrate (graph) this situation in Figure 1.8.5. Note that there are four horizontal copies of the real line representing the situation.

$$R \xrightarrow{f} R \xrightarrow{g} R \xrightarrow{h} R \qquad \bullet$$

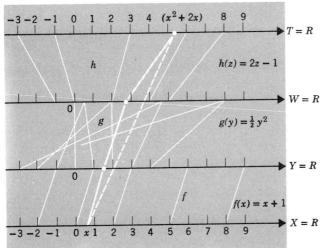

figure 1.8.5

55

fundamentals

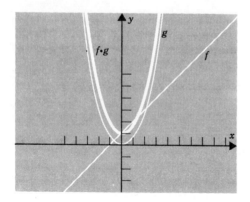

figure 1.8.6

In Figure 1.8.6, the graphs of f, g, and $f \circ g$ of Example 6 are plotted on a *Cartesian* coordinate system: $f(x) = x + 1$ and $g(x) = x^2/2$. Thus, $f \circ g = (x^2/2) + 1$.

exercises

1. Given the functions $f: R \to R$ and $g: R \to R$ defined for each $x \in R$ as follows, find the sum function $f + g$ in each case and sketch its graph.

   (a) $f(x) = 2$, $g(x) = 2x - 1$
   (b) $f(x) = 1 + x$, $g(x) = 1 - x$
   (c) $f(x) = x^2$, $g(x) = 2x + 1$
   (d) $f(x) = 1 - x^2$, $g(x) = (x + 1)^2$
   (e) $f(x) = \sqrt{1 - x}$, $x \le 1$, $g(x) = x^2$
   (f) $f(x) = x^3$, $g(x) = x^{1/2}$, $x \ge 0$
   (g) $f(x) = x^2 + 2x - 3$, $g(x) = x^2 - 3x + 2$
   (h) $f(x) = x^3 - 1$, $g(x) = x - 1$
   (i) $f(x) = \sqrt{x}$, $x \ge 0$, $g(x) = |x + 1|$
   (j) $f(x) = 2x + 1$, $g(x) = [x]$
   (k) $f(x) = (3/2)x - 1$, $g(x) = -[x]$
   (l) $f(x) = x^2 - 1$, $g(x) = x + 1$
   (m) $f(x) = x^2 - 16$, $g(x) = x - 4$
   (n) $f(x) = 9 - x^2$, $g(x) = x + 3$
   (o) $f(x) = x^3 + 1$, $g(x) = x^2 - x + 1$

2. Find the product fg of each of the functions in Exercise 1. Sketch the graph of fg in each case.

3. Find the quotient f/g for each of the functions in Exercise 1. Discuss how the domain must be restricted for f/g to be defined. Sketch a graph of each.

4. Find the composites $f \circ g = f(g(x))$ and $g \circ f = g(f(x))$ for each pair of functions in Exercise 1. Sketch a diagram like Figure 1.8.5 for each.

5. Let X be a set and let $E \subset X$ and $E' = X \backslash E$. Let $f: X \to R$ be any real function. Define two functions f_E and $f_{E'}$ as follows:

$$f_E(x) = \begin{cases} f(x), & \text{for } x \in E \\ 0, & \text{for } x \in E' \end{cases}$$

$$f_{E'}(x) = \begin{cases} 0, & \text{for } x \in E \\ f(x), & \text{for } x \in E' \end{cases}$$

Verify that $f_E + f_{E'} = f$ and $f_E f_{E'} = 0$.

an algebra of functions

6. Suppose f and g are two real functions $f: R \to R$ and $g: R \to R$.

    (a) If $fg(x) = 0$ for each $x \in R$, can we conclude that either f or g is the constant function $0: R \to \{0\}$? Explain.

    (b) If $(f \circ g)(x) = f(g(x)) = 0$ for each $x \in R$, can we conclude that either $f = 0$ or $g = 0$? Explain.

7. Verify that the identity function $i: X \to X$, $i(x) = x$ is an identity for function composition; i.e., $i \circ f = f$.

8. Show by and example that $f \circ g \neq fg$.

9. Consider the function **abs**: $C \to R$ defined in Exercise 13 of the preceding section. Describe **conj** $\circ$ **abs** where **conj** is the conjugation function $C \to C$. What about **abs** $\circ$ **conj**?

10. Illustrate with an example that for real functions f, g, and h we need not have

$$h \circ (f + g) = (h \circ f) + (h \circ g)$$

11. Let a, b, c, be real numbers and f, g, h real functions. The function $af + bg + ch$ is called a *linear combination* of f, g, and h. Find the function $2f - 3g + (1/2)h$ for each of the following functions f, g, and h.

    (a) $f(x) = x$, $g(x) = x^2$, $h(x) = x^3$

    (b) $f(x) = x^4$, $g(x) = x - 1$, $h(x) = 4x^2$

    (c) $f(x) = 3x$, $g(x) = x - 1$, $h(x) = 2x - 1$

    (d) $f(x) = 4x^2$, $g(x) = 1 - x^2$, $h(x) = x^3 + 2$

    (e) $f(x) = [x]$, $g(x) = 1$, $h(x) = [x - 1]$

    (f) $f(x) = 1/x$, $g(x) = \sqrt{1 - x}$, $h(x) = 1 + x^2$

12. Let $X = \{1, 2, 3\}$, and consider the functions σ and θ defined as follows.

$$X \xrightarrow{\sigma} X \xrightarrow{\theta} X$$

$$\sigma(1) = 2, \quad \sigma(2) = 3, \quad \sigma(3) = 1, \quad \theta(1) = 1, \quad \theta(2) = 3, \quad \theta(3) = 2$$

Describe each of the following functions.

    (a) $\sigma \circ \sigma$

    (b) $\theta \circ \theta$

    (c) $\sigma \circ \theta$

    (d) $\theta \circ \sigma$

    (e) $\sigma \circ \theta \circ \sigma$

    (f) $\sigma \circ \sigma \circ \theta$

    (g) $\theta \circ \sigma \circ \theta$

13. Let $f: X \to R$ be defined by $f(x) = 1 - x$, find $f^2(2), f^3(2), f^4(0), [f(2)]^{-1}, [f(3)]^{-2}$.

14. Prove the following concerning composite functions. Let $h = g \circ f: X \to W$ denote the composition of the functions

$$X \xrightarrow{f} Y \xrightarrow{g} W$$

    (a) Then, if both f and g are surjective, so is h.

    (b) Then, if both f and g are injective, so is h.

    (c) Then, if both f and g are bijective, so is h.

    (d) Then, if h is surjective, so is g.

    (e) Then, if h is injective, so is f.

15. Is the set of real functions with the operations of function addition and function multiplication a field (see Appendix I)? Discuss.

fundamentals

1.9. inverse functions

Suppose that f is a function from X to Y. Let $y \in f(X)$ be an element of Y which is the image of one or more elements of X under f. Can the element(s) x in X be determined so that $y = f(x)$?

example 1

Let f be the function given by

$$f(x) = 2x - 1$$

Find x such that $f(x) = 5$, and find x such that $f(x) = y$. To find the first x set $f(x) = 5 = 2x - 1$ and solve for x.

$$5 = 2x - 1$$

$$2x = 6$$

$$x = 3$$

Thus, $f(3) = 5$. To find x such that $f(x) = y$ proceed the same way.

$$y = 2x - 1$$

$$2x = y + 1$$

$$x = \frac{y + 1}{2}$$

This last expression suggests a rule for a function from R to R, f^*, such that $f^*(y) = (y + 1)/2$. Is there always such a function? ●

In this section we discuss this question and the question asked at the beginning. To do this, recall the second definition of a function as a collection of ordered pairs. Thus, if $f \colon X \to Y$, then $f \subset X \times Y$, is the set $\{(x, y) | y = f(x)\}$ with the property that each $x \in X$ appears as the first member of *exactly one* pair in f. Let us reverse the roles of X and Y and consider a correspondence with domain Y and range X by taking f backwards. That is, consider the set

$$f^* = \{(y, x) | y = f(x)\}$$

This set is sometimes called the **inverse of** f. The function described in Example 1 is an illustration of such an f^*.

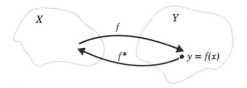

figure 1.9.1

This inverse correspondence, however, is *not always a function* as you can readily see from the next example.

inverse functions

example 2

Suppose $f: X \to Y$ is given by the following formula, where $X = R$ and Y is the set of non-negative real numbers,

$$f(x) = x^2$$

We can see that, when f is considered as a set of ordered pairs, the pairs $(0, 0)$, $(1, 1)$, $(-1, 1)$, $(2, 4)$, $(-2, 4)$ are all members of f. So the pairs $(0, 0)$, $(1, 1)$, $(1, -1)$, $(4, 2)$, and $(4, -2)$ all belong to $f*$. But we note that the number $4 \in Y$ appears as the first element of two different pairs in $f*$, so $f*$ is not a function. The graphs of f and $f*$ are sketched in Figure 1.9.2 with the domain horizontal in each case, as is the convention.

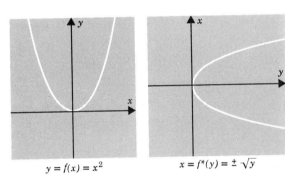

figure 1.9.2

$$y = f(x) = x^2 \qquad\qquad x = f*(y) = \pm \sqrt{y}$$

Note that a vertical line through the domain Y of $f*$ intersects the graph in two places, not just one. ●

example 3

Suppose that X and Y are the finite sets and f and $f*$ are the correspondences given by Table 1.9.1.

table 1.9.1

	f			$f*$	
X	Y		X	Y	
a	α		α	a	
b	β		β	b	
c	γ		γ	c	
d	δ		δ	d	
e	ε		ε	e	

In this case f is a bijection and $f*$ *is* a function.

Let $f: X \to Y$ be the function indicated in the diagram of Figure 1.9.3. The inverse $f*: Y \to X$ can be thought of as retracing the arrow from Y back to X. However, in this figure, we have recopied X on the line above Y so that

fundamentals

it can easily be seen that $f*$ is actually the inverse for the operation of function composition. That is, the composite function $(f* \circ f) = i$ where i denotes the identity function $i: X \to X$ which maps each x to itself. This is because for each $x \in X$, $f*(f(x)) = x$.

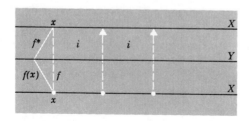

figure 1.9.3

$f* \circ f = i$

Of course, the circumstance illustrated in the next diagram, Figure 1.9.4, can also occur, in which case $f*$ is not a function. ●

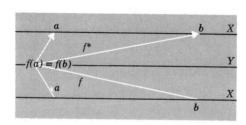

figure 1.9.4

$f*$ is not a function

example 4

Let $X = Y = R$ and let $f: X \to Y$ be the *cube function $f(x) = x^3$*. Graphs of f and $f*$ are sketched in Figure 1.9.5. Here again f is bijective, and $f*$ is a function. ●

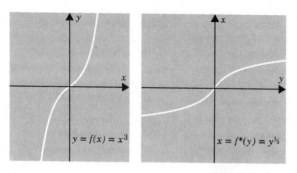

figure 1.9.5

It is fairly obvious from these examples and from the definition of f as a collection of ordered pairs, that the inverse $f*$ will be a function if, and only if, the original function f is one-to-one.

Let $f: X \to Y$ be both one-to-one and onto Y (a bijection). The function $f*: Y \to X$, given by $f*(y) = x$, if and only if, $y = f(x)$, for all $y \in Y$, is called the **inverse** of f, and is usually written f^{-1}.

inverse functions

example 5

Consider the three real functions f, g, and h, each with domain $X = R$, defined by the following formulas.

$$f(x) = 2x \qquad \text{(all } x \in X)$$

$$g(x) = 3 - x \qquad \text{(all } x \in X)$$

$$h(x) = \frac{x^3}{4} \qquad \text{(all } x \in X)$$

Find $f^{-1}(y)$, $g^{-1}(y)$ and $h^{-1}(y)$. Each inverse is a function since each of f, g, and h is one-to-one. Thus,

if

$$y = 2x$$

then

$$x = \frac{1}{2}y \qquad \text{(all } y \in R)$$

Hence, $f^{-1}(y) = (1/2)y$ gives the inverse of $f(x) = 2x$.

If

$$y = 3 - x$$

then

$$x = 3 - y \qquad \text{(all } y \in R)$$

and $g^{-1}(y) = 3 - y$ defines the inverse of g.
Similarly, if

$$y = \frac{x^3}{4}$$

then

$$x = (4y)^{1/3} \qquad \text{(all } y \in R)$$

defines the inverse h^{-1} of h. ●

As another example consider the *"area-of-a-square" function A*. This function is described by the following equation (formula) where a is the area and s the length of the side.

$$a = A(s) = s^2, \qquad s > 0$$

The domain of the function A is the set of all positive real numbers. (We ignore the "square" (dot) whose side has length 0.) *This* function is (both one-to-one and onto) bijective, so its inverse A^{-1} is given by

$$s = A^{-1}(a) = \sqrt{a}$$

The relatively simple idea of a function and its inverse has a minor complication when we deal with functions whose domain and range are both subsets of the real numbers. This complication arises because of a convention which is almost universal. This convention requires that the *domain* of a function lie along the *horizontal X* axis and its *range* along the *vertical Y* axis. Therefore, rather than the more natural $y = f(x)$ and $x = f^{-1}(y)$, frequently we see $y = f(x)$ written for the function f and $y = f^{-1}(x)$ for the related inverse function. For example, the cube function is $y = x^3$, and its inverse, the cube root function, is most often written $y = x^{1/3}$ rather than $x = y^{1/3}$ (see Figure 1.9.6).

fundamentals

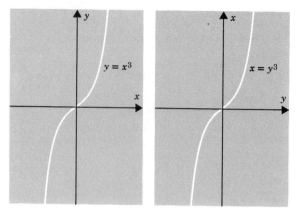

figure 1.9.6

The cube and inverse.

Geometrically, then, because of the convention that the domain of a function be on the X axis and its range on the Y axis, taking the inverse of a function f whose domain and range are subsets of the real numbers means switching the roles of x and y. In other words, we relabel the X axis and the Y axis as in the following figure; that is, to obtain the inverse, in this sense, for a function, we must switch the two axes. Hence, if $y = x^3$, switching x and y gives $x = y^3$ and $y = x^{1/3}$ as its inverse.

Even when f is not one-to-one, we may perform this switching of x and y to obtain the graph of the inverse f^* of f. But the frame of reference in such a graph is different than we are used to. To orient the graph in the conventional way requires a rotation and a reflection. This is illustrated in the series of graphs of Figure 1.9.7.

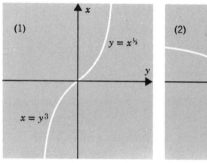

(1) *Unconventional graph.*   **(2)** *Rotated 90° clockwise.*

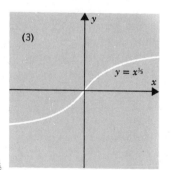

figure 1.9.7

Obtaining a conventional graph of $f^{-1}(x) = X^{1/3}$

(3) *Graph flipped about x axis.*

inverse functions

This turn-flip process can be combined into a single flip about the diagonal line whose equation is

$$y = x$$

As illustrated in Figure 1.9.8, this is the geometric equivalent switching the variables x and y in the algebraic expression for $f(x)$ or $f^{-1}(y)$.

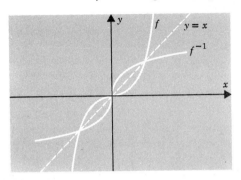

figure 1.9.8

The cube function and its inverse.

example 6

Given a "linear equation" of the form

$$Ax + By = C$$

Unless $B = 0$, this can be reexpressed as a function in the form

$$y = -\frac{A}{B}x + \frac{C}{B}$$

If A is also not zero, we have a bijective function

$$y = h(x) = mx + b$$

where $m = -(A/B)$ and $b = C/B$.

Then it is clear that, upon switching x and y, we have $x = h(y) = my + b$. Solving for y in this expression gives $y = h^{-1}(x) = (1/m)(x - b)$. Therefore,

$$h(x) = mx + b$$

and

$$h^{-1}(x) = \frac{1}{m}(x - b) \quad \bullet$$

example 7

Note that if n is a positive integer, then the set

$$f = \{(x, y) \mid y = x^n\}$$

is always a function, the *power function*. We usually write it as $f(x) = x^n$. Whether f is one-to-one or not depends upon whether n is an *odd* or *even* integer. When n is *odd*, the power function $f(x) = x^n$ is a bijection (both one-to-one and onto R); and hence its inverse is a function. We can write this inverse as $f^{-1}(x) = x^{1/n} = \sqrt[n]{x}$. For the symbol $x^{1/n}$ to have meaning when n is even, however, we must actually consider a restriction of f to only a part of its domain. We also consider only the *image* of this restricted function as the *range*. This new function is a "branch" of f which *is* one-to-one and onto its range (a bijection). That is, we restrict our domain and range to non-negative values of x only, and consider the *new function* $f: R \to R^+$.

63

(1.7.4)

$$f = \{(x, y) \mid y = x^n \text{ and } x \geq 0\}$$

fundamentals

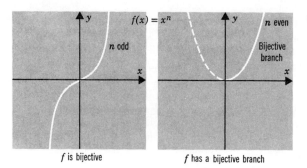

figure 1.9.9

$f(x) = x^n$

n odd

n even

Bijective
branch

f is bijective   *f* has a bijective branch

For this "new" f, since it is bijective in our myopic view, the inverse is a function, and the symbol $x^{1/n} = \sqrt[n]{x}$ now has meaning. We shall discuss this symbol and the idea of a fractional exponent from a slightly different point of view in Chapter 4. The two views complement each other. This idea, restricting the domain of a function f which is not one-to-one to obtain a branch which is, will be exploited when the inverse trigonometric functions are discussed in Chapter 5.

Any real valued function f is said to be **increasing** on an interval $[a, b]$ in its domain if, for each x and y in $[a, b]$, with $x < y$, we have $f(x) < f(y)$. If we have, for x and y in $[a, b]$ with $x < y$, $f(x) \leq f(y)$, we say that f is a **nondecreasing** function on $[a, b]$. **Decreasing** and **nonincreasing** functions are defined analogously. A function is called **monotonic** on $[a, b]$ if it is either nonincreasing on (all of) $[a, b]$ or nondecreasing there. One uses the term **strictly monotonic** on $[a, b]$ if f is either increasing on all of $[a, b]$ or decreasing there. It is an easy exercise to show that a function is injective on $[a, b]$ if it is *strictly monotonic* there. ●

example 8

The linear function $y = 3x + 1$ is increasing everywhere and the greatest integer function is nondecreasing. The function f whose graph is sketched in Figure 1.9.10 is nonincreasing while the function g of Figure 1.9.11 is nondecreasing. ●

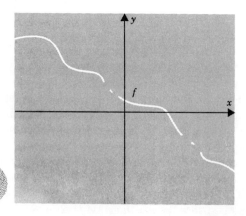

figure 1.9.10

f is nonincreasing

f

64

inverse functions

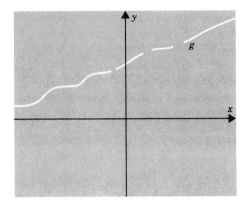

figure 1.9.11

g is nondecreasing

exercises

1. Sketch a graph of the functions $f(x) = x^3$ and $g(x) = 1 + x$ over the interval $[0, 2]$. The sketch graphs of each of the following functions over the same interval.

(a) $f + g$

(b) $f - g$

(c) fg

(d) f/g

(e) $f \circ g$

(f) $g \circ f$

(g) f^{-1}

(h) $1/f$

(i) g^{-1}

(j) $1/g$

(k) $f^{-1} \circ f$

(l) $f \circ f^{-1}$

(m) $f \circ g^{-1}$

2. In each of the functions of Exercise 1, state which are increasing and which are decreasing on $[0, 2]$.

3. Sketch the graph of each of the functions $f: X \to Y$, where $y = f(x)$ is defined by the following formulas, X is given, and $Y = f(X)$.

(a) $y = 2x + 1$, $X = [-2, 2]$

(b) $y = x^2$, $X = [-2, 2]$

(c) $y = x^3$, $X = [-2, 2]$

(d) $y = \dfrac{x^2 - 9}{x - 3}$, $X = [0,3) \cup (3, 5]$

(e) $y = x^{1/2}$, $X = [0, 4]$

(f) $y = x^5$, $X = [-2, 2]$

(g) $y = x^2 + 1$, $X = R$

(h) $y = x^3 + 1$, $X = R$

(i) $y = [x]$, $X = [0, 3]$

(j) $y = |x|$, $X = [0, 3]$

(k) $y = |x|$, $X = [-2, 2]$

(l) $y = \sqrt{1 - x^2}$, $X = [-1, 1]$

(m) $y = x^{1/3}$, $X = [-1, 1]$

(n) $y = x^2 - 1$, $X = [-2, 2]$

(o) $y = x^3 - x^2$, $X = [-2, 2]$

(p) $y = \begin{cases} 0, & \text{when } x = 0 \\ \dfrac{1}{x}, & \text{when } x \neq 0, \end{cases}$ $X = R$

4. Which of the functions f in Exercise 3 have inverse functions f^{-1} on the set Y? Find $f^{-1}(y)$ in each such case. Then find $f^{-1}(x)$ for each of these.

5. Each of the following functions is bijective. Describe the inverse function.

(a) $f(x) = x - 1$; $f^{-1}(x) =$

(b) $g(a) = a^3 + 1$; $g^{-1}(a) =$

(c) $H(x) = |2x|$, $x > 0$; $H^{-1}(x) =$

(d) $Q(x) = x^4$, $x \leq 0$; find $Q^{-1}(x)$

(e) $C(x) = 1 - x^2$, $x \in [0, 1]$; find $C^{-1}(x)$

fundamentals

6. In each of the following functions, select a suitable bijective branch and then find the inverse function $f^{-1}(x)$ for that branch

(a) $f(x) = 1 - x^2$ (e) $f(x) = x^4$

(b) $f(x) = x^2 - 4$ (f) $f(x) = x^4 - 1$

(c) $f(x) = x^2 + 2x + 1$ (g) $f(x) = x^4 - 2x^2 + 1$

(d) $f(x) = x^2 - 6x + 9$ (h) $f(x) = 1 - x^6$

7. Given a real function $f: X \rightarrow R$, does $f^{-1} = 1/f$? Explain.

8. Prove that if f is a function with domain and image the same finite set, then f is also an injection. Such functions are called *permutations*.
Hint: Can any element be the image of two different elements without decreasing the size of the range?

9. Les $S = \{x_1, x_2, x_3,\}$. List *all* the functions which have domain and range S; i.e., list all the permutations of S.

10. Is the greatest integer function injective? Is it monotonic? What about $y = |x|$ for $x \geq 0$?

11. Show that every function f which is strictly monotonic on $[a, b]$ is injective on $[a, b]$.

12. Let $f(x)$ and $g(x)$ be as in Exercise 1. Let $h(x) = [x]$ and $a(x) = |x|$. Write algebraic expression for each of the following.

(a) $a(f(x))$ (d) $h(g(x))$

(b) $h(f(x))$ (e) $f^{-1}(h(x))$, $(x \in [0, 4])$

(c) $a(g(x))$ (f) $g^{-1}(h(x))$

13. Show that if $f: X \rightarrow Y$ is any bijection, then the following are true.

(a) The composite function $f^{-1} \circ f$ is the identity function on X.

(b) The composite function $f \circ f^{-1}$ is the identity function on Y.

14. Prove that if $f: X \rightarrow Y$ and $g: Y \rightarrow X$ are two functions such that:

(i) $g \circ f$ is the identity function on X

(ii) $f \circ g$ is the identity function on Y

then the following are true:

(a) both f and g are bijections

(b) $g = f^{-1}$ and $f = g^{-1}$

15. If f and g are both injective functions, is $f + g$ injective? (Proof.) What about fg, $f - g$ and $f \circ g$?

16. With the notation of Exercise 12,

(a) show $f \circ a = a \circ f$

(b) show $h \circ g = g \circ h$

(c) For $x \geq 0$ show $(f \circ g)^{-1} = g^{-1} \circ f^{-1}$

17. Let S be the set of all bijective functions (one-to-one correspondences) $f: R \rightarrow R$. With function addition as addition and function composition as "multiplication" does $S, +, \circ$ form a field (see Appendix I)? Discuss.

quiz for review

PART I.   True or False

Problems 1–10 refer to the following sets.

Z = the set of all the integers
R = the set of all real numbers
Q = the set of all rational numbers
$A = \{x \in R: -5 \le x \le 5\}$
$B = \{x \in Q: 0 \le x \le 3\}$
$C = \{x \in Z: -5 < x < 0\}$

Each of the following is either true or false. Mark your answer sheet a (true) or b (false).

1. $0 \in C$

2. $A \cap C = \{-4, -3, -2, -1\}$

3. $A \cup C = A$

4. A is the closed interval, $[-5, 5]$

5. C is the open interval, $(-5, 0)$

6. $0 \in Z \cap B$

7. $A \cap B = B$

8. $B \cap C = \varnothing$

9. The pair $(1, 1) \in B \times B$

10. The pair $(\pi/2, -1) \in B \times C$

11. For any set X, $X \cap \tilde{X} = \varnothing$

12. $\tilde{\varnothing} = U$

13. If k^2 is an odd natural number and k is a natural number, then k is odd.

14. $\sqrt{x^2} = x$ for every real number x

15. $|x - 3| = x - 3$ for every real number x

16. If $|x - 1| \le 2$, then $-1 \le x \le 3$.

17. $|z|$ is a real number for every complex number z

18. If $z = a + bi$, then $\bar{z} = a - bi$.

19. $z\bar{z}$ is real if and only if z is a real number

20. The set of all x such that $|x - 2| < 0.1$ is a neighborhood of -2.

21. Every one-to-one function is surjective.

22. If $f: X \to X$ is an injection and X is a finite set, then f is a bijection.

23. If $x = y$ implies $f(x) = f(y)$, then f is injective.

24. If $f: R \to R$ and $g: R \to R$ are any functions, then $fg \ne gf$.

25. If $f: R \to R$ is an injection, then f^{-1} exists with domain $f(R)$.

PART II

In Problems 1–3, identify each shaded region of the given Venn diagram as one of
(a) $A \cap B$, (b) $A \cup B$, (c) $A \backslash B$, (d) $B \backslash A$, (e) $\varnothing$.

1.

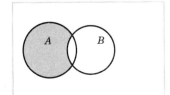

fundamentals

2.

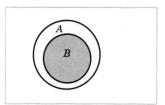

3.

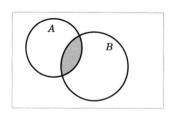

4. Let $f: X \rightarrow Y$ be a function. The set X is called

    (a) the image of f (d) injective
    (b) the domain of f (e) surjective
    (c) the range of f

5. Let $f: X \rightarrow Y$ be a function. The set Y is called

    (a) the image of f (d) injective
    (b) the domain of f (e) surjective
    (c) the range of f

6. Let $f: X \rightarrow Y$ be a function with $f(X) = Y$. Then f is called

    (a) the image of f (d) injective
    (b) the domain of f (e) surjective
    (c) the range of f

7. Let $f: X \rightarrow Y$ be a function. We call f injective if, and only if,

    (a) Y is the range of f (d) $f(x) = f(y)$ imples $x = y$
    (b) $f(X) = Y$ (e) $f(x) \in X$
    (c) Each x in X has a unique image in Y

8. Let p denote the postage stamp function, with domain $D = (0, 640)$ (ounces) $p: D \rightarrow Z$. Let r be the current rate.

    (a) Z is the image of p (d) p is surjective
    (b) p is injective
    (c) $p(x) = rx$ (e) $p(x) = \begin{cases} rx, & \text{if } x \in Z \cap D \\ r[x + 1], & \text{if } x \in D/Z \end{cases}$

Problems 9–13 refer to the following diagrams.

    (a)

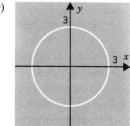

    (b)

X	Y
Jones	A
Smith	B
Williams	D
Zundell	C

quiz for review

(c)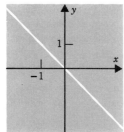

(d)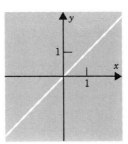

(e)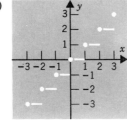

9. The graph of the identity function $x(t) = t$ looks like:

10. The graph of the greatest integer function $[t]$ looks like:

11. The graph of the negation function $n(t) = t$ looks like:

12. Which function has finite range and is bijective?

13. Which is not a function?

Suppose $f: R \to R$, with $f = t + 2$ and $g: R \to R$ with $g(x) = (3/2)t^2$. Match each of the functions denoted in the Problems 26–30 with the appropriate value at the real number t.

Function		Value at t
14. $(f + g)(t)$	(a)	$(3/2)t^2 + 2$
15. $f(g(t))$	(b)	$t - 2$
16. $(fg)(t)$	(c)	$(3/2)(t + 2)^2$
17. $(3/2)f^2(t)$	(d)	$(3/2)t^3 + 3t^2$
18. $f^{-1}(t)$	(e)	$(3/2)t^2 + t + 2$

19. If $z = 3 - 2i$ then $|z| =$
   (a) 3   (b) -2 (c) 5   (d) 13   (e) $\sqrt{13}$

20. Which of the following is a neighborhood of $1/2$?
   (a) $[0, 1]$ (b) $(1/2, 1)$ (c) $[0, 1/2]$ (d) $(0, 1)$ (e) $[-1, 2]$

21. $1/(1 + i) =$
   (a) $1 - i$ (b) $1 + (1/i)$ (c) $(1/2) - (i/2)$ (d) $-i$ (e) is undefined

22. The solution set for $|x - (1/2)| > 5$ is
   (a) $((11/2), \infty)$ (b) $(-\infty, (9/2))$ (c) $R \setminus [-(9/2), (11/2)]$
   (d) $\{(9/2), (11/2)\}$ (e) any real number except -4.5 and 5.5

23. If z and w are complex numbers then it is *not true* that
   (a) $\overline{z + w} = \bar{z} + \bar{w}$ (b) $\overline{(zw)} = (\bar{z})(\bar{w})$ (c) $\overline{(z/w)} = \bar{z}/\bar{w}$
   (d) $\overline{(\bar{z} - \bar{w})} = (\overline{z - w})$ (e) $(\bar{z})(\bar{w})$ is always real

fundamentals

24. Solve the equation $|3x + 1| = |x^2 - 3|$ for x a real number

   (a) $x = 4$ and $x = -1$ (b) $x = -(3/2) - (\sqrt{17}/2), x = -(3/2) + (\sqrt{17}/2)$
   (c) $x = -4, x = 1$ (d) The numbers in both (a) and (b) are solutions.
   (e) The numbers in both (b) and (c) are solutions.

25. If $f(x) = x^3$ and $g(x) = x - 1$, then

   (a) $f(g(x)) = x^3 - 1$ (b) $f^{-1}(g(x)) = \sqrt[3]{x - 1}$ (c) $f(g^{-1}(x)) = x^3 + 1$ ·
   (d) $f^{-1}(g^{-1}(x)) = x$ (e) $g(f^{-1}(x)) = \sqrt[3]{x - 1}$

sophisticated counting principles

②

At an early age, each of us learned to count things. In fact, counting preceded our personal development of the abstract idea of a number and gave meaning to it. As the title states, this chapter will not review ways to count sheep or pieces of candy—unless they are involved in committees or something. Rather, we develop here some sophistication in the kind of counting that is involved in many abstract mathematical situations.

In Exercise 14 and 15 of Section 1.7, you were asked to attempt to apply some sophisticated counting techniques. It was pointed out in these exercises that the process of counting the objects (members) in a set can be so interpreted as to involve one-to-one functions (injections) either *from* or *to* the set of natural numbers (positive integers). In this chapter, we shall examine such functions in greater detail. We begin by discussing an important property of the set of positive integers.

2.1. mathematical induction

If a sophisticated approach to counting involves functions whose domain or whose range is the set of positive integers N, it is then necessary to take a longer look at this set. In this section we shall exploit a fairly obvious property of the set N to produce a powerful tool known as **mathematical induction.** This tool is an important technique for proof which not only will be used in several places in this text, but which is a standard part of the equipment of any mathematician.

This technique exploits a property of the set of positive (non-negative) integers which is not shared by the rational or real numbers in their usual

sophisticated counting principles

ordering on the number line. The set N of positive integers, ordered under the $<$ relation as real numbers, satisfies:

WELL-ORDERING PRINCIPLE: *Every nonempty subset of N has a least (smallest) element.*

That is, given any set of positive (or non-negative) integers, we can always find the smallest integer in the set. This is not true for every subset of the reals as is easily seen from the *open* interval $(1, 6)$. There is no *smallest* real number x in this interval; neither is there a smallest rational number r. But 2 is the smallest positive integer in the set.

The technique of mathematical induction is easily derived from this principle. Suppose we have some statement $P(n)$ which purports to be true for every positive integer. For example, $P(n)$ might be any one of the following statements:

(2.1.1) "The sum of the first n odd positive integers is n^2."
(2.1.2) "We always have $2^n > n$, for every positive integer n."
(2.1.3) "For every pair of real numbers x and y and each positive integer n, $(xy)^n = x^n y^n$."

To prove any such statement $P(n)$, we let S denote the set of those positive integers for which the statement is true, and let T denote the set of those positive integers for which the statement fails. Our task is to show that $T = \emptyset$ (is empty); that is, to show that S contains *every* positive integer. This is equivalent to saying that $P(n)$ *is true for every positive integer.*

Suppose that T is not empty. Then it has a least element r. Each of the statements (2.1.1)–(2.1.3) above is true when $n = 1, 1 \in S$; vis, $1 = 1^2, 2^1 > 1$, and $(xy)^1 = x^1 y^1$, so $1 \notin T$. Thus, in these cases, $r > 1$.

Each of the statements in our examples has an additional property. *Whenever a positive integer k is in S, so is the next positive integer, k + 1.* We verify this for each example as follows

(2.1.1) If $k \in S$, then $1 + 3 + 5 + \cdots + 2k - 1 = k^2$. Hence, $(1 + 3 + 5 + \cdots + 2k - 1) + (2k + 1) = k^2 + 2k + 1 = (k + 1)^2$. Thus, if the sum of k odd positive integers is $k^2 (k \in S)$; then the sum of $k + 1$ odd positive integers is $(k + 1)^2 (k + 1 \in S)$.

(2.1.2) If $k \in S$, then $2^k > k$. So $2^{k+1} = 2^k \cdot 2 \geq k \cdot 2$, or $2^{k+1} \geq 2k = k + k$. But $k \geq 1$, so $2^{k+1} \geq k + 1$. Thus, when $k \in S$, so is $k + 1$ in S.

(2.1.3) If $k \in S$, then $(xy)^k = x^k y^k$. Consider $(xy)^{k+1}$. We have $(xy)^{k+1} = (xy)^k (xy) = x^k y^k xy = x^k \cdot x \cdot y^k \cdot y = x^{k+1} y^{k+1}$, and $k + 1 \in S$.

In each of these examples we have verified that the truth set S for the statement P has the following two properties

(a) 1 is in S; and
(b) whenever an integer k is in S, so is the next integer, $k + 1$, in S.

What does this tell us about the complement T of S? From (a) we have the *least element* r of T is larger than 1; therefore, $r - 1$ is a positive integer less than r, so $r - 1 \notin T$. Hence, $r - 1 \in S$. But if $r - 1 \in S$, property (b) insists that the next integer, $r - 1 + 1 = r$, is in S. Hence, $r \in S \cap T$. But $S \cap T$ is empty. This must mean that T was not a nonempty set to begin with.

72

mathematical induction

We have, therefore, demonstrated the induction principle: *A statement P(n) is true for every positive integer n provided its truth set has properties (a) and (b), above.* This is equivalent to saying that *P(n) is true for every positive integer n provided:*

(a)   *P(1) is true.*
(b)   *If P(k) is true, then P(k + 1) is true.*

As an additional example of a proof using mathematical induction consider the following.

example 1

If a line segment of unit length (length 1) is given, then a line segment of length $\sqrt{n}$ can be constructed using a straight edge and compass only, for every positive integer *n*.

This is clearly a statement of the form $P(n)$. Since $\sqrt{1} = 1$, the statement $P(1)$ is true. Now suppose that $P(k)$ is true. That is, suppose that a segment of length $\sqrt{k}$ is constructible when a unit segment is given. To construct a segment of length $\sqrt{k + 1}$ one proceeds as follows.

Construct the segment of length $\sqrt{k}$ as segment AB (see Figure 2.1.1). Then at point A erect a perpendicular to the segment AB. Locate the point C on this perpendicular so that AC has unit length. Then the hypotenuse BC of the right triangle ABC has length $\sqrt{k + 1}$ from Pythagoras's theorem.

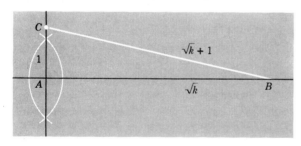

figure 2.1.1

Therefore if $P(k)$ is true, so is $P(k + 1)$. Hence $P(n)$ is true for every positive integer *n*. ●

Often we make use of the induction principle to give *inductive* or *recursive* *definitions*. The idea is to define the first of the items in question, or to define the value at 1 of the symbol being defined. Then define the *n*th item in terms of the preceding one. For example, we define the symbol

$$n!$$

read *n factorial*, by

$$0! = 1! = 1$$

and

$$n! = n(n - 1)!$$

Thus, $2! = 2(1!) = 2$, $3! = 3 \cdot 2! = 3 \cdot 2 = 6$, etc.

sophisticated counting principles

example 2.   summation notation Σ

The capital sigma is used to denote a (finite) sum. Thus, by

$$\sum_{i=1}^{n} a_i$$

we mean the sum $a_1 + a_2 + \cdots + a_n$. This definition given inductively would read

$$\sum_{i=1}^{1} a_i = a_1,$$

and

$$\sum_{i=1}^{n} a_i = \left(\sum_{i=1}^{n-1} a_i\right) + a_n \quad \bullet$$

example 3.   product Π

As the capital sigma is used to denote sums, the capital pi is used to indicate a product. Defined inductively, this would be

$$\prod_{i=1}^{1} a_i = a_1$$

and

$$\prod_{i=1}^{n} a_i = \left(\prod_{i=1}^{n-1} a_i\right) a_n$$

Thus, $\Pi_{i=1}^{n} a_i = a_1, a_2, a_3, \ldots a_n.$ $\bullet$

This kind of definition is frequently written in the latter form, rather than giving the inductive definition formally; nevertheless, all such definitions are in fact inductive whether so stated or not. Indeed, when we define

$$a^n = a \cdot a \cdot \ldots \cdot a, n \text{ factors}$$

we are actually stating the inductive definition

$$a^1 = a$$

and

$$a^n = (a^{n-1})a$$

example 4

Use induction to show that for any finite number n of complex numbers $z_1, z_2, \ldots z_n$, the conjugate of the sum of the z_i's is the sum of the individual conjugates. (The case $n = 2$ is Theorem 1.5.1(i)).

This proposition, using the sigma notation, would read

$$\overline{\sum_{i=1}^{n} z_i} = \sum_{i=1}^{n} \bar{z}_i$$

The proposition is true for $n = 1$ since $\overline{\Sigma_{i=1}^{1} z_i} = \bar{z}_1 = \Sigma_{i=1}^{1} z_i$. Assume that $P(k)$ is true; that is, assume that

$$\overline{\sum_{i=1}^{k} z_i} = \overline{z_1 + z_2 + \cdots z_k} = \bar{z}_1 + \bar{z}_2 + \cdots + \bar{z}_k = \sum_{i=1}^{k} \bar{z}_k$$

mathematical induction

Then consider the conjugate of the sum of $k + 1$ complex numbers.

$$\overline{\sum_{i=1}^{k+1} z_i} = \overline{\sum_{i=1}^{k} z_i + z_{k+1}}$$

The right side of this equation is the conjugate of the sum of two complex numbers, z_{k+1} and

$$w = \sum_{i=1}^{k} z_i$$

Theorem 1.5.1 (i) assures us that

$$\overline{w + z_{k+1}} = \overline{w} + \overline{z_{k+1}}$$

Therefore

$$\overline{\sum_{i=1}^{k+1} z_i} = \overline{\left(\sum_{i=1}^{k} z_i\right) + z_{k+1}} = \overline{\left(\sum_{i=1}^{k} z_i\right)} + \overline{z_{k+1}}$$

But $\overline{w} = \overline{\sum_{i=1}^{k} z_i} = \sum_{i=1}^{k} \overline{z_i}$ by our assumption that $P(k)$ is true. Therefore,

$$\overline{\sum_{i=1}^{k+1} z_i} = \left(\sum_{i=1}^{k} \overline{z_i}\right) + \overline{z_{k+1}} = \sum_{i=1}^{k+1} \overline{z_i}$$

and $P(k + 1)$ is true. Thus the principle of induction has shown that $P(n)$ is true for every positive integer n. ●

You are asked in Exercises 17 and 18 to show that the conjugate of a finite product is the product of the conjugates and, in particular that $(\overline{z^n}) = (\overline{z})^n$ for every complex number z and natural number n.

exercises

For each positive integer n, prove the following by using induction.

1. $1 + 2 + 3 + \cdots + n = \dfrac{n(n + 1)}{2}$

2. $1^2 + 2^2 + 3^2 + \cdots + n^2 = \dfrac{n(n + 1)(2n + 1)}{6}$

3. $1^3 + 2^3 + 3^3 + \cdots + n^3 = \dfrac{n^2(n + 1)^2}{4}$

4. $\dfrac{1}{2} + 1 + \dfrac{3}{2} + 2 + \cdots + \dfrac{n}{2} = \dfrac{n(n + 1)}{4}$

5. $2 + 4 + 6 + \cdots + 2n = n(n + 1)$

6. $1 \cdot 1! + 2 \cdot 2! + \cdots + n \cdot n! = (n + 1)! - 1$

7. $\dfrac{1}{1 \cdot 2} + \dfrac{1}{2 \cdot 3} + \cdots + \dfrac{1}{n(n + 1)} = \dfrac{n}{n + 1}$

8. $x^n - y^n = (x - y)(x^{n-1} + x^{n-2}y + \cdots + xy^{n-2} + y^{n-1})$

9. The sum of the angles in a polygon having $n + 2$ sides is $n(180°)$.

10. $2^n > n$

sophisticated counting principles

11. $2^{n+3} < (n+3)!$

12. For $x \geq -1$, $(1+x)^n \geq 1 + nx$

13. $\dfrac{n^5}{5} + \dfrac{n^3}{3} + \dfrac{7n}{15}$ is an integer

14. $3^{2n} - 1$ is divisible by 8

15. $7^{2n} + 16n - 1$ is divisible by 64

16. $1^3 + 2^3 + \cdots + (n-1)^3 < \dfrac{n^4}{4} < 1^3 + 2^3 + \cdots + n^3$

17. For every complex number z, $(\overline{z^n}) = (\bar{z})^n$, where $\bar{z}$ denotes the conjugate of z.

18. For n complex numbers $z_1, z_2, \ldots, z_n$

$$\overline{\prod_{i=1}^{n} z_i} = \prod_{i=1}^{n} \bar{z}_i$$

19. Let $P(n)$ be the statement $3 + 6 + 9 + \cdots + 3n = \dfrac{3n(n+1)}{2} + (n-1)$. Show that $P(1)$ is true. Is $P(n)$ true for every positive integer n? Why?

20. Let $P(n)$ be the statement $3 + 6 + 9 + \cdots + 3n = \dfrac{3n(n+1)}{2} + 1$. Show that if $P(k)$ is true, so is $P(k+1)$. Is $P(n)$ true for all n?

21. $a + ar + ar^2 + \cdots + ar^{n-1} = \dfrac{a - ar^n}{1 - r}$ (geometric progression).

22. $a + (a+d) + (a+2d) + (a + (n-1)d) = \dfrac{n}{2}(2a + (n-1)d)$ (arithmetic progression).

23. (This is a paraphrase of an example of G. Polya.) Criticize the following induction proof:

> $P(n) = $ *In any collection of n girls, if one of them has red hair, all have red hair.*

Proof. Clearly, $P(1)$ is true. Suppose $P(k)$ is true. We consider a collection of $k + 1$ girls, $\{g_1, g_2, \cdots, g_{k+1}\}$, at least one of whom has red hair. Without loss of generality, we assume that it is g_1. Then, taking the set of girls $\{g_1, g_2, \ldots, g_k\}$, we have a set of k girls at least one of whom has red hair. Hence, since $P(k)$ is true, all have red hair. Then the collection $\{g_2, g_3, \ldots, g_k, g_{k+1}\}$ is a set of k girls at least one of whom has red hair—in fact, $k - 1$ of them do. Since $P(k)$ is true, they all must have red hair. Therefore, g_{k+1} has red hair. Thus, $k + 1$ girls have red hair.

24. Prove that the well-ordering principle, in fact, follows from the induction principle. Thus, we could have taken the induction principle as our axiom. (This is the approach taken by the Italian mathematician G. Peano (1858–1932) who stated five axioms, one of which was the induction principle, from which the number systems of algebra could be deduced.)

25. The following is an alternative formulation of the principle of induction. Prove that it is equivalent to the one given.

Let $P(n)$ be a statement such that
(a) $P(1)$ is true.
(b) $P(n)$ is true whenever $P(k)$ is true for all positive integers $k < n$.
Then $P(n)$ is true for every positive integer n.

26. Prove that $\Sigma_{i=1}^{n} (a_i + b_i) = (\Sigma_{i=1}^{n} a_i) + (\Sigma_{i=1}^{n} b_i)$.
27. Show that if k is any fixed real number then $\Sigma_{i=1}^{n} ka_i = k \Sigma_{i=1}^{n} a_i$ and that $\Sigma_{i=1}^{n} k = kn$.
28. Show that if m and n are any positive integers $\Sigma_{i=1}^{n} a_i + \Sigma_{i=n+1}^{n+m} a_i = \Sigma_{i=1}^{n+m} a_i$.
29. Rewrite each of Exercises 1 thru 7 using the sigma notation.

2.2. counting principles

In this section we study systematic ways of counting the elements in a set. At first glance, this hardly seems a study worth much time or effort. Yet we shall see that the techniques for doing this are indeed important and that the counting tasks are not always trivial. These techniques are important in many branches of mathematics, most particularly to questions of probability (the science in which one studies situations in which the outcome is uncertain, but about which he wishes to express some degree of confidence).

One is frequently interested in a method for enumerating all the possibilities stemming from a given situation. A very useful tool for such an analysis is called a tree diagram. We illustrate this with several examples.

example 1

Suppose we have two urns. The first contains two black balls and one white ball, while the second contains one black ball and two white balls. A typical elementary probability question involves the likelihood of drawing a specific combination of black and white balls by first choosing an urn and then drawing two balls in succession. To solve such a problem, one must be aware of the possibilities. We can illustrate these possibilities by the following three-stage tree (Figure 2.2.1). The three stages are:

(1) select an urn,
(2) draw a ball,
(3) draw a second ball.

In constructing the tree, we consider the possible outcomes. For example, if we selected the first urn (2B and 1W) and draw a black ball first, we may select either color on the second draw. However, if the first ball is white (the only one), the second draw must be a black ball. A similar list is made for the possibilities in choosing the second urn, resulting in the tree diagram of Figure 2.2.1. ●

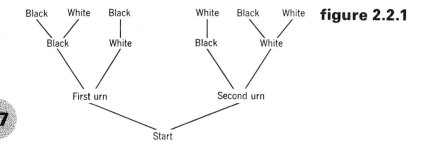

figure 2.2.1

sophisticated counting principles

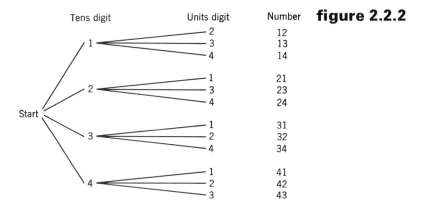

figure 2.2.2

example 2

Consider a set of four discs numbered 1, 2, 3, and 4, respectively. How many different two-digit numbers can be represented by placing two discs side by side? (See Figure 2.2.2.) ●

These two examples lead us to the following theorem. We do not prove the theorem here.

theorem 2.2.1. fundamental counting principle

Suppose two or more actions are performed in order with the first action having m_1 possible outcomes, the second action m_2 possible outcomes, etc. Then the number of possible outcomes of performing this set of actions is the product $m_1 m_2 m_3, \ldots,$ etc.

For additional applications of this theorem, we consider other examples.

example 3

How many four-letter "words" can be formed using the letters a, b, c, and d only? (A "word" need not be sensible in English.)

m_1 = number of ways to select the first letter $m_1 = 4$
m_2 = number of ways to select the second letter $m_2 = 4$
m_3 = number of ways to select the third letter $m_3 = 4$
m_4 = number of ways to select the fourth letter $m_4 = 4$

Thus, there are $(4)(4)(4)(4) = 4^4$ possible "words." ●

example 4

A certain six-man committee will elect three officers: a chairman, vice chairman, and secretary (distinct people). How many different possible sets of officers are there?

Number of possible choices for chairman $m_1 = 6$
Number of possible choices for vice chairman $m_2 = 5$
Number of possible choices for secretary $m_3 = 4$

Thus, there are $6 \cdot 5 \cdot 4 = 120$ possible sets of officers. ●

counting principles

In the preceding section we gave an inductive definition of the number $n!$. We repeat the definition here. Let n be any non-negative integer; $n!$, read "n factorial," is defined by

$$0! = 1$$
$$1! = 1$$
$$2! = 1 \cdot 2$$
$$3! = 1 \cdot 2 \cdot 3$$
$$\vdots \qquad \vdots$$
$$n! = 1 \cdot 2 \cdot 3 \cdots (n - 1)(n)$$

You can easily show that $n! = n(n - 1)!$ and that, for n and r positive integers $r < n$,

$$n! = r! \, (r + 1) \, (r + 2) \cdots (n)$$

We leave these as exercises. Thus,

$$5! = 5 \cdot 4 \cdot 3 \cdot 2 \cdot 1 = 120$$

and

$$7! = 7 \cdot 6 \cdot 5! = (42)(120) = 5040$$

Let S be an arbitrary set. A **partition** $[A_1, A_2, \ldots, A_k]$ of S is a subdivision of S into disjoint subsets A_i which exhaust S; that is, every element of S must be in one, and only one, of the subsets. Thus, $S = A_1 \cup A_2 \cup \cdots \cup A_k$ and $A_1 \cap A_j = \varnothing$ (empty) for $i \neq j$. The subsets A_i in the partition of S are often called **cells**.

example 5

Suppose $S = \{0,1,2, \ldots ,9\}$ is the set of digits. Then the subsets $A_1 = \{0,2,4,6,8\}$ and $A_2 = \{1,3,5,7,9\}$ make up a partition of S. Similarly, $[B_1 = \{0,1,2\}, B_2 = \{3,4,5\}, B_3 = \{6,7,8\}, B_4 = \{9\}]$ is also a partition of S. However, $[C_1 = \{1,3,5,7\}, C_2 = \{1,2,3,4\}, C_3 = \{5,6,8,9\}]$ is not a partition, even though $S = C_1 \cup C_2 \cup C_3$. (Why?) Another partition of S is given by the sets $[A_{11} = \{0,2,4\}, A_{12} = \{6,8\}, A_{21} = \{1,3,5\}, A_{22} = \{7,9\}]$. We note that this last partition is in a sense finer than the first. It is a **refinement** of the first.

●

We turn now to a way to count partitions.

example 6

Suppose a certain school has a traveling basketball squad of 12 players. The hotel assigns the squad two rooms for three people and three rooms for two people. In how many different ways may the coach assign players to rooms? (He is not concerned with the arrangement of the players in the rooms.)

We use the symbol $\begin{pmatrix} 12 \\ 3,3,2,2,2 \end{pmatrix}$ for this number. If the coach listed all the possibilities (i.e., room 1, bed 1; room 1, bed 2; room 1, bed 3; room 2, bed 1; etc.), he could make $12 \cdot 11 \cdot 10 \cdots 3 \cdot 2 \cdot 1 = 12!$ different assignments. But since he is unconcerned with which bed a given player in a room occupies,

sophisticated counting principles

there are too many outcomes listed. We divide by the number of ways in which the players may choose beds once they are assigned to a room. Thus, by 3! for room one; 3! for room two; and 2! for each of the succeeding rooms.

Hence, there are $\begin{pmatrix} 12 \\ 3,3,2,2,2 \end{pmatrix} = \dfrac{12!}{3!3!2!2!2!} = (11)(10)(9)(8)(7)(6)(5)$ possible room assignments. ●

The special symbol introduced in this example is defined, in general, as follows: *Let $n, r_1, r_2, \ldots, r_k$ be positive integers with $n = r_1 + r_2 + \cdots + r_k$. If we wish to partition a set of n elements into k subsets containing respectively, $r_1, r_2, \ldots, r_k$ elements each, then there are*

$$\begin{pmatrix} n \\ r_1, r_2, \ldots, r_k \end{pmatrix} = \frac{n!}{r_1!, r_2!, \ldots, r_k!}$$

possible partitions. Note that this symbol is defined only if $r_1 + r_2 + \cdots + r_k = n$. The symbol

$$\begin{pmatrix} n \\ r_1, r_2, \ldots, r_3 \end{pmatrix}$$

is called *the multinomial coefficient* because it represents the coefficient of $x_1^{r_1} x_2^{r_2} \cdots x_k^{r_k}$ in the expansion of the multinomial $(x_1 + x_2 + \cdots + x_k)^n$.

The special case where $k = 2$, that is, where we want to partition a set of n elements into two subsets, is important. In this case we put r elements in the first set and $n - r$ elements in the second. The number of such partitions is

$$\frac{n!}{r! \, (n-r)!}$$

This number also represents the number of ways r elements can be selected from a set of n elements. We are, then, partitioning a set of n elements into two subsets: *the chosen r and the rejected $n - r$.* The multinomial symbol $\begin{pmatrix} n \\ r, n-r \end{pmatrix}$ is always shortened in this case to

$$\begin{pmatrix} n \\ r \end{pmatrix}$$

Notice that $\begin{pmatrix} n \\ r \end{pmatrix} = \begin{pmatrix} n \\ n-r \end{pmatrix}$. Why? The symbol $\begin{pmatrix} n \\ r \end{pmatrix}$ is called *the binomial coefficient.*

example 7

If State plays ten football games in a season, in how many different ways can they end with a 5:3:2 record (5 wins, 3 losses, 2 ties)?

We partition the set of ten opposing teams into three subsets: those whom State beats, those who beat State, and those with whom State ties. Thus, there are $\begin{pmatrix} 10 \\ 5,3,2 \end{pmatrix} = \dfrac{10}{5! \, 3! \, 2!} = 2520$ ways to do this. ●

example 8

This example is basic in the study of probability. Suppose a coin is thrown 5 times. There are 2^5 possible outcomes since on each throw there are two possibilities, a head or a tail. Each set of tosses of the coin results in a partition

counting principles

of the numbers 1, 2, 3, 4, 5 in the following way. In the first subset, put the numbers corresponding to a toss on which a head appeared and in the second the number of tosses on which a tail appeared. If we want three numbers in the first subset and two in the second subset, there are $\binom{5}{3} = 10$ different ways this works. ●

We prove the following theorem concerning the binomial coefficient:

theorem 2.2.2

If n and r are non-negative integers with $r \le n - 1$, then

$$\binom{n}{r} + \binom{n}{r+1} = \binom{n+1}{r+1}.$$

proof.

$$\binom{n}{r} + \binom{n}{r+1} = \frac{n!}{(n-r)!\,r!} + \frac{n!}{(n-(r+1))!\,(r+1)r!}$$

Factoring this expression yields

$$\binom{n}{r} + \binom{n}{r+1} = \frac{n!}{(n-(r+1))!\,r!}\left[\frac{1}{n-r} + \frac{1}{r+1}\right]$$

$$= \frac{n!}{(n-(r+1))!\,r!}\left[\frac{n+1}{(n-r)(r+1)}\right]$$

$$= \frac{(n+1)n!}{(n-r)!\,(r+1)!}$$

$$= \frac{(n+1)n!}{[(n+1)-(r+1)]!\,(r+1)!}$$

$$= \binom{n+1}{r+1}$$

The formula in Theorem 2.2.2 expresses the relationship between binomial coefficients which is known as "*Pascal's triangle.*" There are many relations which involve the numbers in this triangle, several of which were developed by the French mathematician Blaise Pascal (1623–1662). He was not, however, the originator of the triangle. The oldest currently known reference to it occurs in the work of the Chinese algebraist Chu Shi-ki in 1303.

```
                1
              1   1
            1   2   1
          1   3   3   1
        1   4   6   4   1
      1   5  10  10   5   1
    1   6  15  20  15   6   1
```

sophisticated counting principles

In the next section we shall see that the nth row of the triangle contains the coefficients in the expansion of the binomial $(a + b)^n$, n a positive integer.

exercises

1. Find:

(a) $\binom{3}{2}$ (d) $\binom{32}{4}$ (g) $\binom{4}{2,0,2}$

(b) $\binom{6}{4}$ (e) $\binom{550}{549}$ (h) $\binom{6}{3,1,2}$

(c) $\binom{7}{2}$ (f) $\binom{5}{1,2,2}$ (i) $\binom{2}{1,1,1}$

2. Interpret $\binom{n}{0}$ and $\binom{n}{n}$.

3. How many four-digit numbers can be formed (repetition of digits allowed) from the digits 0,1,2,3?

4. Some license plates contain two letters and four numerals. How many different combinations are possible?

5. New California license plates contain three numerals followed by three letters. If we disregard the fact that certain combinations of letters are avoided, how many different license plates can be issued?

6. How many committees of four individuals can be selected from a group of ten people?

7. In how many different ways can five books be arranged on a shelf?

8. Suppose a coin is tossed seven times. How many possible outcomes are there? How many of these result in three heads and four tails?

9. In the coin toss of Exercise 8, how many outcomes result in a head on the first toss and also a head on the last toss?

10. How many subsets does a set of n elements have?

11. In how many different ways can a student answer a multiple-choice examination if the exam has twenty questions, each with five possible responses, (a), (b), (c), (d), and (e)?

12. How many different individual sets of initials can there be among a group of people if each person has three initials?

13. There are four different routes from town A to town B and three different routes from town B to town C. There are also two roads from town A to town C that bypass town B.

     (a) Sketch a diagram.
     (b) How many possible routes may one take from town A to town C?

14. (a) Let $X = \{a,b,c\}$. How many different functions $f: X \rightarrow X$ exist? List them. How many are bijections?
     (b) Let $X = \{a,b,c,d\}$. How many different functions $f: X \rightarrow X$ exist? List them. How many are bijections?
     (c) Let X be a set of n elements. How many different functions $f: X \rightarrow X$ exist? How many bijections?

15. From the set $A = \{1,2,3,4,5,6\}$, in how many different ways may one select two numbers whose sum is seven?

the binomial theorem

16. How many different possible bridge hands (13 cards) can be dealt from a deck (52 cards)?

17. Prove that if $p > 2$ is a prime, then the integer $\binom{p}{2}$ is divisible by p.

18. Write 10 rows of Pascal's triangle.

19. Use the results of Exercise 10 to show that

$$\binom{n}{0} + \binom{n}{1} + \binom{n}{2} + \cdots + \binom{n}{n} = 2^n$$

20. Using the definition, show that if $n \geq r \geq k \geq s$, then

$$\binom{n}{r}\binom{r}{k}\binom{k}{s} = \binom{n}{s}\binom{n-s}{k-s}\binom{n-k}{r-k}$$

2.3. the binomial theorem

From your previous mathematical experience you have, no doubt, learned that the following identities are true whenever $a, b \in R$.

$$(a + b)^2 = a^2 + 2ab + b^2$$

$$(a + b)^3 = a^3 + 3a^2b + 3ab^2 + b^3$$

$$(a + b)^4 = a^4 + 4a^3b + 6a^2b^2 + 4ab^3 + b^4$$

If we add $(a + b)^0 = 1$ and $(a + b)^1 = a + b$ to the top, we note that the coefficients can be arranged in the form of Pascal's triangle.

$$
\begin{array}{ccccccccc}
 & & & & 1 & & & & \\
 & & & 1 & & 1 & & & \\
 & & 1 & & 2 & & 1 & & \\
 & 1 & & 3 & & 3 & & 1 & \\
1 & & 4 & & 6 & & 4 & & 1 \\
\end{array}
$$

While it would be possible by successive multiplication of $(a + b)$ to obtain $(a + b)^n$ for any positive integer n, we are saved from this labor by means of the binomial theorem. This theorem is stated as follows.

theorem 2.3.1

If a and b are real numbers and n is a positive integer, then

$$(a + b)^n = \sum_{r=0}^{n} \binom{n}{r} a^{n-r}b^r$$

The Σ sign, of course, means we take the sum

$$\binom{n}{0}a^n + \binom{n}{1}a^{n-1}b + \binom{n}{2}a^{n-2}b^2 + \cdots + \binom{n}{r}a^{n-r}b^r + \cdots + \binom{n}{n}b^n$$

83 We give a proof of this theorem which uses mathematical induction (on n) and the results of the theorem of the previous section. Notice that the formula

sophisticated counting principles

indicate: that the binomial expansion is a function whose domain is the set of non-negative integers n. A slightly more elegant, but a little less simple proof is suggested in Exercise 10.

Induction proof.

If $n = 1$, then $(a + b)^1 = a + b = \binom{1}{0}a + \binom{1}{1}b$. We assume that the theorem is true when $n = k$; that is,

$$(a + b)^k = \sum_{r=0}^{k} \binom{k}{r} a^{k-r} b^r$$

It must be shown that, based upon this assumption, it follows that

$$(a + b)^{k+1} = a^{k+1} + \binom{k+1}{1}a^k b + \cdots + \binom{k+1}{r}a^{k+1-r}b^r + \cdots + b^{k+1}.$$

Now, $(a + b)^{k+1} = (a + b)(a + b)^k$

$$= (a + b)\sum_{r=0}^{k}\binom{k}{r}a^{k-r}b^r$$

$$= a\sum_{r=0}^{k}\binom{k}{r}a^{k-r}b^r + b\sum_{r=0}^{k}\binom{k}{r}a^{k-r}b^r$$

$$= a^{k+1} + \binom{k}{1}a^k b + \cdots + \binom{k}{k-1}a^2 b^{k-1} + ab^k$$

$$+ a^k b + \cdots + \binom{k}{k-2}a^2 b^{k-1} + \binom{k}{k-1}ab^k + b^{k+1}$$

In the last expression above we notice that, except for the a^{k+1} and b^{k+1} terms, each of the other powers $a^{k+1-r}b^r$, $r = 1,2,\ldots,k$, appears twice: once with coefficient $\binom{k}{r}$ and once with coefficient $\binom{k}{r-1}$. Thus, we have, upon combining these and using the result of Theorem 2.2.2,

$$\left[\binom{k}{r} + \binom{k}{r-1}\right]a^{k+1-r}b^r = \binom{k+1}{r}a^{k+1-r}b^r$$

Or, in other words,

$$(a + b)^{k+1} = \sum_{r=0}^{k=1}\binom{k+1}{r}a^{k+1-r}b^r$$

This proves the theorem.

example 1

Find $(x - y)^5$ by use of the binomial theorem. We have

$$(x - y)^5 = \sum_{r=0}^{5}\binom{5}{r}x^{5-r}(-y)^r$$

the binomial theorem

Thus,

$$(x - y)^5 = \binom{5}{0}x^5 + \binom{5}{1}x^4(-y) + \binom{5}{2}x^3(-y)^2 + \binom{5}{3}x^2(-y)^3 + \binom{5}{4}x(-y)^4$$
$$+ \binom{5}{5}(-y)^5$$
$$= x^5 - 5x^4 + 10x^3y^2 - 10x^2y^3 + 5xy^4 - y^5 \quad \bullet$$

example 2

Find $(1.01)^6$. We write

$$(1.01)^6 = (1 + .01)^6 = \sum_{r=0}^{6} \binom{6}{r} 1^{6-r}(0.01)^r$$

$$= \sum_{r=0}^{6} \binom{6}{r}(.01)^r$$

$$= 1 + 6(.01) + 15(.01)^2 + 20(.01)^3$$
$$+ 15(.01)^4 + 6(.01)^5 + (.01)^6$$

$$= 1 + .06 + .0015 + .00002 + .00000015$$
$$+ .0000000006 + .000000000001$$

$$= 1.061520150601 \quad \bullet$$

example 3

Find the fifth term in the expansion of $(3a^2 - 2b)^7$ without writing out the entire expansion. The $r + 1$st term, for $r = 0,1,2, \ldots ,7$, in this expansion has the form

$$\binom{7}{r}(3a^2)^{7-r}(-2b)^r$$

Thus, when $r = 0$, we obtain the first term; $r = 1$, the second term; etc. Hence, the fifth term is

$$\binom{7}{4}(3a^2)^3(-2b)^4 = 35(27a^6)(16b^4) = 15120a^6b^4. \quad \bullet$$

exercises

1. In each of the following, use the binomial theorem to expand the expression, then simplify:

(a) $(a + b)^7$
(b) $(1 + b)^6$
(c) $(1 - b)^5$

(d) $(a - b)^8$
(e) $(p - q)^7$
(f) $(3c - d)^4$

(g) $(2x^2 - 3y)^3$
(h) $(1 - .02)^3$
(i) $(1 + .001)^3$

2. Without writing the entire expansion, find the specified term in each of the following. Simplify the expression.

(a) the fourth term of the expansion of $(7a - 2b)^6$

85

sophisticated counting principles

(b) the third term of the expansion of $\left(\dfrac{1}{2}a + \dfrac{1}{3}b\right)^7$

(c) the third term of the expansion of $\left(x + \dfrac{1}{x}\right)^8$

(d) the fourth term of the expansion of $\left(\dfrac{1}{x^2} - 2x^4\right)^6$

3. Using the binomial theorem expansion of $(1 + 1)^n$, show that

$$\sum_{r=0}^{n} \binom{n}{r} = 2^n$$

4. Show that

$$\sum_{r=0}^{n} (-1)^r \binom{n}{r} = 0$$

5. Show that, if n is even, $\binom{n}{0} + \binom{n}{2} + \cdots + \binom{n}{n} = 2^{n-1}$.

6. Use the appropriate number of terms to find an approximation accurate to four decimal places for the following.

(a) $(.99)^4$ (d) $(1.2)^{10}$ (g) $(99)^5$ find exact value

(b) $(1.1)^7$ (e) $(2.1)^6$ (h) $(102)^5$ find exact value

(c) $(.98)^5$ (f) $(1.8)^6$

7. Find the term which involves x^2 in the expansion of each of the following binomials.

(a) $(x - y)^6$ (c) $(3x^{-3} + 2x^2)^6$

(b) $\left(\dfrac{1}{x} - x\right)^4$ (d) $(x^{1/2} + x^{-1/2})^{12}$

8. An argument similar to that given for the binomial theorem leads to the *multinomial theorem*. If p and n are positive integers, then $(a_1 + a_2 + \cdots + a_p)^n$ is the sum of terms of the form

$$\binom{n}{r_1, r_2, \ldots, r_p} a_1^{r_1}, a_2^{r_2}, \ldots, a_p^{r_p}$$

where we choose all possible sets $\{r_1, r_2, \ldots, r_p\}$ of p non-negative integers whose sum is n. Use this to find the following.

(a) $(x + y + z)^2$ (d) $(a + b + c + d)^2$

(b) $(a + b + c)^3$ (e) $(2w + 3x^{1/2} - y^{-1} + z^2)^2$

(c) $(u - v + w)^2$ (f) $[(a + ib) - (c + id)]^3$

9. How many terms are there in the expansion of:

(a) $(x + y + z)^6$?   (c) $(a + 2b + 5c + d)^5$?

(b) $(a + b + c + d)^4$?   (d) $(r + s + t + u + v)^6$?

10. Prove the binomial theorem by using only the methods of Section 2.2; that is, count the number of ways in the expansion of $(a + b)^n$ by multiplication, using the distributive law, that one can obtain $a^{n-r}b^r$, for $r = 0,1,2,\ldots,n$.

86

2.4. sequences & progressions

After looking at several specialized examples of functions involving the positive integers, as we did in the previous sections of this chapter, we turn to a more general study of all those functions

$$a: N \to R$$

whose domain is the set of positive integers.

Consider the following list of numbers

$$2, 4, 8, 16, 32, \ldots, 2^n, \ldots$$

Such a list, consisting of one number after another in succession, is called a (infinite) *sequence*. The numbers are called the *terms* of the sequences; 2 is the *first term*, 4 the *second term*, etc. The example is a sequence in which each term is a positive power of 2. The expression 2^n gives the *nth term* of this sequence and indicates a rule for finding any term once its number (position) in the sequence is specified.

As an example consider the sequence whose *n*th term is $1/n$. The first term would be $1/1 = 1$; the second term is $1/2$; the third term is $1/3$; etc. We would write the sequence in the form

$$1, \frac{1}{2}, \frac{1}{3}, \frac{1}{4}, \ldots, \frac{1}{n}, \ldots$$

You notice that in the above discussion we have informally described a particular class of functions. The domain of each of these functions (sequences) is the set of positive integers (natural numbers) N. The range is contained in the set R of real numbers, and the *n*th term is the rule of association. Formally then, a sequence is defined in the following way.

A (real) **sequence** *a is any function*

$$a: N \to R$$

whose domain is the set N of positive integers (and whose range is the set R of real numbers).

The "*n*th term" a_n of the sequence is a specialized way of writing the function notation $a(n)$. The list, as in the above examples, is a sort of function table. In general the sequence a could be written

$$a_1, a_2, a_3, \ldots, a_n, \ldots$$

Here are some examples of sequences written in this way. In each case the first five terms—values of $a(1) = a_1$, $a(2) = a_2$, a_3, a_4, and a_5—are listed, and then the *n*th term $a_n = a(n)$.

(1) $1, 2, 6, 24, 120, \ldots, n!, \ldots$

(2) $1, 4, 9, 16, 25, \ldots, n^2, \ldots$

(3) $1, \dfrac{1}{2}, \dfrac{1}{4}, \dfrac{1}{8}, \dfrac{1}{16}, \ldots, \dfrac{1}{2^{n-1}} \ldots$

(4) $3, 5, 7, 9, 11, \ldots, 3 + 2(n - 1), \ldots$

sophisticated counting principles

(5) $1, -1, 1, -1, 1, \ldots, (-1)^{n-1}, \ldots$

(6) $1, 1, 2, 3, 5, \ldots, a_n = a_{n-1} + a_{n-2}, \ldots$

Sequence number (6) is known as the **Fibonacci sequence** after the thirteenth-century mathematician Leonardo of Pisa (Fibonacci) who was an early European expositor of arithmetic and elementary algebra. The sequence is generated by the formula

$$a_1 = a_2 = 1$$

and

$$a_n = a_{n-1} + a_{n-2}, \qquad \text{for } n \geq 3$$

This description of the Fibonacci sequence illustrates the *recursive* way in which sequences may be defined. We discussed this procedure in the section on induction.

As is the case with many functions, a sequence is most economically defined by simply stating a formula for the nth term a_n. Each of the above examples could have been listed in that way only.

The story is often told that when the mathematician Gauss was a young school boy, his teacher, in an effort to gain some free time, set the pupils the task of adding up all the numbers from 1 to 100. He supposed that this would keep the class occupied for a little while. He was surprised when, after about three minutes, Gauss appeared at his desk with the correct answer.

This story illustrates a frequent problem encountered with sequences; viz, finding the sum s_m of the first m terms of the sequence. Gauss's problem involved the sequence (progression)

$$1, 2, 3, 4, \ldots, n, \ldots 100$$

This is a fairly simple example of a class of sequences called *arithmetic progressions. An* **arithmetic progression** *is a sequence given by*

$$a_n = a_{n-1} + d$$

for each $n > 1$. Here d is a fixed number called the common difference.

In the cited example, we have $a_1 = 1$ and $d = 1$.

Each of the following is an arithmetic progression. We specify a_1 and d

(1) $2, 4, 6, 8, 10, \ldots$ $\qquad\qquad$ $a_1 = 2, \qquad d = 2$

(2) $3, 1, -1, -3, -5, \ldots$ $\qquad$ $a_1 = 3, \qquad d = -2$

(3) $1, 1\frac{1}{2}, 2, 2\frac{1}{2}, 3, \ldots$ $\qquad\quad$ $a_1 = 1, \qquad d = \frac{1}{2}$

It is seen by induction that

$$a_n = a_1 + (n - 1)d$$

This follows since $a_1 = a_1 + (1 - 1)d$; and if $a_k = a_1 + (k - 1)d$,

$$a_{k+1} = a_k + d = a_1 + (k - 1)d + d$$

$$= a_1 + (k + 1 - 1)d$$

sequences & progressions

To return to the problem of Gauss's teacher. To find the sum s_m of the first m terms of an arithmetic progression we write these terms down twice as follows.

$$s_m = a_1 + [a_1 + d] + [a_1 + 2d] + \ldots + [a_1 + (m-2)d] +$$
$$[a_1 + (m-1)d]$$

and in reverse order,

$$s_m = [a_1 + (m-1)d] + [a_1 + (m-2)d] + [a_1 + (m-3)d] +$$
$$\ldots + [a_1 + d] + a_1$$

Upon adding both sides we have,

$$2s_m = [2a_1 + (m-1)d] + [2a_1 + (m-1)d] + \ldots + [2a_1 + (m-1)d]$$
$$= m[2a + (m-1)d]$$

Thus, $s_m = (m/2)[2a + (m-1)d]$ is the formula for the sum s_m of the first m term of an arithmetic progression.

Gauss's answer was

$$s_{100} = \frac{100}{2}[2(1) + 100 - 1)1] = 50(2 + 99)$$

$$= (50)(101) = 5050$$

This formula can also be proved by mathematical induction (see Section 2.1).

theorem 2.4.1

For each natural number m and fixed number d, the sum s_m of the first m terms of an arithmetic progression a_n with common difference d and first term a_1 is

$$s_m = \frac{m}{2}(a_1 + a_m)$$

proof. The proof is given above, since $a_m = a_1 + (m-1)d$.

example 1

The fourth term of an arithmetic progression is -16 and the eighth term is -40. Find the sum of ten terms. To do this we note that

$$a_4 = a_1 + 3d = -16$$

and

$$a_8 = a_1 + 7d = -40$$

Thus,

$$a_1 + 3d = -16$$

$$a_1 + 7d = -40$$

sophisticated counting principles

One easily computes that $a_1 = 2$ and $d = -6$. Hence,

$$s_{10} = \frac{10}{2}[2(2) + 9(-6)]$$

$$= 5(-50)$$

$$= -250 \quad \bullet$$

Suppose that a and b are real numbers, and that x_1, x_2, ..., x_k are k real numbers such that

$$a, x_1, x_2, \ldots, x_k, b$$

are consecutive terms of some arithmetic progression. Then the numbers x_1, x_2, ... x_k, are called **arithmetic means** *between a and b. Clearly, the arithmetic mean, $(a + b)/2$, of a and b satisfies this definition for $k = 1$.*

To compute the numbers x_1, x_2, ..., x_k for $k > 1$ we notice that a is the first term of the progression and b is the $(k + 2)$nd term. Thus,

$$b = a + (k + 1)d$$

Therefore, $d = (b - a)/(k + 1)$. It follows, then, that

$$x_1 = a + \left[\frac{b-a}{k+1}\right], \; x_2 = a + 2\left[\frac{b-a}{k+1}\right], \cdots, x_k = a + k\left[\frac{b-a}{k+1}\right]$$

example 2

Insert five arithmetic means between 3 and 7. We see that $k = 5$, so $d = (7 - 3)/6 = 2/3$. The desired means are

$$x_1 = 3 + \frac{2}{3} = 3\frac{2}{3}, \; x_2 = 4\frac{1}{3}, \; x_3 = 5, \; x_4 = 5\frac{2}{3}, \; x_5 = 6\frac{1}{3} \quad \bullet$$

A second important class of sequences are the geometric progressions. Unfortunately, we are unable to supply an interesting historical anecdote to introduce this class of sequences.

A **geometric progression** *is a sequence with first term a_1 and succeeding terms given by*

$$a_n = a_{n-1}r$$

where r is a fixed real number called the **common ratio.**

While they are technically qualified as geometric progressions we, henceforth, specifically exclude from our consideration, sequences with $r = 0$ or 1 and those with $a_1 = 0$. Aside from these trivial examples, the following are also geometric progressions.

(1) $2, 4, 8, 16, 32, \ldots, 2^n, \ldots$; $a_1 = 2, r = 2$
(2) $1, -1, 1, -1, 1, \ldots, (-1)^{n-1}, \ldots$; $a_1 = 1, r = -1$
(3) $1, \frac{1}{3}, \frac{1}{9}, \frac{1}{27}, \frac{1}{81}, \ldots, (\frac{1}{3})^{n-1}, \ldots$; $a_1 = 1, r = \frac{1}{3}$
(4) $\sqrt{2}, 2, \sqrt{8}, 4, \sqrt{32}, \ldots, 2^{n/2}, \ldots$; $a_1 = \sqrt{2}, r = \sqrt{2}$
(5) $5, -15, 45, -135, \ldots, 5(-3)^{n-1}, \ldots$; $a_1 = 5, 4 = -3$

sequences & progressions

It is left to you to use induction to show that for a geometric progression

$$a_n = a_1 r^{n-1}$$

for each natural number n.

To find the sum of the first m terms of a geometric progression we write the sum formally.

$$S_m = a_1 + a_1 r + a_1 r^2 + \cdots + a_1 r^{m-1}$$

Then

$$rS_m = a_1 r + a_1 r^2 + \cdots + a_1 r^{m-1} + a_1 r^m$$

Subtracting these two expressions gives

$$S_m - rS_m = a_1 + 0 + 0 + \cdots + 0 - a_1 r^m$$

or

$$S_m(1 - r) = a_1(1 - r^m)$$

Thus, for $r \neq 1$,

$$S_m = \frac{a_1}{1 - r}(1 - r^m)$$

This gives one proof of the following theorem. It may also be proved by induction (see Section 2.1).

theorem 2.4.2

For each natural number m and fixed real number $r \neq 1$, the sum s_m of the first m terms of a geometric progression with first term a_1 and common ratio r is

$$S_m = \frac{a_1}{1 - r}(1 - r^m)$$

example 3

Compute the sum of the first six terms of the geometric progression $a_n = 2^n$.
Here $a_1 = 2$ and $r = 2$. Thus

$$S_6 = \frac{2}{1 - 2}(1 - 2^6)$$

$$= \frac{2}{1}(1 - 64) = (-2)(-63) = 126 \quad \bullet$$

Suppose a and b are real numbers, and that $x_1, x_2, \ldots, x_k$ are k real numbers such that

$$a, x_1, x_2, \ldots, x_k, b$$

are consecutive terms of some geometric progression. Then the numbers x_1, $x_2, \ldots, x_k$ are called **geometric means** *between a and b.*
When $k = 1$ we have *the geometric mean $\sqrt{xy}$ of x and y* satisfying this definition.

sophisticated counting principles

Generally, we have a progression with first term a and $(k + 2)$nd term b. Thus,

$$b = ar^{k+1}$$

and

$$r = \left(\frac{b}{a}\right)^{1/(k+1)}$$

Hence,

$$x_1 = ar, \; x_2 = ar^2, \; \ldots, \; x_k = ar^k$$

example 4

Insert three geometric means between 3 and 48.

We have $r = \left(\dfrac{48}{3}\right)^{1/4} = 16^{1/4} = 2$

Thus,

$$x_1 = 6, \; x_2 = 12, \; x_3 = 24 \quad \bullet$$

Suppose, as a parting shot, we consider the behavior of s_m when r is suitable and m is very large. We have

$$s_m = \frac{a_1}{1 - r}(1 - r^m)$$

$$= \frac{a_1}{1 - r} - \frac{a_1 r^m}{1 - r}$$

The last of these two terms is the only one involving m. We recall that if $|r| < 1$ and m is very large, r^m is a very small number. In fact, one can show that r^m can be made as small as we please for a suitably large value of m. Hence s_m "approaches" the value of

$$s = \frac{a_1}{1 - r}, \text{ provided } |r| < 1$$

example 5

Consider the (infinite) repeating decimal expansion

$$k = 0.1111 \ldots$$

At any stage we can consider the resulting number (when we stop) to be the sum s_m of the geometric progression $a_n = .1(1/10)^{n-1}$. Here $r = 1/10$ and $a_1 = 1/10$. Thus the number k approaches the value

$$s = \frac{1/10}{1 - (1/10)} = \frac{1/10}{9/10} = \frac{1}{9}$$

In fact k is the decimal expansion of $1/9$. $\quad \bullet$

sequences & progressions

exercises

1. Find the 30th term for each of the arithmetic progressions given below.

   (a) 2, 4, 6, 8, . . .

   (b) 1, 4, 7, 10, . . .

   (c) 0, −3, −6, −9, . . .

   (d) −2, 4, 10, 16, . . .

   (e) $-\dfrac{1}{6}, \dfrac{1}{6}, \dfrac{1}{2}, \dfrac{5}{6}, \ldots$

   (f) 6, 3, 0, −3, . . .

   (g) $-\dfrac{1}{3}, -1\dfrac{2}{3}, -3, -4\dfrac{1}{3}, \ldots$

   (h) 8, 5, 2, −1, . . .

   (i) $a, b, (2b - a), (3b - 2a), \ldots$

   (j) $u, (v + 2a), (2v + 3a), (3v + 4a), \ldots$

2. Write a formula for the nth term of each of the arithmetic progressions in Exercise 1.
3. Find the sum of the first 15 terms in each of the progressions in Exercise 1.
4. Insert three arithmetic means between 2 and 12.
5. Find the sum of the even numbers between 49 and 101.
6. The third term of an arithmetic progression is −4 and the ninth term is 14. Find the sum of the first ten terms.
7. Find the fifth term for each of the following geometric progressions.

   (a) 1, 2, 4, . . .

   (b) 1, 3, 9, . . .

   (c) 1, −3, 9, −27, . . .

   (d) 24, 8, $\dfrac{8}{3}$, . . .

   (e) $2\dfrac{1}{4}, 1\dfrac{1}{2}, 1, \ldots$

   (f) $1, \dfrac{1}{2}, \dfrac{1}{4}, \dfrac{1}{8}, \ldots$

   (g) $1, \dfrac{1}{3}, \dfrac{1}{9}, \dfrac{1}{27}, \ldots$

   (h) $\sqrt{2}, \sqrt{6}, 3\sqrt{2}, \ldots$

   (i) 0.4, 0.04, 0.004, . . .

   (j) $a, b, \dfrac{b^2}{a}, \ldots$

8. For each of the geometric progressions in Exercise 7, write a formula for the nth term.
9. Find the sum of the first five terms of each of the progressions in Exercise 7.
10. Find the arithmetic mean and the geometric mean of 3 and 27.
11. Find the first and second terms of a geometric progression whose third term is 18 and whose fifth term is 40.5.
12. Find the "infinite" sum of each of the following geometric progressions.

   (a) $1, \dfrac{1}{2}, \dfrac{1}{4}, \dfrac{1}{8}, \ldots$

   (b) $1, \dfrac{1}{3}, \dfrac{1}{9}, \dfrac{1}{27}, \ldots$

   (c) $1, -\dfrac{1}{2}, \dfrac{1}{4}, -\dfrac{1}{8}, \ldots$

   (d) $1, \dfrac{1}{\sqrt{3}}, \dfrac{1}{3}, \ldots$

   (e) $2\dfrac{1}{4}, 1\dfrac{1}{2}, 1, \ldots$

   (f) 0.2, 0.02, 0.002, . . .

   (g) 0.4, 0.04, 0.004, . . .

   (h) 0.39, 0.0039, 0.000039, . . .

   (i) 0.01, 0.0001, 0.000001, . . .

   (j) 0.04, 0.0004, 0.000004, . . .

sophisticated counting principles

13. Find the rational number p/q whose decimal expansion is

(a) 0.4444 . . .   (e) 0.040404 . . .
(b) 0.2222 . . .   (f) 1.4444 . . .
(c) 0.393939 . . .   (g) 1.9999 . . .
(d) 0.3999 . . .   (h) 3.142857142857142857 . . .

14. Show that the reciprocals of the terms of a geometric progression also form a geometric progression.

15. A harmonic progression is a sequence the reciprocals of whose terms form an arithmetic progression. Find the seventh term of the harmonic progression

$$\frac{3}{2}, \frac{4}{3}, \frac{6}{5}, \ldots$$

16. A ball is dropped from a height of 24 feet. On each rebound it rises to a height one-third less than the height from which it fell. How far does it rise on the fourth rebound?

17. If a person saves $1.00 the first week, $2.00 the second week, $3.00 the third, and so on, how much will he have saved (exclusive of interest) in a year?

quiz for review

PART I.   True or False

1. Every sequence is a function whose domain is the set of positive integers.

2. Every sequence is a bijection from the integers to the natural numbers.

3. *Every* subset of a well-ordered set has a least element.

4. Mathematical induction is a deductive method for proof.

5. If a statement $P(n)$ has the property that $P(k)$ is true for some k, then $P(k + 1)$ must also be true.

6. $\binom{n}{0} = 1$ for every positive integer n.

7. $1 + 2 + 3 + \cdots + n = \dfrac{n(n + 1)}{2}$ for every positive integer n.

8. $(n!)\,(k!) = (nk)!$

9. $\binom{n}{r} = \binom{n}{n - r}$

10. $\binom{n}{r} = \binom{n + 1}{r + 1}$

11. $\binom{n}{k} + \binom{n}{k - 1} = \binom{n + 1}{k}$

12. 1.4, 1.44, 1.444, 1.4444, . . . are terms of a geometric progression.

13. 1, 3, 5, 7, . . . are terms of an arithmetic progression.

14. The sum of m terms of an arithmetic progression is $s_m = a_1 + (m - 1)d$; a_1 the first term and d the common difference.

15. The sum of m terms of a geometric progression is $s_m = \dfrac{a(1 - r^m)}{1 - r}$, $r \neq 1$.

16. $\binom{3}{0} > \binom{2}{0}$

17. $(n + 1) + n! = (n + 1)!$

18. $\binom{n}{n} = n$

19. $\binom{5}{2, 2, 2} = \dfrac{5!}{2!\ 2!\ 2!}$

20. 1.4, 1.44, 1.444, 1.4444, . . . approaches the rational number 13/9.

PART II

Chose the best response.

1. $\binom{4}{2} =$ (a) 4!   (b) 2!   (c) $\dfrac{4!}{2!}$ (d) 6   (e) 1

2. $\binom{7}{5} =$ (a) 42   (b) 21   (c) 7!   (d) 5!   (e) 1

3. $\binom{6}{2, 2, 2} =$ (a) $\binom{6}{2}$ (b) 90   (c) $\dfrac{6!}{8!}$ (d) 6!   (e) is not defined

4. A certain club has fifteen members. How many different nominating committees are possible if three members are to make up the committee?

   (a) 2800   (b) 3!   (c) $\dfrac{15!}{3!}$ (d) 455   (e) 15!

5. The sum of the first n integers, $1 + 2 + \cdots + n$, equals

   (a) n^2 (b) $\dfrac{1}{6}(n(n + 1)(2n + 1))$ (c) $\dfrac{n}{2}(n + 1)$ (d) n^3 (e) 5050

6. The nth term of the geometric progression $\sqrt{3}$, 3, $\sqrt{27}$, 9, . . . is

   (a) 3   (b) $(\sqrt{3})^n$ (c) $(\sqrt{3})^{n-1}$ (d) 3^n (e) 3^{n-1}

7. In giving a proof by mathematical induction of a proposition $P(n)$, n a natural number, one must show

   (a) that $P(1)$ is true.
   (b) that $P(k)$ is true.
   (c) that if $P(k)$ is true, then $P(k + 1)$ is true.
   (d) that $P(1)$ is true; and if $P(k)$ is true, then $P(k + 1)$ is true.
   (e) that $P(1)$, $P(k)$, and $P(k + 1)$ are all true.

8. With eight football games to play in a season, in how many ways may a team end the season with three wins, three losses, and two ties?

   (a) $(3)(3)(2)$ (b) $\binom{8}{32}$ (c) $\binom{8}{3, 3, 2}$ (d) $\dfrac{8!}{5!\ 3!}$ (e) it can't be done

9. How many straight lines are determined by six points if no three of the points are collinear?

   (a) 6   (b) 3   (c) 90   (d) 30   (e) 15

10. An infinite sequence is defined recursively by

$$a_1 = 1, \quad a_{n+1} = na_n$$

then $a_4 =$ (a) 4   (b) 4!   (c) 12   (d) 1   (e) 0

11. $\Sigma_{i=1}^{3}\ i =$ (a) 1   (b) 3   (c) i^3 (d) 5   (e) 6

sophisticated counting principles

12. $\Pi_{i=1}^{3} 2^i =$ (a) 8   (b) 3   (c) 2^3 (d) 32   (e) 64

13. $\sum_{r=0}^{5} \binom{5}{r} (-2)^r x^{5-r} y^r =$ (a) $5!$ (b) $(x + y)^5$ (c) $(x - 2y)^5$

                                                 (d) $(y - x)^5$ (e) $(y - 2x)^5$

14. $\sum_{r=0}^{6} \binom{6}{r} =$ (a) $6!$ (b) 0   (c) 2^6 (d) $(a + b)^6$ (e) $\binom{6}{0}$

15. The sixth term in the expansion of $(x - y)^8$ is

     (a) $56 \, x^5 y^3$ (b) $28 \, x^2 y^6$ (c) $-56 \, x^3 y^5$ (d) $-28 \, x^6 y^2$

     (e) $\binom{8}{6} x^6 y^2$

polynomials & rational functions

3

In the first chapter, the general concept of a function

$$f: X \to Y$$

was discussed. There we met with varied examples of some special functions which occur in the physical and mathematical worlds. In this chapter, two large classes of functions will be described: the polynomials and the rational functions.

Polynomial functions are in one sense the simplest real functions. They have been studied carefully for several centuries, and because of man's familiarity with them are most often used for models of the natural world. You will discover that the polynomials have many algebraic properties which are analogous with those of the integers. Polynomials are, roughly, the integers of the function family.

Just as rational numbers are expressible as quotients of integers, the second class we discuss, the rational functions, consists of functions which are expressed as quotients of polynomials. These functions, too, have interesting and important properties which make them useful in building models of natural phenomena.

3.1. polynomials & polynomial functions

97 Technically, there is a difference between a polynomial (in a single indeterminate x) and a polynomial function. However, a detailed description of this difference need not really concern us here. We shall point out how each poly-

polynomials & rational functions

nomial gives rise to a polynomial function and conversely. These concepts will then be used interchangably as the need arises.

A (real) **polynomial** is an expression of the form

$$(3.1.1) \qquad p(x) = a_n x^n + a_{n-1} x^{n-1} + \cdots + a_1 x + a_0$$

where n is a non-negative integer, and the coefficients a_i, $i = 1, 2, \ldots, n$, are real numbers (or perhaps complex numbers, integers, rational numbers, etc.). The only *exponents* allowed for x in a polynomial are natural numbers or 0. The number a_0 is called *the constant term*; it is the coefficient of x^0. If $a_n \neq 0$ it is called *the leading coefficient* of the polynomial $p(x)$. The non-negative integer n, the exponent of the highest power of x, is the *degree* of $p(x)$, $n = \deg p(x)$. In the special case where $a_i = 0$ for each $i = 0, 1, 2, \ldots, n$; i.e., each coefficient is zero, we have *the zero polynomial*. The zero polynomial does not have a degree.

The following is a list of some polynomials and their degrees.

(1) $p(x) = x^2 + 2x + 1$
   $\deg p(x) = 2$ (p is called *quadratic*)
(2) $p(x) = 3x + 1$
   $\deg p(x) = 1$ (p is called *linear*)
(3) $p(x) = 1 - x^3$
   $\deg p(x) = 3$ (p is called *cubic*)
(4) $p(x) = 2$
   $\deg p(x) = 0$ (p is a *constant polynomial*)
(5) $p(x) = 2x^4 - 2x^3 - 3$
   $\deg p(x) = 4$ (p is called *quartic*, or *biquadratic*)

The following functions are *not* polynomials. Why?

(6) $p(x) = \dfrac{1}{x} - 1 = x^{-1} - 1$

(7) $p(x) = x^2 - \sqrt{x} + 3x - 7$

(8) $p(x) = \dfrac{x^2 + x + 1}{x + 1}$

(9) $p(x) = [x]$

(10) $p(x) = 2^x$

For most of the results of this chapter it will be immaterial whether the coefficients a_i of the polynomial $p(x)$ are real numbers, complex numbers, rational numbers, etc. All that we require is that they possess certain algebraic properties. (They must belong to a Field—see Appendix I.) Henceforth we shall assume that the coefficients of $p(x)$ belong to a certain algebraic system F which you may think of as the real numbers or the complex numbers, etc., the set of all polynomials with coefficients in F, is symbolized by $F[x]$. Then $p(x) \in F[x]$ means $p(x)$ is such a polynomial.

Any polynomial $p(x) \in F[x]$ determines a function

$$p: F \to F$$

polynomials & polynomial functions

as follows. Let

$$p(x) = a_n x^n + a_{n-1} x^{n-1} + \cdots + a_1 x + a_0$$

First look upon the function x as the identity map on the set F:

$$x(t) = t$$

for each $t \in F$. Thus, by a direct application of function multiplication, ax is the function

$$ax(t) = at$$

for each $t \in F$.

Recall the definition of f^k, for f a function and k a positive integer. It follows that, for each such k, $a_k x^k$ is the function

$$a_k x^k(t) = a_k t^k$$

for each $t \in F$.

Thus, the polynomial $p(x)$ can be thought of as defining a function which is made up of a finite sum of functions each of which is a multiple of a constant function and a power of the identity function. Therefore, for each $t \in F$,

$$p(t) = a_n t^n + a_{n-1} t^{n-1} + \cdots + a_1 t + a_0$$

is an element of F.

In this way, we see that each polynomial in $F[x]$ determines a function from F to F. Such functions are called *polynomial functions*. The converse is also true.

When $F = R$, the real numbers, we call $p(x) \in F[x]$ a *real polynomial*. Likewise, when $F = C$, the complex numbers, we call $p(x) \in F[x]$ a *complex polynomial*.

example 1

Let $p(x)$ be the real cubic polynomial

$$p(x) = x^3 - 2x^2 + x - 1$$

Then $p(x)$ determines polynomial function $p : R \to R$ given by

$$p(t) = t^3 - 2t^2 + t - 1, \text{ for each real number } t.$$

In particular $p(0) = 0^3 - 2(0)^2 + 0 - 1 = -1$, $p(1) = 1^3 - 2(1)^2 + 1 - 1 = -1$ $p(2) = 2^3 - 2(2)^2 + 2 - 1 = 1$, etc. ●

The above example also illustrates the facts contained in the following theorem.

theorem 3.1.1

Let $f(x) \in F[x]$ be given by

$$f(x) = a_n x^n + a_{n-1} x^{n-1} + \cdots + a_1 x + a_0$$

Then, for $0, 1 \in F$, we have $f(0) = a_0$, and $f(1) = \sum_{i=1}^{n} a_i$.

proof. Clearly, $f(0) = a_n \cdot 0 + a_{n-1} \cdot 0 + \cdots + a_1 \cdot 0 + a_0 = a_0$, and $f(1) = a_n(1) + a_{n-1}(1) + \cdots + a_1(1) + a_0 = a_n + a_{n-1} + \cdots + a_1 + a_0$, as desired.

One can always insert zero coefficients in a polynomial, or leave them out as he wishes. Thus, $x^3 + 1$ may be written as $x^3 + 0x^2 + 0x + 1$, or even as $0x^5 + 0x^4 + x^3 + 0x^2 + 0x + 1$. Therefore, in any discussion involving a finite collection of polynomials, one may, if he wishes, assume that the same powers of x appear. We shall do just this in considering adding polynomials. First, however, we define *equal polynomials*. Equal polynomials give rise to equal polynomial functions.

Suppose $p(x)$ and $q(x)$ are in $F[x]$. One may write

(3.1.2)
$$p(x) = a_n x^n + a_{n-1} x^{n-1} + \cdots + a_1 x + a_0, \text{ and}$$
$$q(x) = b_n x^n + b_{n-1} x^{n-1} + \cdots + b_1 x + b_0$$

As was mentioned above, we do not insist that a_n or b_n be non-zero. Thus, $\deg p(a) \leq n$ and $\deg q(x) \leq n$.

The polynomials $p(x)$ and $q(x)$ are equal *if and only if the coefficient of corresponding powers of x are equal*; i.e., $a_i = b_i$ for all $i = 0, 1, 2, \ldots, n$.

To add $p(x)$ and $q(x)$, their sum is defined as follows.

$$p(x) + q(x) = (a_n + b_n)x^n + (a_{n-1} + b_{n-1}) + \cdots + (a_1 + b_1)x + (a_0 + b_0)$$

You are invited to prove that, with this definition, the addition of polynomials satisfies the usual properties of function addition. Notice particularly that the zero polynomial is the zero function and the negative of the polynomial $p(x)$ is $-p(x) = -a_n x^n - a_{n-1} x^{n-1} - \cdots - a_1 x - a_0$, etc.

theorem 3.1.2

Let $p(x)$ and $q(x)$ be polynomials in $F[x]$. Then $\deg\left(p(x) + q(x)\right) \leq \max \{\deg p(x), \deg q(x)\}$.

proof. Let $p(x)$ and $q(x)$ be as in Equation (3.1.2) with $\deg p(x) = k$ and $\deg q(x) = l$. Let $m = \max \{k, l\}$. Then $a_m + b_m = 0$ if and only if $a_m = -b_m$. But $a_t = 0$, and $b_t = 0$ for all $t > m$. Thus, $a_t + b_t = 0$ for all $t > m$, so $\deg p(x) + q(x) \leq m$.

example 2

Let $p(x) = x^3 - 1$ and $q(x) = 1 + x^2 - x^3$. Then $p(x) + q(x) = (1 - 1)x^3 + (0 + 1)x^2 + (0 + 0)x + (-1 + 1) = x^2$. Note that $\deg x^2 = 2$, while $\deg p(x) = \deg q(x) = 3$. ●

To define polynomial multiplication, proceed as follows. Let $p(x)$ and $q(x)$ be as in (3.1.2) and *define*

$$p(x) \cdot q(x) = s(x) = c_h x^h + c_{h-1} x^{h-1} + \cdots + c_1 x + c_0$$

where

$$c_k = \sum_{i+j=k} a_i b_i$$

polynomials & polynomial functions

This means, to obtain c_k, sum over all non-negative integers i and j which add up to k. Thus,

$$c_0 = a_0 b_0$$
$$c_1 = a_0 b_1 + a_1 b_0$$
$$c_2 = a_0 b_2 + a_1 b_1 + a_2 b_0$$
$$c^3 = a_0 b_3 + a_1 b_2 + a_2 b_1 + a_3 b_0$$

etc.

example 3

Let $p(x) = 2x^2 - x + 1$, and let $q(x) = x^3 - 3x^2 - x + 2$. Then

$$p(x)q(x) = 2x^5 + (-1 - 6)x^4 + (1 + 3 - 2)x^3 + (-3 + 1 + 4)x^2$$
$$+ (-1 - 2)x + 2$$
$$= 2x^5 - 7x^4 + 2x^3 + 2x^2 - 3x + 2$$

Notice that $p(x) + q(x) = x^3 - x^2 - 2x + 3$. ●

In this example the formulas in the definitions were used. We state the following theorem which considerably shortens the labor involved in multiplication. It states that polynomial multiplication distributes over addition.

theorem 3.1.3

For $f(x)$, $g(x)$, and $h(x)$ in $F[x]$.

$$f(x)[g(x) + h(x)] = f(x)q(x) + f(x)h(x)$$

proof. Use the definitions on each side and compute. The details are left as Exercise 12.

example 4

Let $p(x) = 3x^3 + 2x^2 - x - 1$ and $q(x) = x^2 - x - 1$. Compute $p(x)q(x)$. Write $(3x^3 + 2x^2 - x - 1)(x^2 - x - 1)$, and use distributivity as follows.

$$p(x)q(x) = (3x^3 + 2x^2 - x - 1)x^2 + (3x^3 + 2x^2 - x - 1)(-x) +$$
$$(3x^3 + 2x^2 - x - 1)(-1)$$
$$= 3x^5 + 2x^4 - x^3 - x^2 - 3x^4 - 2x^3 + x^2 + x - 3x^3 - 2x^2 + x + 1$$
$$= 3x^5 - x^4 - 6x^3 - 2x^2 + 2x + 1$$ ●

In Exercise 13 you are invited to supply the proof of the following theorem.

theorem 3.1.4

Let $p(x)$ and $q(x)$ be polynomials in $F[x]$. Then $\deg p(x)q(x) = \deg$ $p(x) + \deg q(x)$.

polynomials & rational functions

exercises

1. Give the following collection of functions. Identify each as a *polynomial* or *not a polynomial*. State the degree of each polynomial and explain why each of the nonpolynomials fails to be one.

(a) $x^2 + x + 1$

(b) $1 - x^2$

(c) $1/x$

(d) $(\sqrt{2})x - 1$

(e) 3

(f) $\sqrt{x^2 + 1}$

(g) $1/x + x - x^2 - x^3$

(h) $1/(x^3 + x + 1)$

(i) x

(j) 0

(k) $\sqrt{x^3 + x + 1}$

(l) $x^7 - x^4 + x - 1$

2. For each of the following pairs of polynomials $p(x)$ and $q(x)$ find $p(x) + q(x)$, find $\deg\big(p(x) + q(x)\big)$.

(a) $p(x) = x + 1$, $q(x) = x - 1$

(b) $p(x) = x^2 - x + 3x + 1$, $q(x) = x^3 - x^2 + 2x - 1$

(c) $p(x) = x^5 - 3x - 1$, $q(x) = x^2 - 7x + x^3 - x^4$

(d) $p(x) = x^6 - 1$, $q(x) = 1 - x^2 + x^4 - x^6$

(e) $p(x) = x^4 + x^2 - 3x - 1$, $q(x) = x^5 - x^4 + x^2 + x + 1$

3. Find $p(x)q(x)$ for the polynomials in Exercise 2. Find $\deg p(x)q(x)$.

4. For each $p(x)$ in Exercise 2, find $p(0)$, $p(1)$, $p(1/2)$, $p(-1)$.

5. For each $q(x)$ in Exercise 2 find $q(0)$, $q(1)$, $q(-2)$, and $q(-1)$.

6. Consider each of the following polynomials in $C[x]$, (with complex coefficients). Find $f(x) + g(x)$, $\deg\big(f(x) + g(x)\big)$ in each case.

(a) $f(x) = x + i$; $g(x) = x - i$

(b) $f(x) = x^2 + x + 1$; $g(x) = x - i$

(c) $f(x) = x^2 + x - 1$; $g(x) = x^2 - ix + 3i$

(d) $f(x) = x - (1 + i)$; $g(x) = x - (1 - i)$

(e) $f(x) = x - (\sqrt{3}/2 - 1/2i)$; $g(x) = x - (\sqrt{3}/2 + 1/2i)$

(f) $f(x) = x^3 - ix + i$; $g(x) = x - \sqrt{2}i$

(g) $f(x) = i\sqrt{3}x^2 - 1$; $g(x) = 1 + i\sqrt{3}x^2$

(h) $f(x) = x - (1 + i)$; $g(x) = x^2 + ix - 1$

7. For each of the polynomials in Exercise 6 find $f(x)g(x)$ and $\deg f(x)g(x)$.

8. For each of the polynomials $f(x)$ in Exercise 6 find $f(0)$, $f(1)$, $f(-1)$, $f(i)$, and $f(-i)$.

9. Give examples of two polynomials $p(x)$ and $q(x)$ each of degree 3 such that the degree of $p(x) + q(x)$ is

(a) 3   (b) 2   (c) 1   (d) 0

10. Show that polynomial addition gives rise to polynomial function addition which obeys the laws in Section 1.8.

11. Prove that if $p(x)$ and $q(x)$ are real polynomials such that $p(x)q(x) = 0$, then either $p(x)$ or $q(x)$ is the zero polynomial.
Hint: Consider the degrees, and use Theorem 3.1.4.

12. Prove Theorem 3.1.3 by working out the details.

13. Prove Theorem 3.1.4.

14. Show that polynomial multiplication gives rise to multiplication of polynomial functions which satisfies the properties of function multiplication in Section 1.8.

15. Let $p(x) = 5x^3 - 6x^2 + cx - c$. Determine the real number c so that $p(0) = 0$.

102

16. For $p(x) = 5x^3 - 6x^2 + cx + c$. Determine a real number c so that $p(1) = 1$.

17. For the linear polynomial $f(x) = mx + b$ determine m and b so that $f(0) = 2$ and $f(1) = 3$.

3.2. graphs of polynomial functions

In this section, we shall consider only those polynomials which have real coefficients; that is $F[x]$ is $R[x]$. Thus, since such polynomials give rise to real functions, their graphs will be subsets of the plane R^2. For each polynomial $p(x)$ in $R[x]$ its graph is the collection of all points (x,y) in the plane such that $y = p(x)$. We shall describe how to sketch graphs for various polynomials.

First consider the graph of the *zero polynomial* $y = 0$. This graph is, of course, the set of points

$$\{(t,0) | t \in R\}$$

which is *the x axis*.

Polynomials of *degree zero* are of the form

$$p(x) = b$$

for some *non-zero* real number b. Thus the graph of the constant polynomial is the graph of $y = b$, which is a horizontal line intersecting the y axis at $(0,b)$; namely the set of points

$$\{(t,b) | t \in R\}$$

Consider now polynomials of *degree 1*. All have the form

$$f(x) = mx + b$$

where m and b are real numbers with $m \neq 0$. It is easily seen that each such polynomial has a zero (crosses the x axis) at the number $-b/m$. This is, in fact, the only zero. Polynomials of degree 1 are called *linear polynomials*. In Chapter 1 we noticed that such polynomials have graphs which are straight lines, but these lines are not horizontal. In Chapter 7 this fact will be proved. Therefore, the graph of the polynomial

$$f(x) = mx + b$$

is a straight line passing through the points $(-b/m, 0)$ and $(0,b)$. These points are known as the **x intercept** and the **y intercept**, respectively, of the line.

example 1

Sketch the graph of $y = 3x - 1$. The graph will be a straight line determined by the points $(0, -1)$ and $(1/3, 0)$. This graph is sketched in Figure 3.2.1. ●

An equation of the form $ay + cx = d$ with $a \neq 0$, although not a polynomial in x, can be related to a linear polynomial. In Chapter 7 we shall see that each such equation has a graph which is a straight line. Given

$$ay + cx = d, a \neq 0$$

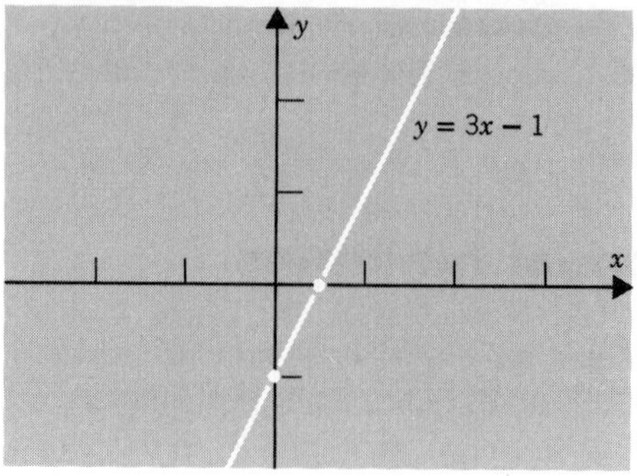

polynomials & rational functions

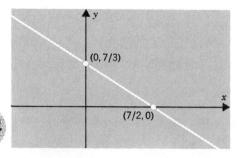

figure 3.2.1

$$y = 3x - 1$$

First solve for y to obtain

$$y = -\frac{c}{a}x - \frac{d}{a}$$

which now has the form $y = p(x)$, where $p(x)$ is a linear polynomial. The graph of the equation is then the graph of

$$p(x) = \frac{-c}{a}x - \frac{d}{a}$$

example 2

Sketch the graph of the linear equation $2x + 3y - 7 = 0$.
First note that if

$$2x + 3y - 7 = 0, \text{ then}$$
$$3y = -2x + 7, \text{ or}$$
$$y = -2/3x + 7/3$$

Hence, the graph is a straight line passing through the points $(0, 7/3)$ and $(7/2, 0)$. The graph is sketched in Figure 3.2.2. ●

Consider now polynomials of degree two. The most general *quadratic* polynomial has the form

$$p(x) = ax^2 + bx + c$$

for a, b, and c real numbers, and $a \neq 0$. The graphs of these functions $y = p(x)$ always have the same form. Such curves are called *parabolas* and are studied

figure 3.2.2

$(0, 7/3)$

$(7/2, 0)$

graphs of polynomial functions

in detail in Chapter 7. We make some remarks here, however, which enable us to sketch these graphs.

A real function $f: R \to R$ is said to *have an absolute maximum at the real number r* provided

$$f(t) \le f(r)$$

for every $t \in R$. The **absolute maximum** is the number $f(r)$. Similarily f is said to *have an absolute minimum at a real number r* provided

$$f(t) \ge f(r)$$

for every $t \in R$. In this case, $f(r)$ is the **absolute minimum** of f. An **absolute extremum** for f is either an *absolute maximum* or an *absolute minimum*. We shall see that every quadratic polynomial function has an absolute extremum which is an absolute maximum when $a < 0$ and is an absolute minimum when $a > 0$. This is formally stated in the following theorem.

theorem 3.2.1

A quadratic polynomial function

$$p(x) = ax^2 + bx + c, \, a \ne 0$$

has an absolute maximum at $r = -b/2a$ when $a < 0$, or an absolute minimum at $r = -b/2a$ if $a > 0$. In either case the value of the absolute extremum is

$$f(r) = \frac{-(b^2 - 4ac)}{4a}$$

proof. For each $t \in R$,

$$p(t) = at^2 + bt + c$$

$$= a\left(t^2 + \frac{b}{a}t\right) + c$$

Upon completing the square on the right, this becomes

$$p(t) = a\left(t^2 + \frac{b}{a}t + \left(\frac{b}{2a}\right)^2\right) + c - \frac{b^2}{4a}$$

$$= a\left(t + \frac{b}{2a}\right)^2 + \frac{4ac - b^2}{4a}$$

Now, since the first term on the right side, $a[t + (b/2a)]^2$, is always a times the square of a real number, its sign depends upon the sign a. Therefore, when $a < 0$, for every real number t,

$$p(t) \le \frac{4ac - b^2}{4a}$$

Similarly when $a > 0$, for every real number t

$$p(t) \ge \frac{4ac - b^2}{4a}$$

polynomials & rational functions

Thus, the extreme value of $p(t)$ is $(4ac - b^2)/4a$, which is the value of $p(-b/2a)$.

The point $(-b/2a, (b^2 - 4ac)/2a)$ is the extreme value of the graph of the quadratic polynomial. It is known as the *vertex* of the parabola, $y = p(x)$.

The number $\Delta = b^2 - 4ac$ will shortly be seen to have other consequences for the graph of a quadratic function and is called the **discriminant** of the quadratic.

example 3

Find the extreme value of the polynomial function $p(x) = 5x^2 + 6x - 1$. Sketch its graph.

It is clear that $a = 5$, $b = 6$, and $c = -1$. Therefore, $b^2 - 4ac = 36 - 4(5)(-1) = 56$. Hence, the extreme value is $-56/10 = -28/5$. The vertex of the parabola

$$y = 5x^2 + 6x - 1$$

is at the point $(-3/5, -28/5)$.

By direct substitution, we see that $p(0) = -1$, $p(1) = 10$, $p(-1) = -2$, $p(-2) = 7$. Since $p(0)$ and $p(1)$ have opposite signs the graph of this parabola ought to cross the x axis between 0, and 1. Similarly, it also ought to cross the x axis between -1 and -2. The sketch in Figure 3.2.3 approximates the graph of $p(x)$ under these assumptions.

To find the number r such that $p(r) = 0$ proceed as follows.
Set

$$p(r) = 5r^2 + 6r - 1 = 0$$

Then

$$r^2 + \frac{6}{5}r = 1/5$$

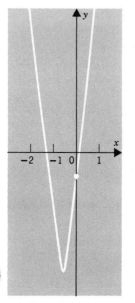

figure 3.2.3

graphs of polynomial functions

Adding $(3/5)^2 = 9/25$ to both sides of this equation we have

$$r^2 + \frac{6}{5}r + \left(\frac{3}{5}\right)^2 = 1/5 + 9/25 = 14/25$$

The left member of the resulting equation is a perfect square. That is, $r^2 + (6/5)r + (3/5)^2 = [r + (3/5)]^2$. Making this substitution gives the expression

$$\left(r + \frac{3}{5}\right)^2 = \frac{14}{25}$$

Taking square roots of both sides results in the equation

$$r + \frac{3}{5} = \pm\frac{\sqrt{14}}{5}$$

Thus, $r = -(3/5) + (\sqrt{14}/5)$ and $r = -(3/5) - (\sqrt{14}/5)$ will make $p(r) = 0$. These numbers are called the **zeros** or **roots** of $p(x)$.

Note that the right member of the equation above is $\pm(\sqrt{\Delta}/2a)$ for the given quadratic, where $\Delta = b^2 - 4ac$, is its discriminant. ●

A quadratic polynomial may have two, one, or no real zeros. Its graph may thus (respectively) cross, touch, or fail altogether to cross the x axis. This is determined by its discriminant $\Delta = b^2 - 4ac$.

theorem 3.2.2

The quadratic function

$$p(x) = ax^2 + bx + c$$

$a \neq 0$, has two, one, or a no real zero accordingly as the discriminant

$$\Delta = b^2 - 4ac$$

is positive, zero, or negative, respectively. When they exist the zeros t are given by the formula

$$t = \frac{-b \pm \sqrt{b^2 - 4ac}}{2a}$$

proof. Set $p(t) = 0$. Thus, if for some $t \in R$

$$at^2 + bt + c = 0$$

it follows upon dividing both sides of the equation by $a \neq 0$ that

$$t^2 + \frac{b}{a}t = -\frac{c}{a}$$

Completing the square on the left yields

$$t^2 + \frac{b}{a}t + \frac{b^2}{4a^2} = \frac{b^2}{4a^2} - \frac{c}{d}$$

or

$$\left(t + \frac{b}{2a}\right)^2 = \frac{b^2 - 4ac}{2a}$$

polynomials & rational functions

The number on the right side is the square of a real number, hence is non-negative. Therefore if

$$\Delta = b^2 - 4ac < 0$$

There is no real number t for which $p(t) = 0$ on the other hand, if

$$\Delta = b - 4ac \geq 0$$

then

$$t + \frac{b}{2a} = \pm \frac{\sqrt{b^2 - 4ac}}{2a}$$

$$t = \frac{-b \pm \sqrt{b^2 - 4ac}}{2a} = \frac{-b \pm \sqrt{\Delta}}{2a}$$

Therefore, if $\Delta = 0$, $t = -b/2a$ is the only zero for $p(x)$. If $\Delta > 0$ then p has two zeros $(-b + \sqrt{\Delta})/2a$ and $(-b + \sqrt{\Delta})/2a$.

In Figure 3.2.4 we sketch these cases both for positive a and for negative a, respectively.

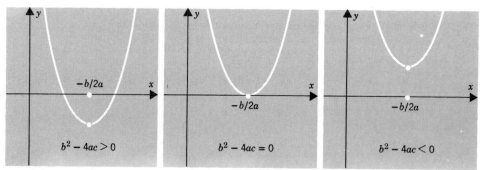

$b^2 - 4ac > 0$ $b^2 - 4ac = 0$ $b^2 - 4ac < 0$

Possible cases for a > 0.

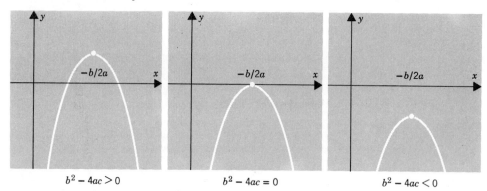

$b^2 - 4ac > 0$ $b^2 - 4ac = 0$ $b^2 - 4ac < 0$

Possible cases for a < 0.

figure 3.2.4

example 4

108 Sketch the graphs of each of the following quadratic polynomials.

$$f(x) = 4x^2 - 12x + 9, \qquad g(x) = 1 - x^2, \qquad \text{and} \qquad h(x) = x^2 + x + 1$$

graphs of polynomial functions

For each polynomial we compute the discriminant as follows.

$$\Delta_f = (-12)^2 - 4(4)(9) = 0$$
$$\Delta_g = (0)^2 - 4(-1)(1) = 4$$
$$\Delta_h = (1)^2 - 4(1)(1) = -3$$

Thus, f has one real zero, g has two real zeros, and h has no real zeros. The zero for f is the point $(3/2,0)$. The zeros for g are the points $\left(\pm\dfrac{\sqrt{\Delta}}{2a}, 0\right) = (\pm 1, 0)$.

We also know that the extremum for each function occurs at $(-b/2a, -\Delta/4a)$. These points are:

$$\begin{array}{llll} & \text{for } f & \text{at} & (3/2,0), \text{ a minimum} \\ & \text{for } g & \text{at} & (0,1), \text{ a maximum} \\ \text{and} & \text{for } h & \text{at} & (-1/2,3/4), \text{ a minimum} \end{array}$$

By determining a few more points on each graph as listed in the following table, we are able to sketch in Figure 3.2.5 the graph of each curve. ●

t	0	1	-1	$\dfrac{1}{2}$	$-\dfrac{1}{2}$	2	3
$f(t)$	9	1	25	4	16	1	9
$g(t)$	1	0	0	$\dfrac{3}{4}$	$\dfrac{3}{4}$	-3	-8
$h(t)$	1	3	1	$1\dfrac{3}{4}$	$\dfrac{3}{4}$	7	-12

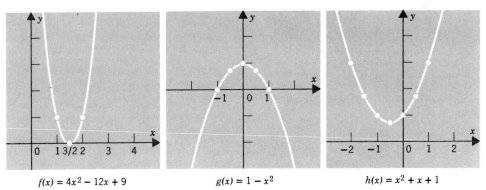

$f(x) = 4x^2 - 12x + 9$　　　　$g(x) = 1 - x^2$　　　　$h(x) = x^2 + x + 1$

figure 3.2.5

The **y intercept** of the graph of any polynomial $p(x)$ occurs when $x = 0$. It is the point $(0,c)$ for any quadratic. The **x intercepts** of the graph, if any, occur at the zeros of the polynomial.

polynomials & rational functions

In Chapter 1, we considered the graph of the general power functions;

$$f(x) = x^n, \quad n \text{ a positive integer.}$$

These are, of course, special types of polynomial functions, each of which passes through the origin. The shapes of their graphs were given in Figure 1.7.9.
As the degree of the general polynomial

$$p(x) = a_n x^n + a_{n-1} x^{n-1} + \cdots + a_1 x + a_0$$

increases, the graph becomes more complicated. These graphs are always smooth, however, with a number of humps. Frequently, the number of humps is related to the degree, but of course $p(x) = x^{50}$ only has one flat hump. The following example illustrates a fifth degree polynomial with several humps.

example 5

Sketch the graph of $f(x) = x(x - 1)(x + 1)(x - 2)(x + 2)$. We have written $p(x)$ in factored form so that you can easily see that its graph will cross the x axis at 0, $+1$, -1, $+2$, and -2. By plotting additional points its graph is determined to be somewhat as that shown in Figure 3.2.6.

A more complete description and more accurate graph of this polynomial, as well as others of this type, requires techniques of the calculus. The exact determination of the number and location of the humps for such functions was,

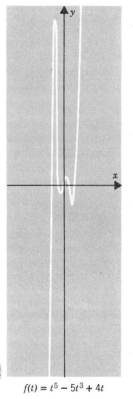

figure 3.2.6

$f(t) = t^5 - 5t^3 + 4t$

graphs of polynomial functions

t	0	1	-1	2	-2	1/2	$-1/2$	3/2	$-3/2$	$-5/2$	3	-3
$f(t)$	0	0	0	0	0	$1\frac{13}{32}$	$-1\frac{13}{32}$	$-3\frac{9}{32}$	$27\frac{17}{32}$	$-29\frac{17}{32}$	120	-120

in fact, one of the things which motivated the development of differential calculus. ●

exercises

1. Sketch the graphs of each of the following constant polynomials.

(a) 3
(b) -3
(c) 1/2
(d) $-1/2$

(e) π
(f) $\sqrt{2}$
(g) $\sqrt{-2}/2$
(h) $-\pi$

2. Sketch the graph of each of the following linear polynomials by determining the x intercept and the y intercept of each.

(a) $x - 1$
(b) $x + 1$
(c) $1 - x$
(d) $1 + 2x$

(e) $1 - 2x$
(f) $3x + 4$
(g) $x/2 - y = 1$
(h) $\frac{1}{2}x - 3$

(i) $3 - \frac{1}{2}x$
(j) $\pi x + 2$
(k) $\sqrt{2}x - 1$
(l) $1 - \pi^2 x$

3. Find the x *intercept*, the y *intercept*, and sketch the graph of each of the following linear equations.

(a) $x + y = 1$
(b) $x - y = 1$
(c) $2x - y = 3$
(d) $x - 2y = 4$

(e) $2x - 3y = 6$
(f) $3x + 2y = 6$
(g) $x/2 - y = 1$
(h) $x - y/2 = 3$

(i) $5x + 7y = 9$
(j) $3x - 8y = 12$
(k) $x/2 + y/3 = 1$
(l) $x/3 - y/2 = 1$

4. Find the x and y intercepts for each of the following quadratic polynomials.

(a) $x^2 - 2x + 1$
(b) $x^2 - 1$
(c) $x^2 - 9$
(d) $9 - x^2$

(e) $x^2 + x - 2$
(f) $x^2 - x + 6$
(g) $16 - 9x^2$
(h) $8 - 2x^2$

(i) $2x^2 + x - 1$
(j) $2x^2 - x - 3$
(k) $6x^2 - x - 2$
(l) $1 + 5x - 6x^2$

5. Find the absolute extremum for each of the polynomials in Exercise 4.

6. Sketch the graph of each of the polynomials in Exercise 4.

7. Sketch the graphs of each of the following polynomials.

(a) x^2
(b) $-x^2$
(c) x^3
(d) $-x^3$

(e) x^4
(f) x^5
(g) x^6
(h) x^7

polynomials & rational functions

8. Sketch the graph of each of the following cubics.

(a) $x^3 - \cdot 1$
(b) $1 - x^3$
(c) $x^3 - x$
(d) $x^3 - 4$
(e) $x^3 - 2x^2 + x$

(f) $6x^3 - x^2 - 2x$
(g) $x^3 - 3x^2 + 3x - 1$
(h) $x^3 - x^2 - 4x + 4$
(i) $x^3 - x^2 - 9x + 9$
(j) $(2x - 1)(x + 1)(3x + 1)$

9. Sketch the graph of each of the following polynomials.

(a) $x^2 + x + 1$
(b) $x^2 + 1$
(c) $1 - x - x^2$
(d) $6 + x - x^2$
(e) $x^2 - 2x + 2$

(f) $x^2 + 3x + 3$
(g) $4 - x - x^2$
(h) $1 - x + x^2$
(i) $1 - x^4$
(j) $1 - x^6$

10. Sketch the graph of the fourth degree polynomial

$$p(x) = (1 - x)^4$$

11. Sketch the graph of the fourth degree polynomial

$$p(x) = (x - 1)(x + 2)(x - 3)(2x + 1)$$

3.3. algebra of polynomials: divisibility

At the beginning of this chapter we remarked that the polynomials were the integers of the function family. This is particularly evident in a consideration of their algebraic properties. The definitions for addition and multiplication of polynomials were given in the first section. That these are analogous to the same operations on the integers is most easily seen when we consider the fact that positional notation in writing integers is similar to polynomials in ten; for example,

$$2576 = 2(10)^3 + 5(10)^2 + 7(10) + 6$$

You are invited to compare the addition and multiplication of integers with that of polynomials in this light.

If a polynomial is written as the product of other polynomials these are called **factors** *of the original polynomial.* Thus, since

(3.3.1) $$x^5 - 5x^3 + 4x = x(x^2 - 1)(x^2 - 4)$$

it has x, $(x^2 - 1)$, $(x^2 - 4)$, as factors. For a polynomial $p(x)$ with real coefficients, every non-zero real number a is a factor of $p(x)$ since

(3.3.2) $$a\left(\frac{1}{a}p(x)\right) = p(x)$$

We are usually not so much concerned with factors of this type. They are referred to as *trivial factors*. We are, however, concerned with polynomial factors whose degree is one or larger. We are interested in factoring a given polynomial as completely as possible.

Again the integers form an interesting analogy. You will recall that an integer $p > 1$ is called *prime* if when $p = ab$, then it must be true that one factor, say a, is $+1$ or -1; thus, $b = p$ or $b = -p$. In a similar way, *a poly-*

algebra of polynomials: divisibility

nomial $p(x)$ is called **prime** *or* **irreducible** *if it has degree at least one, and in a factorization*

$$p(x) = a(x)b(x)$$

one of the factors, say $a(x)$, is a polynomial of degree zero; i.e., a nonzero real number. For example, $x - 1$ is a prime polynomial since any factorization must contain a factor whose degree is less than 1. The polynomial $x^2 - 1$ is not prime since $x^2 - 1 = (x - 1)(x + 1)$.

The question of whether a polynomial $p(x) \in F[x]$ is irreducible or not is not absolute for those whose degree is larger than one. It depends upon the set F from which the coefficients come. Thus

$$x^2 + 1$$

is an irreducible *real* polynomial but can be factored when considered as a *complex* polynomial; thus,

$$x^2 + 1 = (x + i)(x - i)$$

example 1

Given the polynomials in (3.3.1) and the factorization there. The polynomial x is irreducible but the other two factors are not. A prime factorization is given by

$$(3.3.3) \qquad x^5 - 5x^3 + 4x = x(x - 1)(x + 1)(x - 2)(x + 2)$$

The determination of a prime factorization for a given polynomial is often a trial-and-error process. We will be aided by several theorems in this and subsequent sections. It is also true, although we do not prove it here, that the factorization of any polynomial is unique except perhaps for the order in which the factors are written. Thus (3.3.3) is the only way in which the polynomial $x^5 - 5x^3 + 4x$ can be written as the product of irreducible polynomial factors, except for possible trivial factors and the order of the factors. ●

If a polynomial $g(x)$ is a factor of a polynomial $f(x)$, we say that $f(x)$ is **divisible** *by $g(x)$.* Thus, $x^2 - 4$ is divisible by $x - 2$ and not divisible by $x - 1$; just as integers are divisible by some integers and not by others. On the other hand, given two integers a and b we can always find two other integers q and r such that

$$b = aq + r$$

with $0 \le r < |a|$. For example, given $b = 75$ and $a = 12$. Divide 12 into 75 as follows

$$
\begin{array}{r}
6 \\
12\overline{)75} \\
72 \\
\hline
3
\end{array}
$$

Thus, the *quotient* q is 6 and the *remainder* r is 3. That is

$$75 = 12(6) + 3$$

113 The division of polynomials can be done in a similar manner. Consider the example which follows.

polynomials & rational functions

example 2

Divide, using long division, the polynomial $f(x) = x^3 + 3x^2 - x - 1$ by the polynomial $g(x) = x^2 + 1$. The division goes as follows.

$$
\begin{array}{r}
x + 3 \\
x^2 + 1{\overline{\smash{\big)}\,x^3 + 3x^2 - x - 1}} \\
\underline{x^3 + x } \\
3x^2 - 2x - 1 \\
\underline{3x^2 + 3} \\
-2x - 4
\end{array}
$$

Therefore, the *quotient* is $(x + 3)$, and the *remainder* is $-2x - 4$. The division process stopped when the degree of the remainder was smaller than the degree of the divisor, $g(x)$. Thus,

$$f(x) = g(x)(x + 3) - 2(x + 2) \quad \bullet$$

The divisibility of polynomials is, therefore, analogous with divisibility in the integers.

theorem 3.3.1

If $f(x)$ and $g(x)$ are polynomials in $F[x]$ then there exist polynomials $q(x)$ and $r(x)$ in $F[x]$, with $r(x)$ equal to the zero polynomial or with $\deg r(x) < \deg g(x)$, such that

$$f(x) = g(x)q(x) + r(x)$$

The proof of this theorem is actually nothing more than an application of the long division process as in the example above. If $\deg f(x) < \deg g(x)$ then there is really nothing to prove, since $q(x) = 0$ and $r(x) = f(x)$ will suffice. Hence, one assumes that $\deg f(x) = m \geq n = \deg g(x)$ and performs an induction on the degree of $f(x)$. The details are left as Exercise 5.

Just as the long division process for the integers has a shorter version, so does that for polynomials. That is to say, we did not perform the division of 75 by 12 in the extended form

$$1(10) + 2{\overline{\smash{\big)}\,7(10) + 5}}$$

The method of shortening the long division of polynomials is completely analogous to the usual method of long division for integers. The process is know as the *method of detached coefficients*. Thus, we may write the division in Example 1 as follows. Notice that the powers of x are omitted, since they, in reality (like powers of 10 in an integer), serve largely as place holders. Take care, however, that missing power of x, those with coefficient zero, are supplied with the zero.

$$
\begin{array}{r}
1 + 3 \\
1 + 0 + 1{\overline{\smash{\big)}\,1 + 3 - 1 - 1}} \\
\underline{1 + 0 + 1 } \\
3 - 2 - 1 \\
\underline{3 + 0 + 3} \\
-2 - 4
\end{array}
$$

algebra of polynomials: divisibility

Displayed above are the coefficients of the quotient $q(x) = x + 3$ and of the remainder $r(x) = -2x - 4$.

Of particular interest is division by linear divisors of the form $x - a$. In the next section, we shall see that the method of detached coefficients can be shortened even further for such divisors when use is made of the two theorems presented there.

example 3

Use the method of detached coefficients to divide $x^3 - 4x^2 + 5x - 1$ by $x - 3$. Write

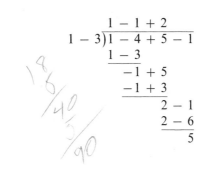

$$
\begin{array}{r}
1 - 1 + 2 \\
1 - 3\overline{)1 - 4 + 5 - 1} \\
\underline{1 - 3} \\
-1 + 5 \\
\underline{-1 + 3} \\
2 - 1 \\
\underline{2 - 6} \\
5
\end{array}
$$

Therefore,

$$x^3 - 4x^2 + 5x - 1 = (x - 3)(x^2 - x + 2) + 5 \quad \bullet$$

exercises

1. Express each of the following polynomials as the product of polynomials which are prime over the reals R. (Factor them.)

(a) $x^2 - 9$
(b) $16 - 4x^2$
(c) $x^3 - 1$
(d) $x^3 + 1$
(e) $x^8 - 1$

(f) $x^2 - 5x - 6$
(g) $x^2 - 5x + 4$
(h) $x^3 - x^2 - 9x + 9$
(i) $x^3 - x^2 - 4x + 4$
(j) $x^2 + x + 1$

2. Given the polynomials $f(x)$ and $g(x)$ as follows. Use long division to find the quotient and the remainder $r(x)$ so that $f(x) = g(x)q(x) + r(x)$.

(a) $f(x) = x^2 - 9,$ $g(x) = x + 3$
(b) $f(x) = x^3 - 1,$ $g(x) = x^2 + x + 1$
(c) $f(x) = x^3 - x^2 + 4,$ $g(x) = x^2 - 4x$
(d) $f(x) = 2x^4 - x^3 + 7x + 3,$ $g(x) = x^2 + 2x - 5$
(e) $f(x) = 3x^3 - 7x^2 + 2x + 1,$ $g(x) = x^2 + x + 1$
(f) $f(x) = 4x^5 + 10x^2 - 1,$ $g(x) = x^2 + x - 1$
(g) $f(x) = x^3 + 1,$ $g(x) = x^2 - x + 1$
(h) $f(x) = 3x^4 - 1,$ $g(x) = x^2 - 1$
(i) $f(x) = 6x + 9,$ $g(x) = 9x^2 + 6$
(j) $f(x) = 8x,$ $g(x) = 7x^2$

3. For each of the following pairs of polynomials $f(x)$ and $g(x)$, use the method of detached coefficients to find the quotient and the remainder when $f(x)$ is divided by $g(x)$.

(a) $f(x) = x^3 + 2x^2 + 3x + 4$ $g(x) = x - 5$
(b) $f(x) = x^3 - 1$ $g(x) = x - 1$

polynomials & rational functions

$$
\begin{aligned}
&\text{(c)} \quad f(x) = x^3 - x^2 + 4 && g(x) = x + 2 \\
&\text{(d)} \quad f(x) = 5x^5 - 7x^2 + 3x && g(x) = x - 6 \\
&\text{(e)} \quad f(x) = 8x^3 - 12x^2 + 6x - 3 && g(x) = 2x + 1 \\
&\text{(f)} \quad f(x) = 3x^4 + 7x^3 + 10x^2 + 12x + 1 && g(x) = 3x + 7 \\
&\text{(g)} \quad f(x) = x^5 - 2x^3 + x^2 - 4 && g(x) = x^2 + 9 \\
&\text{(h)} \quad f(x) = x^6 - 1 && g(x) = x^3 - 1 \\
&\text{(i)} \quad f(x) = x^6 - 1 && g(x) = x^4 + x^2 + 1 \\
&\text{(j)} \quad f(x) = x^6 - 1 && g(x) = x^2 + x + 1
\end{aligned}
$$

4. Show that every polynomial of degree one in $F[x]$ is irreducible (prime) no matter what the field F of coefficients is.

5. Use Exercise 25 of Section 2.1 to complete the proof of the division algorithm. To do this note that if $a_m \neq 0$, $b_n \neq 0$, $m \geq n$, and

$$f(x) = a_m x^m + a_{m-1} x^{m-1} + \cdots + a_1 x + a_0$$

$$g(x) = b_n x^n + b_{n-1} x^{n-1} + \cdots + b_1 x + b_0$$

then the polynomial $f_1(x) = f(x) - (a_m/b_n)x^{m-n}g(x)$ has degree $\leq m - 1$, and, therefore, the division algorithm holds for it.

3.4. two easy theorems & synthetic division

We remarked in Section 3.1 that if $p(x)$ is a polynomial in $F[x]$, it also determines a function p from the set of coefficients F to F. We noticed that this p took 0 to the constant term of $p(x)$ and took $1 \in F$ to the sum of the coefficients of $p(x)$. In this section, we begin by developing a method called **synthetic substitution** for determining $p(t)$ for any $t \in F$. Suppose

$$(3.4.1) \qquad p(x) = a_n x^n + a_{n-1} x^{n-1} + \cdots + a_1 x + a_0$$

Then surely $p(t)$ can be computed directly by substituting t for x in (3.4.1). This process can be quite tedious, however. Consider the following example.

example 1

Let $p(x) = 3x^4 - 7x^3 + 2x^2 + x + 1$. By direct substitution we find

$$p(5) = 3(5)^4 - 7(5)^3 + 2(5)^2 + (5) + 1 = 1056$$

The arithmetic involved is considerable and can be done in a slightly different way if we write

$$p(5) = (\{[3(5) + (-7)](5) + 2\}(5) + 1)(5) + 1$$

More generally for any $t \in F$, $p(t)$ would appear as

$$p(t) = 3t^4 - 7t^3 + 2t^2 + t + 1$$
$$(3.4.2)$$
$$= \{[(3t - 7)t + 2]t + 1\}t + 1 \quad \bullet$$

The expression in (3.4.2) suggests a general algorithm for computing $p(t)$, for any t and any polynomial $p(x)$. We state this algorithm explicitly for an arbitrary fifth degree polynomial. It will then, hopefully, be obvious what to do for any degree n.

116 Given

$$p(x) = a_5 x^5 + a_4 x^4 + a_3 x^3 + a_2 x^2 + a_1 x + a_0$$

two easy theorems & synthetic division

For any $t \in F$ write

$$p(t) = (\{[(a_5 t + a_4)t + a_3]t + a_2\}t + a_1)t + a_0$$

Now set $k_5 = a_5$ and $k_4 = k_5 t + a_4$. Making this substitution yields

$$p(t) = \{[(k_4 + a_3)t + a_2]t + a_i\}t + a_0$$

Set $k_3 = k_4 t + a_3$, so that this becomes

$$p(t) = [(k_3 t + a_2)t + a_1]t + a_0$$

Set $k_2 = k_3 t + a_2$ and $p(t)$ can be written in the form

$$p(t) = (k_2 t + a_1)t + a_0$$

Set $k_1 = k_2 t + a_1$ and $p(t)$ becomes

$$p(t) = k_1 t + a_0$$

Call this last number k_0; $k_0 = k_1 t + a_0$. Thus, $p(t) = k_0$, where k_0 is computed by the above formulas.

For a general polynomial of degree n these formulas are
$$k_n = a_n, \text{ and}$$

(3.4.3)

$$k_i = k_{i+1}t + a_i \text{ for } i < n$$

These formulas are usually written in the following tabular form (again using $n = 5$ as the example).

	a_5	a_4	a_3	a_2	a_1	a_0	$\lfloor t$
(3.4.4)		$k_5 t$	$k_4 t$	$k_3 t$	$k_2 t$	$k_1 t$	
	k_5	k_4	k_3	k_2	k_1	$k_0 = p(t)$	

In this form, it is easier to see that the numbers k_i are obtained by adding the appropriate columns of the table. The first row contains the detached coefficients of $p(t)$, and the second row contains the various products obtained by multiplying t by the previous entry in the last row.

example 2

Let $p(x) = x^6 - 2x^2 + 1$. Use synthetic substitution to compute $p(2)$ and $p(-1)$.
Write the coefficients of $p(x)$ in tabular form, being careful to include the zero coefficients of omitted powers. Then formulas (3.4.3) would give

1	0	0	0	-2	0	1	$\lfloor 2$
	2	4	8	16	28	56	
1	2	4	8	14	28	57	

Thus, $p(2) = 57$. For $p(-1)$ the array would be

1	0	0	0	-2	0	1	$\lfloor -1$
	-1	1	-1	1	1	-1	
1	-1	1	-1	-1	1	0	

Thus, $p(-1) = 0$; that is, -1 is a zero (root) of $p(x)$. ●

polynomials & rational functions

There is a relationship between the substitution of $t \in F$, in a polynomial $p(x)$ to obtain $p(t)$, and division of $p(x)$ by the linear polynomial $x - t$. We have the following important, and easily proved theorem.

theorem 3.4.1.   the remainder theorem

If $p(x) \in F[x]$ and $t \in F$, then the remainder, when $p(x)$ is divided by $x - t$, is the value $p(t)$.

proof.   By the division algorithm we have, on dividing,

$$p(x) = (x - t)q(x) + r(x) \qquad \text{where } r(x) = 0,$$

or degree $r(x) < \deg (x - t)$. Thus, $r(x) = r$ is zero or of degree zero. Therefore, $p(x) = (x - t)q(x) + r$. Now substituting t in p gives

$$p(t) = (t - t)q(t) + r$$

$$= 0 + r$$

Thus, $p(t) = r$ is the remainder.

An immediate consequence of the remainder theorem is the next theorem.

theorem 3.4.2.   the factor theorem

The polynomial $p(x) \in F[x]$ has $x - t$ as a factor if, and only if, $p(t) = 0$.

proof.   By the remainder theorem

$$p(x) = (x - t)q(x) + p(t)$$

Thus, if $p(t) = 0$, $(x - t)$ divides $p(x)$. On the other hand, if $(x - t)$ divides $p(x)$, then the remainder is zero; i.e., $p(t) = 0$.

The remainder theorem allows us to use synthetic substitution to compute the remainder when the polynomial $p(x)$ is divided by the linear polynomial $x - t$, $t \in F$. In fact it does more. The numbers $k_1, \ldots, k_n$ occurring in the synthetic substitution process are, in fact, the coefficients of the quotient $q(x)$; of course, $k_0 = p(t)$ is the remainder $r(x)$. This is apparent from the following long division.

$$
\begin{array}{r}
k_5 x^4 + k_4 x^3 + k_3 x^2 + k_2 x + k_1 = q(x) \\
\hline
x - t \overline{)\, a_5 x^5 + a_4 x^4 + a_3 x^3 + a_2 x^2 + a_1 x + a_0} \\
\underline{k_5 x^5 - k_5 t x^4} \\
k_4 x^4 + a_3 x^3 \\
\underline{k_4 x^4 - k_4 t x^3} \\
k_3 x^3 + a_2 x^2 \\
\underline{k_3 x^3 - k_3 t x^2} \\
k_2 x^2 + a_1 x \\
\underline{k_2 x - k_2 t x} \\
k_1 x + a_0 \\
\underline{k_1 x - k_1 t} \\
k_0 = r(x) = p(t)
\end{array}
$$

Here the k_i are as previously defined.

two easy theorems & synthetic division

Because of this connection, the process of synthetic substitution is most often called **synthetic division**.

example 3

Divide $g(x) = 3x^4 - 2x^3 - x^2 + x - 1$ by $x - 1$, and by $x + 2$. Using an array as (3.4.4) we have

$$
\begin{array}{rrrrr|r}
3 & -2 & -1 & 1 & -1 & 1 \\
  & 3 & 1 & 0 & 1 & \\
\hline
3 & 1 & 0 & 1 & 0 &
\end{array}
$$

Therefore, $3x^4 - 2x^3 + -x^2 + x - 1 = (x - 1)(3x^3 + x^2 + 1)$. For division by $x + 2$ write $x + 2 = x - (-2)$. Hence,

$$
\begin{array}{rrrrr|r}
3 & -2 & -1 & 1 & -1 & -2 \\
  & -6 & 16 & -30 & 58 & \\
\hline
3 & -8 & 15 & -29 & 57 &
\end{array}
$$

and $3x^4 - 2x^3 + x - 1 = (x + 2)(3x^3 - 8x^2 + 15x - 29) + 57$. This shows that 1 is a root of $g(x)$ while $g(-2) = 57$. ●

example 4

Show that $f(x) = 2x + 1$ divides $g(x) = 2x^4 + 3x^3 - 17x^2 - 27x - 9$. One approach is simply to divide $g(x)$ by $f(x)$, using long division. On the other hand if

$$g(x) = f(x)q(x) \text{ for some } q(x)$$

then it is also true that

$$g(x) = (x + 1/2)2q(x)$$

and conversely. The factor theorem tells us that $x + 1/2$ will be a factor of $g(x)$ if, and only if, $-1/2$ is a root of $g(x)$. We may, therefore, use synthetic division as follows.

$$
\begin{array}{rrrrr|r}
2 & 3 & -17 & -27 & -9 & -1/2 \\
  & -1 & -1 & 9 & 9 & \\
\hline
2 & 2 & -18 & -18 & 0 &
\end{array}
$$

Thus, $g(-1/2) = 0$, and $(x + 1/2)$ divides $g(x)$. Therefore, so does $(2x + 1)$. In fact

$$g(x) = (x + 1/2)(2x^3 + 2x^2 - 18x - 18)$$

Therefore, $g(x) = (2x + 1)(x^3 + x^2 - 9x - 9)$. ●

exercises

1. For each of the following polynomials $p(x)$, use the remainder theorem (and synthetic substitution) to determine $p(t)$ for the given t.

(a) $p(x) = x^4 - 3x^3 + x^2 - 1$ $t = 2$

(b) $p(x) = 2x^3 - 4x + 2$ $t = \sqrt{2}$

polynomials & rational functions

(c) $p(x) = x^3 + 2x^2 - 5x + 1$ $t = -3$
(d) $p(x) = x^2 - 3x + 1$ $t = -1$
(e) $p(x) = x^4 - 3x^3 + x^2 - 1$ $t = i$
(f) $p(x) = 2x^3 - 4x + 2$ $t = -i$
(g) $p(x) = x^2 - 3x + 1$ $t = i$

2. Use synthetic division to show that, in each of the following, $f(x)$ is a factor of $g(x)$.

(a) $g(x) = x^2 - 3x + 2$ $f(x) = x - 1$
(b) $g(x) = 2x^2 + 5x - 3$ $f(x) = x + 3$

(c) $g(x) = 6x^2 - 7x + 2$ $f(x) = x - \dfrac{2}{3}$

(d) $g(x) = x^3 - 7x + 6$ $f(x) = x - 2$
(e) $g(x) = 2x^3 - 4x^2 + x - 2$ $f(x) = x - 2$
(f) $g(x) = 2x^4 - x^3 - 2x^2 + x$ $f(x) = 2x - 1$

(g) $g(x) = \dfrac{1}{3}x^3 - \dfrac{2}{9}x^2 + \dfrac{1}{27}x + \dfrac{4}{81}$ $f(x) = 3x + 1$

(h) $g(x) = x^4 + 2ix^3 - ix - 4$ $f(x) = x - i$
(i) $g(x) = x^3 - ix^2$ $f(x) = x - i$

3. For each of the following, use synthetic division to find the quotient polynomial when $s(x)$ is divided by $t(x)$. Indicate any nonzero remainder $r(x)$.

(a) $s(x) = 2x^3 + 11x^2 - 2x - 21$ $t(x) = 2x + 3$
(b) $s(x) = 2x^4 + 3x^3 - 4x^2 + 3x - 1$ $t(x) = 2x - 1$
(c) $s(x) = x^3 - (a + b + c)x^2 + (bc + ca + ab)x - abc$ $t(x) = x - a$
(d) $s(x) = 0.2x^3 - 0.03x + 0.015$ $t(x) = x + 0.2$
(e) $s(x) = x^3 - x^2 + 5$ $t(x) = x - 1 - 2i$
(f) $s(x) = x^2 + 4x - 2$ $t(x) = x - 3 + i$

4. In each of the following, verify that the given number a is a zero of the given polynomial $p(x)$.

(a) $p(x) = 4x^2 - 25$ $a = 5/2$
(b) $p(x) = 2x^4 - 6x^3 + 4x^2 - 17x + 15$ $a = 3$
(c) $p(x) = 2x^4 - 5x^3 - x^2 + 3x + 1$ $a = -1/2$
(d) $p(x) = 3x^3 + 9x^2 - 11x + 4$ $a = -4$
(e) $p(x) = 3x^5 - 2x^4 + 6x^3 - 4x^2 - 9x + 6$ $a = 2/3$
(f) $p(x) = x^4 - 6x^3 + 10x^2 - 6x + 1$ $a = 1$
(g) $p(x) = 3x^3 - 7x^2 - 4x + 2$ $a = 1 - \sqrt{3}$
(h) $p(x) = x^3 + x^2 + 2x + 2$ $a = i\sqrt{2}$
(i) $p(x) = 2x^5 - x^4 + 4x^3 - 2x^2 + 2x - 1$ $a = i$

5. Determine the real number c so that the polynomial

$$h(x) = c^2x^4 - 3c^2 + 1$$

is divisible by $x - 1$.

6. Show that $x + 2$ divides $x^{10} - 1024$.

7. Prove that the polynomial $p(x) = x^4 + 2x^2 + 1$ is not divisible by $x - a$ for any real number a. Is $p(x)$ a prime polynomial in $R[x]$?

8. Prove that for each positive integer n, $x - y$ is a factor of $x^n - y^n$.

120   **9.** Prove that for each positive even integer n, $x + y$ is a factor of $x^n - y^n$.

10. Prove that for each positive odd integer n, $x + y$ is a factor of $x^n + y^n$.

3.5. zeros of polynomial functions

In the second section of this chapter, an informal definition was given that the zero of a polynomial $p(x) \in R[x]$ was the place where its graph crossed the x axis. In a formal and more general sense we shall discuss zeros for polynomials in this section and in the two which follow.

If $p(x) \in F[x]$, *an element (number)* $r \in F$ *is a* **zero** *(or a* **root***) of* $p(x)$ *if and only if* $p(r) = 0 \in F$.

The study of the zeros of polynomials is both useful and traditional. In fact, the solution of the general quadratic was known by the Babylonians about 2000 B.C. They were also able to find the roots of some third and fourth degree polynomials. It took nearly 3500 years, until the sixteenth century, before general formulas analogous with the quadratic formula were found which give the zero of any cubic or quartic (biquadratic) in terms of the coefficients of the polynomial.

In 1824 the famous Norwegian mathematician, N. H. Abel (at age 22) established the fact that, in general, formulas expressing the zeros of polynomials of degree five or higher in terms of the coefficients are impossible. However, the earlier attempts to find such formulas and subsequent developments have amassed a considerable body of knowledge concerning the nature of the zeros of many polynomials. Because the general formulas for the roots of polynomials of degree five and higher cannot exist, and because of the quite complicated nature of the general formulas for degree three and four, this additional theory has important practical application. We shall discuss a small part of it here.

To begin with, consider the zeros of the product of two polynomials.

theorem 3.5.1

If $f(x) \in F[x]$ *with* $f(x) = g(x)h(x)$ *then, for* $r \in F$, $f(r) = 0$ *if, and only if,* $h(r) = 0$, *or* $g(r) = 0$. *That is, a zero of* $f(x)$ *is a zero of at least one of its factors.*

proof. Clearly, if $h(r) = 0$, or if $g(r) = 0$ then $f(r) = g(r)h(r) = 0$. On the other hand, if $f(r) = 0$ then $g(r)h(r) = 0$; thus at least one of $g(r)$ or $h(r)$ must be zero.

The fact expressed by the above theorem is important in finding the roots of many polynomials. In the following example note how it is used to find all the zeros of a particular fifth degree polynomial.

example 1

Find all the roots of $p(x) = x^5 - 13x^3 + 36x$. It is easily seen that $p(x) = x(x^4 - 13x^2 + 36)$. The second factor is of degree four but is a quadratic in x^2 in form. Therefore,

$$x^4 - 13x^2 + 36 = (x^2)^2 - 13(x^2) + 36$$
$$= (x^2 - 4)(x^2 - 9)$$

121

Thus, the zeros of $p(x)$ are the zeros of its factors

$$x, x^2 - 4, \text{ and } x^2 - 9$$

These are 0, 2, -2, 3, and -3. ●

example 2

Find all the roots of $h(x) = x^4 - 3x^2 - 4$. Again here, $h(x)$ is a quadratic in x^2. It factors as follows.

$$x^4 - 3x^2 - 4 = (x^2 - 4)(x^2 + 1)$$

The factor $x^2 - 4$ has 2 and -2 as zeros. Considered as a real polynomial $x^2 + 1$ has no zeros since its discriminant $\Delta = -4$. On the other hand, if we consider $h(x)$ to be a complex polynomial; i.e., $F[x] = C[x]$, then $x^2 + 1$ has roots $-i$ and i. These are obtained by direct application of the quadratic formula (Theorem 3.2.2)

$$t = \frac{-b \pm \sqrt{b^2 - 4ac}}{2a}$$

If this is the case, the roots of $h(x)$ are -2, 2, $-i$, and i. ●

As the above example shows, the existence of roots for a given polynomial $p(x)$ may depend upon the field of coefficients F, as was the case for $h(x)$ in that example. For a given complex polynomial of positive degree, $p(x) \in C[x]$, there always exists a complex number r which is a zero of $p(x)$. This fact is called *the fundamental theorem of algebra*.

theorem 3.5.2

If $p(x) \in C[x]$ and $p(x)$ has degree $n > 0$, then there exists $r \in C$ such that

$$p(r) = 0$$

This theorem was first satisfactorily proved in 1799 by the great German mathematician, C. F. Gauss. Although it was named the fundamental theorem of algebra it is really neither fundamental to all that is now included in algebra, nor can it be proved easily by algebraic methods. Gauss gave four different proofs, the first one for his doctoral dissertation (1799), the second and third published in 1816, and the fourth in 1850. We shall omit these complicated proofs and take the theorem to be true on faith.

As a direct result of Theorem 3.5.2 note that each $p(x) \in C[x]$ of positive degree has a linear factor $x - r$, $r \in C$. This follows upon application of the factor theorem. For this reason, the only prime polynomials in $C[x]$ are linear.

Also note that if $p(x)$ is a polynomial with real, rational, or integeral coefficients it is *a fortiori* a complex polynomial to which the theorem applies. On the other hand, we have no guarantee, as was the case with $x^2 + 1$, that the root will be real, rational, or an integer.

Suppose that $p(x)$ is a complex polynomial. Then using the remark following the theorem, there is some complex number r_1, such that

$$p(x) = (x - r_1)p_1(x)$$

zeros of polynomial functions

Now $p_1(x)$ is in $C[x]$ as well. If the degree of $p(x)$ is $n \geq 1$ then the degree of $p_1(x)$ is $n - 1$. If $n - 1 \neq 0$, we may consider $p_1(x)$ to be the polynomial in question. It has a linear factor $x - r_2$ for some $r_2 \in C$. Thus

$$p(x) = (x - r_1)(x - r_2)p_2(x)$$

where the degree $p_2(x)$ is $n - 2$. Since n is a fixed positive integer, we shall ultimately reach a polynomial $p_n(x)$ of degree zero, i.e., a constant complex polynomial a_n. In fact this a_n is the leading coefficient of $p(x)$.

$$p(x) = (x - r_1)(x - r_2) \cdots (x - r_n)p_n(x)$$
$$= a_n(x - r_1)(x - r_2) \cdots (x - r_n)$$

This proves:

theorem 3.5.3

If $p(x) \in C[x]$ is of degree $n > 0$ there exist n complex numbers r_1, $r_2, \ldots, r_n$ such that each is a root of $p(x)$, and $p(x)$ factors completely into linear factors as

$$p(x) = a_n(x - r_1)(x - r_2) \cdots (x - r_n)$$

example 3

Factor $p(x) = x^4 + 2x^3 + 3x^2 + 4x + 2$ completely into linear factors, given that $x^2 + 2$ divides $p(x)$. Division shows that

$$p(x) = (x^2 + 2)(x^2 + 2x + 1)$$

which can be factored further as

$$p(x) = (x + i\sqrt{2})(x - i\sqrt{2})(x + 1)(x + 1)$$

Therefore, the four complex numbers assured by Theorem 3.5.3 are $r_1 = -i\sqrt{2}$, $r_2 = i\sqrt{2}$, $r_3 = -1$, and $r_4 = -1$. Notice that $r_3 = r_4$. There is nothing in Theorem 3.5.3, or its proof, that guarantees that each of the r_i is distinct. ●

corollary

If $p(x)$ is a polynomial in $C[x]$ of degree $n > 0$, then $p(x)$ has at most n distinct roots.

proof. Exercise 5.

In the above example we could have written the factorization of $p(x)$ as follows.

$$p(x) = (x + i\sqrt{2})(x - i\sqrt{2})(x + 1)^2$$

If $(x - r)^k$ divides a polynomial $p(x)$ while $(x - r)^{k+1}$ does not, we say that r is a root (zero) of $p(x)$ **of multiplicity k.**

Thus, in Example 3, -1 is a root of $x^4 + 2x^3 + 3x^2 + 4x + 2$ of multiplicity two. Each of the other two zeros, $i\sqrt{2}$ and $-i\sqrt{2}$ has multiplicity one.

example 4

Write a polynomial of degree five which has 1 as a root of multiplicity three and 2 as a root of multiplicity two.

polynomials & rational functions

It follows from the factor theorem that

$$p(x) = a(x - 1)^3(x - 2)^2$$

where a is any nonzero constant (polynomial). In particular, if we choose $a = 2$:

$$p(x) = 2(x - 1)^3(x - 2)^2$$
$$= 2(x^5 - 7x^4 + 19x^3 - 25x^2 + 16x - 4)$$
$$= 2x^5 - 14x^4 + 38x^3 - 50x^2 + 32x - 8 \quad \bullet$$

If one counts a root of multiplicity k, k times he immediately sees that Theorem 3.5.3 assures him that a given complex polynomial has exactly $n \geq 1$ complex roots. When $n = 0$, one has a nonzero complex polynomial which has no zeros.

theorem 3.5.4

Every polynomial $p(x) \in C[x]$ of degree n has exactly n zeros in C, provided each zero of multiplicity k is counted as k zeros.

example 5

Suppose $p(x) = x^5 + 6x^4 + 9x^3$. Find all the zeros of $p(x)$.
First note that $p(x)$ has x^3 as a factor. Hence

$$p(x) = x^3(x^2 + 6x + 9)$$

Factoring the second factor yields

$$p(x) = x^3(x + 3)^2$$

Thus, 0 is a root of $p(x)$ of multiplicity 3, and -3 is a zero of multiplicity 2. $\quad \bullet$

Consider again the polynomial of Example 5. If t is any positive real number then surely each integeral power of t is positive as is, in particular, $6t^4$ and $9t^3$. It is, therefore, not possible for $p(t) = 0$ when $t > 0$. This illustrates an important fact concerning the zeros of polynomials with only positive coefficients. Let

$$p(x) = a_n x^n + a_{n-1} x^{n-1} + \cdots + a_1 x + a_0$$

with $n \geq 1$ and each coefficient a non-negative real number. Then it is clear that for every positive real number r, $p(r) > 0$. Hence, $p(x)$ has no positive zero. This proves:

theorem 3.5.5

Let $p(x) \in R[x]$ with $\deg p(x) = n \geq 1$ and each coefficient a_i of $p(x)$ non-negative. Then $p(x)$ has no positive zero.

This theorem assures us that 0 is larger than every real zero of $p(x)$, when $p(x)$ has only non-negative coefficients. *A number b is an* **upper bound** *for the zeros of a polynomial $p(x)$ if $p(r) \neq 0$ for all $r > b$. Similarly, a number c is a* **lower bound** *for the zeros of $p(x)$ if $p(r) \neq 0$ for all $r < c$.* One can use the above theorem and synthetic division to determine upper and lower bounds for the zeros of a polynomial $p(x)$.

zeros of polynomial functions

example 6

Let $f(x) = -3x^3 - x^2 - x - 1$. It is clear that $f(x)$ does not satisfy the hypothesis of Theorem 3.5.5 since its coefficients are all negative. On the other hand,

$$-f(x) = 3x^3 + x^2 + x + 1$$

does satisfy these hypotheses and, therefore, has no positive root r. But, since

$$f(x) = -[-f(x)]$$

and

$$f(r) = -[-f(r)]$$

We see that $f(x)$ can have no positive zero either. ●

example 7

Let $k(x) = 2x^3 + 3x^2 - 4x + 5$. Since there is no way for $k(x)$ or $-k(x)$ to satisfy the theorem, it is possible that $k(x)$ has a positive zero. Suppose we try $r = 1$. By synthetic division we obtain

$$
\begin{array}{rrrr|r}
2 & 3 & -4 & 5 & \underline{1} \\
  & 2 & 5 & 1 & \\
\hline
2 & 5 & 1 & 6 &
\end{array}
$$

Thus, $k(x) = (x - 1)(2x^2 + 5x + 1) + 6$. Now if r is any real number larger than 1 we see that

$$k(r) = (r - 1)(2r^2 + 5r + 1) + 6$$

is larger than 0. Hence, k has no zeros larger than 1. That is, 1 is an upper bound for the zeros of k. It was apparent in the synthetic division process that the quotient polynomial having only positive coefficients could not have a positive zero. Therefore, since the remainder was 6, a positive number, the only hope for a positive zero for k is that the factor $x - 1$ be negative. Hence no zero can be larger than 1. ●

A generalization of the above analysis leads to the following theorem.

theorem 3.5.6

Let $r \geq 0$ be a real number. Let $p(x) \in R[x]$ be a polynomial with real coefficients. Then in the division of $p(x)$ by $x - r$, we have $p(x) = (x - r)q(x) + p(r)$. If $p(r) \geq 0$ and the coefficients of $q(x)$ are nonnegative, $p(x)$ has no zero larger than r.

The proof is left as an exercise as is also the proof of the fact that if in dividing synthetically by the negative real number $-r$, one obtains coefficients of the quotient and remainder with alternating signs, then $-r$ is a lower bound for the zero of $p(x)$.

example 8

Consider again the polynomial $k(x) = 2x^3 + 3x^2 - 4x + 5$ of Example 7.

Set

$$r = -3$$

polynomials & rational functions

then

$$\begin{array}{rrrr|r} 2 & 3 & -4 & 5 & -3 \\ & -6 & 9 & -15 & \\ \hline 2 & -3 & 5 & -10 \end{array}$$

The signs alternate, hence -3 is a lower bound for the zero of $k(x)$. All the zeros must be between -3 and 1.

We have also seen, in the case of $k(x)$, that $k(1) = 6$ while $k(-3) = -10$. Thus, the graph of $p(x)$ must cross the x axis somewhere between -3 and 1, perhaps it crosses more than once. We may plot points, using the remainder theorem and the fact that $k(0) = 5$ to obtain the graph of $k(x)$ and an approximation to the zero as follows:

$$\begin{array}{rrrr|r} 2 & 3 & -4 & 5 & -2 \\ & -4 & 2 & 4 & \\ \hline 2 & -1 & -2 & 9 \end{array}$$

$$\begin{array}{rrrr|r} 2 & 3 & -4 & 5 & -1 \\ & -2 & -1 & 5 & \\ \hline 2 & 1 & -5 & 10 \end{array}$$

$$\begin{array}{rrrr|r} 2 & 3 & -4 & 5 & 2 \\ & 4 & 14 & 20 & \\ \hline 2 & 7 & 10 & 25 \end{array}$$

The graph of $k(x)$ is sketched in Figure 3.5.1. We see that the points $(0,5)$, $(1,6)$; $(2,25)$, $(-1,10)$, $(-2, 9)$, and $(-3, -10)$ all satisfy the equation $y = k(x)$ and lie on the graph. Since $k(-3) = -10$ and $k(-2) = 9$, the graph of $y = k(x)$ must cross the x axis between $x = -3$ and $x = -2$.

This method can be refined by subdividing the interval $[-3,-2]$ and

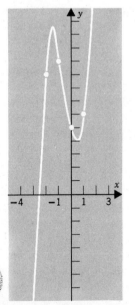

figure 3.5.1

126

zeros of polynomial functions

testing rational numbers such as -2.9, -2.8, etc. In this manner, a rational approximation to the zero of $k(x)$ can be determined to any desired accuracy. ●

exercises

1. Find any polynomial of degree four with the given roots.

(a) $1, -1, i, -i$

(b) $2, -2, 1 + i, 1 - i$

(c) $1, 2, -1, -2$

(d) $1, i, 2i, 1 - i$

(e) $i, 2i, 3i, 0$

(f) $1, 1, 1, 1$

(g) $1, 1, 1, i$

(h) $1, 2, 2, 3$

(i) $0, 0, i, -i$

(j) $i, i, i, -i$

2. Each of the following polynomials has been partially factored. Find all the zeros and their multiplicities. (Use the quadratic formula if necessary.)

(a) $p(x) = (x^2 - 5x + 6)(x + 1)^3$

(b) $p(x) = (x - 1)^2(x^2 - 1)$

(c) $p(x) = 2(x + 1)^2(x^2 + 1)$

(d) $p(x) = (3x + 2)^5$

(e) $p(x) = (x^2 - 25)^3$

(f) $p(x) = 3x^4(x - 1)^3(x + 1)^2$

(g) $p(x) = x^3(x + 1)^2(x^2 + 1)^3$

(h) $p(x) = 4(x^2 - x - 12)^3$

(i) $p(x) = (x^2 + x + 1)^2$

(j) $p(x) = 5(x^2 - x - 1)^3$

3. Use synthetic division to show that 3 is a root of multiplicity 2 of the polynomial $f(x) = x^4 - 6x^3 + 13x^2 - 24x + 36$. Then express $f(x)$ as the product of linear factors.

4. Show that -1 is a root of multiplicity 5 of the polynomial $f(x) = x^6 + 3x^5 - 10x^3 - 15x^2 - 9x - 2$. Find the other root.

5. Prove the corollary of Theorem 3.5.3.

6. Find a cubic polynomial $p(x)$ with -2, 1, and 3 as zeros and $p(0) = 12$.

7. Find a cubic polynomial $p(x)$ with zeros 1, -1, and -2 and $p(0) = -4$.

8. Prove that if b is a positive real number the polynomial $x^n - b$ has a unique positive root.

9. Prove Theorem 3.5.6.

10. Prove the statement concerning lower bounds following Theorem 3.5.6.

11. Approximate the real zero of $k(x) = 2x^3 + 3x^2 - 4x + 5$ to three decimal places by using synthetic division.

12. Find integers which are upper and lower bounds for the zeros of each of the following polynomials.

(a) $3x^3 + 4x^2 - 3x + 1$

(b) $8x^3 - 6x^2 - 5x + 15$

(c) $x^4 + x^2 + 1$

(d) $x^4 - x^2 - 1$

(e) $x^4 - 6x^3 + x^2 + 12x - 6$

(f) $x^4 + 6x^3 - x^2 - 12x - 6$

(g) $2x^5 + x^4 - 2x - 2$

13. Use synthetic division to compute $p(b)$ for enough values of b to sketch a fairly accurate graph $p(x)$ in each of the following. Approximate the real root(s) of p from the graph.

(a) $p(x) = 2x^2 - 5x - 3$

(b) $p(x) = x^2 - 2x - 6$

(c) $p(x) = x^4 + 3x^2 - 4$

(d) $p(x) = x^4 - 3x^2 - 4$

(e) $p(x) = 2x^3 - 3x^2 + 6x - 5$

(f) $p(x) = x^3 - x^2 - 6x$

(g) $p(x) = 2x^4 - 3x^2 + 4x - 5$

(h) $p(x) = x^5 - 3x^4 - 5x^3 + 15x^2 + 4x - 12$

polynomials & rational functions

3.6. rational roots

Our objective in this section is to establish a criterion for discovering all the rational roots of a polynomial

$$(3.6.1) \qquad p(x) = a_n x^n + a_{n-1} x^{n-1} + \cdots + a_1 x + a_0$$

whose coefficients are *rational numbers*. That is, we shall see that any rational root $r = u/v$, u, v integers, $v \neq 0$, of $p(x)$ must have a special relationship to the coefficient of $p(x)$. This will enable us to list *all* the *possible* rational roots of $p(x)$. The number of possibilities will always be finite. Then testing each by synthetic division, all rational roots of $p(x)$ can be determined.

Suppose then that $p(x)$ has rational coefficients with

$$p(x) = \frac{c_n}{b_n} x^n + \frac{c_{n-1}}{b_{n-1}} x^{n-1} + \cdots + \frac{c_1}{b_1} x + \frac{c_0}{b_0}$$

where each of the c_i and b_i are integers and none of the b_i's is zero, $i = 1$, $2, \ldots , n$. Then if b is the least common multiple of $b_0, b_1, \ldots , b_n$, any zero of $p(x)$ is also a zero of

$$bp(x) = d_n x^n + d_{n-1} x^{n-1} + \cdots + d_1 x + d_0$$

This polynomial has integers for coefficients. It is true, therefore, that we may just as well consider polynomials with integral coefficients as far as determining roots is concerned. For example, if

$$f(x) = \frac{5}{3} x^2 + \frac{1}{6} x - \frac{7}{2}$$

we may just as well look for the roots of

$$6f(x) = 10x^2 + x - 21$$

example 1

Let $p(x) = 10x^2 + x - 21$. As one can determine from factoring $p(x)$, or from the quadratic formula, $-3/2$ is a root of $p(x)$

$$p(3/2) = 10(-3/2)^2 + (-3/2) - 21 = 0$$

We write this in the following form.

$$10(-3/2)^2 + (-3/2) = 21$$

One can see that -3 is a factor of the left member of this equation; i.e.,

$$-3 \left[10 \frac{(-3)}{2^2} + \frac{1}{2} \right] = 21$$

and since -3 divides the left member, it must also divide the right member 21. The right member is however, the negative of the constant term of the given polynomial, hence, -3 divides the constant term of $p(x)$. ●

In general, let $p(x)$ be a polynomial with integer coefficients,

$$p(x) = a_n x^n + a_{n-1} x^{n-1} + \cdots + a_1 x + a_0$$

rational roots

Let $r = u/v$ be a rational number, $u, v \in Z$, $v \neq 0$ such that

$$p(r) = a_n(u/v)^n + a_{n-1}(u/v)^{n-1} + \cdots + a_1(u/v) + a_0 = 0$$

Then, transposing the constant term results in

$$a_n(u/v)^n + a_{n-1}(u/v)^{n-1} + \cdots + a_1(u/v) = -a_0$$

But since u is a factor of the left member of this equation it must follow that u divides a_0. Let us also assume that r is in lowest terms; i.e., u and v have no common factors. Note that since $v \neq 0$, v^n is also nonzero. Thus,

$$v^n p(r) = a_n u^n + a_{n-1} u^{n-1} v + \cdots + a_1 u v^{n-1} + a_0 v^n = 0$$

Therefore, by transposing the first term, the following equation results.

$$-a_n u^n = a_{n-1} u^{n-1} v + a_{n-2} u^{n-2} v + \cdots + a_1 u v^{n-1} + a_0 v^n$$

The right member of this equation is divisible by v; the left member must also be divisible by v. That is, v divides $-a_n u^n$. But u and v have no common factors, so v does not divide u^n. It must therefore divide $-a_n$. Thus, v divides the leading coefficient of $p(x)$.

In Example 1 we saw that $r = -3/2$, $u = -3$, $v = 2$. It is clear that in that case u divides the constant term of

$$p(x) = 10x^2 + x - 21$$

while v divides the leading coefficient. The general remarks constitute the proof of the following important theorem.

theorem 3.6.1

Let $p(x)$ be a polynomial with integers as coefficients. Then any rational root $r = u/v$ (in lowest terms), $v \neq 0$, $u, v \in Z$, of $p(x)$ must be such that u divides the constant term a_0 of $p(x)$ and v divides the leading coefficient a_n of $p(x)$.

There are a number of things that this theorem does not say. First of all, it does not guarantee that a particular polynomial with integer coefficients has *any* rational roots. For example, $2x^2 + 2x + 2$ has no rational roots at all, nor does $3x^2 + x + 2$, nor $x^2 + 3x + 1$. It also does not guarantee that every divisor u of a_0 and v of a_n will give rise to rational roots u/v. Witness the polynomial of the following example.

example 2

List all the possible rational roots of

$$f(x) = 3x^3 - 14x^2 - 3x - 10$$

The leading coefficient of $f(x)$ is 3, while the constant term is -10. The possible integral divisors of -10 are ± 1, ± 2, ± 5, and ± 10. The possible integral divisors of v of 3 are ± 1, ± 3. Therefore, the possible rational roots u/v for $f(x)$ are only

$$\pm 1, \pm 2, \pm 5, \pm 10, \pm 2/3, \pm 1/3, \pm 5/3, \text{ and } \pm 10/3$$

129

Although there are 16 possible rational roots, not all of them can be roots of

$p(x)$. To try each of these sixteen rational root candidates by synthetic division is an easy task when compared to the infinite collection of all the rational numbers. In doing so we discover that

$$
\begin{array}{rrrrr|r}
3 & -14 & -3 & -10 & & \underline{5} \\
  & 15 & 5 & 10 & & \\
\hline
3 & 1 & 2 & 0 & &
\end{array}
$$

Thus, $f(5) = 0$, and $f(x) = (x - 5)(3x^2 + x + 2)$. Since the second factor has negative discriminant, $\Delta = b^2 - 4ac = -23$, 5 is the *only* rational root of $f(x)$. The other two roots are the complex numbers $-1/6 - i(\sqrt{23}/6)$ and $-1/6 - i(\sqrt{23}/6)$. ●

example 3

Find all the roots of

$$p(x) = 6x^5 - 7x^4 - 18x^3 + 22x^2 - 3$$

The factors v of $a_n = 6$ are ± 1, ± 2, ± 3, and ± 6. The factors u of $a_0 = -3$ are ± 1, and ± 3. Therefore, the possible rational roots of $p(x)$ are

$$\pm 1, \ \pm 3, \ \pm 1/2, \ \pm 3/2, \ \pm 1/3, \ \text{and} \ \pm 1/6$$

First we test $r = 3$ as follows.

$$
\begin{array}{rrrrrr|r}
6 & -7 & -18 & +22 & 0 & -3 & \underline{3} \\
  & 18 & 33 & 45 & 201 & 603 & \\
\hline
6 & 11 & 15 & 67 & 201 & 600 &
\end{array}
$$

While we see that 3 is not a root, it is an upper bound for all the real roots of $p(x)$. Now test $r = 1$

$$
\begin{array}{rrrrrr|r}
6 & -7 & -18 & 22 & 0 & -3 & \underline{1} \\
  & 6 & -1 & -19 & 3 & 3 & \\
\hline
6 & -1 & -19 & 3 & 3 & 0 &
\end{array}
$$

$r = 1$ is a rational root and is *not* an upper bound for the real roots of $p(x)$. It may be that there is an additional root between 1 and 3.

Next test $r = 1/2$. Since any root different from 1 of $f(x)$ must be also a root of the quotient $6x^4 - x^3 - 19x^2 + 3x + 3$ obtained by dividing $p(x)$ by $x - 1$, we need only use this quotient. Hence,

$$
\begin{array}{rrrrr|r}
6 & -1 & -19 & 3 & 3 & \underline{1/2} \\
  & 3 & 1 & -9 & -3 & \\
\hline
6 & 2 & -18 & -6 & 0 &
\end{array}
$$

Therefore, $1/2$ is also a root. Next test $r = -1/3$ in the quotient $6x^3 + 2x^2 - 18x - 6$, thus,

$$
\begin{array}{rrrr|r}
6 & 2 & -18 & -6 & \underline{-1/3} \\
  & -2 & 0 & 0 & \\
\hline
6 & 0 & -18 & 0 &
\end{array}
$$

rational roots

We see that $f(x) = (x - 1)(x - 1/2)(x + 1/3)(6x^2 - 18)$. The two remaining roots of $p(x)$ are roots of $6x^2 - 18$. These are the roots of $6(x^2 - 3)$; viz., $\pm\sqrt{3}$. Neither of these is rational. Note too that, as we speculated earlier, $\sqrt{3}$ is between 1 and 3. ●

example 4

Find all the roots of the polynomial

$$p(x) = 4x^4 + 27x^3 + 46x^2 + 61x + 30$$

Since this polynomial has integers as coefficients, we list first the possible rational roots. These are: ± 30, ± 15, ± 10, ± 5, ± 3, ± 2, ± 1, $\pm 15/2$, $\pm 5/2$, $\pm 3/2$, $\pm 1/2$, $\pm 15/4$, $\pm 5/4$, $\pm 3/4$, and $\pm 1/4$. This is a rather formidable number of possibilities to test. We note first, however, that the number can immediately be cut in half since $p(x)$ can have *no positive roots* (why?). Then testing -10 by synthetic division results in alternating signs in the quotient and remainder, so -10 is a lower bound for the roots of $p(x)$.

Then test -5 as follows:

4	27	46	61	30	$\underline{}-5$
	-20	-35	-55	-30	
4	7	11	6	0	

Thus -5 is a root of $p(x)$. It is however not a double root. If we test -2 in the quotient $q(x) = 4x^3 + 7x^2 + 11x + 6$, we get

4	7	11	6	$\underline{}-2$
	-8	2	-26	
4	-1	13	-20	

Therefore, -2 is a lower bound for the rest of the roots of $p(x)$.
Testing -1 by synthetic division reveals

4	7	11	6	$\underline{}-1$
	-4	-3	-8	
4	3	8	-2	

Thus, $q(-1) = -2 < 0$.
Testing $-1/2$ by synthetic division reveals

4	7	11	6	$\underline{}-1/2$
	-2	$-5/2$	$17/4$	
4	5	$-17/2$	10 1/4	

Hence, $q(-1/2) = 10\ 1/4 > 0$, and q has a root between -1 and $-1/2$. This might be $-3/4$, a possible rational root. It, in fact, is as we see from the following.

4	7	11	6	$\underline{}-3/4$
	-3	-3	-6	
4	4	8	0	

two easy theorems & synthetic division

Furthermore, any additional roots of $p(x)$ $\big($or $q(x)\big)$ must be roots of $h(x) = 4(x^2 + x + 2)$. Since this is a quadratic with negative discriminant, it has only complex roots. They are obtained from the quadratic formula; viz.,

$$t = \frac{-1 + \sqrt{1 - 4(2)(1)}}{2} = -1/2 \pm i\frac{\sqrt{7}}{2}$$

Thus, the roots of $p(x)$ are -5, $-3/4$, $-1/2 + i\sqrt{7}/2$, and $-1/2 - i\sqrt{7}/2$. ●

exercises

1. List all the *possible* rational roots for the following polynomials.

(a) $x^2 - 2x + 1$ $(x-1)(x-1)$
(b) $x^2 + 2x + 1$ $(x+1)(x+1)$
(c) $2x^2 - 2x + 3$
(d) $3x^2 - 4x + 1$
(e) $3x^3 - 4x^2 + x - 1$
(f) $3x^3 - x^2 + x + 2$
(g) $x^4 + 2x^3 + x^2 + x - 2$
(h) $6x^4 - 5x^3 - 5x - 6$
(i) $2x^4 - 13x^3 + 36x^2 - 70x + 75$
(j) $3x^4 - 4x^3 - 8x^2 - 9x - 2$
(k) $12x^4 + 7x^3 + 13x^2 + 7x + 1$
(l) $2x^5 + 9x^4 + 21x^3 + 28x^2 + 21x + 9$

2. Find all of the rational roots for the following polynomials.

(a) $4x + 1$
(b) $3x - 5$
(c) $6x^2 + x - 2$
(d) $6x^2 - 13x - 5$
(e) $6x^2 - 19x + 10$
(f) $3x^2 - 2x + 12$
(g) $6x^3 + 13x^2 - 41x + 12$
(h) $5x^3 + 17x^2 + 11x + 2$
(i) $3x^3 + 5x^2 + 5x + 2$
(j) $6x^4 + 7x^3 - 12x^2 - 3x + 2$
(k) $5x^5 - 7x^4 + 5x^3 - 7x^2 + 5x - 7$
(l) $10x^5 - 11x^4 - 61x^3 + 8x^2 + 12x$

3. Factor each of the following polynomials into irreducible factors with rational coefficients.

(a) $4x + 1$
(b) $x^3 - 3x^2 - 3x + 1$
(c) $x^2 + x + 1$
(d) $x^3 - 2x^2 - x + 2$
(e) $6x^3 + 5x^2 - 7x - 4$
(f) $4x^3 + 4x^2 - 3x + 10$
(g) $2x^3 + 7x^2 + 18x - 22$
(h) $4x^4 - 4x^3 - 17x^2 + 9x + 18$
(i) $x^4 - 4x^3 + 3x^2 - 4x + 4$
(j) $x^4 + 4x^3 + x^2 - 16x - 20$
(k) $x^5 - x^4 - 5x^3 + x^2 + 8x + 4$
(l) $x^5 - x^4 - 5x^3 + 5x^2 + 4x - 4$

4. First find the rational roots for the following polynomials and then use the quadratic formula, if necessary, to determine all the roots.

(a) $3x - 2$
(b) $x^2 + x + 1$
(c) $x^3 - 2x^2 + 3x - 2$
(d) $3x^3 - 7x^2 - 3x + 2$
(e) $2x^3 - 9x^2 + 8x - 2$
(f) $12x^3 + 8x^2 - 13x + 3$
(g) $2x^3 - 7x^2 - 2x + 12$

(h) $x^4 - x^3 - 2x - 4$
(i) $x^4 + x^3 - 2x^2 - 4x - 8$
(j) $3x^5 + x^4 - 39x^3 - 13x^2 + 108x + 36$

5. Find the roots of those polynomials in Exercise 1 (a), (b), (c), and (d).

6. Find all the roots of the polynomials in Exercise 2.

7. Consider all the possible rational roots for the polynomial $p(x) = x^2 - 2$. Show that $p(x)$ has no rational roots, and therefore $\sqrt{2}$ is not a rational number.

8. Use the procedure of Exercise 7 to demonstrate that none of the following is rational.

(a) $\sqrt{3}$

(b) $\sqrt{5}$

(c) $\sqrt{7}$

(d) $\sqrt{10}$

(e) $\sqrt{11}$

(f) $\sqrt{15}$

(g) $\sqrt[3]{2}$

(h) $\sqrt[3]{3}$

(i) $\sqrt[3]{4}$

(j) $\sqrt[3]{5}$

(k) $\sqrt[3]{6}$

(l) $\sqrt[3]{9}$

Hint: Look at $x^3 - 2$.

9. Each of the following cubic polynomials $p(x)$ has a real root. Use the method outlined in Section 3.5 to locate the root in the interval $[a,b]$, when $p(a)$ and $p(b)$ have different signs. Approximate all the real roots of $p(x)$ to two decimal places.

(a) $x^3 + 6x^2 - 3x - 18$
(b) $x^3 - x^2 - 3x + 2$
(c) $x^3 + 3x^2 - 5x - 4$

(d) $x^3 + 5x^2 + x - 15$
(e) $12x^3 - 8x^2 - 21x + 14$

10. How long is the edge of a cube if, after a one-inch thick slice is cut off of one side, the remaining figure has a volume of 100 cubic inches.

11. A rectangular box is to be made from a piece of sheet metal 6 inches by 14 inches. To do so, equal squares are cut from the four corners. The sides are then turned up. Find the size of the square to be cut if the volume of the box is to be 48 cubic inches.

12. Find three consecutive integers $(n, n + 1, n + 2)$ so that the sum of their reciprocals is 47/60.

13. Show that if $p(x)$ has integral coefficients and leading coefficient $a_n = 1$, then its only rational roots must be integeral factors of the constant term a_0.

14. Find all the rational roots of the following polynomials.

(a) $x^5 - 3x^4 + x^3 + 4x$
(b) $x^6 - 3x^5 + 4x^4 - 4x^3 + 3x^2 - x$
(c) $x^5 - 2x^4 - 11x^3 + 12x^2 + 36x$

(d) $2x^5 - 9x^4 + 6x^3 + 20x^2 - 24x$
(e) $4x^5 - 20x^4 + 29x^3 - 16x^2 + 3x$
(f) $27x^5 - 27x^4 + 9x^3 + x^2$

3.7. complex roots

In the previous sections we saw that many real polynomials have roots which are not real numbers. We noticed that the general real quadratic polynomial

$$q(x) = ax^2 + bx + c, \qquad a \neq 0$$

may have a graph which crosses the x axis twice, one, or not at all. Just which of these occurs is indicated by the discriminant

$$\Delta = b^2 - 4ac$$

polynomials & rational functions

Therefore, $q(x)$ may have two real roots, a real root, of multiplicity 2, or no real roots, respectively as $\Delta > 0$, $\Delta = 0$ or $\Delta < 0$. On the other hand, we are assured that all quadratics have two roots in the *complex numbers*.

example 1

Let $q(x) = x^2 + 2x + 2$. Since for this polynomial, $\Delta = 4 - 8 = -4$, its graph will not intersect the x axis, and it has no real roots. Nevertheless, it is true that for

$$r_1 = -1 + i \qquad \text{and} \qquad r_2 = -1 - i$$

$q(r_1) = q(r_2) = 0$. We notice that r_2 is the conjugate of r_1. ●

This is not a coincidence, as we see from the following theorem.

theorem 3.7.1

Let $p(x) \in R[x]$ be any polynomial with real coefficients and degree $n > 0$. If $r = a + bi$ is a complex root of $p(x)$, then the conjugate $\bar{r} = a - bi$ of r is a root of $p(x)$.

proof. The proof is an easy and straightforward application of the properties of complex numbers. Recall that the conjugate of any finite sum of complex numbers is the sum of the conjugate, $\overline{\sum_{i=1}^{n} z_i} = \sum_{i=1}^{n} \bar{z}_i$. Similarly, recall that the same is true for finite products; i.e.,

$$\overline{\prod_{i=1}^{n} z_i} \qquad \prod_{i=1}^{n} \bar{z}_i$$

the conjugate of a product is the product of the conjugates. Apply this to the fact that $p(r) = 0$. If

$$p(x) = a_n x^n + a_{n-1} x^{n-1} + \cdots + a_1 x + a_0$$

then

$$p(r) = 0 = a_n (r)^n + a_{n-1} (r)^{n-1} + \cdots + a_1 (r) + a_0$$

Therefore, the conjugate of $p(r)$ is

$$\overline{p(r)} = \bar{0} = 0 = \overline{a_n (r)^n + a_{n-1} (r)^{n-1} + \cdots + a_1 (r) + a_0}$$

Thus,

$$0 = \bar{a}_n (\bar{r})^n + \bar{a}_{n-1} (\bar{r})^{n-1} + \cdots + \bar{a}_1 (\bar{r}) + \bar{a}_0$$

But, $p(x) \in R[x]$, so the coefficients a_i are real numbers. Hence, $\bar{a}_i = a_i$, for each $i = 0, 1, 2, \ldots, n$. Therefore,

$$0 = a_n (\bar{r})^n + a_{n-1} (\bar{r})^{n-1} + \cdots + a_1 (\bar{r}) + a_0$$

That is, $p(\bar{r}) = 0$ as desired.

example 2

Find all the roots of $f(x) = x^4 - 6x^3 + 11x^2 - 6x + 10$ given that i is a root. The easiest way to proceed is to divide synthetically by $x - i$. This yields

1	-6	11	-6	10	$\underline{i}$
	i	$-1 - 6i$	$6 + 10i$	-10	
1	$-6 + i$	$10 - 6i$	$10i$	0	

complex roots

But if i is a root, since $f(x)$ has real coefficients, it must follow that $-i$ is also a root. Continuing the division (in the quotient), this time by $x - (-i)$, yields

$$
\begin{array}{ccccc}
1 & -6+i & 10-6i & 10 & \underline{\mid -i} \\
 & -i & +6i & -10i & \\
\hline
1 & -6 & 10 & 0 &
\end{array}
$$

So far we know that

$$f(x) = x^4 - 6x^3 + 11x^2 - 6x + 10 = (x - i)(x + i)(x^2 - 6x + 10)$$

Therefore, the other two roots are roots of the quadratic factor. Thus, they are given by the quadratic formula as

$$r = \frac{6 \pm \sqrt{(-6)^2 - 4(1)(10)}}{2(1)} = \frac{6 \pm \sqrt{-4}}{2} = 3 \pm i$$

Hence, i, $-i$, $3 + i$, and $3 - i$ are the roots of $f(x)$. Notice that $3 + i$ and $3 - i$ are also conjugate of each other as was to be expected. ●

It is true that a real number $s = a + 0i$ is also a complex number. Theorem 3.7.1 tells us that if $p(s) = 0$, then $p(\bar{s}) = 0$. But $s = \bar{s}$. This theorem *does not* say that s is a double root (multiplicity two) of $p(x)$.

example 3

Factor $p(x) = x^3 - 1$ over the complex numbers. One quickly sees that $p(1) = 0$ so

$$p(x) = (x - 1)(x^2 + x + 1)$$

But 1 is a root of multiplicity 1. The other two roots of $p(x)$ are roots of $x^2 + x + 1$. They are obtained from the quadratic formula as $-1/2 + i\sqrt{3}/2$, and $-1/2 - i\sqrt{3}/2$. ●

As an immediate consequence of Theorem 3.7.1 we see that every polynomial of odd degree $n > 1$ with real coefficients must have at least one real root. This follows because if $a + bi$ and $a - bi$ are roots of a polynomial $p(x)$ then the factor theorem assures us that $(x - (a + bi))$ and $(x - (a - bi))$ both divide $p(x)$. Since each of these is a prime polynomial, it must follow that their product divides $p(x)$.

$$
\begin{aligned}
p(x) &= [(x - (a + bi))][x - (a - bi)]q(x) \\
&= [(x - a) - bi][(x - a) + bi]q(x) \\
&= [(x - a)^2 + b^2]q(x)
\end{aligned}
$$

Therefore, every real polynomial which has a nonreal root $a + bi$ must be divisible by the quadratic

$$(x - a)^2 + b^2 = x^2 - 2a + (a^2 + b^2)$$

Hence, if the degree of $p(x)$ is odd, it must have a real linear factor.

One sees that each of the above quadratics has a negative discriminant. This fact and the other theorems give us:

polynomaisl & rational functions

theorem 3.7.2

The only polynomials which are irreducible (prime) over the real numbers are linear, or else quadratic with negative discriminant.

example 4

Factor $p(x) = 3x^5 + 8x^4 + 19x^3 + 4x^2 - 22x - 12$ into irreducible real polynomials. What are the roots of $p(x)$?

First try to obtain linear factors $(x - r)$. This will occur if r is a real root. If r is also rational, it must be among the possible rational roots ± 12, ± 6, ± 4, ± 3, ± 2, $\pm 4/3$, $\pm 2/3$, ± 1. Dividing synthetically by $x - r$ we find, in succession

$$
\begin{array}{rrrrrr|}
3 & 8 & 19 & 4 & -22 & -12 \quad \underline{|1} \\
 & 3 & 11 & 30 & 34 & 12 \\
\hline
3 & 11 & 30 & 34 & 12 \quad \underline{|-1} \\
 & -3 & -8 & -22 & -12 \\
\hline
3 & 8 & 22 & 12 \quad \underline{|-2/3} \\
 & -2 & -4 & -12 \\
\hline
3 & 6 & 18 \\
\end{array}
$$

4. 4(1 36

Therefore,

$$p(x) = (x - 1)(x + 1)(x + 2/3)(3x^2 + 6x + 18)$$
$$= (x - 1)(x + 1)(3x + 2)(x^2 + 2x + 6)$$

The quadratic $x^2 + 2x + 6$ has discriminant -20 and is, therefore, irreducible. Thus, the above is the desired factorization.

The roots of $p(x)$ are, of course, 1, -1, $-2/3$, and the conjugate pair of complex roots $-1 \pm i\sqrt{5}$. ●

example 5

Determine a polynomial $f(x)$ with real coefficients such that i and 2 are double roots and that -1 is a root of multiplicity 3.

By the factor theorem $f(x)$ must be divisible by

$$(x - i)^2(x - 2)^2(x + 1)^3$$

Besides this however, since $f(x)$ must have real coefficients, $-i$ must also be a double root. Therefore, such an $f(x)$ is

$$
\begin{aligned}
f(x) &= (x - i)^2(x + i)^2(x - 2)^2(x + 1)^3 \\
&= (x^2 + 1)^2(x^2 - 4x + 4)(x^3 + 3x^2 + 3x + 1) \\
&= (x^4 + 2x^2 + 1)(x^2 - 4x + 4)(x^3 + 3x^2 + 3x + 1) \\
&= (x^6 - 4x^5 + 6x^4 - 8x^3 + 9x^2 - 4x + 4)(x^3 + 3x^2 + 3x + 1) \\
&= (x^9 - x^8 + 3x^6 - x^5 + 5x^4 - 7x^3 + 9x^2 + 8x + 4)
\end{aligned}
$$

This is, of course, not the only possible polynomial since

$$cf(x)$$

complex roots

for any nonzero real number c satisfies the above requirements. If one additionally knew that $f(0) = 8$ then the desired polynomial would be

$$2x^9 - 2x^8 + 6x^6 - 2x^5 + 10x^4 - 14x^3 + 18x^2 + 16x + 8 \quad \bullet$$

exercises

1. Find a polynomial $f(x)$ of degree two which has real coefficients with the given root r and value of $f(0)$.

(a) $r = -3i$, $f(0) = 9$ (g) $r = 2 - 3i$, $f(0) = 13$
(b) $r = 2i$, $f(0) = 12$ (h) $r = 3i$, $f(0) = 6$
(c) $r = 3i$, $f(0) = 27$ (i) $r = 2i$, $f(0) = 8$
(d) $r = 2i$, $f(0) = 4$ (j) $r = 1/2 - i\sqrt{3}/2$, $f(0) = 2$
(e) $r = i\sqrt{5}$, $f(0) = 10$ (k) $r = \sqrt{2}/2 + i\sqrt{2}/2$, $f(0) = 2$
(f) $r = 1 + i$, $f(0) = 2$ (l) $r = \sqrt{3}/2 + i/2$, $f(0) = 3$

2. Find a polynomial of degree three with real coefficients which has the following roots.

(a) $r_1 = 3$, $r_2 = i$ (g) $r_1 = 1$, $r_2 = -i\sqrt{5}$
(b) $r_1 = 1$, $r_2 = -2i$ (h) $r_1 = 2/3$, $r_2 = \sqrt{2} - i$
(c) $r_1 = -2$, $r_2 = 1 + i$ (i) $r_1 = -3/2$, $r_2 = \sqrt{3} + i$
(d) $r_1 = -1$, $r_2 = 2 - 3i$ (j) $r_1 = \sqrt{2}$, $r_2 = -i$
(e) $r_1 = -3$, $r_2 = 3 + 2i$ (k) $r_1 = \sqrt{3}$, $r_2 = i\sqrt{3}$
(f) $r_1 = -5$, $r_2 = i\sqrt{3}$ (l) $r_2 = \sqrt{5}$, $r_3 = -i\sqrt{5}$

3. In each of the following one root of the polynomial is given. Find all the roots.

(a) $x^4 + 4x^3 + 3x^2 + 4x + 2$ $r = i$
(b) $2x^3 - 2x^2 - x - 6$ $r = 2$
(c) $2x^3 - 5x^2 + 6x - 2$ $r = 1 + i$
(d) $2x^3 - 5x^2 + 1$ $r = 1/2$
(e) $3x^3 - 7x^2 - 4x + 2$ $r = 1 - \sqrt{3}$
(f) $x^3 + x^2 - 6x + 4$ $r = -1 + \sqrt{5}$
(g) $x^3 + x^2 + 2x + 2$ $r = i\sqrt{2}$
(h) $x^4 + x^3 - x^2 + 5x + 6$ $r = 1 - i\sqrt{2}$
(i) $x^4 + 2x^2 + 8x + 5$ $r = 1 - 2i$
(j) $x^4 - 8x^3 + 42x^2 - 104x + 169$ $r = 2 + 3i$
(k) $x^5 + 2x^4 + 4x^3 - 8x^2 - 16x - 32$ $r = -1 - i\sqrt{3}$
(l) $x^6 + x^5 + 2x^4 + x^3 + 2x^2 + x + 1$ $r = -1/2 - i\sqrt{3}/2$

4. If $-2i$ is a root of $x^3 + 3x^2 + bx + c$. Find the real numbers b and c.

5. Let $ax^2 + bx + c$ be a quadratic polynomial with complex coefficients. Show that

(a) $(\sqrt{2}/2 + i\sqrt{2}/2)^2 = i$
(b) $(-\sqrt{2}/2 + i\sqrt{2}/2)^2 = -i$
(c) If $az^2 + bz + c = 0$ for some complex number z, then

$$z = \frac{-b \pm \sqrt{b^2 - 4ac}}{2a}$$

6. By the fundamental theorem of algebra the complex polynomial $x^n - z$ has n com-

polynomials & rational functions

plex roots. These are called the **nth roots of z**. Use the methods of this chapter to determine each of the following.

   (a) the cube roots of 1
   (b) the cube roots of -1
   (c) the square roots of i (see Exercise 5)
   (d) the square roots of $-i$ (see Exercise 5)
   (e) the fourth roots of 1
   (f) the fourth roots of -1 (see (c) and (d))

7. Show that $i^4 = 1$, and therefore $i^k = i^{k-4}$ for $k \geq 4$.

8. Show that i is a root of $x^7 + i$. Is $-i$ also a root? Check it. Does this violate Theorem 3.7.1? Why?

9. Factor each of the following polynomials into irreducible polynomials over the real numbers.

   (a) $2x^3 - 3x^2 - 17x + 30$ (e) $12x^3 - x^2 + 7x + 2$
   (b) $6x^3 + 11x^2 - 4x - 4$ (f) $2x^3 - x^2 + 8x - 4$
   (c) $3x^3 + 8x^2 - x - 10$ (g) $x^4 - 3x^3 + x^2 + 4$
   (d) $x^4 + x^3 - 5x^2 - 15x - 18$ (h) $9x^4 + 12x^3 + 7x^2 - 10x + 2$

10. Factor each of the following polynomials into irreducible polynomials over the *rational numbers*.

   (a) $3x^2 - x - 2$ (e) $x^4 + 2x^3 - 2x^2 + 2x + 1$
   (b) $2x^2 - 4$ (f) $x^4 - 5x^2 + 6$
   (c) $2x^2 - 5x - 12$ (g) $2x^4 - 5x^3 - 4x^2 + 15x - 6$
   (d) $x^2 + 3x - 1$ (h) $x^5 + 3x^4 - 8x^2 - 9x - 3$

11. Prove that if r is a complex number which is a root of multiplicity k of a polynomial $p(x)$ with real coefficients then the conjugate $\bar{r}$ is also a root of multiplicity k of $p(x)$.

12. Prove that if $p(x)$ is a polynomial with rational coefficients, and if $a + \sqrt{b}$, where a and b are rational and $\sqrt{b}$ is irrational, is a root of $p(x)$, then so is $a - \sqrt{b}$.

13. Let $p(x) = a_n x^n + a_{n-1} x^{n-1} + \cdots + a_1 x + a_0$ be a polynomial with degree $n \geq 1$. Define the *formal derviative $p'(x)$* as follows.

$$p'(x) = na_n x^{n-1} + (n-1)a_{n-1} x^{n-2} + (n-2)a_{n-2} x^{n-3} + \cdots + 2a_2 x + a_1$$

Compute $p'(x)$ for each of the following polynomials.

   (a) x (f) $x^2 + 2x + 1$
   (b) x^2 (g) $x^2 - 4x + 4$
   (c) $x - r$ (h) $2x^3 - 3x^2 + 4x + 2$
   (d) $3x^2$ (i) $2x^5 + 3x^2 + 7$
   (e) $2x + 1$ (j) $9x^4 + 12x^3 + 7x^2 - 10x + 2$

14. With $p'(x)$ as above it follows that if $p(x) = (x - r)q(x)$, then $p'(x) = (x - r)g'(x) + g(x)$. Use this fact to show that if r is a double root of $p(x)$ then r is a root of $p'(x)$.

3.8. the rational functions

The second large class of functions which we discuss in this chapter is the set of rational functions. Just as (roughly speaking) the rational numbers are the ratios a/b of two integers, $b \neq 0$, the rational functions are ratios of polynomials.

the rational functions

Let $p(x)$ and $q(x)$ be two polynomial functions. For the purposes of this section, we will assume that they are real polynomials. The function

$$f(x) = \frac{p(x)}{q(x)}$$

is then, a real function whose domain D is all the real numbers t except those for which $q(t) = 0$. For example,

$$f(x) = \frac{x + 3}{x^2 - 4}$$

is a rational function. The domain of this function is the set $R \setminus \{2, -2\}$; i.e., all the real numbers except $+2$ and -2.

If $q(x)$ is a (nonzero) constant polynomial, a, then $p(x)/q(x) = (1/a)p(x)$ is a polynomial. Given any polynomial $p(x)$, it can, therefore, be considered as a rational function of the form

$$\frac{p(x)}{1}$$

Hence, the class $R(x)$ of rational functions contains (properly) the class $R[x]$ of polynomials;

$$R[x] \subset R(x)$$

As is the case with the rational numbers where

$$1/2 = 2/4 = -3/-6 = 4/8 = \text{etc.}$$

we may have a rational function

$$f(x) = \frac{\hat{p}(x)}{\hat{q}(x)} = \frac{p(x)r(x)}{q(x)r(x)}$$

That is, the numerator and denominator polynomials could have a common factor $r(x)$. The situation here is, in contrast, to the rational numbers, slightly more complicated, since it is not entirely correct to say that

$$f(x) = \frac{x^2 - 9}{(x + 1)(x - 3)} \quad \text{and} \quad f^*(x) = \frac{x + 3}{x + 1}$$

are equal. The reason for this is that $f^*(3) = 3/2$, while $f(3)$ does not exist. That is, 3 is not in the domain of f, but 3 does belong to the domain of f^*. For every other real number t (except $t = -1$ which is in the domain of neither function) $f(t) = f^*(t)$.

With this caution in mind we shall, in the rest of this section, ignore this difference and assume that f and f^* are the same function. We assume that every rational function

$$f(x) = \frac{p(x)}{q(x)}$$

is already in *lowest terms*; that is that $p(x)$ and $q(x)$ *have no common factors*.

Any thorough study of rational functions can best be done with the methods of calculus. There are, however, a few interesting properties which

polynomials & rational functions

we can discuss here. We shall be interested in sketching graphs of rational functions. In doing so we shall investigate the behavior of the function f *near* the zero of $q(x)$. We shall also see that many have interesting behavior for $|x|$ very large. It is the "asymptotic behavior" of these functions which makes them useful in describing many physical phenomena.

example 1

Sketch the graph of the rational function.

$$h(x) = \frac{2 - 6x}{x^2 + 2x - 3} = \frac{2 - 6x}{(x - 1)(x + 3)}$$

First note the following about h.

(i) $h(1/3) = 0$, since $2 - 6(1/3) = 0$.
(ii) h is not defined at 1 or at -3.
(iii) For each real t, different from 1 or -3, $h(t)$ is a real number.

In particular, let $N_a(1) = (1 - a, 1 + a)$ be a neighborhood of 1.

Let $N_a^0(1)$ be $N_a(1)\backslash\{1\}$, the neighborhood with 1 deleted. For each $t \in N_a^0(1)$, $h(t)$ exists, provided $a < 2$. If t is in this deleted neighborhood of 1 and is larger than 1, then consider $h(t)$. Consider each factor of the numerator and denominator: $2 - 6t$ is near -4; $t - 1$ is positive and small; $t + 3$ is near 4. Thus $h(t)$ is a fraction with a negative numerator and a *small* positive denominator. If t is *quite near*, 1 the denominator is *very small*. Hence the number $|h(t)|$ is *very large*. For example, if $t = 1 + 10^{-9}$, then $h(t)$ is nearly

$$\frac{-4}{\dfrac{4}{10^9}} = -10^9$$

Thus, $|h(t)|$ is near 10^9.

Similarly, when t is in $N_a^0(1)$ but $t < 1$, $h(t)$ is positive and $|h(t)|$ is *very large*. This sort of behavior is described succinctly by saying that $h(x)$ has a **vertical asymptote** at $x = 1$. See Figure 3.8.1.

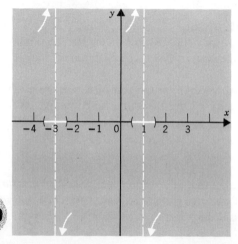

figure 3.8.1

the rational functions

A similar analysis discloses the fact that $x = -3$ is also a vertical asymptote for $h(x)$. In this case when $t < -3$, $h(t)$ is positive, and when $t > -3$ (but $t < 1/3$), $h(t)$ is negative (see Figure 3.8.1).

To complete the sketch of the graph of $h(x)$, we compute the values of $h(t)$ for various values of t and then connect the points $(t, h(t))$ with a smooth curve. Some values of $h(t)$ are given in the following table.

t	-5	-4	$-7/2$	$-5/2$	-2	-1	0	$1/3$	$1/2$	$3/2$	2	3	4
$h(t)$	$\dfrac{8}{3}$	$\dfrac{26}{5}$	$\dfrac{92}{9}$	$\dfrac{-68}{7}$	$\dfrac{-14}{3}$	-2	$\dfrac{-2}{3}$	0	$\dfrac{4}{7}$	$\dfrac{-28}{9}$	-2	$\dfrac{-4}{3}$	$\dfrac{-23}{27}$

The completed sketch is figure 3.8.2. ●

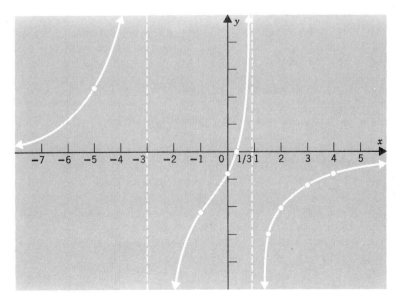

figure 3.8.2

$$h(x) = \frac{2 - 6x}{x^2 + 2x - 3}$$

This example illustrates the following general theorem, the proof of which follows the same analysis given.

theorem 3.8.1

Let $f(x) = p(x)/q(x)$ be a rational function where $p(x)$ and $q(x)$ have no common factors, then $f(x)$ has a vertical asymptote $x = t_0$ at each root t_0 of $q(x)$.

polynomials & rational functions

example 2

Sketch the graph of the rational function

$$k(x) = \frac{x - 1}{(x + 1)(2x - 1)}$$

A (partial) function table for k is given below. We note that according to Theorem 3.8.1, $x = -1$ and $x = 1/2$ are vertical asymptotes for k.

t	-5	-4	-3	-2	$-3/2$	$-1/2$	0	3/4	3/2	1	2	3	4
$k(t)$	$\dfrac{-3}{22}$	$\dfrac{-5}{27}$	$\dfrac{-2}{7}$	$\dfrac{-3}{5}$	$\dfrac{-5}{4}$	$\dfrac{3}{2}$	1	$\dfrac{-2}{7}$	$\dfrac{1}{10}$	0	$\dfrac{1}{9}$	$\dfrac{1}{10}$	$\dfrac{3}{35}$

Plotting the points given in the table and connecting them with a smooth curve gives the graph in Figure 3.8.3. ●

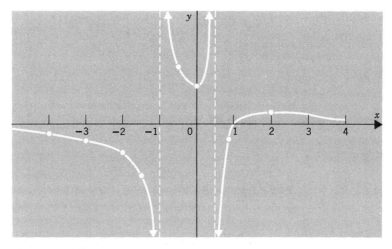

figure 3.8.3

$$k(x) = \frac{x - 1}{(x + 1)(2x - 1)}$$

Observe the additional asymptotic behavior in the two graphs of Examples 1 and 2. As $|t|$ becomes very large, the graphs of these curves approach the x axis. We say, in such cases, that the x axis is a **horizontal asymptote**. In Example 2, note that as t becomes large and positive, symbolized by $t \to \infty$, $k(t) > 0$, but differs from zero by less and less. Similarly, as t becomes large in absolute value but negative, $t \to -\infty$, $k(t) < 0$ but also differs very little from 0. The situation in Example 1 is analogous; as $t \to \infty$, $h(t) \to 0$ from below, and as $t \to -\infty$, $h(t) \to 0$ from above.
It is a theorem that such is the case for every rational function of this type.

the rational functions

theorem 3.8.2

> Let $f(x) = p(x)/q(x)$ be a rational function in lowest terms, with deg $p(x) <$ deg $q(x)$. Then the x axis is a horizontal asymptote for $f(x)$.

example 3

Consider the rational function

$$f(x) = 2 + \frac{2 - 6x}{x^2 + 2x - 3}$$

The vertical asymptotes of $f(x)$ are determined by the proper fractional part of f, as in Example 1. In fact we have, with $h(x)$ as in Example 1,

$$f(x) - h(x) = 2$$

Thus for every real number t in their common domain, $f(t)$ and $h(t)$ differ only by the constant polynomial 2.

While the vertical asymptotes of $f(x)$ are $x = 1$ and $x = -3$, the same as those of $h(x)$, the horizontal asymptote for $f(x)$ is the line $y = 2$. This follows readily from the fact that as $t \to \infty$, $f(t) = 2 + h(t)$ "approaches" $2 + 0$ or 2. Such is also the case when $t \to -\infty$. ●
In general if $f(x)$ is a rational function

$$f(x) = \frac{p(x)}{q(x)}$$

and the degree of $p(x) \geq$ degree $q(x)$, one may always use the divisibility properties of polynomials,

$$p(x) = s(x)q(x) + r(x), \qquad r(x) = 0, \qquad \text{or} \qquad \text{deg } r(x) < \text{deg } q(x)$$

to write $f(x)$ as a mixed fraction

$$f(x) = s(x) + \frac{r(x)}{q(x)}$$

Here $s(x)$ is a polynomial and $(r(x))/(q(x))$ is a "proper fractional function," that is, deg $r(x) <$ deg $q(x)$.

The vertical asymptotes of $f(x)$ are, according to Theorem 3.8.1, the zeros of $q(x)$. The function $f(x)$ will have a horizontal asymptote if and only if, $s(x)$ is a constant polynomial $s(x) = a$. This is because

$$f(x) - \frac{r(x)}{q(x)} = s(x)$$

and as $|t|$ becomes large $(r(x))/(q(x))$ approaches zero, according to Theorem 3.8.2. Therefore $f(x)$ approaches $s(x) = a$.

When $s(x)$ is not constant, it is still true that the difference between $f(x)$ and $s(x)$ becomes small

$$f(x) - s(x) = \frac{r(x)}{q(x)} \to 0$$

as $t \to \infty$ and $t \to -\infty$. Thus $f(x)$ is **asymptotic** to the graph of $s(x)$.

polynomials & rational functions

example 4

Sketch the graph of

$$f(x) = \frac{x^2}{x-1}$$

Since the degree of $x - 1$ is less than the degree of x^2, divide and write

$$f(x) = x + 1 + \frac{1}{x-1}$$

Since $x - 1$ has a zero at 1, the line $x = 1$ is a vertical asymptote for f.
Since $f(x) - (x + 1) = 1/(x - 1)$, and since as $t \to \infty$, $1/(t - 1) \to 0$, and is positive for $x > 1$, the graph of $f(x)$ is asymptotic to, and above, the graph of $x + 1$. As $t \to -\infty$, $1/(t - 1) \to 0$, and is negative. Thus, in this case the graph of f is asymptotic to, and below, the graph of $x + 1$. We sketch this in Figure 3.8.4.

The graphs of the asymptotes $y = x + 1$ and $x = 1$ are the dotted straight lines in the figure. One must analyze and plot points $(t, f(t))$ as in the previous examples to sketch the graph of $f(t)$. ●

t	-5	-4	-3	-2	-1	0	1/2	3/2	2	3	4	5
$f(t)$	$-\dfrac{25}{6}$	$-\dfrac{16}{5}$	$-\dfrac{9}{4}$	$-\dfrac{4}{3}$	$-\dfrac{1}{2}$	0	$-1/2$	9/2	4	9/2	$\dfrac{16}{3}$	$\dfrac{25}{4}$

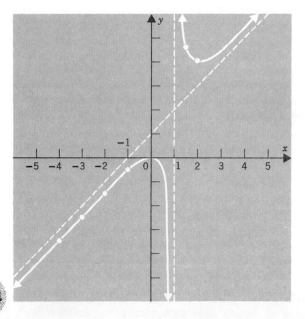

figure 3.8.4

144

the rational functions

We have done little more than sketch the graphs of rational functions in this section. However, the graph of any function reveals many of its properties. Other properties, such as the location of vertices, maximum and minimum values, etc. are studied extensively in most calculus courses.

exercises

1. Find the vertical asymptotes for each of the following rational functions.

(a) $f(x) = \dfrac{1}{x}$

(f) $f(x) = \dfrac{3}{5 - x}$

(b) $f(x) = \dfrac{1}{x + 1}$

(g) $f(x) = \dfrac{x}{x + 1}$

(c) $f(x) = \dfrac{4}{2x - 1}$

(h) $f(x) = \dfrac{x}{(x - 2)^2}$

(d) $f(x) = \dfrac{1}{(x - 1)^2}$

(i) $f(x) = \dfrac{x^2}{2x^2 - 1}$

(e) $f(x) = \dfrac{2}{x^2 - 1}$

(j) $f(x) = \dfrac{x - 2}{x - 1}$

2. Find the horizontal asymptotes for each of the functions in Exercise 1.

3. Sketch the graphs of each of the functions in Exercise 1 by making a partial function table, plotting the points and connecting these points with a smooth curve.

4. Find the asymptotes of each of the following rational functions.

(a) $f(x) = \dfrac{x^2}{(x - 1)^2}$

(f) $f(x) = \dfrac{3x - 2}{x^2 - 3x - 4}$

(b) $f(x) = \dfrac{x^2 - 4}{x}$

(g) $f(x) = \dfrac{x^3 + 1}{x^3 - 6x^2 + 11x - 6}$

(c) $f(x) = \dfrac{2x}{x^2 + x - 2}$

(h) $f(x) = \dfrac{x + 1}{x^2 - x + 1}$

(d) $f(x) = \dfrac{x^2 - 1}{x - 2}$

(i) $f(x) = \dfrac{x^2 - 5x + 4}{x^2 - 8x + 15}$

(e) $f(x) = \dfrac{x^2 + 1}{x^2 - x + 1}$

(j) $f(x) = \dfrac{x^2 - 7x + 6}{x^2 - 8x + 15}$

5. By plotting points $(t, f(t))$ and connecting them with a smooth curve, use the information of Exercise 4 to sketch the graphs of each of the functions in Exercise 4.

6. It is clear that the graph of a rational function does not cross its vertical asymptote. Can it cross any other type of asymptote? Explain.

7. Determine real numbers A, B, and C so that for every real number t different from 1, -1, and 2

$$\frac{t^2 + t - 3}{(t - 1)(t + 1)(t - 3)} = \frac{A}{t - 1} + \frac{B}{t + 1} + \frac{C}{t - 3}$$

145

polynomials & rational functions

quiz for review

PART I.   True or False

In Problems 1–10 identify the degree of the given function if it is polynomial.

      (a) 0   (b) 1   (c) 3   (d) 4   (e) not a polynomial

1. x

2. π

3. $2x^4 - 3$

4. $\dfrac{3}{x}$

5. $x^3 - 3x^2 + 2x + 1$

6. $x^3 - 3x + 1 - x^{-2}$

7. $\sqrt{1 - 2x^3}$

8. $3x + 4$

9. $\dfrac{6x}{x^2 - 1}$

10. $x^4 - 7x^3 + 3x + 2$

In Problems 11–15 consider the following graphs.

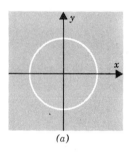

(a)

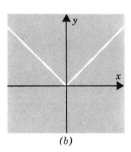

(b)

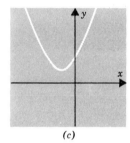

(c)

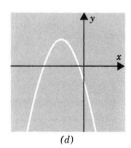

(d)

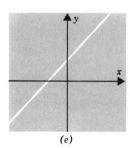

(e)

11. Which of these graphs might be that of the polynomial function $f(t) = at^2 + bt + c$, with $a < 0$?

12. Which of these graphs might be that of the polynomial function $f(t) = at^2 + bt + c$, with $a > 0$?

13. Which graph is the graph of a bijective function?

14. Which of these graphs might be that of the polynomial function $f(t) = mt + b$?

15. Which is not the graph of a function?

In Problems 16–25, the statement is either (a) always true, or (b) false.

16. If f is a polynomial with integer coefficients, and $f(a/b) = 0$, a/b in lowest terms, $a, b \in Z$, $b \neq 0$, then a divides the leading coefficient of f and b divides the constant term of f.

17. The zero function, $f(x) = 0$ for all $x \in R$, gives rise to a polynomial, the zero polynomial.

18. The real numbers may be considered as polynomials of degree zero.

quiz for review

T *19.* Let f be a polynomial with real coefficients and degree n at least 1. Then, if $f(b) = 0$, $f(x) = (x - b)q(x)$.

T *20.* Let f be as in Problem 19 and let k be any real number. Then there exists a real polynomial $s(x)$ such that $f(x) = (x - k)s(x) + f(k)$.

F *21.* The polynomial $x^3 + x + 1$ has no real roots.

F *22.* The discriminant of the quadratic $x^2 - x + 1$ is positive.

T *23.* If $p(x)$ is any complex polynomial, then there exists some complex number r such that $p(r) = 0$.

T *24.* If $p(x)$ is any complex polynomial and $p(r) = 0$, then $p(\bar{r}) = 0$, where $\bar{r}$ is the conjugate of r.

T *25.* The polynomial $x^4 + 2x^2 + 1$ has no positive roots.

26. Let p and q be real polynomials with deg $p = m$ and deg $q = n$. Then deg $(p + q)$
    (a) $= m + n$ (b) $= mn$ (c) $\leq$ max $\{m, n\}$ (d) $\leq$ min $\{m, n\}$
    (e) $\leq m - n$

27. For p and q as in Problem 26, deg $(pq) =$
    (a) $m + n$ (b) mn (c) max $\{m, n\}$ (d) min $\{m, n\}$ (e) $m - n$

28. The function $p(x) = x^3 + 27$
    (a) has no real root   (b) has two real roots   (c) has three real roots
    (d) has no positive roots   (e) has only complex roots

29. Given the polynomial $h(x) = 2x^2 + 3x + 1$, with discriminant Δ. Then
    (a) $\Delta = 0$ (b) $\Delta > 0$ (c) $\Delta < 0$ (d) h has no negative roots
    (e) h has no real roots.

30. Given that f is a polynomial with real coefficients and with i and $1 - i$ as roots. Then f could be
    (a) $x^2 + 1$ (b) $x^2 - x$ (c) $x^4 - 2x^3 + 3x^2 - 2x + 2$
    (d) $x^2 - x + (1 + i)$ (e) $x^2 - 1$

31. The number 5 is a zero of which of the following polynomials?
    (a) $x^2 - 5x + 1$ (b) $3x^3 - 2x + 5$ (c) $x^2 + x + 5$ (d) $x^2 - 5$
    (e) $x^3 - 3x^2 - 9x - 5$

32. The x axis is a horizontal asymptote for the graph of the rational function $f = p/q$
    (a) provided $p(0) \neq 0$ (b) provided $p(0) = 0$ (c) provided deg $p \geq$ deg q
    (d) provided $p = qs + r$ and deg $r > 0$ (e) provided deg $p <$ deg q.

33. The rational function $f(x) = (x^2 - 6x + 9)/(x^2 + 6x + 9)$ has
    (a) a vertical asymptote $x = -3$ and a horizontal asymptote $y = 1$ (b) vertical asymptote $x = 3$ (c) horizontal asymptote $y = 3$ (d) vertical asymptote $x = 0$ (e) no horizontal asymptotes

34. The number 2 is a zero of which of the following polynomials?
    (a) $x^2 + 3x + 2$ (b) $2x^3 - 2x - 1$ (c) $x^2 - x - 2$ (d) $x^2 - 2$
    (e) $x^3 - 2x^2 + x + 2$

35. Which of the following is *not* a rational function?
    (a) 3   (b) x (c) $x^{1/2}$ (d) $(x^3 - 1)/(x^3 + 1)$ (e) $\sqrt{3}/\pi x^2$

PART II

1. Find the roots of the polynomial $x^5 - 2x^4 + x^3 + 2x^2 - 2x$ by first finding the rational roots. Use synthetic division.

2. Given that $-1/2 + i(\sqrt{3}/2)$ is a root of $x^6 + x^5 + 2x^4 + x^3 + 2x^2 + x + 1$ of multiplicity 2. Find all the roots and their multiplicities.

polynomials & rational functions

3. Write a polynomial with real coefficients having -2 as a root of multiplicity 3 and $-2i$ a root of multiplicity 2.

4. Given the quadratic $ax^2 + bx + c$. Derive the *quadratic formula*

$$x = \frac{-b \pm \sqrt{b^2 - 4ac}}{2a}$$

5. Show that $[(\sqrt{2/2} + i(\sqrt{2/2})]^2 = i$. Use this fact to find four fourth roots of -1.

exponential & logarithmic functions

4

Just as is the concept of *number*, the function concept is an abstraction of a phenomenon from the real world. To be an effective scholar, one must have a fairly extensive catalog of functions available to him. In the first chapter, in addition to introducing the general concept, several special functions were described. In the chapter which just preceded this one, two large classes of functions were examined: The polynomial and the rational functions. In each case, the properties of certain individual members of these classes were examined in some detail.

In this chapter, two additional classes of functions will be described. As their properties are examined, you should bear in mind that it is these particular properties which make it possible for one to describe many natural phenomena in mathematical terms. The functions in these two classes are in some ways more mathematically sophisticated than were the polynomials and rational functions, although in other respects they have somewhat nicer behavior.

4.1. exponents, rational & irrational

Before actually studying the exponential and the logarithmic functions, it is necessary to make a small detour. In this section, we lay the necessary groundwork upon which these functions are erected. You will recall that if $b \neq 0$ is a real number and n is a positive integer, then the symbol

$$b^n$$

exponential & logarithmic functions

indicates the product of n factors of b. Thus, if b is 3 and n is 5, 3^5 means

$$(3)(3)(3)(3)(3)$$

A mathematically precise (inductive) definition would be as follows.

$$b^1 = b$$

and for $n > 1$,

$$b^n = b(b^{n-1})$$

For example,

$$3^5 = 3 \cdot 3^4 = 3(3 \cdot 3^3) = (3)(3)(3 \cdot 3^2) = (3)(3)(3)(3 \cdot 3^1) = (3)(3)(3)(3)(3)$$

On the basis of this definition, it is quite easy to prove (using induction) the following **laws of exponents**.

For $a, b \in R$, $a, b \neq 0$, and $m, n \in Z$, $m > 1$, $n > 1$

(4.1.1) $$b^n \cdot b^m = b^{m+n}$$

(4.1.2) $$(b^m)^n = b^{mn}$$

(4.1.3) $$\frac{b^m}{b^n} = \begin{cases} b^{m-n}, & \text{if } m > n \\ 1, & \text{if } m = n \\ \dfrac{1}{b^{n-m}}, & \text{if } n > m \end{cases}$$

(4.1.4) $$(ab)^m = a^m b^m$$

(4.1.5) $$\left(\frac{a}{b}\right)^m = \frac{a^m}{b^m}$$

These laws of exponents are very useful in calculations. In fact, they form the basis for the algorithms of ordinary arithmetic.

example 1

Consider the following applications of these laws of exponents.

(a) $(a^2)(a^3) = a^{2+3} = a^5$

(b) $(a^3)(a^4) = a^{3+4} = a^7$

(c) $(16)(32) = (2^4)(2^5) = 2^{4+5} = 2^9$

(d) $(20)(300) = (2)(10)(3)(10^2) = (2)(3)(10)(10^2) = (6)(10^3) = 6000$

(e) $\dfrac{a^4}{a^2} = a^{4-2} = a^2$

(f) $\dfrac{a^3}{a^7} = \dfrac{1}{a^4}$

(g) $\dfrac{16}{32} = \dfrac{2^4}{2^5} = \dfrac{1}{2}$

(h) $\dfrac{300}{20} = \dfrac{(3)(10^2)}{(2)(10)} = \dfrac{(3)(10)}{(2)} = 15$

(i) $(30^3) = [(3)(10)]^3 = (3^3)(10^3) = (27)(1000) = 27000$ ●

exponents, rational & irrational

There is something in the nature of man that drives him to further generalization. If the laws of exponents work so well, can they not be extended to handle cases where the exponent is something other than a positive integer? The answer to this question is the major content of this section.

Notice first that a superficial statement of (4.1.3) might read,

$$\frac{b^n}{b^m} = b^{n-m}$$

Then you might expect such statements as

$$\frac{2^3}{2^5} = 2^{3-5} = 2^{-2} \quad \text{and} \quad \frac{5^3}{5^3} = 5^{3 \cdot 3} = 5^0$$

On the other hand, the appropriate statements are

$$\frac{2^3}{2^5} = \frac{1}{2^2} \quad \text{and} \quad \frac{5^3}{5^3} = 1$$

The definition of the symbol b^n when n is a *negative* integer and when $n = 0$ is motivated by these results. If 5^0 is to equal 1, we are forced to *define*, for $b \neq 0$

(4.1.6) $$b^0 = 1$$

Similarly, if 2^{-2} ought to be $1/2^2$, we should define, for $b \neq 0$

(4.1.7) $$b^{-m} = \frac{1}{b^m} \qquad m \in Z$$

The statement (4.1.7) is stronger than we advertised, since m is *any* integer.

example 2

Simplify each of the following expressions. Use the definitions and the laws of exponents.

(a) $\dfrac{a^5}{a^5} = a^{5-5} = a^0 = 1$

(b) $a^{-3} = \dfrac{1}{a^3}$

(c) $\dfrac{5x^2}{25x} = \dfrac{1}{5}x^{2-1} = \dfrac{x}{5}$

(d) $\dfrac{a^3 b^2}{a^4 b} = a^3 a^{-4} b^2 b^{-1} = a^{-1} b^1 = \dfrac{b}{a}$

(e) $\dfrac{2^3 c^2}{2c^4} = 2^3 2^{-1} c^2 c^{-4} = 2^2 c^{-2} = \dfrac{4}{c^2}$

(f) $\dfrac{1}{b^{-6}} = b^{-(-6)} = b^6$ ●

Having defined b^n for *each integer* n, positive, negative, or zero, we push forward to ask about giving a meaning to the symbol

$$b^r$$

exponential & logarithmic functions

where r is a rational number, $r = m/n$, m, $n \in Z$, $n \neq 0$. Again, we desire to maintain the laws of exponents—at least so far as is possible. There, however, we have a problem. Consider the following demonstration that $2 = -2$.

$$2 = (\sqrt{2})(\sqrt{2}) = \sqrt{4} = \sqrt{(-2)^2} = (\sqrt{-2})(\sqrt{-2}) = (i\sqrt{2})(i\sqrt{2}) =$$
$$i^2(\sqrt{2})(\sqrt{2}) = -1(\sqrt{2})(\sqrt{2}) = -2$$

This is an absurd conclusion which we do not want to accept. There is an error somewhere. It occurs at the fifth stage

$$\sqrt{(-2)^2} \neq \sqrt{-2}\sqrt{-2}$$

The rule for multiplying square roots

$$\sqrt{a}\sqrt{b} = \sqrt{ab}$$

is *not true* when both a and b are negative.

With the above caution in mind, let us proceed to define, for $b > 0$ and n a positive integer, the symbol $b^{1/n}$. *Define*

(4.1.8)
$$b^{1/n} = \sqrt[n]{b}$$

This definition is chosen because the law of exponents (4.1.2) would then yield the result $(b^{1/n})^n = b^{(n)(1/n)} = b^1 = b$, which is, of course, the meaning of $\sqrt[n]{b}$. Again a cautionary note: $(-2)^2 = 4$ as well as $(2)^2 = 4$. What then is $4^{1/2}$? Is it 2, -2, or both? *The symbol* $\sqrt{4}$ *means the unique positive real number whose square is 4*; thus, 2. To obtain the negative square root -2 of 4 it is necessary to write $-\sqrt{4}$. When *both* roots of 4 are desired the symbol $\pm\sqrt{4}$ is needed. Therefore, $4^{1/2} = 2$. Any less rigid regulation leads to the absurd conclusion that $-2 = 2$ as we stated earlier. In Exercise 10(a) you are asked to show, among other things, that for b a *positive* real number there is, in fact, a unique *positive* square root of b, symbolized by $\sqrt{b}$.

When n is an *odd* positive integer equation (4.1.8) can be used to define $b^{1/n}$ for *negative* real numbers b; e.g.,

$$(-8)^{1/3} = -2$$

since $(-2)^3 = -8$. Bear this possible relaxation in mind as we proceed with some additional definitions stated for positive b only.

Let $b > 0$ be a real number and let $r = m/n$ be a rational number, m, $n \in Z$, $n > 0$. Define

$$b^{m/n} = (\sqrt[n]{b})^m$$

It can be verified, then, that the following laws of exponents hold.

theorem 4.1.1

Let a, b be positive real number. Let r, s be rational numbers. Then

(i) $(b^r)(b^s) = b^{r+s}$

(ii) $(b^r)^s = b^{rs}$

(iii) $\dfrac{b^r}{b^s} = b^{r-s}$

exponents, rational & irrational

$$\text{(iv)} \quad (ab)^r = a^r b^r$$

$$\text{(v)} \quad \left(\frac{a}{b}\right)^r = \frac{a^r}{b^r}$$

proof. Exercise 13.

example 3

Consider the following statements.

(a) $9^{1/2} = 3$

(b) $27^{1/3} = 3$

(c) $81^{3/4} = [81^{1/4}]^3 = 3^3 = 27$

(d) $625^{-3/4} = (625^{1/4})^{-3} = \dfrac{1}{5^3} = \dfrac{1}{125}$

(e) $x^{1/2}x^{3/2} = x^{1/2+3/2} = x^2$

(f) $\dfrac{(u^{-1}v^2)^3}{u^{1/2}v^{1/2}} = \dfrac{u^{-3}v^6}{u^{1/2}v^{1/2}} = u^{-3-1/2}v^{6-1/2} = \dfrac{v^{11/2}}{u^{7/2}}$ ●

Each of these applies the laws of exponents given by the theorem. You should decide which ones are used. The final expression in (f) of Example 3 may also be written in the following forms.

$$\frac{v^{11/2}}{u^{7/2}} = \left(\frac{v^{11}}{u^7}\right)^{1/2} = \sqrt{\frac{v^{11}}{u^7}} \cdots$$

Which of these forms is to be used is largely a matter of personal taste or computational expediency.

Finally, we wish to extend the results of Theorem 4.1.1 to include the case where r and s are any real numbers. It is here that a certain mathematical sophistication is required. We shall not give a formally precise definition, but shall indicate the ideas involved. To do this we indicate how one meaningfully defines 3^π.

To accomplish this we need the results contained in Theorem 4.1.2. These facts are about what would be expected and have some interest on their own.

theorem 4.1.2

Let $b > 0$ be a real number and let r and s be two positive rational numbers with $r < s$. Then

(i)   *if $b > 1$, $b^r < b^s$*
(ii) *if $b = 1$, $b^r = b^s$*
(iii) *if $0 < b < 1$, $b^r > b^s$*

The proof of this theorem involves considering a number of separate cases. Because it is quite involved we omit it. However, Exercises 9—12 outline the steps.

To define 3^π, first notice that π, being irrational, can be approximated by rational numbers to any desired accuracy.

$$\pi = 3.14159 \ldots$$

exponential & logarithmic functions

Therefore,

$$3 < \pi < 4$$
$$3.1 < \pi < 3.2$$
$$3.14 < \pi < 3.15$$
$$3.141 < \pi < 3.142$$
$$3.1415 < \pi < 3.1416$$

$$\vdots \qquad \qquad \vdots$$

Apply Theorem 4.1.2 using the numbers less than π in the display above for r and those larger than π for s. Hence,

$$27 = 3^3 < 3^4 = 81$$
$$30.13 \doteq 3^{3.1} \quad < 3^{3.2} \quad \doteq 33.63$$
$$31.49 \doteq 3^{3.14} \quad < 3^{3.15} \doteq 31.84$$
$$31.52 \doteq 3^{31.41} < 3^{3.141} \doteq 31.56$$

$$\vdots \qquad \qquad \vdots$$

The symbol $\doteq$ indicates "approximately equal to." The display given here indicates how two sequences of real numbers are generated. One is *increasing*

(I) $\qquad\qquad 3^3, \ 3^{3.1}, \ 3^{3.14}, \ 3^{3.141}, \ 3^{3.1415}, \ldots$

the other *decreasing*

(II) $\qquad\qquad 3^4, \ 3^{3.2}, \ 3^{3.15}, \ 3^{3.142}, \ 3^{3.1416}, \ldots$

Each term of the second sequence is, however, larger than the corresponding term of the first sequence. The differences between corresponding terms

$$3^4 - 3^3, \ 3^{3.2} - 3^{3.1}, \ 3^{3.15} - 3^{3.14}, \ 3^{3.142} - 3^{3.141}, \ 3^{3.1416} - 3^{3.1415}, \ldots$$

is also a decreasing sequence which approaches zero. By the *completeness* of the real numbers (see the Appendix) there exists a unique real number k which each of sequences (I) and (II) approaches. Thus,

$$3^3 < k < 3^4$$
$$3^{3.1} < k < 3^{3.2}$$
$$3^{3.14} < k < 3^{3.15}$$
$$3^{3.141} < k < 3^{3.142}$$
$$3^{3.1415} < k < 3^{3.1416}$$

$$\vdots \qquad \qquad \vdots$$

This unique real number $k = 31.5 \ldots$ is the number we must assign to 3^{π} to preserve the laws of exponents and the results of Theorem 4.1.2, when r and s are arbitrary real numbers.

In general then, to define b^r for any positive *real* number r and $b > 1$, select any increasing sequence $r_1, r_2, \ldots, r_n, \ldots$ of positive *rational* numbers and any decreasing sequence $s_1, s_2, \ldots, s_n, \ldots$ of positive rational numbers satisfying

$$r_n < r < s_n$$

exponents, rational & irrational

for all $n = 1, 2, \ldots$. Further, select these sequences such that the difference $s_n - r_n$ approaches zero as n becomes large. Then it follows *that the unique real number k determined by these sequences satisfies*

$$b^{r_n} < k < b^{s_n}, \text{ for all } n = 1, 2, \ldots,$$

This k is assigned as the value of b^r.

If r is a negative real number and $b > 1$ set

$$b^r = \frac{1}{b^{-r}}$$

since $-r$ is positive. If b is 1, then $1^r = 1$ for every real number r. If $0 < b < 1$, then $1/b > 1$, so $(1/b)^r$ has been defined for every real number r. Therefore, define

$$b^r = \frac{1}{(1/b)^r}$$

We have thus defined b^r for $b > 0$ and r *any real number.* This definition was made to be compatible with the laws of exponents originally discovered for b^n, when n is a positive integer. In this process of generalization, we have been required to be more restrictive of b then at the outset, yet the laws of exponents are preserved for a positive real number b, called the *base*, and any real number *power* of b. These results are condensed as the following theorem.

theorem 4.1.3

For real numbers a, b, r, and s with a > 0 and b > 0, it is true that

(i) $(b^r)(b^s) = b^{r+s}$

(ii) $(b^r)^s = b^{rs}$

(iii) $\dfrac{b^r}{b^s} = b^{r-s}$

(iv) $(ab)^r = a^r b^r$

(v) $\left(\dfrac{a}{b}\right)^r = \dfrac{a^r}{b^r}$

Furthermore, for r < s, then

(vi) *if* $b > 1$, $b^r < b^s$

(vii) *if* $b = 1$, $b^r = b^s = 1$

(viii) *if* $0 < b < 1$, $b^r > b^s$

The proof is quite involved. It traces in general terms the procedure we used to define 3^π. The machinery necessary to give a good proof is not part of our equipment at present.

example 4

Use the laws of exponents to simplify each of the following.

(a) $\sqrt[6]{27a^3} = (3^3 a^3)^{1/6} = (3a)^{3/6} = (3a)^{1/2} = \sqrt{3a}$

(b) $(3\sqrt[3]{2})^4 = [(3)(2^{1/3})]^4 = 3^4\, 2^{4/3} = (81)(2)(2^{1/3}) = 162\sqrt[3]{2}$

155

exponential & logarithmic functions

(c) $(5^{\sqrt{2}})^{2\sqrt{2}} = 5^{2\cdot 2} = 5^4 = 625$

(d) $\left[\left(\dfrac{1}{2}\right)^{0.5}\right]^{-2} = \left(\dfrac{1}{2}\right)^{-2(0.5)} = \left(\dfrac{1}{2}\right)^{-1} = 2$

(e) $\dfrac{\sqrt[3]{8xy^2}}{\sqrt[4]{16x^2y^2}} = \dfrac{8^{1/3}x^{1/3}y^{2/3}}{16^{1/4}x^{2/4}y^{2/4}} = \dfrac{2x^{1/3}y^{2/3}}{2x^{1/2}y^{1/2}} = x^{-1/6}y^{1/6} = \sqrt[6]{\dfrac{x}{y}}$ ●

exercises

1. Simplify the following expressions. Leave the answer in exponential form.

(a) $(3^4)(3^{-3})(3^2)$

(b) $(2^2)(2^5)(2^{-6})$

(c) $(5^{-1})(5^3)(5^{-2})$

(d) $(3^3)(2^2)(6^3)$
(e) $(2^4)(5^{-2})(10^3)$
(f) $(7^{-3})(14^2)(2^4)$

(g) $\left(\dfrac{2}{3}\right)^4 (2)^{-2}(3)^3$

(h) $\left(\dfrac{3}{5}\right)^{-2}\left(\dfrac{5}{3}\right)^2 (3)^3$

(i) $\left[\left(\dfrac{5}{2}\right)^2\left(\dfrac{2}{5}\right)^{-3}(3)^3\right]^2$

(j) $[2^2 - 1][2^{-2} - 1][2^2 + 1]$
(k) $[(2)(10) + 1][3(10^2) + 5(10) + 2]$
(l) $[(12^4)(2^{-3})(8)^{2/3}(2^2)]^3$

2. Evaluate each of the following.

(a) $16^{1/2}$
(b) $25^{1/2}$
(c) $32^{1/5}$
(d) $81^{-1/4}$
(e) $144^{-1/2}$

(f) $8^{2/3}$
(g) $125^{-2/3}$
(h) $16^{-3/4}$
(i) $(2.25)^{3/2}$
(j) $(.09)^{1/2}$

3. Find, by inspection, a real number r which satisfies each of the following

(a) $2^r = 128$

(b) $2^r = \dfrac{1}{32}$

(c) $3^r = \dfrac{1}{81}$

(d) $10^r = 1000$

(e) $10^r = .01$
(f) $10^r = 1$

(g) $16^r = 4$

(h) $125^r = 5$

(i) $\left(\dfrac{1}{16}\right)^r = 2$

(j) $3^r = 9\sqrt{3}$

(k) $32^r = 4\sqrt{2}$
(l) $125^r = 1$

4. Simplify the following expressions. For appropriate real numbers a, b, etc.

(a) $(a^{1/2})(a^{2/3})$
(b) $(b^{3/5})(b^{1/2})$

(c) $(u^{2/3})^6$

(d) $(v^{1/3})(v^{3/4})$
(e) $(27^{2/3})(2^3)$

(f) $\dfrac{2a}{a^{1/3}}$

5. Simplify each of the following, where a, b, etc., are appropriate real numbers. Use only positive exponents to express the answers.

(a) $(a^2a^{-3}a^5)^{1/2}$

(b) $(b^4b^{-1}b^{-3}b^6)^{1/3}$

exponents, rational & irrational

(c) $\dfrac{a^{-3}b^{2/3}}{a^{-4}b^{-2}}$

(h) $\dfrac{1}{(a^4a^{-2}a^{-3/2})^{-3}}$

(d) $\dfrac{(a^{-1}b^2)^3}{a^{1/2}b^{1/2}}$

(i) $(a + a^{-1})(a - a^{-1})$

(e) $\dfrac{b^5}{b^{2/3}}$

(j) $(a + a^{-1})(a^{-1} - a^{-2})$

(f) $\left(\dfrac{ab^2}{3c^{1/2}}\right)^3$

(k) $(1 + b)(1 + b^{-1} + b^{-2})$

(g) $\dfrac{a^3a^{-1/2}a^{3/2}}{a^2a^{-3}}$

(l) $(1 + b^{1/2} + b)(1 - b^{1/2} + b)$

6. Simplify the following. Leave the answer in exponential form, using only positive exponents. Here a, b, etc., are positive real numbers.

(a) $\sqrt[12]{a^3}$

(g) $(\sqrt{2} + 1)(\sqrt{2} - 1)$

(b) $\sqrt[8]{b^2}$

(h) $(2 + \sqrt{3})(2 - \sqrt{3})$

(c) $\sqrt[6]{a^3}$

(i) $(\sqrt{2b})^5$

(d) $\sqrt[6]{8}$

(j) $\sqrt[3]{\sqrt[4]{8a^3}}$

(e) $\sqrt[4]{9a^2}$

(k) $\dfrac{\sqrt[3]{9a^2b}}{\sqrt{2b}}$

(f) $\sqrt[6]{27b^3}$

(l) $\dfrac{\sqrt{3ab}}{\sqrt[3]{4a^2b^2}}$

7. Simplify the following, assuming that a, b, etc. are positive real numbers. Express the answer using only positive exponents.

(a) $\sqrt{3\sqrt{3}}$

(g) $\dfrac{a^{-1} + b^{-2}}{a^{-2} + b^{-1}}$

(b) $\sqrt{2\sqrt{2\sqrt{2}}}$

(h) $\dfrac{a^{-2} + b^{-1}}{a^{-2} + b^{-2}}$

(c) $\sqrt[3]{\sqrt{a}}$

(i) $\dfrac{a^{-3}b^{-3}}{a^{-3} + b^{-3}}$

(d) $\sqrt{\dfrac{a}{\sqrt[3]{a}}}$

(j) $\dfrac{(a^2 - b^2)^{-1}}{(a + b)^{-3}}$

(e) $(\sqrt[4]{ab^3})(\sqrt[5]{a^5b})(\sqrt[3]{a^2b^2})^{-1}$

(k) $\dfrac{2a^{-1} + 3b^{-1}}{4a^{-2} - 9b^{-2}}$

(f) $\left(\sqrt[3]{b^{9/2}}\sqrt{b^{-3}}\right)\left(\sqrt{(\sqrt[3]{b^{-7}})(\sqrt[3]{b})}\right)^{-1}$

(l) $\dfrac{a^{-1} - 5b^{-1}}{a^{-2} - 25b^{-2}}$

8. Show that for each real number r and $b > 0$ and $h \neq 0$

$$\frac{b^{r+h} - b^r}{h} = b^r\left(\frac{b^h - 1}{h}\right)$$

9.

(a) Let b be a real number and n a positive integer. Show that

$$b^n - 1 = (b - 1)(b^{n-1} + b^{n-2} + \cdots + b + 1)$$

(b) Use part (a) to prove that if (i) $b = 1$ $b^n = 1$, (ii) if $b > 1$ then $b^n > 1$ and (iii) if $0 < b < 1$, then $0 < b^n < 1$.

Hint: The second factor in part (a) is always positive.

10.

(a) Use the methods of Chapter 3 to show that if $b > 0$ and n is a positive integer then the polynomial $x^n - b$ has a unique positive root. This real number is denoted by $b^{1/n}$.

(b) Use the results of Exercise 9 to show that if r is a positive real number $r \leq 1$, and $b > 1$, then $r^n \leq 1$ and

$$f(r) = r^n - b \leq 1 - b < 0$$

(c) Use the results of parts (a) and (b) to prove that if $b > 1$ then $b^{1/n} > 1$.

(d) If $0 < b < 1$ and $r \geq 1$ is any real number, show that, since $r^n \geq 1$ by Exercise 9

$$f(r) = r^n - b \geq 1 - b > 0$$

(e) Use part (d) to establish that $0 < b^{1/n} < 1$ when $0 < b < 1$.

11. Let m and n be positive integers. Use Exercise 9 and 10 to show that if $r = m/n$ is a positive rational number and b is any positive real number, then the following are true.

(a) If $b > 1$, then $b^r > 1$

(b) If $b = 1$, then $b^r = 1$

(c) If $0 < b < 1$, then $b^r < 1$

12. Complete the proof of Theorem 4.1.2 by using the results of Exercise 11. Consider the ratio

$$\frac{b^s}{b^r} = b^{s-r}$$

13. Use the laws of exponents for the integers to prove Theorem 4.1.1.

14. If a and b are positive real numbers show that $0 \leq a \leq b$ if and only if $\sqrt{a} \leq \sqrt{b}$.

15. If a and b are real numbers, is it true that $\sqrt[3]{a} < \sqrt[3]{b}$ if, and only if, $a < b$?

16. Entertain yourself by using mathematical induction to prove the laws of exponents (Equations (4.1.1)–(4.1.5)) when the exponents are positive integers.

4.2.   the exponential functions

The exponential functions are descriptive of much of nature. Such ideas as growth, decay, normal distribution, etc., will be seen to be described by such functions. We shall discuss the properties of exponential functions in this section and devote Section 4.5 to application of them.

Let $b > 0$ be a real number. For each such *base b define the real function*

$$\exp_b: R \to R$$

by the rule

$$\exp_b(x) = b^x$$

for each $x \in R$. That this is indeed a function is the result of the discussion of real exponents in Section 4.1. This function is called *the* **exponential function** *of base b.*

158   Before discussing the properties of $\exp_b$, it is meaningful to know that these functions form a class separate from the class of polynomial function.

the exponential functions

(Otherwise there would not be much point in studying them.) If $b = 1$, then $1^x = 1$ is a constant polynomial. However, if $b \neq 1$, then $\exp_b$ is not constant since

$$\exp_b(0) = b^0 = 1 \quad \text{and} \quad \exp_b(1) = b^1 = b \neq 1$$

Suppose that $\exp_b$ were a nonconstant polynomial $p(x)$. Then the degree of $p(x)$ is $n \geq 1$. We have, for each $x \in R$,

$$p(x) = \exp_b x = b^x$$

Now, square both sides. This results in

$$p^2(x) = b^{2x} = \exp_b 2x = p(2x)$$

Hence, for every real number x, $p^2(x) = p(2x)$. But $p^2(x)$ is a polynomial of degree $2n$, while $p(2x)$ has degree n. This implies that $2n = n$ or $n = 0$ contrary to the fact that $\exp_b x$ is not constant. This contradiction proves that $\exp_b$ is *not a polynomial* when $b \neq 1$. We shall, in the rest of this chapter, exclude the function $\exp_1$ from consideration.

As an immediate consequence of Theorem 4.1.3 some interesting properties of $\exp_b x$ are obtained.

theorem 4.2.1

Let b be a positive real number different from 1. Then

> (i) *If $b > 1$, $\exp_b x$ is an increasing function*
> (ii) *If $b < 1$, $\exp_b x$ is a decreasing function*

example 1

Sketch the graphs of $y = \exp_2 x$ and of $y = \exp_{1/2} x$. Notice first that $\exp_2(0) = 2^0 = 1$, and $\exp_{1/2}(0) = (1/2)^0 = 1$. Thus, both curves pass through the point $(0,1)$. Also note that, for every real number r,

$$\exp_{1/2}(-r) = (1/2)^{-r} = 2^r = \exp_2(r)$$

so there is a certain symmetry between the two graphs. Construct the following partial function tables, plot the points as in Figure 4.2.1, and then connect them with smooth curves to obtain the desired graphs. ●

x	0	1	-1	2	-2	3	-3	4	-4
2^x	1	2	1/2	4	1/4	8	1/8	16	1/16
$(1/2)^x$	1	1/2	2	1/4	4	1/8	8	1/16	16

As can be seen in this example, and as an immediate consequence of the fact that strictly monotonic functions are injective, we have the following corollary to theorem 4.2.1.

exponential & logarithmic functions

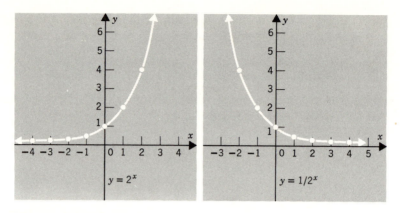

$y = 2^x$

$y = 1/2^x$

figure 4.2.1

Exponential functions ●

theorem 4.2.2

If $b \neq 1$ is a positive real number, the function

$$\exp_b \colon R \to R$$

is one-to-one (injective). Furthermore, if the range of $\exp_b$ is taken as the non-negative real numbers R^+, the function

$$\exp_b \colon R \to R^+$$

is a bijection.

This theorem will take on added importance in the next several sections.

example 2

Suppose $\exp_3 t = 1/81$. Find t. But $1/81 = 3^{-4}$. Therefore,

$$\exp_3 t = \exp_3(-4)$$

Since the exponential functions are all one-to-one, it must follow that $t = -4$. ●

The asymptotic behavior of the exponential functions is also worth noting. First of all, if $b > 1$ it is quite apparent, from Theorem 4.1.3, that b^r approaches ∞ as r approaches ∞ and that b^r approaches zero as r approaches $-\infty$.

If $0 < b < 1$, then $a = b^{-1} > 1$. Therefore, $a^r = b^{-r}$ approaches ∞ as r does and approaches 0 as r approaches $-\infty$. Hence,

$$b^r = \left(\frac{1}{a}\right)^r$$

approaches 0 as r approaches ∞ and approaches ∞ as r approaches $-\infty$. We have proved the following theorem.

theorem 4.2.3

The x axis is a horizontal asymptote for the graph of the exponential function $y = \exp_b(x)$, $b \neq 1$ a positive real number. In particular, (i) if

the exponential functions

$b > 1$, $\exp_b(x)$ *approaches zero as x approaches* $-\infty$, *and* (ii) *if* $0 < b < 1$, $\exp_b(x)$ *approaches zero as x approaches* ∞.

Although the *exponential function* is defined by

$$\exp_b(x) = b^x$$

if A is a nonzero real number the function $A \cdot \exp_b$ is also frequently referred to as *an exponential function*.

$$A \exp_b(x) = Ab^x$$

example 3

Sketch the graph of $y = (1/2)3^x$.

To sketch this graph we determine the values in the following table, plot the points (x, y) and connect them with a smooth curve as in Figure 4.2.2. ●

x	0	-1	-1	2	-2	3	4
3^x	1	3	1/3	9	1/9	27	81
$y = \frac{1}{2}3^x$	1/2	3/2	1/6	9/2	1/18	27/2	81/2

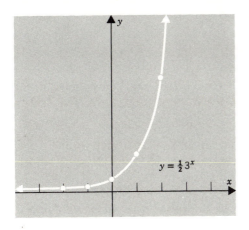

figure 4.2.2

$$y = \frac{1}{2}3^x$$

Up to this point in our discussion, we have not singled out any particular number b as the base of an exponential function preferred above others. The use of $\exp_2(x)$ and $\exp_3(x)$ in the examples was purely for convenience. However, there is an irrational number

$$e = 2.718281828459 \ldots$$

such that $\exp_e(x)$ has particularly nice properties. Some of these will only

exponential & logarithmic functions

become apparent in applications and later study. Nevertheless, just as the irrational number π arises in nature, so does this number e which is called *the natural base*.

The number e may be defined as that real number approached by the function

$$f(x) = (1 + x)^{1/x}$$

as x approaches zero, i.e., as x is in a deleted neighborhood $N_a^o(0)$ of 0. The choice of the letter e to represent this number is, no doubt in honor of the mathematician Leonard Euler who in 1748 published a comprehensive discussion of it in his *Introductio in Analysin Infinitorum*.

The number e can also be arrived at, and approximated rationally, by considering the sequence of rational numbers

$$a_1, a_2, a_3, \ldots$$

where, for each n

$$a_n = 1 + \frac{1}{1!} + \frac{1}{2!} + \frac{1}{3!} + \cdots + \frac{1}{n!}$$

This sequence is sketched in Figure 4.2.3.

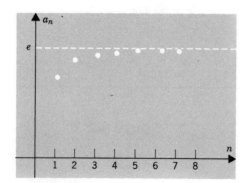

figure 4.2.3

Because e is the preferred *natural base* it is conventional to omit it and write $\exp_e(x) = e^x$ simply as

$$\exp(x)$$

In the Exercises (1) you are asked to sketch graphs for $y = e^x$, $y = e^{-x}$, etc. You will be helped in these sketches by Table 2 in the Appendix. This table gives four-place approximations to the values of $\exp(t) = e^t$ and of $\exp(t) = e^{-t} = 1/e^t$ for values of the real number t between 0 and 4 at intervals of length 0.05, and from 4 to 10 at various other intervals. From it we find, for example, that

$$\exp(3.7) = 40.447$$

and

$$\exp(-2.25) = 0.1054$$

Any exponential function $\exp_b(x)$ can be expressed in terms of the natural exponential

$$\exp(x)$$

the exponential functions

Since $\exp(x) = e^x$ is a bijection from R to R^+, for any positive real number b there exists a real number c such that $b = e^c = \exp(c)$. Thus,

$$\exp_b(x) = b^x = (e^c)^x = e^{cx} = \exp(cx)$$

example 4

Use Table 2 to find $2^{1.10}$. From the body of Table 2 we see that $2 \doteq \exp(0.07)$. Therefore,

$$\exp_2(1.10) = \exp[(0.7)(1.10)] = \exp(.77) \doteq 2.2 \quad \bullet$$

example 5

The real function

$$p(x) = \frac{1}{\sqrt{2\pi}} \exp\left(-\frac{x^2}{2}\right)$$

is used in statistics as the description of the *normal distribution*. Sketch its graph.

By carefully plotting points using Table 2 you can verify that this curve has the well-known bell shape shown in Figure 4.2.4. $\quad\bullet$

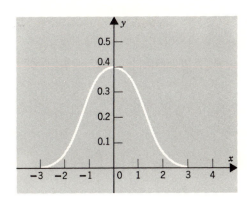

figure 4.2.4

We conclude this section with a theorem which is nothing more than a restatement of the laws of exponents (Theorem 4.1.4) in terms of the exponential functions $\exp_b$. The properties are interesting, however. The proof is left as an exercise.

theorem 4.2.4

Let $b \neq 1$ be a positive real number. For all real numbers x and y,

(i) $\quad \exp_b(x + y) = \exp_b(x) \times \exp_b(y)$

(ii) $\quad \exp_b(x - y) = \dfrac{\exp_b(x)}{\exp_b(y)}$

(iii) $\quad \exp_b(xy) = [\exp_b(x)]^y = [\exp_b(y)]^x$

(iv) $\quad \dfrac{1}{\exp_b(x)} = \exp_b(-x)$

exponential & logarithmic functions

exercises

1. Sketch graphs for each of the following.

   (a) $y = \exp_3(x)$

   (b) $y = \exp_{1/3}(x)$

   (c) $y = \exp_{0.1}(x)$

   (d) $y = e^x$

   (e) $y = e^{-x}$

   (f) $y = \exp(x^2)$

   (g) $y = \exp(-x^2)$

   (h) $y = 10^x$

   (i) $y = 2^{x^2}$

   (j) $y = 2^{-x^2}$

2. Use the same set of axes to sketch the graphs of $y = 2^x$, $y = 3^x$, and $y = 4^x$.

3. Use the same set of axes to sketch the graphs of $y = \exp(x - 3)$, and $y = \exp(3 - x)$.

4. Sketch the graph of the function $f(x) = (1 + x)^{1/x}$ on the interval $[-1,1]$. (Note that $f(0)$ is not defined.)

5. Use Table 2 to approximate each of the following as accurately as possible

   (a) $e^{1.1}$

   (b) $3^{1.05}$

   (c) $3^{-1.1}$

   (d) $e^{-2.05}$

   (e) $\exp(-2.3)$

   (f) $\exp(3.6)$

   (g) $\exp(-3)$

   (h) $\exp(3e)$

   (i) $\exp(-2e)$

   (j) $\exp(12)$

6. Use Table 2 to approximate each of the following.

   (a) $\exp_{1.284}(4)$

   (b) $\exp_3(2)$

   (c) $\exp_2(3)$

   (d) $\exp_2(1.05)$

   (e) $\exp_3(-0.10)$

   (f) $\exp_2(-1.05)$

7. Let $f(x) = c \exp(kx)$. Sketch the graph for each of the following given values of c and k. (Use Table 2.)

   (a) $c = 2, k = 2$

   (b) $c = 100, k = .01$

   (c) $c = 1/2, k = 2$

   (d) $c = 0.1, k = .1$

8. If $f(x) = c \exp(kx)$, what is true about $f(x + y)$ and $f(x - y)$ with respect to $f(x)$ and $f(y)$.

9. A function $f: R \to R$ is called *even* if its graph is symmetric with respect to the y axis. This is true whenever

$$f(-x) = f(x)$$

for every real number x. Which of the following are even functions?

   (a) $y = x^2$

   (b) $y = x^3$

   (c) $y = x^2 + 1$

   (d) $y = e^x$

   (e) $y = e^{-x}$

   (f) $y = e^{x^2}$

   (g) $y = e^{-x^2}$

   (h) $y = x^2 e^{x^2}$

10. Show that if $b \neq 1$ is any postive real number and c is any real number $\exp_b(cx) = [\exp_b(x)]^c$.

11. Prove Theorem 4.2.4

   (a) part (i)

   (b) part (ii)

   (c) part (iii)

   (d) part (iv)

12. Sketch the graphs of the following functions.

   (a) $y = 2^x + 2^{-x}$

   (b) $y = 2^x - 2^{-x}$

the logarithmic functions

(c) $y = \dfrac{e^x + e^{-x}}{2}$ (e) $y = \dfrac{e^x - e^{-x}}{e^x + e^{-x}}$

(d) $y = \dfrac{e^x - e^{-x}}{2}$ (f) $y = \dfrac{e^x + e^{-x}}{e^x - e^{-x}}$

13. The graph of the function

$$h(x) = k\left[1 - \exp\left(\frac{x}{d}\right)\right]$$

is know as the *Heines Curve* in mental testing. Sketch it for $k = 500$ and $d = 10$.

14. Use the fact that $\exp_b\colon R \to R^+$ is one-to-one (injective) to determine r in each of the following equations.

(a) $\exp_2(r) \doteq 64$ (g) $\exp(r) = \exp(2r - 3)$
(b) $\exp_3(r) = 81$ (h) $\exp_{10}(3r) = \exp_{10}(r^2 - 4)$
(c) $\exp_{10}(r) = 0.001$ (i) $\exp(r^2 - 6) - \exp(-r) = 0$

(d) $\exp_{1/2}(r) = 32$ (j) $\left(\dfrac{1}{27}\right)^r = 9$

(e) $\exp_{1/3}(r) = 243$ (k) $\exp_{32}(r) = \exp_2\left(\dfrac{3}{2}\right)$

(f) $\exp_{1/5}(r) = \dfrac{1}{125}$ (l) $\exp(-0.2r) = .150$

4.3. the logarithmic functions

Let b be any positive real number different from 1. In the last section it was noted that the function

$$\exp_b(x) = b^x$$

is monotonic and, therefore, one-to-one for each such b. Therefore, considered as a function from the real numbers R to the non-negative real numbers R^+

$$\exp_b\colon R \to R^+$$

is a bijection. It must then have an inverse $\exp_b^{-1}$ which is also a bijective function

$$\exp_b^{-1}\colon R^+ \to R$$

The existence of inverses for every bijective function was discussed in Section 1.9.

example 1

Suppose $\exp_3(2t) = 81$. Find t. Since $\exp_3$ is a bijection, we have $3^{2t} = 81$, or $3^{2t} = 3^4$. Therefore, $2t = 4$ and $t = 2$. Thus,

$$2t = \exp_3^{-1}(81) = 4 \quad \bullet$$

The inverse of the exponential is an important function in its own right. This function is called the **logarithm**, abbreviated **log**.

$$\log_b(x) = \exp_b^{-1}(x)$$

exponential & logarithmic functions

The idea of a logarithm actually predates the concept of the exponential function. The idea originated in the late sixteenth and early seventeenth centuries, as a computational tool. Several mathematicians considered a coordination of arithmetic and geometric progressions, mainly to help them work with complicated trigonometric tables. In 1614 a Scottish "laird," John Napier, published his work on logarithms.

Napier's main idea was to construct two sequences of numbers related to each other in such a way that when one sequence increased in arithmetic progression, the second would decrease as a geometric progression. This would allow the product of two numbers in the second sequence to correspond with the sum of two numbers in the first. In this way, complicated multiplication could be found by addition.

Napier was, however, dissatisfied with his system, and together with Henry Briggs of London, worked out a system involving the usual number base ten. After Napier's death, Briggs, in 1624, published a table of logarithms of 14 decimal places for the integers from 1 to 20,000 and from 90,000 to 100,000. The gap was filled in 1627 by the publication of a complete table by the Dutch mathematician, Ezechiel de Decker.

In the next section, we shall elaborate further on the method of using logarithms as computation aids. While this aspect of logarithms was once extremely important, the advent of high-speed electronic computers has made it much less significant now. Of course, the ordinary slide rule still uses the logarithm concept for computation.

Logarithms as real functions have other uses, however, as we shall see in the last section. Among these is the concept of pH in chemistry. The chief importance of the logarithm lies in the fact that as a function it is the inverse of the exponential function.

Therefore, if $b \neq 1$ is a positive real number $\log_b$ is $\exp_b^{-1}$. Thus,

$$y = \log_b(x)$$

if, and only if,

$$x = b^y$$

The symbol $y = \log_b(x)$ is read "the *logarithm to the base b of x is y*."

example 2

Sketch the graphs of the two functions

$$y = \log_2(x) \text{ and } y = \exp_2(x)$$

The graph of $y = 2^x$ was sketched in Figure 4.2.1 and is reproduced as part of Figure 4.3.1. Since $y = \log_2(x)$ is the inverse it can easily be obtained by reflecting the graph of $y = 2^x$ in the line $y = x$. ●

The analysis used in this example illustrates how to sketch the graph of $y = \log_b(x)$, simply as the inverse of the exponential graph $y = \exp_b(x)$ for any base b.

As a direct consequence of its definition as the inverse of the exponential the logarithm has the following two properties which we state as a theorem.

the logarithmic functions

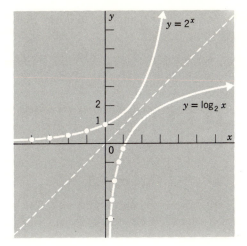

figure 4.3.1

$y = log_2 x$ *as the inverse of* $y = 2^x$

theorem 4.3.1

If $b \neq 1$ *is a positive real number, then for every positive real number s, and every real number t*

(i) $\exp_b \log_b(s) = b^{\log_b b^s} = s$, *and*
(ii) $\log_b \exp_b(r) = \log_b(b^r) = r$

These statements are, of course, of the form $f(f^{-1}(s)) = s$ and $f^{-1}(f(r)) = r$. Yet their rather obvious nature when looked at in this light in no way diminishes their importance.

example 3

Find $\log_5(125)$, $\log_2(1/128)$, and $\log_3(3\sqrt{3})$.

To find these we note that $125 = 5^3$, $1/128 = 2^{-7}$ and $3\sqrt{3} = 3^{3/2}$. Therefore,

$$\log_5(125) = \log_5(5^3) = 3$$

$$\log_2\left(\frac{1}{128}\right) = \log_2(2^{-7}) = -7$$

$$\log_3(3\sqrt{3}) = \log_3(3^{3/2}) = 3/2 \quad \bullet$$

As one can see from Figure 4.3.1 and can expect from its definition, the logarithm as a function has the following properties.

(i)   *The domain of* $\log_b$ *is the non-negative real numbers* R^+.
(ii)   *The image of* $\log_b$ *is all the real numbers, R.*
(iii) *The logarithm is a bijection from* R^+ *to R; that is, it is both one-to-one and onto all of R.*

As a consequence of (iii), for any positive real numbers r and s it is true that

$$\log_b(r) = \log_b(s)$$

implies

$$r = s$$

167

exponential & logarithmic functions

A second consequence also follows. If t is any real number, there exists a unique postive real number s so that

$$\log_b(s) = t$$

This real number s is, of course, the real number b^t.

(iv)   When $b > 1$, $\log_b$ is a strictly increasing function and when $0 < b < 1$, $\log_b$ is strictly decreasing.

A sketch of $y = \log_{1/2}(x)$ is given in Figure 4.3.2.

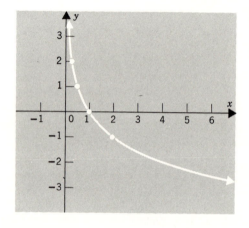

figure 4.3.2

$y = \log_{\frac{1}{2}} x$

example 4

Solve the following equations for x.

(a) $\log_5(x) = -3$
(b) $\log_{1/2}(x) = 4$
(c) $\log_4(x - 2) = 3$

Each of these equations is solved by making use of the fact that the logarithm is a bijection. For (1), if

$$\log_5(x) = -3,$$

then

$$x = 5^{-3}, \text{ or } x = \frac{1}{125}$$

To solve (b) note that if

$$\log_{1/2}(x) = 4, \text{ then}$$

$$x = (1/2)^4 = \frac{1}{64}$$

Finally to solve (c) note that if

$$\log_4(x - 2) = 3, \text{ then } x - 2 = 4^3, \text{ or}$$

$$x - 2 = 64$$

Hence, $x = 66$.   ●

the logarithmic functions

It is easily seen from the definition of the logarithm that

$$\log_b(1) = 0 \text{ and } \log_b(b) = 1$$

for every positive real base b different from 1.

The following theorem states the **laws of logarithms**. The first three are really only direct translations of the laws of exponents.

theorem 4.3.2

Let b be a positive real number $b \neq 1$. Then for every pair of positive real numbers r and s

(i) $\log_b(rs) = \log_b(r) + \log_b(s)$
(ii) $\log_b(r/s) = \log_b(r) - \log_b(s)$
(iii) $\log_b(r^t) = t \log_b(r)$ *for every real number t*

(iv) *if $a > 0$, and $a \neq 1$, then* $\log_a(r) = \dfrac{\log_b(r)}{\log_b(a)}$

proof. Set $m = \log_b(r)$ and $n = \log_b(s)$. Then $b^m = r$ and $b^n = s$. For (i), $rs = (b^m)(b^n) = b^{m+n}$. Thus $\log_b(rs) = m + n$. Therefore, $\log_b(rs) = \log_b(r) + \log_b(s)$.

For (ii) note that

$$\frac{r}{s} = \frac{b^m}{b^n} = b^{m-n}$$

Hence, $\log_b(r/s) = m - n = \log_b(r) - \log_b(s)$.

For (iii) if t is any real number $r^t = (b^m)^t = b^{mt}$. Thus,

$$\log_b(r^t) = mt = t(\log_b r)$$

To prove (iv) if $k = \log_a(r)$, then $a^k = r$. Thus, $\log_b(r) = \log_b(a^k) = k \log_b(a)$. Therefore,

$$\log_b(r) = \log_a(r) \log_b(a)$$

The fact that $a \neq 1$ means that $\log_b(a) \neq 0$. Dividing the above expression by $\log_b(a)$ gives the desired result. This completes the proof of the theorem.

The results of this theorem express those properties which originally led to the development of logarithms as computational aids. In the next section, we shall use them for just such a purpose. Part (iv) is called a *change of base formula* and allows us to use a single table of logarithms plus some computation to obtain the logarithm of a number to any desired base.

example 5

If $\log_e(2) = 0.6931$ find $\log_e(16)$ and $\log_e(0.5)$.

To find these, note that $16 = 2^4$. Thus $\log_e(16) = 4 \log_e(2) = 4(0.6931) = 2.7724$. For the second one, $0.5 = 1/2 = 2^{-1}$. Thus, $\log_e(0.5) = -\log_e(2)$ or -0.6931. ●

exponential & logarithmic functions

example 6

Given $\log_e(3) = 1.0986$. Find $\log_2(3)$. To do this, use the value of $\log_e(2)$ from Example 5 and the change of base formula. Thus,

$$\log_2(3) = \frac{\log_e(3)}{\log_e(2)} = \frac{1.0986}{0.6931} = 1.5851$$

The entries in the table are approximations accurate to the third decimal place. Therefore, our results are also approximations. ●

In the two examples above, we have considered logarithms to the natural base e, where e is the irrational number approximated by 2.718,281, ... In advanced science and technology this base is the most common. In most such cases $\log_e(r)$ is written simply as **log r** or **ln r** and is called the **natural logarithm**.

Because our number system is based on the number ten, when logarithms are used strictly for computational purposes, the base used is generally $b = 10$. In elementary text books $\log_{10}(r)$ is also frequently written as $\log(r)$. This is called the **common logarithm**. As you can see the symbol $\log(r)$ can be ambiguous. However, it is usually fairly easy from the context to discover which base the author is using whether $\log_{10}(r)$, or $\log_e(r)$.

example 7

Given that $\log_{10}(2) = 0.3010$ find $\log_{10}(3.20)$. To do this we note that $3.20 = (32)(10^{-1})$ and $32 = 2^5$. Thus $\log_{10}(3.20) = \log_{10}(2^5)(10^{-1})$. If we apply the laws of logarithms, we obtain

$$\log_{10}(3.20) = \log_{10}(2^5)(10^{-1}) = \log_{10}(2^5) + \log_{10}(10^{-1})$$

$$= 5(0.3010) + (-1)$$

$$= 1.5050 - 1$$

$$= 0.5050$$

●

exercises

1. Sketch the graphs of each of the following function pairs. Use the same set of co-ordinate axes for each pair.

(a) $y = e^x$, $y = \log_e(x)$ (d) $y = \left(\frac{1}{3}\right)^x$, $y = \log_{1/3}(x)$

(b) $y = 3^x$, $y = \log_3(x)$ (e) $y = e^{-x}$, $y = \log_{1/e}(x)$

(c) $y = 10^x$, $y = \log_{10}(x)$ (f) $y = \left(\frac{1}{5}\right)^x$, $y = \log_{1/5}(x)$

2. Find each of the following logarithms without resorting to tables.

(a) $\log_2(32)$ (g) $\log_4(4\sqrt[3]{4})$

(b) $\log_3(27)$ (h) $\log_5(\sqrt[4]{125})$

(c) $\log_5(625)$ (i) $\log_{0.5}\left(\frac{1}{8}\right)$

(d) $\log_{10}(0.001)$ (j) $\log_{0.2}(0.008)$

(e) $\log_{10}(100,000)$ (k) $\log_e(e^{-1.7})$

(f) $\log_{1/2}(8)$ (l) $\log_e(e^{0.004})$

the logarithmic functions

3. Find the number x in each of the following.

(a) $x = 2^{\log_2 5}$

(f) $x = \left(\dfrac{1}{2}\right)^{\log_2 5}$

(b) $x = 3^{\log_3 7}$

(g) $x = 2^{3\log_2 3}$

(c) $x = e^{\ln 3}$

(h) $x = \left(\dfrac{1}{10}\right)^{\log_{10} 2}$

(d) $x = 10^{\log_{10} 2}$

(i) $x = e^{-\ln 2}$

(e) $x = 10^{2\log_{10} 5}$

(j) $x = 10^{-\log_{10} 3}$

4. Given $\log_{10}(2) = 0.3010$, and $\log_{10}(3) = 0.4771$, find the following.

(a) $\log_{10}(8)$

(f) $\log_{10}(2700)$

(b) $\log_{10}(80)$

(g) $\log_{10}(6)$

(c) $\log_{10}(.08)$

(h) $\log_{10}(600)$

(d) $\log_{10}(27)$

(i) $\log_{10}(0.006)$

(e) $\log_{10}(2.7)$

(j) $\log_{10}(2.4)$

5. Given $\log_a 2 = .7$ and $\log_a 3 = 1.1$, find the following.

(a) $\log_a 16$

(f) $\log_a(a/3)$

(b) $\log_a 81$

(g) $\log_a(a/2)$

(c) $\log_a 2/3$

(h) $\log_a(2a)$

(d) $\log_a 6$

(i) $\log_a(3a)$

(e) $\log_a 3/2$

(j) $\log_a \sqrt[3]{2}$

6. Solve each of the following equations for the real number x.

(a) $\log_5(x + 5) = 1$

(f) $\log_e(3\sqrt{x}) + \log_e(3) = \log_e(5)$

(b) $\log_7(2 - x) = 2$

(g) $\log_2(x + 2) + \log_2(x) = 3$

(c) $\log_e(x^2 - 1) = -1$

(h) $\log_{10}(x^2) = -5$

(d) $\log_3(x + 3) + \log_3(x) = \log_3(28)$

(i) $\log_{10}(x) + \log_{10}(x + 3) = 1$

(e) $\log_2(x^2 - x - 2) = 2$

(j) $\log_3(x + 2) = 3 - \log_3(x + 2)$

7. Show that for a and b positive real numbers each different from 1

$$[\log_a(b)][\log_b(a)] = 1$$

8. Use the results of Exercise 7, if necessary, $\log_{10} 2 = 0.3010$, and $\log_{10} 3 = 0.4771$ to compute each of the following.

(a) $\log_2(3)$

(f) $\log_3 10$

(b) $\log_3(2)$

(g) $\log_2 10$

(c) $\log_4(9)$

(h) $\log_{2/3}(5)$

(d) $\log_{10} 5$

(i) $\log_{\sqrt{2}}(25/9)$

(e) $\log_{10} 0.125$

(j) $\log_{\sqrt{3}} \sqrt{2}$

9. Solve each of the following inequalities.

(a) $\log_2(x + 1) < 3$

(f) $2 < \log_{10} x < 3$

(b) $1 < \log_3(x - 3) < 2$

(g) $1 < \log_e x < 2$

(c) $0 < \log_{10} x < 1$

(h) $\log_{10}|2x - 1| < 1$

(d) $0 < \log_e x < 1$

(i) $\log_9 \sqrt{x} > 1/2$

(e) $1 < \log_{10} x < 2$

(j) $\log_e \sqrt{x} < 1$

10. Prove that if $b \neq 1$ is a positive real number, and $r > 0$ is any real number,

$$\log_b(1/r) = \log_{1/b}(r)$$

exponential & logarithmic functions

11. Show that, with b and r as in Exercise 10,

$$\log_b(\sqrt{1 + r^2} + r) = -\log_b(\sqrt{1 + r^2} - r)$$

12. Show that, with b and r as in Exercise 10,

$$\log_b \frac{r - \sqrt{r^2 - 1}}{r + \sqrt{r^2 - 1}} = \log_b(r - \sqrt{r^2 - 1})^2$$

13. Sketch the graphs of each of the following functions which involve logarithms. Identify the domain and image for each of them.

(a) $f(x) = \log_3(1 - x)$ (d) $f(x) = \log_2 x^2$
(b) $f(x) = \log_3(1 + x)$ (e) $f(x) = |\log_2 x|$
(c) $f(x) = \log_3 |1 - x|$ (f) $f(x) = \log_e(-x)$

14. Prove each of the following.

(a) If L is a real function such that

$$L(xy) = L(x) + L(y)$$

for x, y any real numbers, then $L(1) = 0$.
(b) If E is a real function such that for any real numbers x and y

$$E(x + y) = E(x)E(y)$$

then $E(0) = 1$.

4.4.   common logarithms

As we explained in the previous sections, logarithms developed historically as an aid to computation. While the need for such computational aid has diminished in recent years with advances in electronic computer development, the use of logarithms in computations has not completely disappeared. (Not everyone has ready access to a large computer.) A handy pocket computer, the slide rule, is in fact, a logarithmic machine.

In this section, we shall use the properties of the logarithm to actually compute. Since our number system is based on the number ten we shall, for the most part, use $\log_{10}$ here. In this section, only common logarithms will be written without the base. Thus, in this section

$$\log x = \log_{10} x$$

example 1

Compute the number

$$N = \frac{(302)(27.5)}{(0.0051)}$$

Let us first of all approximate the size of the answer N as follows.

$$302 = (3.02)(10^2); \quad 27.5 = (2.75)(10^1)$$

and $0.0051 = (5.1)(10^{-3})$. Thus,

$$N = \frac{(302)(27.5)}{(0.0051)} = \frac{(3.02)(10^2)(2.75)(10^1)}{(5.1)(10^{-3})} = \frac{(3.02)(2.75)(10^3)}{(5.1)(10^{-3})}$$

$$N = \frac{(3.02)(2.75)}{(5.1)}10^6$$

Hence the number N will be approximately $[(3)(2\frac{3}{4})/5]10^6$ or 1.6 million. ⬤

We shall return to this computation after making some remarks about questions raised in it.

The method employed here illustrates the fact that any positive real number r can be written in the form

$$r = (t)(10^n)$$

where $1 \leq t < 10$ and n is an integer (positive, negative or zero). For example, $15 = (1.5)(10^1)$; $720 = (7.2)(10^2)$; 9 billion $= (9)(10^9)$; $0.013 = (1.3)(10^{-2})$; $3.2 = (3.2)(10^0)$; Avogadro's number $= (6.02)(10^{23})$; etc.

By the laws of logarithms (Theorem 4.3.2) the logarithm of r to the base 10 may be written in the form

$$\log r = \log(t) + \log(10^n)$$

$$= n + \log(t)$$

Hence, to find the logarithm (base 10) of any real number it is only necessary to know the logarithms of numbers between 1 and 10.

For r and t as above, $r = (t)(10^n)$, the number

$$\log t$$

is called the *mantissa* of log r. The integer n is called the *characteristic* of log r. Therefore,

$$\log r = \text{characteristic} + \text{mantissa}$$

The previous examples would have logarithm as follows.

$\log 15 = 1 + \log 1.5$ $\log 0.013 = -2 + \log 1.3$
$\log 720 = 2 + \log 7.2$ $\log 3.2 = 0 + \log 3.2$
$\log 9$ billion $= 9 + \log 9$ $\log(\text{Avogadro's Number}) = 23 + \log 6.02$

The characteristic of log r can easily be obtained from the position of the decimal point in the number r. The mantissa of log r is obtained from a table. Such a table is most often called a *logarithm table*. A table of four place mantissa appears in the Appendix (Table 3).

If we return to Example 1 and use Table 3, we have

$$\log\left[\frac{(302)(27.5)}{(0.0051)}\right] = \log 302 + \log 27.5 - \log 0.0051$$

$$= 2 + \log 3.02 + 1 + \log 2.75 - (3 + \log 5.1)$$

$$\doteq 6 + (0.4800) + (0.4393) - (0.7076)$$

$$\doteq 6.2117$$

exponential & logarithmic functions

The numbers obtained from the table are only approximations and the $\doteq$ sign indicates this.

The number N desired in Example 1 is such that

$$\log N \doteq 6.2117$$

We, therefore, know that $N \doteq (K)10^6$ where K is a number such that

$$\log K \doteq 0.2117$$

That is, $K = 10^{0.2117}$ and hence $N = 10^{6.2117}$. If we return to the *body* of Table 3, we find that

$$1.62 < K < 1.63$$

since $\log 1.62 \doteq 0.2095$ and $\log 1.63 \doteq 0.2122$. Hence, $N \doteq 1.6$ million, as stated earlier. After a few other remarks, we shall discuss a method, called *linear interpolation*, for getting a closer approximation to K and, thus, to N.

example 2

Find $\log 0.0312$. Since $0.0312 = (3.12)(10^{-2})$ we know that the characteristic is -2. Therefore,

$$\log(0.0312) = -2 + \log(3.12)$$

The mantissa, $\log 3.12$, obtained from Table 3 is found to be approximately 0.4942. Therefore,

$$\log 0.0312 \doteq -2 + 0.4942$$

$$\doteq -1.5058$$

However, in many cases it is inconvenient to work with negative mantissas, since for every real number t in the interval $[1,10]$ $\log t$ is non-negative. For this reason we prefer to leave the logarithm of every real number with a non-negative decimal part. Thus, -0.5058 is not the mantissa of $\log 0.0312$ even though it is the decimal fractional part of its logarithm. For convenience sake, such logarithms are generally written in the standard form

$$\log 0.0312 \doteq 0.4942 - 2$$

This could also be written, and frequently is, with -10 at the end. By adding and subtracting 8 from the above, one obtains

$$\log 0.0312 \doteq 0.4942 + 8 - 8 - 2$$

$$\doteq 8.4942 - 10.$$

One could also write $7.4942 - 9$, $205.4942 - 207$, or any other such number, so long as the decimal part remains positive and so long as the characteristic is -2. ●

example 3

Given that $\log N = 3.8774$, find N. It is immediately clear that

$$N = 10^{3.8774}$$

common logarithms

But to obtain an approximation to N in usable form, we note that

$$N = (10^{0.8774})(10^3)$$

The mantissa 0.8774 of log N can be located in the body of Table 3 and found to be the logarithm of 7.54. Thus

$$10^{0.8774} \doteq 7.54$$

so

$$N = (7.54)(10^3) = 7540 \quad \bullet$$

Table 3 has four place rational approximations to the logarithms of numbers between 1 and 10 in intervals of .01; that is, for three-digit numbers. Should one desire to obtain the logarithm for a number with *four* digits an approximation method called *linear interpolation* is available. We illustrate this with an example.

example 4

Find log 163.3. To do this, one must locate a mantissa in Table 3; that is, find log 1.633. Since $\log_{10}$ is an increasing function, it is clear that this mantissa will be between log 1.630 and 1.640. Table 3 indicates that log 1.630 = 0.2122 while log 1.640 = 0.2148. This situation is indicated graphically in Figure 4.4.1. The figure is distorted for purposes of illustration. In actual fact, the graph of $\log_{10}$ would lie very close to the line L connecting the points $P_1(1.63, 0.2122)$ and $P_2(1.64, 0.2148)$.

The number log 1.633 is the y coordinate of the point Q on the curve, but can be closely approximated by the y coordinate of the point on the line L whose coordinate is 1.633. By using similar triangles, this number y is easily computed.

If d is the length of the segment indicated in the figure, we have the proportion

$$\frac{d}{0.0026} = \frac{.03}{.1}$$

or

$$d = \frac{(.03)(.0026)}{.1} = (3)(26)10^{-5} = 78 \times 10^{-5} = 0.0078$$

Therefore, log 1.633 $\doteq$ log 1.63 + d = 0.2122 + .0078 = 0.21298. Since these

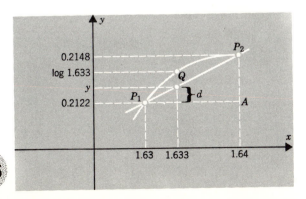

figure 4.4.1

Linear interpolation

mantissas are only four place rational approximations, anyway, always round off the result to four places (or to the accuracy of your table). Thus,

$$\log 163.3 \doteq 2.0230.$$

Since we are primarily interested in ratios, the above proportion could have been simplified to

$$\frac{d}{0.0026} = \frac{3}{10} = .3$$

So $d = (.3)(0.0026) = .0078$, as before. In the next example we make use of this simplification as well as a tabular form for writing the data rather than sketching the graph. ●

example 5

Approximate $\log 0.007637$. The characteristic of $\log 0.007637$ is -3. To find the mantissa, $\log 7.367$ we use Table 3.

$$10 \left\{ {}^7\!\left\{ \begin{array}{l} \log 7.630 \doteq 0.8825 \\ \log 7.637 \doteq ? \end{array} \right\}^d \atop \log 7.640 \doteq 0.8831 \right\} 0.006$$

where the proportional differences are indicated by the symbols alongside the braces. Hence,

$$\frac{d}{.0006} = \frac{7}{10}$$

or

$$d = (.7)(.0006) = .00042 \doteq .0004$$

Therefore, the mantissa $\log\ 7.637 \doteq 0.8825 + 0.0004 = 0.8829$, and $\log 0.007637 \doteq 0.8829 - 3$ or $7.8829 - 10$. ●

The process of linear interpolation can also be used in reverse to approximate N when the four-place mantissa of $\log N$ is not exactly contained in Table 3. This is illustrated in Example 6.

example 6

Approximate N to four significant figures if $\log N = 4.9554$.

From Table 3, we locate the mantissa 0.9554 between the number 0.9552 $\doteq \log 9.02$ and 0.9557 $\doteq \log 9.03$. Thus

$$.01 \left\{ {}^x\!\left\{ \begin{array}{l} \log 9.02 \doteq 0.9552 \\ \log t \quad\ \doteq 0.9554 \end{array} \right\}.0002 \atop \log 9.03 \doteq 0.9557 \right\}.0005$$

Thus, the following proportion

$$\frac{x}{.01} = \frac{.0002}{.0005} = \frac{2}{5}$$

or

176

$$x = \frac{2}{5}(.01) = .004$$

common logarithms

Therefore, $t \doteq 9.02 + 0.004 = 9.024$; $N = t(10^4)$ so

$$N \doteq 90240 \quad \bullet$$

Sometimes the number N is called the *antilogarithm* of log N. In this example, since log $N = 4.9554$, $N =$ antilog (4.9554). However, this usage is archaic since antilog is merely $\exp_{10}$. Thus, when log $N = 4.9554$ we know that $N = \exp_{10}(4.9554)$.

In the computation of Example 1 you will recall that we had the following.

$$.01 \left\{ \begin{array}{l} x \left\{ \begin{array}{l} \log 1.62 \doteq 0.2095 \\ \log K \doteq 0.2117 \end{array} \right\} .0012 \\ \\ \log 1.63 \doteq 0.2122 \end{array} \right\} .0027$$

Therefore,

$$x = \frac{12}{27}(.01) \doteq .0044, \text{ and}$$

$$K = 1.624$$

so that $N \doteq 1,620,000$. Since our data had but three significant figures, this is every bit as accurate as 1,624,000 would be, and not so fraudulent.

example 7

Use logarithms to perform the following computations to four significant figures

$$N = \sqrt[5]{\frac{(127.3)^3(0.06784)}{(37.65)}}$$

The laws of logarithms give us

$$\log N = \frac{1}{5}\{3 \log(127.3) + \log(0.06784) - \log(37.65)\}$$

By using interpolation and the entries in Table 3, each of these logarithms is obtained. They are arranged in tabular form to facilitate greater accuracy.

$$\log 127.3 \doteq 2.1048$$

$$\begin{array}{ll} & 3 \log 127.3 \doteq 6.3144 \\ & \log 0.06784 \doteq 0.8315 - 2 \\ \hline + & = 7.1459 - 2 \\ - & \log 37.65 \doteq 1.5758 \\ \hline & 5 \log N \quad \doteq 5.5701 - 2 \\ & 5 \log N \quad \doteq 3.5701 \end{array}$$

Therefore, log $N = 0.7140$, and $N \doteq \exp_{10} 0.7140$. This value is obtained from Table 3, again by interpolation, as

$$N \doteq 5.174 \quad \bullet$$

exercises

1. Use Table 3 to obtain the following logarithms.

(a) log 1.01

(b) log 10.9

(c) log 268,000

(d) log 513

exponential & logarithmic functions

        (e) log 7710   (h) log 0.912

        (f) log 0.031   (i) $\log(5.91 \times 10^{-23})$

        (g) log 0.00243   (j) $\log(6.02 \times 10^{23})$

2. Use Table 3 to find N, given log N as follows.

(a) log N = 0.4048   (f) log N = 0.5478 − 5

(b) log N = 1.5478   (g) log N = 0.5289 − 4

(c) log N = 2.9547   (h) log N = 0.8149 − 1

(d) log N = 3.8768   (i) log N = 0.1072 − 2

(e) log N = 4.7782   (j) log N = 0.6590 − 3

3. Use Table 3 and interpolate, if necessary, to obtain each of the following logarithms.

(a) log 0.01357   (f) log 2718

(b) log 1.106   (g) log 138.4

(c) log 42.42   (h) log 0.4531

(d) log 0.00005382   (i) log e

(e) log 769.4   (j) log π

4. Use Table 3 and linear interpolation, if necessary, to find N given log N as follows.

(a) log N = 2.7367   (f) log N = 9.6565 − 10

(b) log N = 8.4330 − 10   (g) log N = 0.6078 − 5

(c) log N = 5.1843   (h) log N = 0.9924 − 3

(d) log N = 1.0013   (i) log N = −2.4771

(e) log N = 0.7117 − 2   (j) log N = −1.6990

5. Using logarithms base 10 approximate the value of each of the following to three significant figures.

(a) (0.819)(328)   (f) $\sqrt{2}$

(b) (0.0318)(5.02)   (g) $\sqrt[5]{625}$

(c) (1970)(62)   (h) $\left(\dfrac{4}{3}\right)\pi$

(d) $(23.1)^{1.05}$ (i) $\sqrt[5]{27}$

(e) $\sqrt[3]{79}$ (j) $\sqrt[3]{5}$

6. Using $\log_{10}$ approximate each of the following to four significant figures.

(a) $\dfrac{(1973)(428)}{10.04}$ (f) $(27.04)^{1.03}$

(b) $\dfrac{1.542}{47.08}$ (g) $\sqrt[5]{8041}$

(c) $\dfrac{(36.49)(0.4771)}{8075}$ (h) $(1.015)^{210}$

(d) $\dfrac{(4014)(686.5)}{(2.18)^2}$ (i) $\sqrt{\pi}$

(e) $\sqrt[20]{782,400}$ (j) $\sqrt{e}$

7. Use $\log_{10}$ to approximate each of the following to four significant figures.

(a) πe (f) 2^e

(b) π/e (g) 2^π

(c) $e^{7/2}$ (h) $\sqrt{2^{\sqrt{2}}}$

(d) $2^{\sqrt{2}}$ (i) π^e

(e) 3^π (j) e^π

8. The volume of a right circular cylinder with radius r and altitude h is $V = \pi r^2 h$. Approximate the volume of a cone whose altitude is 8.26 inches and whose radius is 3.41 inches.

9. The volume of a sphere of radius r is $V = (4/3)\pi r^3$. Find the volume of the sphere whose radius is 7.13 inches.

10. The surface area of a sphere of radius r is given by

$$S = 4\pi r^2$$

Find the surface area of the sphere in Exercise 9.

11. The pressure p, in pounds per cubic foot, and the volume v, in cubic feet, of a certain gas are given by the formula

$$pv^{1.6} = 200$$

Approximate the pressure p if the gas is contained in the sphere of Problem 9.

12. The theoretical formula for the period of oscillation T of a certain physical system is

$$T = 2\pi \left(\sqrt{\frac{b^2 + 3a^2}{2ag}} \right)$$

where g is the accelleration due to gravity. Given $g = 32$ ft/sec^2 $a = 12.04$ cm and $b = 10.45$ cm. Find T.

13. The chemical pH of a solution is defined as the negative of the logarithm base 10, of the hydrogen ion concentration, $[H^+]$, in moles per liter:

$$pH = -\log[H^+]$$

If a fluid has a pH $= 7.5$ what is the molecular concentration of hydrogen ions $[H^+]$? (In moles per liter.)

14. In 1916 DuBois and DuBois devised the following empirical formula for the surface area S of the human body.

$$S = 0.007184(W)^{0.425}(H)^{0.725}$$

where W is the weight in kilograms and H is the height in centimeters, calculate the body surface area of a man 2 meters tall who weighs 70 kilograms.

15. Examine a slide rule. Explain how logarithms are used in determining the spacing of the numbers on the C and D scales and how multiplications and divisions using the scales are applications of the laws of logarithms.

16. Locate some log-log paper. Explain why empirical data plotted on such paper will obey a power law $y = Ax^n$, when the curve connecting the data approximates a straight line.

17. Make a crude slide rule using log-log paper and perform some simple computations.

18. Show that the characteristic of log N for any positive real number N is the value of the greatest integer function at log N; i.e., $[\log N]$. Show that the mantissa is log $N - [\log N]$.

19. Explain why $\log_e 2$ and $\log_e 20$ would not be expected to have the same decimal part. Show that $\log_e 20 - \log_e 2$ is not an integer.

4.5.   applications

Some elementary applications of logarithmic functions were indicated in the last section and in the exercises of that section. In this section, several very important applications of exponential and logarithmic functions will be discussed. We shall be concerned primarily with functions of the form

$$f(x) = Ae^{kx}$$

exponential & logarithmic functions

where A and k are given real numbers. Of particular interest is the fact that if one considers f to be a function of time, then at two given times t_0 and $t_1 = t_0 + h$, where h is a positive length of time, the ratio of $f(t_1)$ to $f(t_0)$ becomes

$$\frac{f(t_1)}{f(t_0)} = \frac{f(t_0 + h)}{f(t_0)} = \frac{Ae^{k(t_0 + h)}}{Ae^{kt_0}} = \frac{Ae^{kt_0}e^{kh}}{Ae^{kt_0}} = e^{kh}$$

Hence the ratio of $f(t_1)$ to $f(t_0)$ depends only on the length of elapsed time h, and not on any particular value of t_0. This phenomenon is typical of such things as growth of a population, radioactive decay, measurement of solute concentration, etc. We shall describe some of these applications here.

In these applications, considerable use is made of Tables 2 and 4, in the Appendix, as well as Table 3. Table 2 was discussed in Section 4.2. The methods of linear interpolation described in Section 4.4 may also be adapted for use with both e^x and e^{-x} in this table, if desired. A brief description of the use of Table 4 is given here. It is a logarithm table similar to Table 3. However, Table 4 is for $\log_e x$, called natural logarithms. If our number system were base e, then the comments made in Section 4.4 about the characteristics and mantissa of $\log_e x$ for a number x would also apply here. Such is *not the case*, however, and $\log_e 12.2 - \log_e 1.22$ is not an integer as it would be base 10. It is true, of course, that $\log_e e^k - \log_e e^n = \log_e e^{k-n} = k - n$, so that $\log_e 2e^4$, and $\log_e 2e$ have the same "decimal" part (mantissa?) and differ only by the integer 3.

example 1

Use Table 4 to find the following logarithms.

(i) $\log_e 2$
(ii) $\log_e 583$
(iii) $\log_e 8.744$
(iv) $\log_e 2e^2$
(v) $\log_e 10$

The logarithms for (i) and (ii) are read directly from the table as

(i) $\log_e 2 \doteq 0.6931$ and (ii) $\log_e 5.83 \doteq 1.7630$.

To find (iii) we must use linear interpolation. A glance at the graph of $y = \log_e x$ justifies this approach. From the table we have

$$.01\left\{\begin{array}{l}.004\left\{\begin{array}{l}\log_e 8.740 \doteq 2.1679 \\ \log_e 8.744 \doteq ?\end{array}\right\}d \\ \log 8.750 \doteq 2.1691\end{array}\right\}.0012$$

Thus,

$$\frac{4}{10} = \frac{d}{.0012} \text{ or } d = (.4)(.0012) \doteq (.00048)$$

Therefore, $\log_e 8.744 \doteq 2.1684$.

To find (iv) we note that $\log_e 2e^2 = \log_e 2 + \log_e e^2$. Therefore,

$$\log_e 2e^2 = 2 + \log_e 2 \doteq 2.6931$$

applications

To find (v) we first write $10 = ke^2$ since $e^2 \doteq 7.3891$ and $e^3 \doteq 20.086$. Thus, $k = 10/e^2 \doteq 1.353$, and $10 \doteq (1.353)e^2$. Therefore, $\log_e 10 = \log_e(1.353)e^2 = 2 + \log_e(1.353) = 2.3026$. One can also calculate $\log_e 10$ using Table 3 and Theorem 4.3.2(iv) as

$$\log_e 10 = \frac{1}{\log_{10} e} = \frac{1}{0.4343} = 2.3026 \quad \bullet$$

Let us now turn to the applications of exponential functions.

growth of a bacterial population

Suppose that we have a bacteria population consisting of N_0 bacertia in a suitable culture medium. In a given generation time, each member of the population will have divided to produce two daughter cells, so that the population will have doubled to $2N_0$. A second time interval will increase the population to $4N_0$. After n generations the population will be $2^n N_0$. Provided that there are sufficient nutrients in the culture medium for the population to thrive, we can get a good estimate of the size of the population N_k after k generation times from the formula

$$N_k = N_0 2^k$$

For example, suppose that a culture contains 1,000 cells and has a generation time of one hour. Then in 20 hours there are

$$N_{20} = (1,000)2^{20}$$

bacteria present. Thus

$$N_{20} = (1,000)(1,024)(1,024)$$
$$= 1,048,576,000$$

In general, the generation time varies from culture to culture and also depends on the bacteria. If k denotes the generation time, the number of bacteria present at any given instant t is

$$N_t = N_0 2^{t/k}$$

In terms of the base e, if $e^a = 2^{1/k}$, the formula is

$$N_t = N_0 e^{at}$$

As an additional example, suppose that the bacteria in a given culture were counted after 2 hours of growth and found to be 800. They were counted again after 4 hours of growth. This time there were 2,500 bacteria present. To determine the original number N_0 present and the generation time k, we first find a and N_0 from the formula. Thus

$$N_2 = 800 = N_0 e^{2a}$$
$$N_4 = 2,500 = N_0 e^{4a}$$

Therefore,

$$\frac{N_4}{N_2} = \frac{2,500}{800} = \frac{N_0 e^{4a}}{N_0 e^{2a}} = e^{2a}$$

exponential & logarithmic functions

so that

$$e^{2a} = 3.125$$

Substituting this value for e^{2a} in the formula, we obtain

$$N_2 = 800 = N_0 e^{2a} = N_0 3.125$$

Therefore, $N_0 = (800) \div (3.125) = 256$. Since $e^{2a} = 3.125$, $2a = \log_e 3.125 \doteq 1.1394$, from Table 4. Thus, $a \doteq 0.5697$. The estimate for this culture would be given, then, by the formula

$$N_t = 256 \exp(0.57t)$$

To find the generation time k, we note that $e^a = 2^{1/k}$, hence taking $\log_e$ of both sides we have

$$a = \frac{1}{k} \log_e 2$$

Solving this for k gives

$$k = \frac{1}{a} \log_e 2$$

The generation time for this particular culture is, therefore,

$$k \doteq \frac{1}{0.5697} (0.6931)$$

$$\doteq 1.216 \text{ hours}$$

decay of a radioactive material

Every atom with atomic number greater than 83 is radioactive. That is, they undergo a spontaneous loss of tiny particles from the nucleus. This loss is known as radioactive decay, because the loss results in a decrease in the mass and the charge of the atom. This process changes an atom of one element into an atom of another element. The process continues, in sequence, until a stable atom is reached. The first step in the decay of radium is the loss of an alpha particle (Helium nucleus) which changes the original *radium* to the new element *radon*. If a given amount of a radioactive substance is observed over a period of time, the amount of that substance present decreases exponentially.

Let A_0 denote the amount of the radioactive substance initially present. Then the amount $A(t)$ which is present after a time interval t is given by the formula

$$A(t) = A_0 \exp(-kt)$$

where k is a constant depending upon the nature of the radioactive substance.

The rates of decay of radioactive materials are compared by half-lives. The **half-life** of a substance is the time required for one half of the atoms to decay. The time is always the same for the given material, regardless of the mass of the material originally present, or upon the pressure or temperature (under usual conditions) of the surroundings. The half-life τ does not appear in the

applications

formula given above, but if we set $A(\tau) = 1/2\, A_0$ we get

$$A(\tau) = 1/2A_0 = A_0 \exp(-k\tau),$$

or

$$1/2 = \exp(-k\tau).$$

Hence,

$$-k\tau = \log_e 1/2 \text{ or } \tau = -1/k \, \log_e(1/2)$$

Since $-1/k \, \log_e(1/2) = 1/k[-\log_e(1/2)] = 1/k[\log_e(1/2)^{-1}] = 1/k \, \log_e 2$, we see that $\tau = 1/k \, \log_e 2$, is the half-life. The constant k in our formula is, therefore,

$$k = \frac{\log_e 2}{\tau}$$

where τ is the half-life of the radioactive substance.

For example, the element no. 90, thorium (Z) has a half-life of 1.65×10^{10} years. What fraction of thorium atoms decay per year? The decay constant k is, from the above formula,

$$k = \frac{\log_e 2}{\tau} = \frac{0.693}{(1.65)(10^{10})} = 4.2 \times 10^{-11}$$

Thus 1 atom of thorium for every $1/k = 2.4 \times 10^{10}$ atoms present will decay per year. ●

An important material in establishing age of objects is carbon 14 which has a half-life of 5760 years. Suppose a source originally contained a million atoms. How many of these remain after 20 years? We have

$$k = \frac{0.693}{5670}$$

so

$$A(2) = 10^6 \exp\left[-\frac{0.693}{5760} 20\right]$$

$$= 10^6 \exp(-0.024)$$

Table 2 isn't very helpful in this case. We must interpolate in it to get

$$\exp(-0.024) \doteq 0.9763$$

Hence about 976,300 atoms remain after 20 years.

compound interest

Suppose that an amount P (principal) of money is invested at a rate of interest r, which is added at the end of the year. Then after one year on deposit, the amount of money will be the original principal P plus the amount of interest Pr (assume r is a percent; i.e., $5\% = .05 = r$). The amount on deposit for a second year would be

$$P(1 + r)$$

so that, after two years on deposit, the amount of money will have grown to

$$P(1 + r) + P(1 + r)r = P(1 + r)^2$$

exponential & logarithmic functions

In this way, we see that if the money is left on deposit for t years the amount will be

$$P(1 + r)^t$$

In most cases, interest is compounded more often than once a year. If it is compounded n times per year the rate is r/n per period, and the number of periods when interest is paid in t years becomes nt. Therefore, the total amount of money after t years is

$$P(1 + r/n)^{nt}$$

For example, suppose $200 is invested at 5% interest compounded quarterly. If all the money is left on deposit for 6 years, how much will the account contain at the end of this time? The formula gives us

$$A = \$200\left(1 + \frac{.05}{4}\right)^{24}$$

$$A = \$200(1 + (.0125))^{24} = 200(1.0125)^{24}$$

To compute this number we use logarithms. We have, by interpolation,

so
$$\log_{10} 1.0125 = 0.0054$$
and
$$24\,\log_{10} 1.0125 = 0.1296$$
Therefore,
$$\underline{\log_{10} 200 \qquad\quad = 2.3010}$$
$$\log A \qquad\qquad = 2.4306$$
or

$$A \doteq \$269.50 \qquad \bullet$$

A recent approach is, for some financial institutions, to offer *continuous interest*. In the formula

$$A = P(1 + r/n)^{nt}$$

if n becomes larger and larger, the amount A also increases, but only slightly. When n approaches ∞, the situation approaches continuous interest. To see this we set $h = r/n$, so that $n = r/h$. Substituting this, we have

$$A = P(1 + h)^{(r/h)t} = P[(1 + h)^{1/h}]^{rt}$$

Now, as n approaches ∞, $h = r/n$ will approach zero.

One recalls, from Section 4.2, that as h approaches zero the number

$$(1 + h)^{1/h}$$

approaches e. Thus, for continuous interest the formula is

$$A = Pe^{rt} = P\exp(rt)$$

To contrast this with interest compounded annually, suppose that $500 is invested at 6% interest for 20 years. If interest is compounded annually, the amount of money after this time is

$$A = 500(1 + .06)^{20}$$

$$= 500(1.06)^{20} = \$1603.56$$

applications

If the interest were compounded continuously, the amount after 20 years would be

(handwritten: type 1) Given A&K find P(t) for same t)

$$A = 500 \exp[(20)(.6)] = 500e^{1.2}$$

(handwritten: type 2) $f(t_1) = \#$ $P(t_2) = \#$ find A&K)

$$= 500(3.3201)$$

$$= \$1660.05$$

The difference is $56.49 over twenty years.
The exercises which follow contain problems applying the three principles just desribed as well as additional applications of exponential functions.

(handwritten: $P(t) = Ae^{kt}$)

exercises

1. A population of 1,000 bacteria is grown in a culture in which they divide with a generation time of 1 hour. Find the number of bacteria present after 5 hours.

2. Suppose that after one hour, 0.1% of the bacteria in the culture of Exercise 1 are removed and placed in a different nutrient. In this new environment the generation time is 30 minutes. How many cells should there be in this second culture after 4 additional hours?

3. In a certain bacterial culture there are 420,000 bacteria at the end of two days and 5,650,000 present at the end of four days. Compute the following.

    (a)   the number present at the beginning
    (b)   the number present after one day
    (c)   the number present after three days
    (d)   the number of days required for there to be 42,000,000 bacteria

4. A certain town had a population of 30,000 in 1950. The population in 1960 was 70,000. If growth continues at this rate, what should the 1970 census figure be? 1980?

5. Suppose a certain source contained 2 million atoms of carbon 14 (half-life 5670 years). How much will remain after 150 years?

6. The half-life of radium is 1620 years. After 405 years, how much radium remains from a sample of 10 grams?

7. How much of the ten gram sample of radium in Exercise 6 will be left after 810 years?

8. Phosphorus-32 has a half-life of 14.2 days. How long will it take a given quantity to lose:

    (a) 1% of its activity   (b) 90% of its activity

9. A nerve cell is injected with a quantity A_0 of potassium-42 whose half-life is 12.45 hours. After 1 hour, the amount of potassium-42 present has been reduced to 89.5% of its original value. Has the cell lost any of the original material injected?

10. A bacterial population is increasing so that at time t the population has size $A \exp(t/k)$. A second population has size $(1/2)A \exp(2t/k)$ at the same time t. Which population is the larger at the time:

    (a) $t = 0.1k$
    (b) $t = 0.5k$
    (c) $t = k$
    (d)   At what time will the populations be equal?

11. A sample of a certain radioactive substance has decreased so that after 1 year, 80% of the original amount is present. What is its half-life?

exponential & logarithmic functions

12. If a man invests $600 at 5% interest, determine the amount of money on deposit after 7 years if the interest is compounded:

    (a) annually   (d) monthly
    (b) semiannually   (e) daily
    (c) quarterly   (f) continuously

13. Suppose that you just inherited $1,000 which has been accumulating for 50 years at 4% interest compounded semiannually. How much was originally invested?

14. I shall need $5,000 in fifteen years. How much should I invest in a bond, which pays $5\frac{1}{2}\%$ compounded annually, to be sure that that amount will be available?

15. If an inductance of L henrys and a resistance of R ohms are connected in series with an E.M.F. of E volts, the current of I amperes flowing in the circuit, in t seconds, is given by the formula

$$I = \frac{E}{R}\left[1 - \exp\left(-\frac{R}{L}t\right)\right]$$

Determine the approximate current flow given $t = .02$ sec., $L = 0.1$, $R = 6$, and $E = 180$.

16. Beer's law states that when a light is allowed to pass through a solution, the emerging light intensity I is given by the formula

$$I = I_0 10^{-kcd}$$

where I_0 is the intensity of the incident light, c is the concentration of the solution in moles per liter, d is the thickness of the liquid, and k is the extinction coefficient for the solute for light of a given wavelength. Suppose $c = 0.01$, $d = 5.0$, I_0 is 150 lumens, and $k = 10$. What is the intensity of the transmitted light?

17. Given an incident light of 100 lumens, use the formula in Exercise 16 to determine how thick the solution was which reduced the intensity to 80 lumens if $k = 20$ and $c = .005$.

18. The law of cooling states that the rate at which the temperature of a body warmer than its surroundings decreases is proportional to the difference in temperature between the body and its surroundings. This difference at any time t is given by the formula

$$D(t) = D_0 e^{-rt}$$

where D_0 denotes the initial temperature difference and r is the rate of cooling for the given substance. Suppose that an object originally at 100°C is kept at a constant room temperature of 23°C for 30 minutes. After this time, its temperature is 80°C.

    (a) Find r for this object.
    (b) What will the temperature be after 45 minutes?
    (c) What will the temperature be after 1 hour?
    (d) When will the temperature be 23°C?

19. In mathematical statistics the *density function* of a normal distribution is given as

$$f(x) = \frac{1}{(\sqrt{2\pi})\sigma} \exp\left[-\frac{1}{2\sigma^2}(x - \mu)^2\right]$$

where μ is the mean (average) and σ is the standard deviation. Use Table 2 to sketch a graph of $f(x)$ for $\mu = 0$ and $\sigma = 1$.

20. Compute the value of $f(\mu)$, $f(\mu + \sigma)$, $f(\mu - \sigma)$, $f(\mu + 2\sigma)$, $f(\mu + 3\sigma)$ in the density function of Exercise 19. Discuss the fact that $f(\mu + n\sigma)$ approaches the x axis as an asymptote as n approaches ∞.

quiz for review

quiz for review

1. Let t be a real number such that $5^t = 1/125$. Then $t =$
   (a) 3 (b) -3 (c) 1/3 (d) $-1/3$ (e) 2

2. Let r and s be positive rational numbers. Then
   (a) if $r < s$ and $b < 1$, $b^r < b^s$ (d) if $r < s$ and $b = 1$, $b^r > b^s$
   (b) if $r < s$ and $b > 1$, $b^r < b^s$ (e) if $r < s$ and $b > 1$, $b^r > b^s$
   (c) if $r < s$ and $b = 1$, $b^r < b^s$

3. Let $b \neq 1$ be a positive real number, then $\exp_b(\log_b p) =$
   (a) b (b) $\log_b p$ (c) $\exp_b p$ (d) p (e) $\log_p \exp_p b$

4. Let $b \neq 1$ be a positive real number, then the function $\exp_b, R \to R^+$ is
   (a) strictly decreasing (b) strictly increasing (c) constant
   (d) injective (e) asymptotic to the line $y = x$.

5. Let $b \neq 1$ be a positive real number then
   (a) $\exp_b^{-1} b^a = 1$ (b) $\exp_b^{-1} b^a = b$ (c) $\exp_b^{-1} x = \log_b x$
   (d) $\log_b b^a = 1$ (e) $\log_b \exp_b 1 = b$

6. Suppose $\exp_3 t = 1/81$, then
   (a) $t = 1/4$ (b) $t = -4$ (c) $\log_3 t = 1/81$ (d) $t^3 = 1/81$
   (e) $(1/81)^t = 3$

In Problems 7–11, match the expression in the left-hand column with an equivalent expression from the right-hand column. Here $b \neq 1$ is a positive real number.

7. $\log_b (pq)$ (a) $(\exp_b p)^q$
8. $\log_b(p + q) =$ (b) $q \log_b p$
9. $\exp_b(p + q) =$ (c) $(\exp_b p) \times (\exp_b q)$
10. $\exp_b(pq) =$ (d) $\log_b p + \log_b q$
11. $\log_b(p^q) =$ (e) none of the above

In Problems 12–16 given $\log_{10} 4 = 0.6021$. The requested logarithms as indicated are:
  (a) $8.6990 - 10$ (b) 1.3010 (c) -1.6990 (d) 1.6021 (e) 2.2042

12. $\log_{10} 40$
13. $\log_{10} 0.02$
14. $\log_{10} 20$
15. $\log_{10} 0.05$
16. $\log_{10} 160$

17. Given that $\log_{10} e = 0.4343$, then $\log_e 10 =$
   (a) 2.3026 (b) 1.4329 (c) $9.5671 - 10$ (d) -1.5671 (e) e

18. Given that the number of bacteria present in a certain culture at any time t is given by $W(t) = 200(2^{3t})$, find t when there are 1,600 bacteria present.
   (a) $t = 0$ (b) $t = 1$ (c) $t = 2$ (d) $t = 3$ (e) need a table

19. An important material in establishing the age of an object is carbon-14, which has a half-life of 5,760 years. Suppose a given source originally contained 100,000 atoms of carbon-14. How many of these remain after 20 years?
   (a) $10^5 \exp[(-5760)(20)]$ (b) $20 \exp[10^5]$ (c) $2(10^5) \exp[(20)/(-5760)]$
   (d) $(10^5) \exp_2\left(-\dfrac{20}{5760}\right)$ (e) $(10^5) \exp_2\left(\dfrac{5760}{20}\right)$

187   Given the following table and $\log_{10} 4$ as given before Problem 12, find the following in Problems 20–22.

exponential & logarithmic functions

$\log_{10} t$

t	0	1	2	3
2.5	0.3979	0.3997	0.4014	0.4031
2.6	0.4150	0.4166	0.4183	0.4200
2.7	0.4314	0.4330	0.4346	0.4362
2.8	0.4472	0.4487	0.4502	0.4518
2.9	0.4624	0.4639	0.4654	0.4669

20. $\log_{10} 0.002626 =$
    (a) $0.4183{-}3$ (b) -3.42 (c) $7.4235{-}10$ (d) $0.4193{-}3$
    (e) -3.4183

21. $\exp_{10} 2.4330$
    (a) 2.71   (b) 24.33   (c) e^{10} (d) 271   (e) $e^{2.433}$

22. $\sqrt[3]{16} =$
    (a) 2.007   (b) 0.4014   (c) 4.014   (d) 2.52   (e) need a different table

23. If $\log_5(x + 5) = 1$ then
    (a) $x = 1$ (b) $x = 0$ (c) x is any real number   (d) $x = 10$
    (e) $5^{x+5} = 1$

24. For positive real numbers a and b, $(\log_a b)(\log_b a) =$
    (a) 1   (b) 0   (c) ab (d) $1/\log_{10} a$ (e) e

25. $10^{2\log_{10} 5} =$
    (a) 10   (b) 25   (c) 7   (d) 500   (e) need a table

trigonometric functions

5

In Chapters 1 and 2 we considered several special elementary functions. In Chapters 3 and 4 we studied classes of elementary real functions; viz., the polynomials, the rational functions, exponential and logarithmic functions. In this chapter, we shall discuss another class of real functions. The trigonometric functions are in fact an extremely important class of real functions, with many and varied applications; however, rather than looking at them strictly as real functions $f: R \rightarrow R$, we shall first consider them in their historical context as functions on sets of angles.

The sciences of geography and astronomy require their own special mathematics. It is interesting that this branch, trigonometry, should have a significance which transcends its origin and should have important application to present-day technology. As was the case with the logarithmic functions, the original ideas have lived on as they grew in their usefulness and applicability.

The origins of trigonometry are quite obscure. In the Rhind papyrus, one of the oldest known mathematical records of Egypt (1650 B.C.), there are several problems which appear to involve the cotangent of the dihedral angles at the base of a pyramid. In the remarkable Cuniform tablet Plimpton 322, which tells us something about Babylonian Mathematics, there seems to be a table of secants. However, the actual study of trigonometry as a mathematical discipline, so far as we know, was largely the work of Hipparchus (146 B.C.) and Ptolemy (100–178 A.D.) among others. Ptolemy, about 150 A.D., published much of it in his *Almagest*.

trigonometric functions

5.1.   angles & arcs

The usual textbook introductions to trigonometry are frequently criticized as "not good mathematics," because they rely heavily on the concept of *angle* and/or *arc length* without defining either of these. Each of these concepts is, in fact, possible to place on a rigorous mathematical basis but to do so requires the concepts of calculus. We shall assume that, at least intuitively, you know what an angle is and what the circumference and length of an arc of a circle are. In this section, we digress long enough to consider measuring angles.

One usually considers an angle to be formed by the intersection of two rays, l_1 and l_2 (half-lines), in the plane. We shall regard the angle to have been formed by one of the rays revolving about the point of intersection while the other ray remained fixed. The fixed ray is called the **initial side** and the rotating ray is called the **terminal side** of the angle. The fixed point is called the **vertex** of the angle.

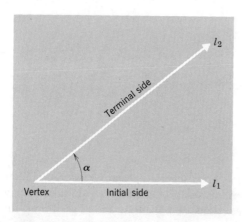

figure 5.1.1

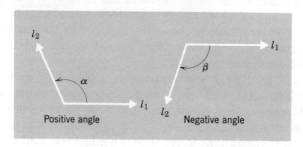

figure 5.1.2

Positive angle.
Negative angle.

If the rotation is in a counterclockwise direction, the angle is said to be a **positive angle.** *When the rotation is clockwise, the angle is* **negative.**

Angles have magnitude as well as direction. The magnitude of the angle is a numerical measure of the amount of rotation. As we shall shortly see, this involves measuring the arc of a circle.

Every angle is congruent to an angle in standard position. (We shall henceforth call congruent angles "equal.") *An angle is in* **standard position** *if it is placed in a coordinate plane* $X \times Y$, *with its vertex at the origin and its initial side along the positive x axis.*

190

angles & arcs

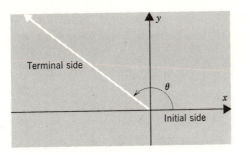

figure 5.1.3

Angle in standard position.

   With the angle in standard position, the terminal side cuts the unit circle ($\{(x,y)\,|\,x^2 + y^2 = 1\}$) in a point, $P:(x,y)$, and cuts off an arc of length s. We use the arc length to measure the angle (see Figure 5.1.4).

   The most familiar measure for the angle θ is the **degree**. *One degree is 1/360 of the circumference of the circle*. Hence, an angle of one degree, $1°$, subtends an arc which is 1/360 of the circumference. In Figure 5.1.5 we have illustrated some of these angles.

   Degrees (°) are subdivided into minutes (′) and seconds (″); there are 60 minutes in a degree and 60 seconds in a minute.

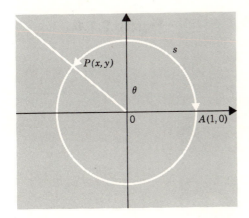

figure 5.1.4

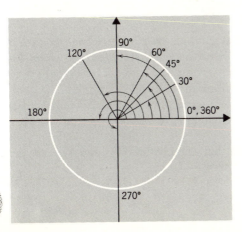

figure 5.1.5

trigonometric functions

example 1

Find $(1/4)(75°\ 29'\ 24'')$. Since $75°$ is not evenly divisible by 4, we write

$$\frac{1}{4}(75°\ 29'\ 24'') = \frac{1}{4}(72°\ 209'\ 24'')$$

$$= \frac{1}{4}(72°\ 208'\ 84'')$$

$$= 18°\ 52'\ 21'' \quad \bullet$$

The degree is $1/360$ of the circumference of *any* circle, so it does not have a constant length in the sense of measurement along a straight line.

example 2

A particle moves in a circular path with a radius of 4 ft, and travels 3/4 of the distance around the circle. How large is the angle which subtends this arc, and what is the distance traveled?

We assume that the particle begins at point $A:(4,0)$ and ends at point $P:(0,-4)$ (see Figure 5.1.6). Since a complete revolution involves $360°$, the angle involved is $3/4(360°)$ or $270°$. Since the circumference of the circle is 8π, the arc has length $(3/4)(8\pi) = 6\pi$. $\quad \bullet$

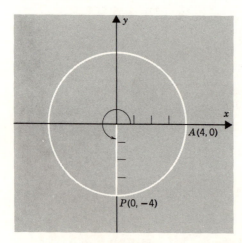

figure 5.1.6

A generally more useful angular measure is the *radian. An angle measuring* **one radian** *is defined as that central angle which subtends an arc of length equal to the radius of the circle.* Thus, in the unit circle, $C = \{(x,y)|x^2 + y^2 = 1\}$, an angle of one radian cuts off an arc of length 1.

Since the circumference of a circle of radius r is $2\pi r$ units and each angle of 1 radian cuts off an arc of length r, there are 2π radians in one complete revolution.

We obtain from the fact that one revolution is equal to 360 degrees and also to 2π radians

$$2\pi = 360°$$

or

$$\pi = 180°$$

angles & arcs

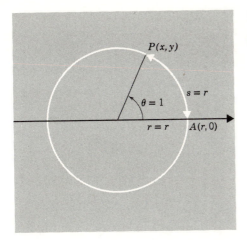

figure 5.1.7

$\theta = 1$ *radian.*

Thus, 1 radian is $180°/\pi$, or approximately $57°\,17'\,45''$. We also have $1° = \pi/180$ radians, or approximately 0.0175 radian. The following figure illustrates some common angles measured in radians. The numbers in parentheses are the rational approximations to the given number.

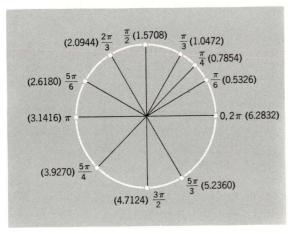

figure 5.1.8

On a circle whose radius is r, an angle of θ radians will subtend an arc of length $r\theta$. We have, then, the *arc length formula*.

$$s = r\theta \text{ (for } \theta \text{ in radians)}$$

example 3

On a circle of radius 5 inches an angle of $\pi/6 = 30°$ subtends an arc of

$$s = 5\frac{\pi}{6} = \frac{5\pi}{6} \text{ inches}$$

The particle in Example 2 moved through an angle of $(3/4)(2\pi) = 3\pi/2$ radians; hence, the distance traveled was

$$s = 4\frac{3\pi}{2} = 6\pi \text{ ft} \quad \bullet$$

193

trigonometric functions

example 4

A freeway curve is to be a circular arc, what radius should be used if the direction is to change 25° in a distance of 120 ft.

Here we must find the radius r of a circle in which a central angle of 25° subtends an arc of 120 ft. From $s = r\theta$ we get

$$r = \frac{s}{\theta}$$

when θ is in radians. To change 25° to radians we take

$$25 \cdot \frac{\pi}{180} = \frac{5\pi}{36} \text{ radians}$$

Thus,

$$r = \frac{120 \text{ ft}}{5\pi/36} = \frac{864}{\pi} \text{ ft, or approximately 275 ft} \quad \bullet$$

From the formula $s = r\theta$, relating arc length with the angle subtended, we can relate linear speed to angular speed. The usual measure of angular speed is in revolutions per minute (rpm). If the terminal side of the angle moves through an angle θ at a constant speed in time t, then its **angular speed** is

$$\omega = \frac{\theta}{t}$$

A point on the end of the terminal side moves a linear distance s during the same time t, so its *linear speed* is $s/t = v$. Thus, if we divide both sides of the equation, $s = r\theta$, by t we obtain

$$v = r\omega$$

This formula relates linear and angular speed. Note that ω is in *radians* per unit time.

example 5

If an automobile tire is 36 inches in diameter and is revolving at the rate 374 rpm, how fast is the car traveling?

First, 374 rpm $= 374\ (2\pi)$ radians per minute, then $v = 18(374)(2\pi) = 12{,}464$ inches per minute. To express this quantity in miles per hour, we divide by $(5280)(12)$ inches per mile and multiply by 60 minutes per hour. Thus,

$$v = \frac{(12{,}464)(60)}{(5280)(12)} = 40.1 \text{ mph} \quad \bullet$$

exercises

1. Change to radians (a multiple of π radians).

(a) 45°	(e) 210°	(i) 40°	(m) 2 revolutions
(b) 30°	(f) 300°	(j) 18°	(n) 4 revolutions
(c) 120°	(g) 330°	(k) 15°	(o) 1.3 revolutions
(d) 135°	(h) 240°	(l) 12°	(p) 3.6 revolutions

194

angles & arcs

2. Change to degrees, minutes, and seconds.

(a) $\dfrac{\pi}{6}$ (e) $\dfrac{3\pi}{4}$ (i) $\dfrac{\pi}{9}$ (m) 2.3

(b) $\dfrac{\pi}{4}$ (f) $\dfrac{2\pi}{3}$ (j) $\dfrac{\pi}{12}$ (n) 2

(c) $\dfrac{\pi}{2}$ (g) $\dfrac{7\pi}{6}$ (k) $\dfrac{\pi}{8}$ (o) 4.1

(d) $\dfrac{5\pi}{6}$ (h) $\dfrac{5\pi}{4}$ (l) $\dfrac{7\pi}{64}$ (p) 1.3 revolutions

3. For the given central angle θ, find the length of arc subtended on a circle whose radius is the given r.

(a) $\theta = \dfrac{\pi}{4}$, $r = 4$ (f) $\theta = 2.384$, $r = 4.832$ meters

(b) $\theta = \dfrac{\pi}{2}$, $r = 2$ (g) $\theta = .30°$, $r = 6$ in.

(c) $\theta = .683$, $r = 39.7$ ft   (h) $\theta = 27°$, $r = 10$ ft
(d) $\theta = 1.32$, $r = 4.01$ cm   (i) $\theta = 33°\,29'$, $r = 10$ ft
(e) $\theta = 2.54$, $r = 13.6$ cm   (j) $\theta = 33°\,29'$, $r = 100$ ft

4. For the given arc length s, find the central angle in radians which subtends this arc on the circle whose radius r is given.

(a) $s = \pi$ in., $r = 2$ in.   (e) $s = 4.8$ cm, $r = 7.3$ cm

(b) $s = \dfrac{\pi}{4}$ in., $r = 3$ in.   (f) $s = 613$ yds, $r = 218$ yds

(c) $s = 5\pi$ in., $r = 10$ in.   (g) $s = 23$ cm, $r = 106$ cm

(d) $s = \dfrac{\pi}{8}$ ft, $r = 6$ ft   (h) $s = 774$ in., $r = 95$ in.

5. How many radians are there between the hands of a clock at 12:20?

6. Through how many radians does the minute hand of a clock move in 39 min?

7. A highway curve is laid out as an arc of a circle of radius 1620 ft. What central angle is subtended in 45 sec by a car that is traveling 40 mph?

8. A locomotive with drive wheels 4.5 ft in diameter is moving at 54 mph. Find the angular velocity of a drive weel in radians per second.

9. The pendulum of a hall clock is 25 in. long and swings through an arc of 20° each second. How far does its tip move in 1 sec?

10. The minute hand of a clock is 2 in. long. How far does its tip move in an hour?

11. Find the linear speed in feet per second of the tip of a two-ft propeller that makes 2,400 rpm.

12. If the tip of the minute hand of a clock moves 630 cm in an hour, how long is the hand?

13. A pulley is moved by a belt. If the pulley turns through an angle of 390° while a point on the belt moves 15 ft, what is the radius of the pulley to the nearest tenth of a foot?

14. A bucket is being drawn from a well by pulling a rope over a pulley. If the bucket is raised 73 in. while the pulley is turned through 3.2 revolutions, find the radius of the pulley.

15. A car is traveling 63 mph. Find the angular speed of the wheels if they are 28 in. in diameter.

16. The tires of a bicycle are 26 in. in diameter and are moving at 4 rps. How fast is the bicycle moving?

17. What is the speed in miles per hour of a point on the equator due to the rotation of the earth? Assume the radius of the earth to be 4000 mi.

5.2. the trigonometric functions—angles

Let θ be any angle in *standard position* on a coordinate system. Remember that this means that the initial side of θ lies along the positive x axis and that its vertex is at the origin. Select any point $P:(x, y)$ on the terminal side of θ. The distance r from the origin to P is given by

$$r = \sqrt{x^2 + y^2}$$

We illustrate in Figure 5.2.1.

The trigonometric functions of the angle θ may be defined in terms of the numbers x, y, and r as follows.

sine: $\sin \theta = \dfrac{y}{r}$ $\qquad\qquad$ *cosecant*: $\csc \theta = \dfrac{1}{\sin \theta} = \dfrac{r}{y},\ y \neq 0$

cosine: $\cos \theta = \dfrac{x}{r}$ $\qquad\qquad$ *secant*: $\sec \theta = \dfrac{1}{\cos \theta} = \dfrac{r}{x},\ x \neq 0$

tangent: $\tan \theta = \dfrac{\sin \theta}{\cos \theta} = \dfrac{y}{x},\ x \neq 0$ $\quad$ *cotangent*: $\cot \theta = \dfrac{1}{\tan \theta} = \dfrac{x}{y},\ y \neq 0$

We shall shortly see that these six functions are, in fact, real functions, i.e., functions from R to R. First, however, a few comments are in order about the definitions given above. At first glance it appears that the number $\sin \theta$ might depend not only on the number θ, but also upon *where* one chooses the point P on the terminal side of an angle of measure θ. This, however, is not the case as the properties of similar triangles verifies. Thus, the ratios y_1/r_1 and y_2/r_2 are the same for a given θ, so long as $P_1(x_1, y_1)$ and $P_2(x_2, y_2)$ both lie on the

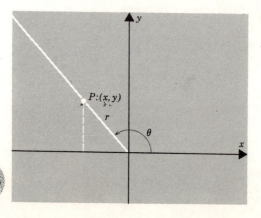

figure 5.2.1

Angle in standard position.

the trigonometric functions—angles

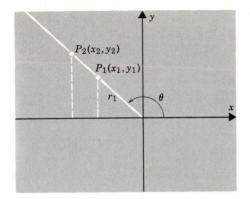

figure 5.2.2

Sine and cosine depend only on θ.

terminal side of the angle whose measure is θ. (We shall, henceforth, shorten this last statement to simply "the angle θ.")

This same comment applies to the other functions as well.

example 1

Compute the six trigonometric functions for an angle of $30° = \pi/6$ radians. From plane geometry, you will recall the theorem that *in a right triangle with angles of 30° and 60°, the side opposite the 30° angle is half the length of the hypotenuse.* Let $\theta = \pi/6$ measure an angle in standard position. Let $P(x, y)$ be selected on the terminal side of θ so that its distance from the origin $r = \sqrt{x^2 + y^2} = 2$. Then $y = 1$ and $x = \sqrt{3}$. The values of the trigonometric functions of θ are then easily computed as follows (see Figure 5.2.3).

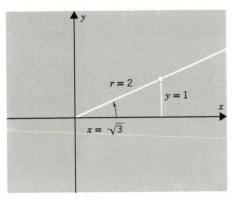

figure 5.2.3

$$\sin \frac{\pi}{6} = \frac{1}{2} \qquad \csc \frac{\pi}{6} = 2$$

$$\cos \frac{\pi}{6} = \frac{\sqrt{3}}{2} \qquad \sec \frac{\pi}{6} = \frac{2\sqrt{3}}{3}$$

$$\tan \frac{\pi}{6} = \frac{\sqrt{3}}{3} \qquad \tan \frac{\pi}{6} = \sqrt{3}$$

trigonometric functions

example 2

Compute the six trigonometric functions of an angle which measures $-60° = (-\pi/3)$.

If an angle of $-60°$ is placed in standard position, its terminal side lies in the fourth quadrant (see Figure 5.2.4). Select P on the terminal side so that $r = 2$. Then the coordinates of P are $(1, -\sqrt{3})$. (Why?) Thus

$$\sin\left(-\frac{\pi}{3}\right) = -\frac{\sqrt{3}}{2} \qquad \csc\left(-\frac{\pi}{3}\right) = -\frac{2\sqrt{3}}{3}$$

$$\cos\left(-\frac{\pi}{3}\right) = \frac{1}{2} \qquad \sec\left(-\frac{\pi}{3}\right) = 2$$

$$\tan\left(-\frac{\pi}{3}\right) = -\sqrt{3} \qquad \cot\left(-\frac{\pi}{3}\right) = -\frac{\sqrt{3}}{3}$$

●

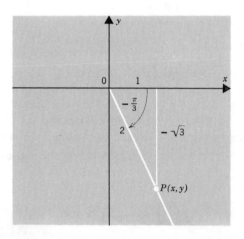

figure 5.2.4

example 3

Compute the value of the sine and cosine for the quadrant angles. For this purpose select $P(x, y)$ to lie on the unit circle $x^2 + y^2 = 1$, so that $r = 1$. When this is done, the definitions of sine and cosine simplify to the following. (Why?)

$$\sin\theta = y \qquad \text{and} \qquad \cos\theta = x$$

Using this information we compute the values given in the following table. (See Figure 5.2.5.) ●

θ in radians	0	$\pi/2$	π	$3\pi/2$	2π
θ in degrees	0	90°	180°	270°	360°
$\sin\theta$	0	1	0	-1	0
$\cos\theta$	1	0	-1	0	1

the trigonometric functions—angles

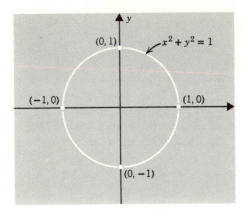

figure 5.2.5

example 4

Find all angles θ for which $\sin \theta = 0$. One sees from the definition that $\sin \theta$ will be zero if and only if the terminal side of θ lies along the x axis ($y = 0$). Thus, $\sin \theta = 0$ provided $\theta = 0°$, or $\theta = 180°$ or θ is a multiple of these (coterminal with one of them). In radians, θ is some integral multiple of π.

$$\theta = n\pi \quad \bullet$$

In defining the trigonometric functions, since $P(x, y)$ can be selected anywhere on the terminal side of θ, there is a certain economy if it is chosen to lie on the unit circle as was done in Example 3. When this is done, we note that, since $r = 1$, the coordinates of P are

$$x = \cos \theta \quad \text{and} \quad y = \sin \theta$$

The equation of the unit circle $x^2 + y^2 = 1$ leads us to the first basic **trigonometric identity**: For any θ,

(5.2.1) $$\cos^2 \theta + \sin^2 \theta = 1$$

Two other relations also follow. First, divide both sides of (5.2.1) by $\cos^2 \theta$; it becomes

$$1 + \frac{\sin^2 \theta}{\cos^2 \theta} = \frac{1}{\cos^2 \theta}$$

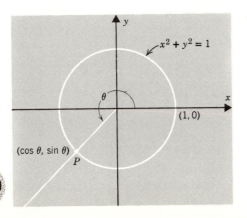

figure 5.2.6

trigonometric functions

or

(5.2.2) $$1 + \tan^2 \theta = \sec^2 \theta$$

Of course, such division can be accomplished only when $\cos \theta \neq 0$; i.e., when the terminal side of θ does not lie along the y axis. But a quick look at the definitions of the tangent and secant shows that those values of θ, for which $\cos \theta = 0$, fail to be in the domain of either *tangent* or *secant*. Thus, equation (5.2.2) holds for all θ in the domains of these two functions; namely, all $\theta \neq \pi/2 + k\pi$, k an integer.

A similar discussion shows that, for all θ for which $\sin \theta \neq 0$; we have

(5.2.3) $$\cot^2 \theta + 1 = \csc^2 \theta$$

The excluded values of θ are integral multiples of π; i.e., $\theta \neq k\pi$.

exercises

1. Complete the following function tables. Use a cross ($\times$) if the function is not defined for a particular value.

	θ	0	$\pi/6$	$-\pi/6$	$\pi/4$	$-\pi/4$	$\pi/3$	$-\pi/3$	$\pi/2$	$-\pi/2$
(a)	$\sin \theta$									
(b)	$\cos \theta$									
(c)	$\tan \theta$									
(d)	θ in degrees									

2. Complete the following function tables. Use a cross ($\times$) if the function is not defined for a particular value.

	θ	0	$\pi/6$	$-\pi/6$	$\pi/4$	$-\pi/4$	$\pi/3$	$-\pi/3$	$\pi/2$	$-\pi/2$
(a)	$\cot \theta$									
(b)	$\sec \theta$									
(c)	$\csc \theta$									
(d)	θ in degrees									

the trigonometric functions—angles

3. Complete the following function tables. Use a cross ($\times$) if the function is not defined for a particular value.

	θ	$2\pi/3$	$3\pi/4$	$5\pi/6$	π	$7\pi/6$	$5\pi/4$	$4\pi/3$	$3\pi/2$	$5\pi/3$	$7\pi/4$	$11\pi/6$
(a)	$\sin\theta$											
(b)	$\cos\theta$											
(c)	$\tan\theta$											
(d)	$\cot\theta$											
(e)	$\sec\theta$											
(f)	$\csc\theta$											
(g)	θ in degrees											

4. Construct an angle in standard position which satisfies the given condition(s). Find the value of the trigonometric functions of that angle.

    (a) $P(12, 5)$ is on the terminal side
    (b) $P(-4, -3)$ is on the terminal side
    (c) $P(15, -8)$ is on the terminal side
    (d) $P(-6, 8)$ is on the terminal side
    (e) $P(x, 8/17)$ lies on the unit circle, $x > 0$
    (f) $P(-12/13, y)$ lies on the unit circle, $y > 0$
    (g) $P(x, y)$ lies on the unit circle, $x, y > 0$

5. State in which quadrants the terminal side of an angle θ may lie if the given condition is to be satisfied.

    (a) the value of each function is positive
    (b) $\cos\theta$ and $\sec\theta$ are positive, the rest are negative
    (c) $\tan\theta$ and $\cot\theta$ are positive, the rest are negative
    (d) $\sin\theta$ and $\csc\theta$ are positive, the rest are negative
    (e) $\sin\theta \tan\theta > 0$
    (f) $\cos\theta \tan\theta > 0$
    (g) $\sin\theta > 0, \cos\theta < 0$
    (h) $\sin\theta < 0, \cos\theta > 0$

6. Determine which statements are true and which are false. Demonstrate that your answers are correct.

    (a) $\sin 30° + \sin 60° = \sin 90°$
    (b) $\sin 30° + \cos 60° = 1$
    (c) $2\sin 30° = \sin 60°$
    (d) $2\cos \pi/6 = \cos \pi/3$
    (e) $\sin \pi/4 + \cos \pi/4 = 1$
    (f) $\sin 120° - \sin 90° = \sin 30°$

    (g) $\cos (90° - 30°) = \sin 30°$
    (h) $\cos 120° + \cos 90° = -\sin 30°$
    (i) $\sin (90° + 30°) = \cos 30°$
    (j) $\cos\theta \sec\theta = 1$
    (k) $\tan\theta \cot\theta = 1$
    (l) $\sin\theta = \tan\theta \cos\theta$

trigonometric functions

7. Show that any point P in the x, y plane can have its coordinates (x, y) given as

$$(r \cos \theta, r \sin \theta)$$

where r is the distance from the origin to P and θ is the angle OP makes with the positive x axis.

5.3. the trigonometric functions & right triangles

In this section, we shall discuss some of the applications of the trigonometric functions. We are concerned first with those applications which involve finding the dimensions of a right triangle, so that the definitions of the proceeding section are adequate. Let us consider some examples.

example 1

Suppose that a boat travels in the direction 25° north of due east at 20 mph for 4 hours. How far north and how far east of its starting point is it at the end of this time?

Refer to Figure 5.3.1. Suppose that the boat begins at point O which we represent as the origin of a coordinate system. The position of the boat at the end of 4 hours is the point P on the terminal side of an angle of 25°. The length of OP is $20(4) = 80$ miles. The distance that P is north of O is the length of the segment AP, the y coordinate of P. The distance OA, or the x coordinate of P, is the distance east. Now

$$\sin 25° = \frac{y}{r} = \frac{AP}{OP}$$

and

$$\cos 25° = \frac{OA}{OP}$$

Thus,

$$AP = OP \sin 25° = 80 \sin 25°$$

and

$$OA = OP \cos 25° = 80 \cos 25°$$

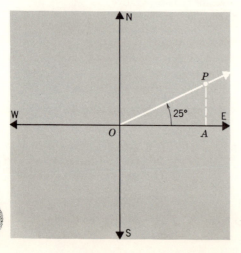

figure 5.3.1

the trigonometric functions & right triangles

Hence, the desired distances are expressed in terms of the values of the sine and cosine functions of the angle 25°. From Table 6 in Appendix II we determine that

$$OA = 80\,(0.9063) = 72.504 \text{ miles east and}$$

$$AP = 80\,(0.4226) = 33.808 \text{ miles north}$$

Hence, the boat is located 33.81 miles north and 72.50 miles east of its original position. ●

Note that in the above example, the angle in question was, fortuitously, in standard position. Suppose instead, that the compass bearing of the path of an airplane is given to be N 15° W. The 15° angle in question will not be in standard position if one sketches the graph in the usual way (Figure 5.3.2).

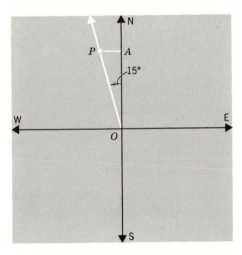

figure 5.3.2

Notice, however, that 15° is an acute angle of the right triangle OAP, where P is any point on the flight path, and A is its projection on the y axis (north-south axis).

Let us consider the general problem of the trigonometric functions of acute angles in right triangles. Given a right triangle ABC with right angle γ at C. Denote the (measure of the) angle at A by α and at B by β. Let a, b, and c be the lengths of the sides opposite angle A, B and C, respectively.

To compute the trigonometric functions of α in terms of the lengths of the sides of the triangle, we introduce a coordinate system in the plane so that A

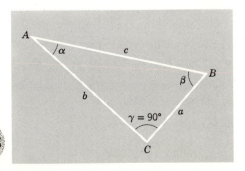

figure 5.3.3

trigonometric functions

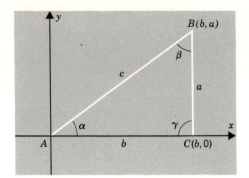

figure 5.3.4

is at the origin and AC lies along the positive x axis. The point B lies in the first quadrant and has coordinates (b, a) (see Figure 5.3.4). Angle α is in standard position.

Thus, the trigonometic functions of α are:

(5.3.1)
$$\sin \alpha = \frac{a}{c} = \frac{\text{side opposite}}{\text{hypotenuse}}$$

(5.3.2)
$$\cos \alpha = \frac{b}{c} = \frac{\text{side adjacent}}{\text{hypotenuse}}$$

(5.3.3)
$$\tan \alpha = \frac{a}{b} = \frac{\text{side opposite}}{\text{side adjacent}}$$

A similar result occurs when the coordinate system is introduced in such a way that β is in standard position. For acute angles in right triangles, then, we may use statements (5.3.1)–(5.3.3) directly without worrying about the orientation of the triangle on a coordinate system.

Let us now return to the question of the airplane whose path has bearing N 15° W.

example 2.

Suppose an airplane flies 400 mph for 2 hours in a direction N 15° W from city O. At the end of this time, how far west of city A will the plane be if A is located directly north of O and directly east of the plane? (See Figure 5.3.5.)

Find $\overline{AP}$, which is the side opposite the 15° angle. Since we can determine that $OP = (400 \times 2) = 800$, if follows that

$$\sin 15° = \frac{AP}{OP}, \text{ or } AP = 800 \sin 15°$$

From the table of trigonometric functions determine that $\sin 15° = 0.2588$, so $AP = 800 \,(0.2588) = 207.04$ miles (see Figure 5.3.5, p. 205). ●

Since the side opposite one acute angle of a right triangle is the side adjacent to the other—its complement—it is clear that

$$\sin \beta = \sin (90° - \alpha) = \cos \alpha$$

and

$$\cos \beta = \sin (90° - \alpha) = \sin \beta$$

the trigonometric functions & right triangles

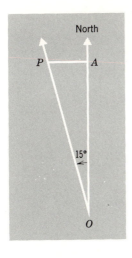

figure 5.3.5

This relation between sine and cosine demonstrates that the name "*cosine*" is derived from *complementary sine* i.e., the sine of the complement.

This relationship is exploited in the tables of trigonometric functions given in the Appendix (Table 6). To locate the sine, cosine, tangent, or cotangent of α, for $0 \leq \alpha \leq 45°$, (or $0 \leq \alpha \leq \pi/4$) one locates θ in one of the two left-hand columns and reads the desired value in the column *headed* sin α, cos α, tan α, or cot α. Thus,

$$\sin 35° = \sin 0.6109 = 0.5736$$
$$\cos 35° = 0.8192$$
$$\tan 35° = 0.7002$$

If $45° \leq \alpha \leq 90°$ ($\pi/4 \leq \alpha \leq \pi/2$), to find the values of the functions of α, one locates the value of α in one of the two right-hand columns and reads the appropriate value using the *footings* of each column. Thus,

$$\sin 0.9076 = 0.7880$$
$$\cos 1.0036 = 0.5373, \text{ etc.}$$

The heading of the column in which we found sin 0.9076 is cos α. We found sin 0.9076 = sin 52° by actually finding the cosine of its complement cos 38°. The table does this for us. To find the value of the functions of numbers α larger than $\pi/2$ we may also use this same table. We shall discuss how in subsequent sections.

One may use the method of linear interpolation described in connection with the logarithm function, to handle values which do not actually appear in the table.

example 3. angle of elevation and angle of depression

An observer is located on a mountain peak which rises 7,320 feet from the plane of the valley below. He observes two villages which lie in a line with the mountain. The angle of depression of the one is 45° and of the other is 20°. How far apart are the towns? Refer to Figure 5.3.6.

Suppose the observer is located at point A, B is the base of the mountain, and T_1 and T_2 are the towns. We want to find the length $T_1 T_2$, given the other

trigonometric functions

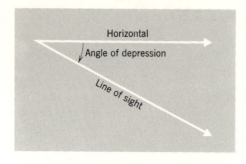

Angle of depression.

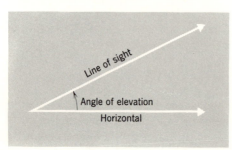

Angle of elevation.

figure 5.3.6

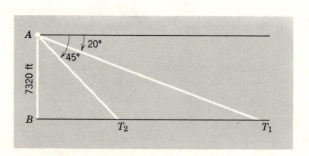

information. Since the angle of depression of T_2 is 45°, so is the (interior) angle BAT_2. Thus the length of BT_2 is also 7320 ft.

The interior angle BAT_1 measures 70°, so the length of BT_1 is 7320 tan 70° = 7320 cot 20°. Why? Hence BT_1 is (7320)(2.747) = 20,108 ft. Hence, the towns are 20,108 − 7320 = 12,788 ft = 2.4 miles apart. ●

exercises

1. Use the tables to find the indicated value.

(a) sin 20°	(f) sin 30°	(k) cos 69°
(b) cos 17°	(g) cos 60°	(l) tan 81°
(c) tan 41°	(h) tan 45°	(m) cot 5°
(d) cot 35°	(i) cot 15°	(n) sec 15°
(e) cos 50°	(j) sin 73°	(o) csc 15°

the trigonometric functions & right triangles

2. Use the tables to find a real number t in the interval $[0, \pi/2]$ which satisfies the given statement.

(a) $\cot t = 0.1763$ (f) $\cos t = 0.2079$ (k) $\sec t = 1.179$
(b) $\sin t = 0.9336$ (g) $\tan t = 0.6009$ (l) $\csc t = 1.103$
(c) $\cos t = 0.9272$ (h) $\sin t = 0.3256$ (m) $\csc t = 1.003$
(d) $\tan t = 1.590$ (i) $\cos t = 0.3987$ (n) $\sec t = 28.65$
(e) $\sin t = 0.6202$ (j) $\tan t = 1.621$ (o) $\sec t = 343.8$

3. Let triangle ABC be a right triangle with $\gamma = 90°$. Without using the tables, find the indicated dimension from the data given.

(a) $\alpha = 25°$, find β (f) $c = 26$, $\alpha = 30°$, find a
(b) $\beta = 71°$, find α (g) $c = 18\sqrt{2}$, $\alpha = 45°$, find a
(c) $a = 16$, $\alpha = 45°$, find b (h) $a = 38$, $c = 76$, find α
(d) $a = 30$, $\beta = 30°$, find b (i) $b = 28\sqrt{3}$, $c = 56$, find β
(e) $a = 25$, $b = 25$, find α (j) $c = 48$, $\beta = 60°$, find b

4. In each of the following, triangle ABC is a right triangle. From the given information determine sides a, b, and c and angles α and β $(\gamma = 90°)$.

(a) $\alpha = 60°$, $c = 3$ (f) $\alpha = 23°$, $c = 2.3$
(b) $\alpha = 45°$, $c = 2$ (g) $\beta = 37°$, $a = 12.9$
(c) $\alpha = 30°$, $c = 6$ (h) $\alpha = 52°$, $b = 1500$
(d) $\beta = 30°$, $a = 8$ (i) $\beta = 69°$, $a = 5.01$
(e) $\beta = 45°$, $b = 4$ (j) $\beta = 58°\,10'$, $b = .002$

5. An airplane flies at 320 mph on a heading of $60°$ (compass reading) from a point P. After flying for two hours, how far east is the plane from the north-south axis through P?

6. In square feet, calculate the area of a plot of land that is bounded as follows: beginning at a marked surveyor's slug in the sidewalk, go south 180 ft; then, N $30°\,10'$ E, 210 ft; then, due west to the starting point. Find the length of the north front as well.

7. Find the base and altitude of an isosceles triangle with a vertex angle of $65°$ and whose equal sides are 415 ft.

8. An A-frame cabin is to be constructed which has a height of 25 ft at the center and is 42 ft wide at the base. What angle does the roof make with the floor?

9. Given the A-frame of Exercise 8, what length is available for a second floor if one needs an 8 ft high ceiling for the bottom floor?

10. If the approach ramp for a certain bridge must gain 10 ft in elevation, and a desired angle of elevation is $3.5°$, how far from the end of the bridge must the ramp begin?

11. A surveyor (who just "happened" to have his instruments with him) was 126 ft from a geyser when it erupted. The angle of elevation of the maximum height the geyser reached was $43°$. What was this maximum height?

12. A motor boat crossed a river from west to east at a speed of 20 mph. The river current was 5 mph and flowed directly south. What would have been the speed of the boat in still water? What was its compass heading as it crossed?

13. A man pulls a rope attached to a sled with a force of 100 lbs. The rope makes an angle of $27°$ with the ground. Find the effective pull tending to move the sled along the ground.

14. There is a wind of 35 mph coming from $320°$ (from N). Find the air speed and heading necessary to give an airplane a groundspeed of 250 mph and a path of $50°$ (from N).

trigonometric functions

5.4. the trigonometric functions—real numbers

The definitions of the trigonometric functions given in the preceding section have relied heavily on geometry and on angles. This is the way in which these functions originated. The angle approach has many applications in surveying, navigation, mechanics, etc. However, the trigonometric functions are also useful in the study of sound, alternating current, etc. These things have no natural relationship to angles. For this reason, we shall now turn to an analytic approach, given in terms of function composition. This approach could be taken for the definitions of the trigonometric functions and the preceding development as an interpretation of these functions. Our development involves the unit circle. For this reason, these functions are often called *circular* functions.

Consider three new functions, two projection maps, and the wrapping function.

The Projection p_x. The function p_x has as its domain the set $R^2 = R \times R$ of ordered pairs of real numbers. Its range is the set R of real numbers.

We define

$$p_x: R^2 \to R$$

by the rule

$$p_x(x, y) = x$$

A geometric interpretation of p_x is easily given. Let $Q(x, y)$ be any point in the plane, then p_x projects the point Q onto its x coordinate.

The Projection p_y.

$$p_y: R^2 \to R$$

is defined by the rule

$$p_y(x, y) = y$$

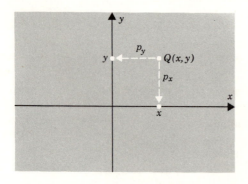

figure 5.4.1

Thus, p_y projects $Q(x, y)$ onto its y coordinate.

Our third function is a bit more complicated and requires some intuition. ***The wrapping function w*** has the set R of real numbers as its domain and the set $R \times R$ as its range. The image of a real number t under w is a point (x, y) of the unit circle $x^2 + y^2 = 1$ such that the length of the arc $\overset{\frown}{AQ}$ from $A(1, 0)$ to $Q(x, y)$ is t. *Thus,*

$$w(0) = (1, 0)$$

the trigonometric functions—real numbers

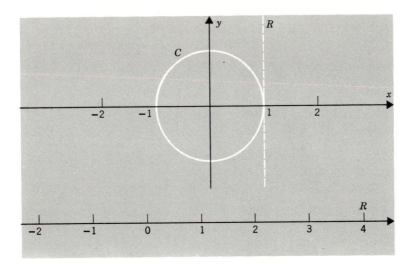

figure 5.4.2

and for each $t \in R$, $t \neq 0$, $w(t) = (x, y)$, *where* $x^2 + y^2 = 1$, *and the arc from* $w(0)$ *to* $w(t)$ *has length* t.

It is fairly easy to see that the image of the interval $(0, 2\pi)$ of R is the entire circumference of the unit circle, and that the entire set R is wrapped around the unit circle—hence the name *wrapping function*. We note that $w(2n\pi) = w(0) = (1, 0)$, while $w((2n + 1)\pi) = w(\pi) = (-1, 0)$, and so forth. Figure 5.4.3 gives an illustration of w.

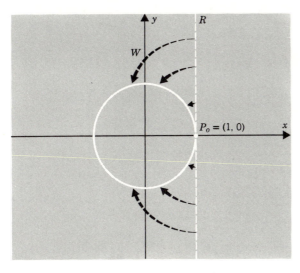

figure 5.4.3

The wrapping function has an interesting property, *periodicity*. We state this formally in the following definition and theorem:

A function f with domain X contained in the real numbers R is called **periodic** *with* **period** p *provided that, for each* $t \in X$

$$f(t + p) = f(t)$$

trigonometric functions

theorem 5.4.1

The wrapping function $w: R \rightarrow R^2$ *is periodic with period* 2π.

This theorem follows readily from the fact that the circumference of the unit circle is 2π. Thus, if $w(t)$ is the point (a, b), then $w(t + 2\pi)$ is also (a, b).

You can verify other properties of the wrapping function in the exercises. The sine and cosine functions can now be described in terms of the wrapping function and the projections. p_x and p_y as follows. *For any real number t,*

$$\sin t = (p_y \circ w)(t) = p_y(w(t)) = y$$

and

$$\cos t = (p_x \circ w)(t) = p_x(w(t)) = x$$

In this way, without actually referring to angles, we see that sine and cosine are functions from R into R. It is also fairly easy to see that the *image* of each is the interval $[-1, 1]$ of the real line.

As a consequence of this view of the sine and cosine function we see that, for each real number t, the image of t under the wrapping function is

$$w(t) = (\cos t, \sin t)$$

The other trigonometric functions are defined in terms of sine and cosine, exactly as was done in the earlier definitions. Thus, *if t is a real number, then*

$$\tan t = \frac{\sin t}{\cos t} \qquad \cot t = \frac{1}{\tan t}$$

$$\sec t = \frac{1}{\cos t} \qquad \csc t = \frac{1}{\sin t}$$

Since $w(t)$ is a point on the unit circle $x^2 + y^2 = 1$, the three fundamental

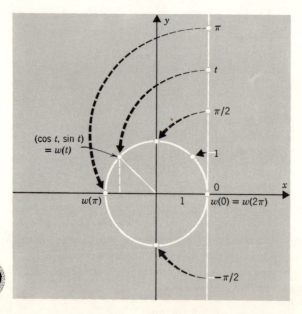

figure 5.4.4

the trigonometric functions—real numbers

identities of section 5.2 hold for all real numbers t (except as noted before). We have the theorem:

theorem 5.4.2

Let t be a real number, then

(i) $\sin^2 t + \cos^2 t = 1$, *all* t;
(ii) $1 + \tan^2 t = \sec^2 t$, $t \neq \pi/2 + k\pi$; $k \in Z$
(iii) $1 + \cot^2 t = \csc^2 t$, $t \neq k\pi$; $k \in Z$

To determine the zeros of the trigonometric functions, we refer to the unit circle and the point $w(t)$. One of the functions will have a zero whenever one of the coordinates of $w(t)$ is zero; that is, when $w(t)$ is located on one of the coordinate axes.

Thus, $\sin t = 0$ when $w(t)$ is on the x axis; i.e., when $t = n\pi$. The zeros of cosine occur when $w(t)$ is on the y axis: $\cos t = 0$ when $t = \pi/2 + n\pi$. The zeros of the tangent correspond to those of the sine, and the cotangent has the same zeros as the cosine. The following table summarizes these results.

Functions	Set of Zeros
sine and tangent	$\{t = n\pi \mid n \in Z\}$
cosine and cotangent	$\{t = \pi/2 + n\pi \mid n \in Z\}$
secant and cosecant	ϕ

As a final remark, we note that, if t is a real number, we can construct an angle θ_t in standard position which measures t radians. Then the point $w(t)$ corresponds with the point at which the terminal side of θ_t cuts the unit circle. Hence, the definitions of the trigonometric functions of real numbers given in this section correspond with those for angles given in the preceding section (see Figure 5.4.5).

figure 5.4.5

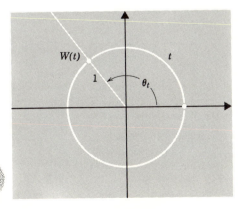

trigonometric functions

exercises

draw pic of wrap. func (handwritten margin note)

1. Let w be the wrapping function. Show that for each $t \in R$, if $w(t) = (a, b)$, then $w(-t) = (a, -b)$.

2. Show that if $w(t) = (a, b)$, $t \in R$, then each of the following is true.

    (a) $w(t + \pi) = (-a, -b)$ (c) $w(t + \pi/2) = (-b, a)$
    (b) $w(t - \pi) = (-a, -b)$ (d) $w(\pi/2 - t) = (b, a)$

3. Show that each of the following statements is true for every real number t for which the given functions are defined.

    (a) $\cos^2 t = 1 - \sin^2 t$ (e) $\cos^2 t - \sin^2 t = 2\cos^2 t - 1$
    (b) $\sin^2 t = 1 - \cos^2 t$ (f) $\tan^2 t = \sec^2 t - 1$
    (c) $(\cos t + \sin t)^2 = 1 + 2\sin t \cos t$
    (d) $\cos^2 t - \sin^2 t = 1 - 2\sin^2 t$ (g) $\dfrac{\csc t}{\sec t} = \cot t$

4. Show that each of the following statements is true for every real number t for which the given functions are defined.

    (a) $\sin t \sec t = \cot(\pi/2 - t)$ (e) $\dfrac{1}{\sec t + \tan t} = \sec t - \tan t$

    (b) $\cos^2 t - \sin^2 t = \dfrac{1 - \tan^2 t}{1 + \tan^2 t}$ (f) $\sin^4 t - \cos^4 t = 1 - 2\cos^2 t$

    (c) $\dfrac{1 + \tan t}{1 - \tan t} = \dfrac{\cos t + \sin t}{\cos t - \sin t}$ (g) $\cos t = \cos(-t)$

    (d) $\dfrac{1 - \cos t}{1 + \cos t} = (\csc t - \cot t)^2$

5. Explain why neither sine nor cosine has an inverse which is a function.

6. Show that for any real number t

    (a) $|\sin t| \le 1$ (b) $|\cos t| \le 1$

7. Is $\tan t$ a periodic function? Explain.

8. Suppose $f(x) = \sin x$ and $g(x) = 2x$ discuss the functions

    (a) $f \circ g$ (b) $g \circ f$

9. Show that for any real number t

    (a) $|\sec t| \ge 1$ (b) $|\csc t| \ge 1$

10. Refer to Figure 5.4.6 in which OA, OB, and OP are radii of the unit circle (each is 1 unit) QB and TA are tangents and CP and DP are half cords. Show that the trigonometric functions of θ can be represented by the six lengths given.

    (a) $\sin \theta = CP$ (d) $\csc \theta = OQ$
    (b) $\cos \theta = DP$ (e) $\sec \theta = OT$
    (c) $\tan \theta = AT$ (f) $\cot \theta = BQ$

11. Prove Theorem 5.4.2.

basic properties of trigonometric functions

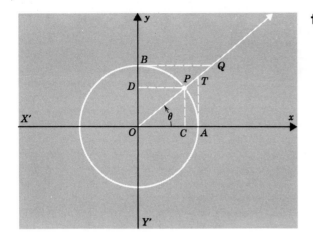

figure 5.4.6

5.5. basic properties of trigonometric functions

As important as the uses of trigonometry in navigation and mechanics are, the trigonometric functions have far wider applications. These applications depend upon the analytic properties these functions—as functions—possess. In this section, we shall examine some of these properties. We shall formally demonstrate a number of them.

One notes, almost immediately, that since the wrapping function is periodic, we cannot hope to have the trigonometric functions be one-to-one (injective). Also, although some uncritical people might like to have it so, the trigonometric functions are surely *not linear*. That is, $\cos(t_1 + t_2)$ is hardly ever equal to $\cos t_1 + \cos t_2$. Just what is true about these functions will be investigated here. First, we turn to a property which is based upon the symmetry of the unit circle.

theorem 5.5.1

For each real number t

$$\text{(i)} \quad \sin(-t) = -\sin t$$
$$\text{(ii)} \quad \cos(-t) = \cos t$$

This follows from the fact that for every real number t, the points $w(-t)$ and $w(t)$ are symmetric with respect to the x axis (see Figure 5.5.1 for an example). Therefore, $\sin(-t) = p_y(w(-t)) = -p_y(w(t))$. By the same token, $\cos(-t) = p_x(w(-t)) = p_x(w(t))$.

These properties are also possessed by the simple polynomial functions (power function)

$$f(x) = x^n$$

n a positive integer. When n is even, $f(x) = x^{2k} = f(-x)$, and when n is odd, we have $f(-x) = (-x)^{2k+1} = (-x)(-x)^{2k} = -(x^{2k+1}) = -f(x)$. By analogy one says that the sine is an *odd function* and the cosine is an *even function*.

213

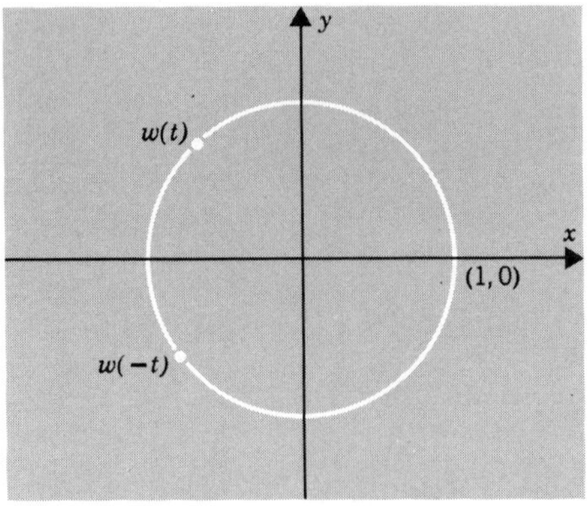

trigonometric functions

figure 5.5.1

In the exercises, you are asked to consider the other trigonometric functions with respect to "evenness" or "oddness."

In Section 5.3. we noticed that in right triangles with acute angles α and β

$$\sin \beta = \sin (90° - \alpha) = \cos \alpha$$

and

$$\cos \beta = \sin (90° - \alpha) = \sin \beta$$

These facts are not only true for acute angles in right triangles, but hold for every real number t. Consider the following theorem.

theorem 5.5.2

Let t be any real number, then

    (i) $\sin (\pi/2 - t) = \cos t$
    (ii) $\cos (\pi/2 - t) = \sin t$

The proof of this theorem depends on the fact that if w is the wrapping function, and $w(t) = (a, b)$, then $w(\pi/2 - t) = (b, a)$ (see Exercise 4.2 (d)). Since $\sin t = b$ and $\cos t = a$ it follows that $w(\pi/2 - t) = (\sin t, \cos t)$. Hence,

$$\sin (\pi/2 - t) = p_y\big(w(\pi/2 - t)\big) = \cos t$$

and

$$\cos (\pi/2 - t) = p_x\big(w(\pi/2 - t)\big) = \sin t$$

as desired.

Now, since for two real numbers α and β, $\cos (\alpha + \beta) \neq \cos \alpha + \cos \beta$ in general, we establish formulas which do relate the functions of a sum with the functions of each number separately.

theorem 5.5.3

    (i) $\cos (\alpha + \beta) = \cos \alpha \cos \beta - \sin \alpha \sin \beta$
    (ii) $\cos (\alpha - \beta) = \cos \alpha \cos \beta + \sin \alpha \sin \beta$

proof. Let $C = \{(x, y) \mid x^2 + y^2 = 1\}$ be the unit circle, and consider the four points $P(1, 0)$, $A = w(-\alpha)$, $B = w(\beta)$, and $Q = w(\alpha + \beta)$, all on C. If *both* α and β are positive, this is illustrated in Figure 5.5.2. Now both of the arcs $\overset{\frown}{AB}$ and $\overset{\frown}{QP}$ have the same length; viz., $\alpha + \beta$. Thus the chords $\overline{AB}$ and

basic properties of trigonometric functions

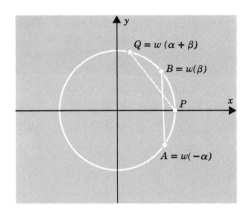

figure 5.5.2

$\overline{PQ}$ have equal lengths as well. Now the coordinates of $P = w(\alpha + \beta)$ are $(\cos (\alpha + \beta), \sin (\alpha + \beta))$, and so the distance formula gives

$$\overline{PQ} = \sqrt{[\cos (\alpha + \beta) - 1]^2 + [\sin (\alpha + \beta)]^2}$$

On the other hand, the same formula gives

$$\overline{AB} = \sqrt{(\cos \beta - \cos \alpha)^2 + (\sin \beta + \sin \alpha)^2}$$

since A has coordinates $(\cos (-\alpha), \sin (-\alpha)) = (\cos \alpha, -\sin \alpha)$, by the previous theorem. The coordinates of B are $(\cos \beta, \sin \beta)$.

Setting $\overline{AB} = \overline{PQ}$ and squaring both sides, results in

$$[\cos (\alpha + \beta) - 1]^2 + [\sin(\alpha + \beta)]^2 = (\cos \beta - \cos \alpha)^2 + (\sin \beta + \sin \alpha)^2.$$

Expanding, this results in

$$\cos^2 (\alpha + \beta) - 2 \cos (\alpha + \beta) + 1 + \sin^2 (\alpha + \beta) =$$
$$\cos^2 \beta - 2 \cos \alpha \cos \beta + \cos^2 \alpha + \sin^2 \beta + 2 \sin \alpha \sin \beta + \sin^2 \beta.$$

We recall that $\cos^2 t + \sin^2 t = 1$, so by appropriate grouping the equation above becomes

$$2 - 2 \cos (\alpha + \beta) = 2 - 2 \cos \alpha \cos \beta + 2 \sin \alpha \sin \beta.$$

This simplifies to

$$cos (\alpha + \beta) = cos\ \alpha\ cos\ \beta - sin\ \alpha\ sin\ \beta,$$

as desired. The second formula, for $\cos (\alpha - \beta)$, follows readily from this one and Theorem 5.5.1. We have

$$\cos (\alpha - \beta) = \cos (\alpha + (-\beta)) = \cos \alpha \cos (-\beta) - \sin \alpha \sin (-\beta).$$

This is the same as $\cos (\alpha - \beta) = \cos \alpha \cos \beta - [(\sin \alpha)(-\sin \beta)]$, which yields the desired result:

$$cos (\alpha - \beta) = cos\ \alpha\ cos\ \beta + sin\ \alpha\ sin\ \beta.$$

To obtain formulas for $\sin (\alpha + \beta)$ and $\sin (\alpha - \beta)$ we refer to Theorem 5.5.2. Thus;

215

$$sin (\alpha + \beta) = cos (\pi/2 - (\alpha + \beta)) = cos\cdot((\pi/2 - \alpha) - \beta).$$

trigonometric functions

But $\cos\left((\pi/2 - \alpha) - \beta\right) = \cos(\pi/2 - \alpha)\cos\beta + \sin(\pi/2 - \alpha)\sin\beta$, from the theorem just proved. Using Theorem 5.5.2 again we obtain,

$$\cos\left((\pi/2 - \alpha) - \beta\right) = \sin\alpha\cos\beta + \cos\alpha\sin\beta.$$

Thus,

$$\sin(\alpha + \beta) = \sin\alpha\cos\beta + \cos\alpha\sin\beta$$

For the $\sin(\alpha - \beta)$, we need only substitute $-\beta$ for β in the preceding formula and use Theorem 5.5.1 to simplify. Thus,

$$\sin(\alpha - \beta) = \sin\alpha\cos\beta - \cos\alpha\sin\beta$$

The so called *double angle* formulas readily follow upon setting $\alpha = \beta$ in the *sum* formulas. Thus,

$$\sin 2\alpha = 2\sin\alpha\cos\alpha$$

and

$$\cos 2\alpha = \cos^2\alpha - \sin^2\alpha$$

We involve Theorem 5.3.1 (a) to obtain

$$\cos 2\alpha = 1 - 2\sin^2\alpha$$

and

$$= 2\cos^2\alpha - 1$$

on substituting $1 - \sin^2\alpha = \cos^2\alpha$, or $1 - \cos^2\alpha = \sin^2\alpha$ in the original formula for $\cos 2\alpha$. These identities are summarized in the following theorem.

theorem 5.5.4

For every two real numbers α and β

    (i) $\sin(\alpha + \beta) = \sin\alpha\cos\beta + \cos\alpha\sin\beta$
    (ii) $\sin(\alpha - \beta) = \sin\alpha\cos\beta - \cos\alpha\sin\beta$
    (iii) $\sin 2\alpha = 2\sin\alpha\cos\alpha$
    (iv) $\cos 2\alpha = \cos^2\alpha - \sin^2\alpha$
    (v) $\cos 2\alpha = 2\cos^2\alpha - 1$
    (vi) $\cos 2\alpha = 1 - 2\sin^2\alpha$

    As you did with elementary arithmetic facts, to be able to successfully apply these mathematical ideas to your subsequent work, you must memorize these *identities* as well as those in Theorem 5.5.3 and the following three for the tangent. The proof of this next theorem is left as an exercise.

theorem 5.5.5

For any two real numbers α, β such that $\cos\alpha \neq 0$, $\cos\beta \neq 0$ we have

    (i) $\tan(\alpha + \beta) = \dfrac{\tan\alpha + \tan\beta}{1 - \tan\alpha\tan\beta}$, *if* $\cos(\alpha + \beta) \neq 0$.

    (ii) $\tan(\alpha - \beta) = \dfrac{\tan\alpha - \tan\beta}{1 + \tan\alpha\tan\beta}$, *if* $\cos(\alpha - \beta) \neq 0$.

    (iii) $\tan 2\alpha = \dfrac{2\tan\alpha}{1 - \tan^2\alpha}$, *if* $\cos 2\alpha \neq 0$

basic properties of trigonometric functions

The proof of (i) is obtained by writing $\tan(\alpha + \beta) = \sin(\alpha + \beta)/\cos(\alpha + \beta)$, using Theorems 5.5.3 and 5.5.4 to expand the right side of this, and then dividing the resulting numerator and denominator by $(\cos \alpha \cos \beta)$. See Exercises 3, 4, and 5.

exercises

1. Determine which statements are true and which are false. Demonstrate that your answers are correct.

(a) $\cos 60° + \cos 30° = \cos 90°$

(b) $2 \tan 30° = \tan 60°$

(c) $2 \sin \dfrac{\pi}{6} = \sin \dfrac{\pi}{3}$

(d) $\sin 60° + \cos 30° = 1$

(e) $\sin(90° - 30°) = \cos 60°$

(f) $\sec \dfrac{\pi}{6} \cos \dfrac{\pi}{6} = \tan \dfrac{\pi}{4}$

(g) $\csc \dfrac{\pi}{4} \sin \dfrac{\pi}{4} = \cot \dfrac{\pi}{4}$

(h) $2 \tan \dfrac{\pi}{8} = \tan \dfrac{\pi}{4}$

(i) $\tan \dfrac{\pi}{8} \cot \dfrac{\pi}{8} = \cot \dfrac{\pi}{4}$

(j) $\sin\left(\pi - \dfrac{\pi}{3}\right) = \sin \dfrac{\pi}{3}$

2. Use the addition and subtraction formulas and the values of the functions of $\pi/6$, $\pi/4$, and $\pi/3$ to compute the desired result. Do not use the tables.

(a) $\cos \dfrac{\pi}{12}$

(b) $\sin \dfrac{\pi}{12}$

(c) $\sin \dfrac{5\pi}{12}$

(d) $\tan \dfrac{\pi}{12}$

(e) $\cot \dfrac{5\pi}{12}$

(f) $\sec \dfrac{5\pi}{12}$

(g) $\csc \dfrac{7\pi}{12}$

(h) $\cot \dfrac{\pi}{12}$

(i) $\tan \dfrac{13\pi}{12}$

(j) $\sin \dfrac{7\pi}{12}$

3. Prove Theorem 5.5.5 (i).

4. Prove Theorem 5.5.5 (ii).

5. Prove Theorem 5.5.5 (iii).

6. Use Theorem 5.5.4 to verify the following identities.

(a) For every real number t, $\sin \dfrac{t}{2} = \pm \sqrt{\dfrac{1 - \cos t}{2}}$

(b) For every real number t, $\cos \dfrac{t}{2} = \pm \sqrt{\dfrac{1 + \cos t}{2}}$

(c) For α and β real numbers such that $\sin \alpha \neq \beta$, $\sin \beta \neq 0$ and $\sin(\alpha + \beta) \neq 0$

$$\cot(\alpha + \beta) = \dfrac{\cot \alpha \cot \beta - 1}{\cot \alpha + \cot \beta}$$

trigonometric functions

7. Show that tangent, cotangent, and cosecant are odd functions while secant is an even function.

8. Show that:

(a) For each real number t, $\sin 3t = 3 \sin t - 4 \sin^3 t$.
(b) For each real number t, $\cos 3t = 4 \cos^3 t - 3 \cos t$.

9. Use the theorems and the results of previous exercises to compute the following.

(a) $\cos 15°$

(b) $\sin 15°$

(c) $\cos 120°$

(d) $\sin \dfrac{\pi}{8}$

(e) $\cos \dfrac{\pi}{8}$

(f) $\sin \dfrac{\pi}{12}$

(g) $\cos \dfrac{3\pi}{8}$

(h) $\sin \dfrac{3\pi}{8}$

(i) If $\cos \alpha = -4/5$ and $\sin \alpha < 0$, compute $\cos \dfrac{\alpha}{2}$.

(j) If $\cos \alpha = 5/13$ and $\sin \alpha < 0$, compute $\cos \dfrac{\alpha}{2}$.

10. Prove that each of the following is an identity.

(a) $\sin x \cos y = \dfrac{1}{2} [\sin (x - y) + \sin (x + y)]$

(b) $\cos x \cos y = \dfrac{1}{2} [\cos (x - y) + \cos (x + y)]$

(c) $\sin x \sin y = \dfrac{1}{2} [\cos (x - y) - \cos (x + y)]$

(d) $\sin x + \sin y = 2 \sin \dfrac{x + y}{2} \cos \dfrac{x - y}{2}$

(e) $\sin x - \sin y = 2 \cos \dfrac{x + y}{2} \sin \dfrac{x - y}{2}$

(f) $\cos x + \cos y = 2 \cos \dfrac{x + y}{2} \cos \dfrac{x - y}{2}$

(g) $\cos x - \cos y = -2 \sin \dfrac{x + y}{2} \sin \dfrac{x - y}{2}$

11. Prove that for each real number t in the domain of the function:

(a) $\tan (-t) = -\tan t$
(b) $\sec (-t) = \sec t$
(c) $\cos t = \sin (t + \pi/2)$
(d) $\sin (t + \pi) = -\cos t$

(e) $\cot (-t) = -\cot t$
(f) $\csc (-t) = -\csc t$
(g) $\cos (x + \pi/2) = -\sin x$
(h) $\cos (x + \pi) = -\sin x$

12. Prove that the following are identities.

(a) $\tan (\pi - x) = -\tan x$
(b) $\sin (\pi - x) = \sin x$
(c) $\cos (x - \pi) = -\cos x$
(d) $\tan (x + \pi) = \tan x$

(e) $\sin (x + \pi) = -\sin x$
(f) $\cos (x + \pi) = -\cos x$
(g) $\tan (x + \pi/2) = -\cot x$
(h) $\tan (x - \pi/2) = \cot x$

218

trigonometric identities

13. Determine which of the following are true and which are false.

(a) $\cos(-x)\sec x = 1$

(d) $\dfrac{-\sin x}{\cos(-x)} = \tan x$

(b) $\tan(-x)\cot(-x) = 1$

(e) $\dfrac{\cos(-x)}{\sin(-x)} = \cot x$

(c) $\dfrac{\sin(-x)}{-\cos x} = \tan x$

(f) $\sin 138° \csc 42° = 1$

14. Prove that the following are identities.

(a) $\sin 3\theta + \cos 6\theta = (1 + 2\sin 3\theta)(1 - \sin 3\theta)$
(b) $4\sin\theta\cos\theta\cos 2\theta = \sin 4\theta$
(c) $\cot 2\theta - \cot 4\theta = \csc 4\theta$
(d) $\cot 2\theta + \tan\theta = \csc 2\theta$
(e) $1 + \tan\theta\tan 2\theta = \sec 2\theta$
(f) $\cot\theta - \tan\theta = 2\cot 2\theta$
(g) $\cot 4\theta + \tan 2\theta = \csc 4\theta$
(h) $\sin 6\theta - \sin 2\theta = 2\sin 2\theta - 4\sin^3 2\theta$

5.6. trigonometric identities

In the exercise of the preceding section, you were asked to demonstrate several important relations among the trigonometric functions. These relationships are statements that certain combinations of trigonometric functions are equal to others. *A statement of the form*

$$f(x) = g(x)$$

that two functions are equal is called an **identity** *if it holds for all values of x in the common domain of f and g. Thus,*

$$(x + 2)^2 = x^2 + 4x + 4$$

is an identity (algebraic) because the square of the real number $x + 2$ is *always* equal to the number $x^2 + 4x + 4$, no matter what real number x is.

We summarize here the important trigonometric identities which were developed in the previous sections.

fundamental identities

$\tan t \cot t = 1$ (1)

$\sec t \cos t = 1$ (2)

$\tan t = \dfrac{\sin t}{\cos t}$ (3)

$\cot t = \dfrac{\cos t}{\sin t}$ (4)

$\sin t \csc t = 1$ (5)

$\sin^2 t + \cos^2 t = 1$ (6)

$1 + \tan^2 t = \sec^2 t$ (7)

$1 + \cot^2 t = \csc^2 t$ (8)

trigonometric functions

other important identities

Sum and Difference Formulas

$$\cos (s + t) = \cos s \cos t - \sin s \sin t \tag{9}$$

$$\sin (s + t) = \sin s \cos t + \cos s \sin t \tag{10}$$

$$\cos (s - t) = \cos s \cos t + \sin s \sin t \tag{11}$$

$$\sin (s - t) = \sin s \cos t - \cos s \sin t \tag{12}$$

$$\tan (s + t) = \frac{\tan s + \tan t}{1 - \tan s \tan t} \tag{13}$$

$$\tan (s - t) = \frac{\tan s - \tan t}{1 + \tan s \tan t} \tag{14}$$

"Double-Angle" Formulas

$$\cos 2s = \cos^2 s - \sin^2 s \tag{15}$$

$$\cos 2s = 1 - 2 \sin^2 s \tag{15a}$$

$$\cos 2s = 2 \cos^2 s - 1 \tag{15b}$$

$$\sin 2s = 2 \sin s \cos s \tag{16}$$

$$\tan 2s = \frac{2 \tan s}{1 - \tan^2 s} \tag{17}$$

"Half-Angle" Formulas

$$\cos \frac{1}{2}\theta = \pm \sqrt{\frac{1 + \cos \theta}{2}} \tag{18}$$

$$\sin \frac{1}{2}\theta = \pm \sqrt{\frac{1 - \cos \theta}{2}} \tag{19}$$

$$\tan \frac{1}{2}\theta = \pm \sqrt{\frac{1 - \cos \theta}{1 + \cos \theta}} \tag{20}$$

$$\tan \frac{1}{2}\theta = \frac{1 - \cos \theta}{\sin \theta} \tag{20a}$$

$$\tan \frac{1}{2}\theta = \frac{\sin \theta}{1 + \cos \theta} \tag{20b}$$

Product and Factor Formulas

$$2 \sin s \cos t = \sin (s + t) + \sin (s - t) \tag{21}$$

$$2 \cos s \sin t = \sin (s + t) - \sin (s - t) \tag{22}$$

$$2 \cos s \cos t = \cos (s + t) + \cos (s - t) \tag{23}$$

$$2 \sin s \sin t = -\cos (s + t) + \cos (s - t) \tag{24}$$

$$\sin s + \sin t = 2 \sin \frac{1}{2}(s + t) \cos \frac{1}{2}(s - t) \tag{25}$$

trigonometric identities

$$\sin s - \sin t = 2 \cos \frac{1}{2}(s + t) \sin \frac{1}{2}(s - t) \qquad (26)$$

$$\cos s + \cos t = 2 \cos \frac{1}{2}(s + t) \cos \frac{1}{2}(s - t) \qquad (27)$$

$$\cos s - \cos t = -2 \sin \frac{1}{2}(s + t) \sin \frac{1}{2}(s - t) \qquad (28)$$

In the exercises which follow, you are asked to verify that other trigonometric relations are identities. To do so, it is necessary to show that the equalities hold for each real number in the domains of the functions involved. The techniques for doing this are much like those employed by research mathematicians in proving conjectures to be theorems. That is, one must begin with the known and by manipulation and insight arrive at the desired result. With trigonometric identities, it is usually safest to work with one side of the equation and reduce it to the other side. If you work with both sides of the proposed identity you are quite likely to fall in the trap of assuming a statement is true because its converse can be shown to be true. Consider the following ridiculous argument—the first part of which is valid.

theorem

If $3 = 5$, then $8 = 8$.

proof.

$3 = 5$; given

$5 = 3$; symmetry of "equals"

Thus

$$8 = 8$$

since, if equals are added to equals, the results are equal.

The argument thus far is valid and the theorem has been proved. The fallacy occurs when we note that $8 = 8$ (the conclusion) is true and then wish to conclude that our hypothesis, $3 = 5$ must, therefore, have been true also. Of course no one would do this, because he "knows" $3 \neq 5$. But suppose that we try to prove an identity as follows.

(5.6.1) Prove:

$$\frac{\sin t}{\cos t} - 1 = \sin^2 t (\sin t \sec t - 1) + \cos t (\sin t - \cos t)$$

Multiply through by $\cos t$. We obtain:

$$(\sin t - \cos t) = \sin^3 t \sec t \cos t - \sin^2 t \cos t + \cos^2 t \sin t - \cos^3 t$$

$$= \sin^3 t - \sin^2 t \cos t + \cos^2 t \sin t - \cos^3 t$$

$$= (\sin t - \cos t)(\sin^2 t + \cos^2 t)$$

(5.6.2) $$\sin t - \cos t = \sin t - \cos t$$

We've reached our $8 = 8$; viz., statement (5.6.2). Can we conclude that

trigonometric functions

$3 = 5$; i.e., that the original statement (5.6.1) is an identity? Although this example is not as transparent as the first one, the fallacy is no less severe.

You are invited to decide, by legitimate means, whether or not equation (5.6.1) is an identity. The following suggestions may be helpful in working with identities.

(1) Usually it is better to work with the most complicated side of the identity.
(2) Perform indicated operations first.
(3) Use the fundamental identities to reduce the number of different trigonometric expressions involved.
(4) Simplify fractions.
(5) Factor whenever possible.
(6) When all else seems to fail, try changing everything to sines and cosines.

exercises

1. Give examples to show that the following are not identities.

   (a) $\sin t + \cos t = 1$

   (b) $\sin^2 t - \cos^2 t = \cos^2 t$

   (c) $\sin t \cot t = 1$
   (d) $1 + \sin^2 t = \cos^2 t$

   (e) $\sin s \sin t = \sin st$

   (f) $\dfrac{\tan t}{\cot t} = 1$

   (g) $\sin (s - t) = \sin s - \sin t$
   (h) $\sin (\csc t) = 1$

2. Express each of the following in terms of sine and/or cosine only.

   (a) $(\sec t + \csc t)^2 \tan t$

   (b) $(\csc t + \cot t)^2$

   (c) $\dfrac{1}{1 + \tan^2 t}$

   (d) $\dfrac{1}{\sec^2 t} - \dfrac{1}{\tan^2 t}$

   (e) $\tan^2 x$

   (f) $\dfrac{1 - \sin^2 x}{\cos x}$

   (g) $\cot (\pi/2 - t)$

   (h) $\left[\sin \dfrac{x}{2} + \cos \dfrac{x}{2} \right]^2$

3. Prove that each of the following is an identity.

   (a) $\dfrac{\cos 3t}{\sin t} + \dfrac{\sin 3t}{\cos t} = 2 \cot 2t$

   (b) $\dfrac{\cos 2t}{\sin t} + \dfrac{\sin 2t}{\cos t} = \csc t$

   (c) $\cos t - \tan s \sin t = \sec s \cos (s + t)$
   (d) $\sin (s + t) \csc t = \cos s + \cot t \sin s$
   (e) $\sin x(\cot x - \tan y) = \cos (x + y) \sec y$

   (f) $\dfrac{\sin (\alpha - \beta)}{\sin \alpha \sin \beta} = \cot \alpha - \cot \beta$

   (g) $\dfrac{\cot \alpha - \tan \beta}{\cot \alpha + \tan \beta} = \dfrac{\cos (\alpha + \beta)}{\cos (\alpha - \beta)}$

   (h) $\dfrac{1 + \tan x}{1 - \tan x} = \dfrac{1 + \sin 2x}{\cos 2x}$

trigonometric identities

4. Prove that each of the following is an identity.

    (a) $(\sin x + \cos x)^2 = 1 + \sin 2x$

    (b) $(\sec x - \tan x)^2 = 1 + 2 \tan^2 x - 2 \sec x \tan x$

    (c) $\sin 4t = 4 \sin t (2 \cos^2 t - 1) \cos t$

    (d) $\cos 3t = 4 \cos^3 t - 3 \cos t$

    (e) $\sin 3t = 3 \sin t - 4 \sin^3 t$

    (f) $\sin 2t - \cos 2t \tan t = \tan t$

    (g) $\cot 2t = \dfrac{\csc t - 2 \sin t}{2 \cos t}$

    (h) $\dfrac{2 \sin t - \sin 2t}{1 - \cos t} = 2 \sin t$

5. Prove that each of the following is an identity.

    (a) $\dfrac{\sin s + \sin t}{\cos s + \cos t} = \tan \dfrac{1}{2}(s + t)$

    (b) $\dfrac{\cos s + \cos t}{\sin s - \sin t} = \cot \dfrac{1}{2}(s - t)$

    (c) $\dfrac{\cos 3t}{\cos t} = 2 \cos 2t - 1$

    (d) $\dfrac{\sin 3t}{\sin t} = 2 \cos 2t + 1$

    (e) $\dfrac{\cos 4t}{\cos 2t} = 2 \cos 2t - \sec 2t$

    (f) $1 - \tan \beta \tan (\alpha - \beta) = \sec (\alpha - \beta) \cos \alpha \sec \beta$

    (g) $\tan (\alpha - \beta) + \tan \beta = \sec (\alpha - \beta) \sin \alpha \sec \beta$

    (h) $\cot \alpha - \cot (\alpha - \beta) = \csc (\beta - \alpha) \sin \beta \csc \alpha$

6. Prove that the following are identities.

    (a) $\sec x = \dfrac{\csc^2 \dfrac{x}{2}}{\cot^2 \dfrac{x}{2} - 1}$

    (c) $\dfrac{\cos x}{\sin \dfrac{x}{2}} = \csc \dfrac{x}{2} - 2 \sin \dfrac{x}{2}$

    (b) $\dfrac{1 + \cot^2 \dfrac{x}{2}}{\left(1 + \cot \dfrac{x}{2}\right)^2} = \dfrac{1}{1 + \sin x}$

    (d) $\dfrac{2 \sin \dfrac{x}{2} - \sin x}{2 - 2 \cos \dfrac{x}{2}} = \sin \dfrac{x}{2}$

7. Identify as true or false and prove your assertion.

    (a) $\cos \pi/2 - \cos \pi/6 = \cos \pi/3$

    (e) $\dfrac{\sin \dfrac{3\pi}{2}}{\sin \dfrac{\pi}{2}} = \sin \pi$

    (b) $\dfrac{\cos 4\pi}{\cos 2\pi} = \cos 2\pi$

    (f) $\sin 21° = \sqrt{\dfrac{1 - \cos 42°}{2}}$

    (c) $\cos 11° = -\sqrt{\dfrac{1 + \cos 22°}{2}}$

    (g) $\tan 144° = \dfrac{2 \tan 72°}{1 - \tan^2 18°}$

    (d) $\sin 2\theta - \sin \theta = \sin \theta$

    (h) $\tan 208° = \dfrac{2 \tan 76°}{\tan^2 76° - 1}$

trigonometric functions

5.7. graphs of the trigonometric functions

In Theorem 5.3.1, we noted that the wrapping functions is periodic with period 2π. It then follows that both the sine and cosine are also periodic with period 2π. We state this fact as a theorem and leave the proof as an exercise.

theorem 5.7.1

For each real number t,

(i) $\sin(t + 2\pi) = \sin t$
(ii) $\cos(t + 2\pi) = \cos t$

Thus, the sine and cosine are periodic functions with period 2π.

It is very useful to consider the graphs of the trigonometric functions. From the, graph one can immediately recognize many of the properties of these functions. The fact that all the trigonometric functions are periodic enables us to more easily sketch their graphs—we shall formally discuss the periodicity of the tangent, cotangent, secant, and cosecant in a subsequent theorem. For the sine and cosine, if we know the graph on an interval of length 2π, we essentially know the complete graph.

For the sine function we first determine its graph on the interval $[0, 2\pi]$. To graph sine in the XY plane, i.e., $y = \sin x$, the following geometric device is useful.

Place an st coordinate system to the left of the y axis so that the s axis and the y axis are parallel, and so that the t axis and the x axis are on the same line. Let $C = \{(s, t) \mid s^2 + t^2 = 1\}$ be the unit circle. For any number $x \in [0, 2\pi]$, locate $w(x)$ on C. The vertical coordinate of $W(x)$ is $\sin x$ which we then use to determine the point $(x, \sin x)$ on the graph of $y = \sin x$. See Figure 5.7.1.

The graph can then be extended to the entire real line using the periodicity of sine. See Figure 5.7.2 for a portion of the graph.

Notice that many of the properties of the sine function can be read from its graph. In particular, the following theorem appears to be true.

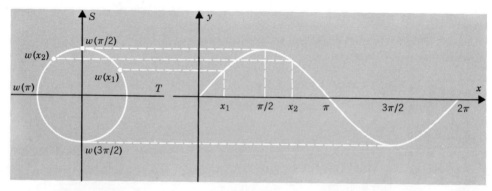

figure 5.7.1

graphs of the trigonometric functions

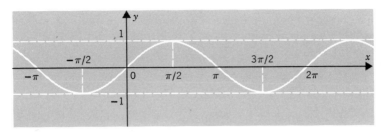

figure 5.7.2

$y = \sin x$.

theorem 5.7.3

A real number x is a zero of the sine function if, and only if, x is a multiple of π; i.e.,

$$\sin x = 0$$

if, and only if, $x = k\pi$, for some $k \in Z$.

To obtain the graph of the cosine function, refer back to Theorem 5.5.2. Thus, for any real number x, $\cos x = \sin (\pi/2 - x)$. From Theorem 5.5.1, we have $\cos (-x) = \cos x$. Combining these,

(5.7.2) $$\cos x = \sin(x + \pi/2)$$

Hence, we obtain the value of $y = \cos x$ if, for each x on the real line, we let y be the vertical distance to the sine curve at the point $x + \pi/2$. Hence, $\cos 0 = \sin \pi/2$, $\cos \pi/2 = \sin \pi$, etc. See this illustrated in Figure 5.7.3.

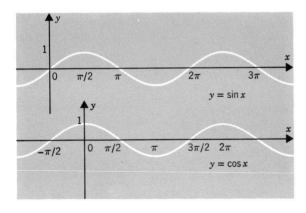

figure 5.7.3

The zeros of the cosine function occur at $s + \pi/2$ where s is a zero of the sine function. Therefore, x is a zero of cosine if $x = k\pi + \pi/2 = (2k + 1)\pi/2$.

theorem 5.7.4

A real number x is a zero of the cosine function if, and only if, x is an odd multiple of $\pi/2$; i.e.,

$$\cos x = 0$$

if, and only if, $x = (2k + 1)\pi/2$, for some $k \in Z$.

The tangent function is also periodic, but its period is π, rather than 2π. To see this we show that, for each real number x for which $\cos x \neq 0$, $\tan (x + \pi) = \tan x$.

$$\tan (x + \pi) = \frac{\sin (x + \pi)}{\cos (x + \pi)} = \frac{-\sin x}{-\cos x} = \tan x$$

Suppose that for some p, satisfying $0 < p \leq \pi$, $\tan (x + p) = \tan x$. Then for $x = 0$

$$\tan p = \tan 0 = 0$$

Therefore, $\sin p = \tan p \cos p = 0$. Thus, p is a zero of the sine function. But $p \in (0, \pi]$, so $p = \pi$. Therefore, the period of the tangent function is π.

In a similar way, one can prove that the cotangent has period π. This, gives the first half of the following theorem. The proof of the rest of the theorem is left to you.

theorem 5.7.4

> *The functions tangent and cotangent are periodic with period π. The secant and cosecant are also periodic functions whose period is 2π.*

To sketch the graph of the tangent function, we also make use of the fact that it is symmetric with respect to the origin.

$$\tan (-x) = -\tan x$$

For $x \in (-\pi/2, \pi/2)$, since $\tan x = \sin x/\cos x$, $\cos x$ is positive, but approaches zero as x approaches either $\pm\pi/2$; thus, the lines $x = -\pi/2$ and $x = -\pi/2$ are vertical asymptotes for the tangent. By plotting points $(x, \tan x)$, for $x \in (-\pi/2, \pi/2)$, we obtain the section of the graph shown in Figure 5.7.4. The tangent has period π, and hence, its graph can be completed by periodicity. See Figure 5.7.5 for a portion of it.

An interesting alternative method to obtain the graph in Figure 5.7.4 is described in Exercise 3.

The zeros of the tangent function are, of course, the same as those of the

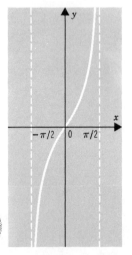

figure 5.7.4

graphs of the trigonometric functions

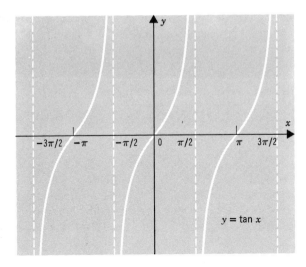

figure 5.7.5

$y = tan\ x$

sine function since a fraction is zero when its numerator is zero and its denominator is not zero.

theorem 5.7.5

A real number x is a zero of the tangent function if, and only if x is a multiple of π; i.e.,

$$tan\ x = 0$$

if, and only if, $x = k\pi$ for some $k \in Z$.

We note that while sine and cosine have only the closed interval $[-1,1]$ for their range, the tangent function maps the open interval $(-\pi/2,\ \pi/2)$ onto the whole real line.

You will be asked in the Exercises to use similar analysis to verify that the graphs of the other trigonometric functions are as they appear in Figures 5.7.6, 5.7.7, and 5.7.8.

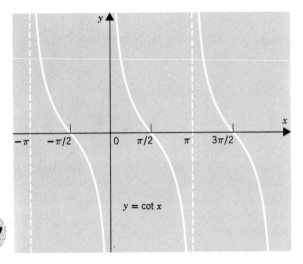

figure 5.7.6

$y = cot\ x$

trigonometric functions

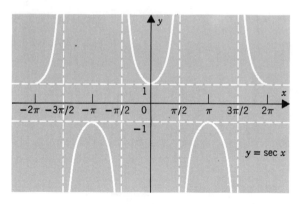

figure 5.7.7

$y = \sec x$

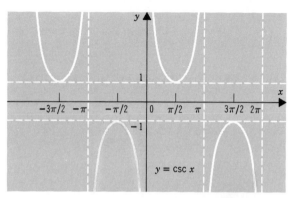

figure 5.7.8

$y = \csc x$

Note that the cotangent function has the whole real line as its range while the range of the secant and cosecant *excludes* the interval $(-1,1)$; i.e.,

$$|\sec x| \geq 1 \quad \text{and} \quad |\csc x| \geq 1, \qquad \text{for all } x$$

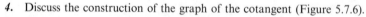

exercises

1. Construct the graph of the specified function by assigning values of x at intervals of $\pi/12$. Make a function table and plot the points of the graph on the interval specified.

(a) $y = \sin x,\ x \in [-3\pi/2,\ \pi/2]$ (d) $y = \cot x,\ x \in [-\pi/2,\ \pi/2]$
(b) $y = \cos x,\ x \in [5\pi/2,\ -\pi/2]$ (e) $y = \sec x,\ x \in [-\pi/2,\ \pi/2]$
(c) $y = \tan x,\ x \in [-\pi,\ \pi]$ (f) $y = \csc x,\ x \in [-\pi,\ \pi/2]$

2. Do the same as Exercise 1 for the following functions.

(a) $y = \sin (x + \pi/2),\ x \in [-\pi,\ \pi]$ (d) $y = \sin (x - \pi/2),\ x \in [-\pi,\ \pi]$
(b) $y = \cos (x + 3\pi/2),\ x \in [-\pi,\ \pi]$ (e) $y = \cos (x - \pi),\ x \in [-3\pi/2,\ \pi/2]$
(c) $y = \sec (x + \pi),\ x \in [-5\pi/2,\ -\pi/2]$ (f) $y = \tan (x - \pi/2),\ x \in [-\pi/2,\ 3\pi/2]$

3. Use Figure 5.4.6 as suggested in Exercise 10 of Section 5.4 to construct the tangent graph as outlined in Figure 5.7.9.

4. Discuss the construction of the graph of the cotangent (Figure 5.7.6).

228 **5.** Discuss the construction of the secant graph (Figure 5.7.7).

6. Discuss the construction of the cosecant graph (Figure 5.7.8).

periodic motion

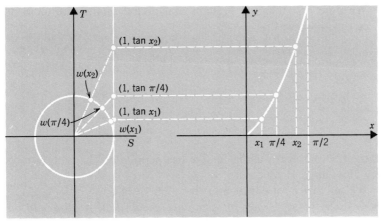

figure 5.7.9 7-9 a,b,c dd

7. Sketch a graph for each of the following functions, and compare it with the graphs of the six trigonometric functions.

(a) $f(x) = -\sin x$
(b) $g(x) = -\cos x$
(c) $h(x) = \sin(-x)$
(d) $k(x) = \cos(-x)$

(e) $t(x) = -\tan x$
(f) $s(x) = -\cot x$
(g) $m(x) = -\sec x$
(h) $n(x) = \csc(-x)$

8. Sketch a graph of each of the following functions.

(a) $y = 2\sin x$
(b) $y = 2\cos x$
(c) $y = (1/2)\sin x$

(d) $y = 5\cos x$
(e) $y = 3\tan x$
(f) $y = 2\sec x$

9. Sketch a graph of each of the following functions.

(a) $y = \sin 2x$
(b) $y = \cos 2x$
(c) $y = \sin x/2$

(d) $y = \cos 5x$
(e) $y = \tan 3x$
(f) $y = \sec 2x$

10. What would the graph of $y = e^t \sin t$ look like? Discuss.

5.8. periodic motion

Much of the present day importance of trigonometry stems from the fact that many natural phenomena are periodic. Problems in light, sound, and electricity, as well as some social phenomena, are periodic. They require periodic functions to describe them adequately.

In previous sections, we have already noted the periodicity of the trigonometric functions. In this section, we shall examine these functions as they apply to periodic phenomena.

Consider the function

(5.8.1) $$f(t) = A\sin(\omega t + \phi)$$

This equation describes what is known as *simple harmonic motion*. Such motion is typical of electromagnetic waves.

trigonometric functions

We note first that the fundamental period of the motion is $\gamma = 2\pi/\omega$. To see this, observe that

$$f\left(t + \frac{2\pi}{\omega}\right) = A \sin\left(\omega\left(t + \frac{2\pi}{\omega}\right) + \phi\right)$$

$$= A \sin(\omega t + \phi + 2\pi)$$

$$= A \sin(\omega t + \phi)$$

$$= f(t)$$

because for each real number t, sine has period 2π. The motion repeats itself every $2\pi/\omega$ units of time.

We see that in one time unit there are $1/(2\pi/\omega) = \omega/2\pi$ cycles. *Thus, the number $\omega/2\pi$, the reciprocal of the period, is called the* **frequency** *of the motion.*

Since $|\sin x| \leq 1$ for all x, we see that $|f(t)| \leq A$. *The number A is called the* **amplitude** *of the motion.*

Now write

$$f(t) = A \sin(\omega t + \phi)$$

$$= A \sin\left[\omega\left(t + \frac{\phi}{\omega}\right)\right]$$

The number ϕ/ω is called the **phase shift** *of the periodic motion.* At $t = x$, $f(t)$ does not equal A times the sine of ωx, but rather A times the value of the sine function at $\omega(x + \phi/\omega)$. The graph of the function is ϕ/ω *out of phase* with the graph of sine function.

This occurs, of course, because when $t = -\phi/\omega$, we have $f(-\phi/\omega) = A \sin \omega(0)$ or $f(-\phi/\omega) = 0$. Thus the zero of f occurs at $t = -\phi/\omega$ rather than at $t = 0$.

We saw in the last section that the cosine graph was shifted $\pi/2$ units to the left of the sine graph. In this case we had $\cos t = \sin(t + \pi/2)$, A and ω both being 1. A *negative* phase shift is a shift to the *right*, while if ϕ/ω is *positive* the shift is to the *left*.

example 1

Consider the function (t in seconds)

$$f(t) = 3 \sin(2t + \pi/3)$$

Determine the amplitude, period, frequency, and phase shift of this motion, and sketch the graph of f.

If we write

$$f(t) = 3 \sin 2(t + \pi/6)$$

We see that $A = 3$, $\omega = 2$ and $\phi/\omega = \pi/6$. Thus, the amplitude is 3. The period $2\pi/\omega = \pi$, and the frequency is $1/\pi$, so the graph repeats itself every π seconds and has a frequency of $1/\pi$ cycles per second. The phase shift is $\pi/6$ units to the left. The graph is sketched in Figure 5.8.1. In fact we sketch, in Figure 5.8.1, the graphs of $y = \sin t$, $y = \sin 2t$, $y = 3 \sin 2t$, and $y = f(t)$ for comparison. ●

periodic motion

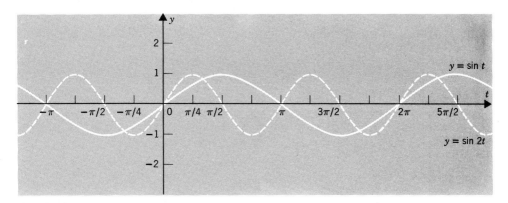

figure 5.8.1(a)

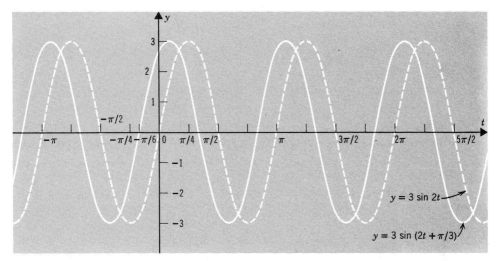

figure 5.8.1(b)

A function of the form

(5.8.2) $$h(t) = A \cos (\omega t + \phi)$$

also describes a harmonic motion with amplitude A and period $2\pi/\omega$. The phase shift is ϕ/ω with respect to the cosine graph. Since $\sin (x + \pi/2) = \cos x$

$$h(t) = A \sin (\omega t + \phi + \pi/2)$$

$$= A \sin \omega \left(t + \frac{\phi + \pi/2}{\omega} \right)$$

Thus, with respect to the sine we have a phase shift of $(\phi + \pi/2)/\omega$.

Consider now a function of the form

(5.8.3) $$k(t) = a \sin \omega t + b \cos \omega t$$

That this represents a harmonic motion as well, is seen as follows. If $a = 0$ and $b = 0$ then $k(t)$ is the constant function O, so

$$k(t) = O \sin \omega t$$

trigonometric functions

Otherwise $a^2 + b^2 > 0$. We consider the numbers

$$x_0 = \frac{a}{\sqrt{a^2 + b^2}} \qquad y_0 = \frac{b}{\sqrt{a^2 + b^2}} \qquad A = \sqrt{a^2 + b^2}$$

The point (x_0, y_0) is easily seen to be a point of the unit circle. Since the wrapping function is a surjection (onto), there is some real number ϕ such that

$$(5.8.4) \qquad \cos \phi = x_0 = \frac{a}{\sqrt{a^2 + b^2}} \qquad \sin \phi = y_0 = \frac{b}{\sqrt{a^2 + b^2}}$$

Substituting these values in (5.8.3), we obtain the following.

$$k(t) = a \sin \omega t + b \cos \omega t$$

$$= \sqrt{a^2 + b^2} \left(\frac{a}{\sqrt{a^2 + b^2}} \sin (\omega t) + \frac{b}{\sqrt{a^2 + b^2}} \cos (\omega t) \right)$$

$$= A(\cos \phi \sin \omega t + \sin \phi \cos \omega t)$$

$$= A(\sin (\omega t + \phi))$$

for each real number t. Hence $k(t)$ is a function describing a harmonic motion with period $2\pi/\omega$, amplitude $A = \sqrt{a^2 + b^2}$, and phase shift ϕ/ω.

example 2

Given the function

$$s(t) = \sqrt{3} \sin \frac{t}{2} - \cos \frac{t}{2}$$

find the amplitude, frequency, period, and phase shift of the motion it describes. Sketch its graph.

Here $\omega = 1/2$, $a = \sqrt{3}$, and $b = -1$. Hence, $a^2 + b^2 = 4$, and so $A = 2$. We set $\cos \phi = \sqrt{3}/2$ and $\sin \phi = -1/2$. This gives

$$s(t) = \sqrt{3} \sin \frac{t}{2} - \cos \frac{t}{2}$$

$$= 2 \left(\frac{\sqrt{3}}{2} \sin \frac{t}{2} - \frac{1}{2} \cos \frac{t}{2} \right)$$

$$= 2 \left(\cos \phi \sin \frac{t}{2} + \sin \phi \cos \frac{t}{2} \right)$$

$$= 2 \sin \left(\frac{t}{2} + \phi \right)$$

The amplitude of s is 2, the period is $2\pi/(1/2) = 4\pi$, the frequency is $1/4\pi$ and the phase shift is $\phi/(1/2) = 2\phi$, where $\cos \phi = \sqrt{3}/2$ and $\sin \phi = -1/2$. The smallest (in magnitude) such ϕ is (see Figure 5.8.2) $-\pi/6$. Thus the phase shift is $-\pi/3$, $\pi/3$ units to the *right*. The graph of $s(t)$ is given in Figure 5.8.3.

$$s(t) = 2 \sin 1/2 (t - \pi/3)$$

Among other things, this analysis has shown that the sum of two functions describing harmonic motion with the same period also describes harmonic motion with that period. ●

232

periodic motion

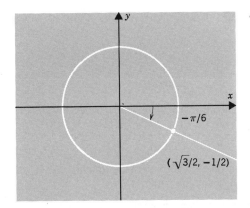

figure 5.8.2

$-\pi/6$

$(\sqrt{3}/2, -1/2)$

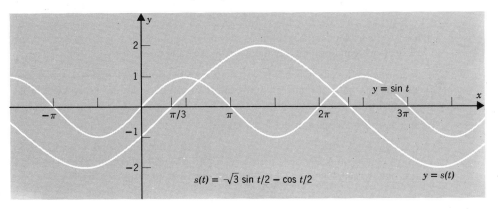

$y = \sin t$

$s(t) = \sqrt{3} \sin t/2 - \cos t/2$

$y = s(t)$

figure 5.8.3

$$s(t) = \sqrt{3} \sin \frac{t}{2} - \cos \frac{t}{2}$$

exercises

1. Find the amplitude, period, frequency, and phase shift, and sketch the graph for each of the following functions.

(a) $f(t) = 3 \sin t$ (d) $f(t) = 5 \sin (2t + \pi/4)$
(b) $f(t) = -1/3 \cos t$ (e) $f(t) = (1/2) \cos (2t - \pi/4)$
(c) $f(t) = \sin (t - \pi/4)$ (f) $f(t) = .01 \sin (t/5 - 0.15)$

2. Do the same as Exercise 1 for the functions.

(a) $f(t) = \sin t + \cos t$ (d) $f(t) = \sqrt{3} \sin t + \cos t$
(b) $f(t) = \sin t - \cos t$ (e) $f(t) = 3 \sin t + 4 \cos t$
(c) $f(t) = \sin t/3 - \sqrt{3} \cos t/3$ (f) $f(t) = 4 \sin t/2 - 3 \cos t/2$

3. Sketch the graphs of each of the following functions.

(a) $f(t) = (1/3) \tan t$ (d) $f(t) = 3 \cot t/2$
(b) $f(t) = -(1/2) \sec 2t$ (e) $f(t) = (1/3) \csc 2t$
(c) $f(t) = 2 \tan t/2$ (f) $f(t) = 2 \tan (t/2 + \pi/4)$

4. Sketch the graphs of each of the following functions. Discuss the periodicity, amplitude, frequency, phase shift, etc.

(a) $f(t) = |\sin t|$

(b) $f(t) = |\cos (2t - \pi/4)|$

(c) $f(t) = t + \sin t$

(d) $f(t) = 2t - \cos t/2$

(e) $f(t) = t \sin t$

(f) $f(t) = e^{-t} \sin t/2$

5. With a simple ac electrical generator, the amount of current (in amperes) at any instant t is expressed by the equation

$$i(t) = A \sin wt$$

where t is time in seconds. State the explicit formula if the amplitude is 20 and the frequency is 60 cycles per second, and sketch the graph.

6. In a simple ac generator, the emf (voltage) is given by the formula

$$e = E_m \sin wt$$

where E_m is the maximum (peak) voltage, t is time in seconds, and w is the angular velocity in radians per second of the generating coil. Sketch and describe $e = 110 \sin 120 \, \pi t$.

7. In an ac circuit, instantaneous power is equal to the product $p = ei$ of instantaneous voltage and instantaneous current. Compute the formula using the expressions in Exercises 5 and 6.

8. Sketch graphs for each of the following.

(a) $i = 10 \sin 60 \, \pi t$

(b) $e = 120 \sin 60 \, \pi t$

(c) $i = 12 \sin (120 \, \pi t - \pi/2)$

(d) $e = 100 \sin (120 \, \pi t - \pi/4)$

(e) $p = (10 \sin 60 \, \pi t)(120 \sin 60 \, \pi t)$

(f) $p = 1000 \sin^2 (120 \, \pi t)$

5.9. inverse trigonometric functions

In Chapter 1, we discussed the concept of an inverse function. We noted that if

$$f: X \to Y$$

were a bijection, then the inverse $f^{-1}: Y \to X$, defined by $f^{-1}(y) = x$ when $y = f(x)$, is also a function. However, if f is not bijective, one obtains an inverse f^* which is not a function. To get a suitable inverse function, in this case, it is necessary to restrict the domain of f, and let f^{-1} be the inverse of a suitable bijective branch.

Here we relate this to the trigonometric functions which, being periodic, are far from bijections.

Consider first the sine function,

$$y = \sin x$$

If we want an inverse function for the sine, we must suitably restrict the domain. A glance at Figure 5.9.1 suggests restricting the sine to the interval $[-\pi/2, \pi/2]$ since on this interval $\sin: [-\pi/2, \pi/2] \to [-1,1]$ is a bijection.

If we, as is conventional, switch the variables x and y, we have the function

(5.9.1) $x = \sin y \qquad y \in [-\pi/2, \pi/2]$

Then the **inverse sine** function is sketched in Figure 5.9.2 and is completely

inverse trigonometric functions

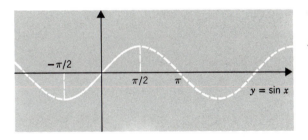

figure 5.9.1

$y = sin\ x$

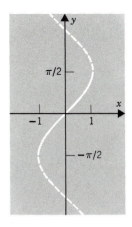

figure 5.9.2

$y = sin^{-1} x$

defined by equation (5.9.1). That is,

$$y = \sin^{-1} x$$

(5.9.2)

$$provided\ x = \sin y\ and\ y \in [-\pi/2, \pi/2]$$

Often the *inverse sine* is called the **arc sine** and $y = \sin^{-1} x$ can be written

$$y = arc\ sin\ x, \qquad -\pi/2 \le y \le \pi/2$$

example 1

Find $\sin^{-1} 1$. The number a equals $\sin^{-1} 1$ if $\sin a = 1$, and if $a \in [-\pi/2, \pi/2]$. A glance at a table or at Figure 5.9.2 shows that $a = \pi/2$. Thus, $\sin^{-1} 1 = \pi/2$. Similarly $\sin^{-1} -1 = -\pi/2$, $\sin^{-1} 0 = 0$, $\sin^{-1} 1/2 = \pi/6$, etc. ●

We emphasize that the domain of the arc sine is the range of the sine function; viz., $[-1, 1]$. The range of the arc sine is the domain of the injective branch of sine chosen; viz., $[-\pi/2, \pi/2]$.

A glance at a graph of the tangent function also suggests that if we wish to consider an inverse, we must restrict our attention to the bijective branch

$$\tan : (-\pi/2, \pi/2) \to R$$

The domain is restricted to the open interval. (Why?) To obtain the inverse for

(5.9.3) $$y = \tan x$$

we switch variables and obtain

235 (5.9.4)

$$x = \tan y$$

$$y \in (-\pi/2, \pi/2)$$

trigonometric functions

Then *the* **inverse tangent (arc tangent)** *is defined by* (5.9.4). That is

$$y = \tan^{-1} x$$

provided (5.9.4) *is true.*

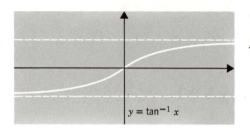

figure 5.9.3

$y = tan^{-1} x$

example 2

Find $\tan^{-1} 1$. Since $\tan \pi/4 = 1$ and $-\pi/2 < \pi/4 < \pi/2$ it follows that

$$\tan^{-1} 1 = \pi/4$$

Similarly, arc tan $(\sqrt{3}) = \pi/3$, arc tan $(0) = 0$, and $\tan^{-1}(-\sqrt{3}) = -\pi/3$, etc. ●

The graph of the inverse tangent is sketched in Figure 5.9.3. Note that the domain is the entire real line, the same as the range of tangent. The range of arc tan is the same as the domain of the bijective branch of tangent chosen; viz., the open interval $(-\pi/2, \pi/2)$.

In Figure 5.9.4, we refer to the graph of the cosine and note that restricting the domain to the closed interval $[-\pi/2, \pi/2]$ *does not* give a bijective branch. For this reason, it seems most convenient to select the branch on the closed interval $[0, \pi]$ instead. This is a bijective branch.

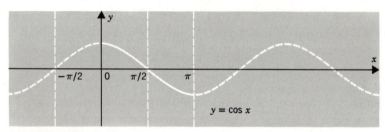

figure 5.9.4

$y = cos\ x$

Once again we interchange x and y and consider

(5.9.5) $x = \cos y$ $y \in [0, \pi]$

Then *the* **inverse cosine (arc cosine)** *is defined by*

$$y = \cos^{-1} x$$

provided (5.9.5) *holds.*

example 3

236 Find $\cos^{-1} 1$. It is clear that $\cos 0 = 1$, and $0 \in [0, \pi]$, so $\cos^{-1} 1 = 0$; similarly $\cos^{-1} -1 = \pi$, arc cos $1/2 = \pi/3$, arc cos $\sqrt{2}/2 = \pi/4$, $\cos^{-1} 0 = \pi/2$, etc. ●

inverse trigonometric functions

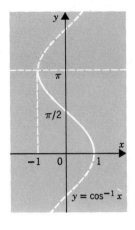

figure 5.9.5

$y = \cos^{-1} x$

$y = \cos^{-1} x$

The graph of the function arc cosine is sketched in Figure 5.9.5.

The domain is the same as the range of the cosine; namely the closed interval $[-1, 1]$. The range of $\cos^{-1}$ is the same as the restricted domain; viz., the closed interval $[0, \pi]$. In summary we have the following table.

Domain	Range
$-1 \le t \le 1$	$-\pi/2 \le \sin^{-1} t \le \pi/2$
$-1 \le t \le 1$	$0 \le \cos^{-1} t \le \pi$
$t \in R$	$-\pi/2 < \tan^{-1} t < \pi/2$

The inverses of the other three trigonometric functions can also be defined, as we suggest in the Exercises.

In many applications it is necessary to express one inverse trigonometric function in terms of another. We consider several examples.

example 4

Find $\sin (\text{arc cos } u)$. To do this set $v = \text{arc cos } u$. Then we know that

$$u = \cos v \qquad \text{and} \qquad 0 \le v \le \pi$$

If we construct an angle v in standard position, we should have a figure similar to Figure 5.9.6 (unless $v = \pi/2$), with v in either the first or second quadrant.

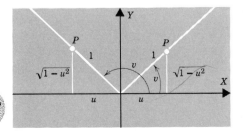

figure 5.9.6

trigonometric functions

Since $u = \cos v$, we may select a point $P(x, y)$ on the terminal side of v in such a way that

$$r = 1 \quad \text{and} \quad x = u; \quad y = \sqrt{1 - u^2}$$

Thus,

$$\sin v = \sqrt{1 - u^2}$$

in any case (even if $v = \pi/2$). ●

example 5

Find the value of $\sec(\tan^{-1} u)$. Proceed as in the previous example to set $v = \tan^{-1} u$ so that $u = \tan v$ and $-\pi/2 < v < \pi/2$. Again sketch a graph of v in standard position, and select a point $P(x, y)$ on the terminal side such that

$$x = 1, \quad y = u$$

Then $r = \sqrt{1 + u^2}$. See Figure 5.9.7 when $v > 0$. Thus, from the figure, $\sec v = \sec(\tan^{-1} u) = 1 + u^2$. ●

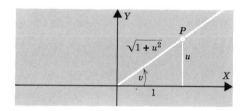

figure 5.9.7

example 6

Show that $\sin^{-1}(3/5) + \cos^{-1}(12/13) = \cos^{-1}(33/65)$. To do this set $\alpha = \sin^{-1}(3/5)$ and $\beta = \cos^{-1}(12/13)$. Then $\sin \alpha = 3/5$ and $\cos \beta = 12/13$. If we set $\gamma = \cos^{-1}(33/65)$ then $\cos \gamma = 33/65$. We wish to show that $\alpha + \beta = \gamma$.

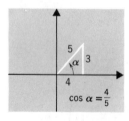

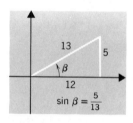

figure 5.9.8

This can be done if $\cos(\alpha + \beta) = \cos \gamma$ *and* $\alpha + \beta$ lies in the same interval, $[0, \pi]$, as does γ. The second criterion is important since, otherwise, we might have, for example

$$\alpha + \beta = \frac{11\pi}{6} \quad \text{while} \quad \gamma = \frac{\pi}{6}$$

and still have $\cos \gamma = \cos(\alpha + \beta)$. However, on $[0, \pi]$, cosine is injective. To check that $\alpha + \beta \in [0, \pi]$ we note that $\cos \beta > 0$, so $\beta \in [0, \pi/2]$; and $\sin \alpha > 0$, so $\alpha \in [0, \pi/2]$. (Explain why.) Thus, $0 \le \alpha + \beta \le \pi$. Now,

$$\cos(\alpha + \beta) = \cos \alpha \cos \beta - \sin \alpha \sin \beta$$

inverse trigonometric functions

We compute $\cos \alpha$ and $\sin \beta$ as in the previous examples. Hence,

$$\cos (\alpha + \beta) = \left(\frac{4}{5}\right)\left(\frac{12}{13}\right) - \left(\frac{3}{5}\right)\left(\frac{5}{13}\right)$$

$$= \frac{48}{65} - \frac{15}{65}$$

$$= \frac{33}{65}$$

Since $\cos (\alpha + \beta) = \cos \gamma$ and $\alpha + \beta \in [0, \pi]$, where the cosine has a injective branch, we have $\alpha + \beta = \gamma$ as desired. ●

exercises

1. Give definitions for arc sec, arc cot, and arc csc.

2. Find the indicated numbers.

    (a) $\cos^{-1} 1/2$
    (b) $\sin^{-1} 1/2$
    (c) $\tan^{-1} -1$
    (d) $\tan^{-1} \sqrt{3}$

    (e) $\cot^{-1} \sqrt{3}$
    (f) $\sec^{-1} \sqrt{2}$
    (g) $\sin^{-1} (-\sqrt{3}/2)$
    (h) $\cos^{-1} \sqrt{3}/2$

    (i) $\arcsin (-\sqrt{2}/2)$
    (j) $\arccos (-\sqrt{2}/2)$
    (k) $\arctan (-\sqrt{3}/3)$
    (l) $\operatorname{arc sec} (-\sqrt{2})$

3. Use the tables backwards to determine each of the following (in radians).

    (a) $\sin^{-1} (.2840)$
    (b) $\cos^{-1} (.2840)$
    (c) $\tan^{-1} (.4245)$
    (d) $\sec^{-1} (1.086)$

    (e) $\arcsin (.8480)$
    (f) $\arctan (.5317)$
    (g) $\operatorname{arc csc} (2.366)$
    (h) $\arccos (-.8910)$

    (i) $\sin^{-1} (-.1219)$
    (j) $\tan^{-1} (-1.780)$
    (k) $\cos^{-1} (-0.6018)$
    (l) $\sec^{-1} (-1.305)$

4. Let u be a positive real number in the appropriate interval. Find:

    (a) $\sin (\arcsin u)$
    (b) $\sec (\arccos u)$
    (c) $\sec (\tan^{-1} u)$

    (d) $\cos (\sin^{-1} u)$
    (e) $\tan (\sin^{-1} u)$
    (f) $\tan (\cot^{-1} u)$

    (g) $\tan (2 \cot^{-1} u)$
    (h) $\sin (2 \cos^{-1} (-u))$
    (i) $\cos (2 \arccos 1/u)$

5. Let u and v be positive real numbers in the appropriate intervals. Find:

    (a) $\sin (2 \arcsin 1/u)$
    (b) $\sin (\pi/2 + \sin^{-1} u)$
    (c) $\tan (\pi - \cot^{-1} u)$
    (d) $\cos (\pi + \tan^{-1} u)$

    (e) $\cos (\arccos u + \arccos v)$
    (f) $\tan (\arctan u + \operatorname{arc cot} v)$
    (g) $\sin (\arcsin u + \arcsin v)$
    (h) $\tan (\cos^{-1} u + \cot^{-1} v)$

6. Suppose $0 < t \leq \pi/2$. Find:

    (a) $\sin^{-1} (\sin t)$
    (b) $\cos^{-1} (\cos t)$
    (c) $\tan^{-1} (\cot t)$
    (d) $\arccos (\sin t)$
    (e) $\arcsin (-\cos t)$

    (f) $\arctan (\cot (\pi - t))$
    (g) $\arcsin (\sin (\pi/2 + t))$
    (h) $\cos^{-1} (\sin (\pi/2 + t))$
    (i) $\sin^{-1} (\cos (\pi - t))$
    (j) $\cos^{-1} (\sin (\pi + t))$

7. Prove that each of the following is true.

    (a) $\tan^{-1} 3 - \tan^{-1} 1 = \tan^{-1} 1/2$
    (b) $\arctan 5/3 - \arctan 1/4 = \pi/4$
    (c) $\arcsin \sqrt{5}/5 + \arcsin (2\sqrt{5}/5) = \pi/2$
    (d) $\sin^{-1} 3/5 + \sin^{-1} \sqrt{2}/10 = \pi/4$
    (e) $\cos^{-1} \sqrt{3}/2 + \sin^{-1} 1/7 = \cos^{-1} 11/14$
    (f) $\sin^{-1} 4/5 - \cos^{-1} 24/25 = \tan^{-1} 3/4$

8. For u a real number as given, verify the following.

(a) $\tan^{-1} u + \tan^{-1}(-u) = 0$, $u \in R$

(b) $\sin^{-1} u + \sin^{-1}(-u) = 0$, $u \in [-1, 1]$

(c) $\sin^{-1} u + \cos^{-1} u = \pi/2$, $u \in [-1, 1]$

(d) $\tan^{-1} u = \sin^{-1}(u/\sqrt{1 + u^2})$, $u \in R$

(e) $\sec^{-1}((1 + u^2)/(1 - u^2)) = 2\tan^{-1} u$, $u > 0$

(f) $\sin^{-1}(u/\sqrt{1 + u^2}) + \sin^{-1}(1/\sqrt{1 + u^2}) = \pi/2$, $u > 0$, $u \neq 1$

(g) $\sec^{-1} u - \cos^{-1}(\sqrt{u^2 - 1}/u) = \sin^{-1}((u^2 - 2)/u^2)$, $u \geq \sqrt{2}$

(h) $\sec^{-1} u + \csc^{-1} u = \pi/2$, $|u| \geq 1$

9. Sketch graphs for the following.

(a) $y = \sec^{-1} x$ (b) $y = \cot^{-1} x$ (c) $y = \csc^{-1} x$

5.10. trigonometric equations

In this section, we apply some of the results of the previous section to solve some conditional equations which involve the trigonometric functions. The techniques for this are best illustrated by examples.

example 1

Solve, for t, the equation

(5.10.1) $$1 - 2\cos t = 2$$

We first find $\cos t = -1/2$; hence, one solution is $t = \cos^{-1}(-1/2)$. That is,

$$t = \frac{2\pi}{3}$$

But $\cos(-2\pi/3) = -1/2$, also, as does $\cos(t + 2k\pi)$, for any solution t. Hence, all the solutions to equation are given by the formulas

$$\frac{2\pi}{3} + 2k\pi$$

$$\frac{-2\pi}{3} + 2k\pi, \qquad k \in Z$$

These can be condensed into $2k\pi \pm (2\pi/3)$, $k \in Z$. ●

We have illustrated a special case of the following general theorem.

theorem 5.10.1

If $|a| \leq 1$, then the equation $\cos x = a$ has solutions given by

$$x_k = 2k\pi \pm \cos^{-1} a$$

for each $k \in Z$. (If $|a| > 1$, $\cos x = a$ has no solutions)

trigonometric equations

proof. It is easily verified that x_k is a solution for each k. To see that all solutions are in this form, we remind ourselves that $\cos^{-1} a$ is in the interval $[0, \pi]$ and that $\cos(-t) = \cos t$, so that $-\cos^{-1} a$ is also a solution. But $-\cos^{-1} a \in [-\pi, 0]$. Because the cosine is injective on both of the intervals $[0, \pi]$ and $[-\pi, 0]$, these are the only solutions in the interval $[-\pi, \pi]$ of length 2π. The periodicity of the cosine then determines that

$$2k\pi + \cos^{-1} a, \quad \text{and} \quad 2k\pi - \cos^{-1} a, \quad k \in Z$$

give all the rest of the solutions.

   A similar analysis yields the results of the next two theorems. The proofs of these are left as exercises.

theorem 5.10.2

If $|a| \leq 1$, then the equation $\sin x = a$ has a solution

$$x_k = k\pi + (-1)^k \sin^{-1} a,$$

for each $k \in Z$. (If $|a| > 1$, $\sin x$ has no solutions.)

theorem 5.10.3

For each real number a the equation $\tan x = a$ has solutions

$$x_k = k\pi + \tan^{-1} a$$

for each $k \in Z$.

example 2

Find all the solutions to the equation

$$\sin t - 1 = \cos^2 t$$

This equation is most easily handled if we substitute $1 - \sin^2 t$ for $\cos^2 t$. It then becomes

$$\sin t - 1 = 1 - \sin^2 t$$

or

$$\sin^2 t + \sin t - 2 = 0$$

This is a quadratic in $(\sin t)$ which factors as

$$(\sin t - 1)(\sin t + 2) = 0$$

Hence, either $(\sin t - 1) = 0$, or $(\sin t + 2) = 0$. In the second case, $\sin t = -2$, which is impossible for any $t \in R$. Hence,

$$\sin t = 1$$

and

$$t = k\pi + (-1)^k \sin^{-1} 1$$

$$= k\pi + (-1)^k \frac{\pi}{2}$$

 gives the solutions, according to Theorem 5.10.2. This can also be written as $((4n + 1)/2)\pi$, where $n \in Z$. ●

trigonometric functions

example 3

Solve $\sin x - \sqrt{3} \cos x = 1$. To solve this equation, we use the method of Section 5.8. We write the left member in the form $A \sin (\omega x + \phi)$. It is now in the form

$$a \sin \omega x + b \cos \omega x$$

with $a = 1$ and $b = -\sqrt{3}$. Thus, $A = \sqrt{a^2 + b^2} = 2$. From this we obtain

$$A(\cos \phi \sin \omega x + \sin \phi \cos \omega x)$$

with $\omega = 1$ (in this case) and

$$\cos \phi = \frac{a}{\sqrt{a^2 + b^2}} = \frac{1}{2} \qquad \text{and} \qquad \sin \phi = \frac{-\sqrt{3}}{2}$$

Since we only need one angle ϕ, it is good enough to select ϕ as any common solution to the equations

$$\cos \phi = \frac{1}{2}$$

$$\sin \phi = \frac{-\sqrt{3}}{2}$$

One such ϕ (the one with smallest magnitude) is $\phi = -\pi/3$. Using this we have $\sin x - \sqrt{3} \cos x = 2 \sin (x - \pi/3) = 1$. Thus,

$$\sin \left(x - \frac{\pi}{3} \right) = \frac{1}{2}$$

From Theorem 5.10.2 we have solutions

$$\left(x - \frac{\pi}{3} \right) = k\pi + (-1)^k \sin^{-1} \frac{1}{2}$$

or

$$\left(x - \frac{\pi}{3} \right) = k\pi + (-1)^k \frac{\pi}{6}$$

Hence,

$$x = k\pi + (-1)^k \frac{\pi}{6} + \frac{\pi}{3}$$

This simplifies, when k is even, $k = 2n$, to

$$x = 2n\pi + \frac{\pi}{2} = \frac{4n + 1}{2}\pi, \qquad n \in Z$$

When k is odd, $k = 2n + 1$, the solution becomes

$$x = (2n + 1)\pi + \frac{\pi}{6}$$

$$= \frac{(12n + 7)}{6}\pi, \qquad n \in Z$$

solving triangles

exercises

1. Find all the solutions in the interval $[0, 2\pi]$ to the following equations.

   (a) $2 \sin \theta - \sqrt{3} = 0$
   (b) $2 \cos \theta + 1 = 0$
   (c) $\tan \theta - \sqrt{3} = 0$
   (d) $\tan^3 \theta - 3 \tan \theta = 0$
   (e) $3 \sec^3 \theta - 4 \sec \theta = 0$

   (f) $\sqrt{3} \tan^2 \theta - 4 \tan \theta + \sqrt{3} = 0$
   (g) $2 \cos^2 \theta - (1 + 2\sqrt{5}) \cos \theta + \sqrt{5} = 0$
   (h) $2 \cos \theta \sin \theta + \cos \theta = 0$
   (i) $\sqrt{3} \cos \theta \tan \theta - \cos \theta = 0$
   (j) $\cos \theta \csc \theta - 2 \cos \theta - \csc \theta + 2 = 0$

2. Find all the solutions to the following equations.

   (a) $\sin^2 x - \sin x = 0$
   (b) $2 \cos^2 x + 3 \cos x = 2$
   (c) $4 \cos^2 x + (2\sqrt{3} - 2) \cos x = \sqrt{3}$
   (d) $2 \sin x \cos x - 2 \sin x - \sqrt{3} \cos x = -\sqrt{3}$
   (e) $\sec x \cot x + \sqrt{3} \sec x = 2 \cot x + 2\sqrt{3}$
   (f) $2 \sin^2 x + 3 \cos x - 3 = 0$
   (g) $4 \tan^2 x = 3 \sec^2 x$
   (h) $1 = \sec x + \tan^2 x$
   (i) $2\sqrt{3} = \sqrt{3} \sec^2 x + 2 \tan x$
   (j) $\csc x + 2 \sin x = 3$

3. Find all the solutions in the interval $[0, 2\pi]$ to the following equations.

   (a) $2 \cos^2 x = 1 - 2 \cos 2x$
   (b) $2 \sin^2 x = 2 + \cos 2x$
   (c) $\sin^2 x - \cos^2 x = -\sin 2x$
   (d) $\cot x - \tan 2x = 0$

   (e) $\cos 2x = 1 - 2 \cos^2 x/2$
   (f) $\cos 2x = 1 - \tan x$
   (g) $\cot 2x = \tan x$
   (h) $\tan (x/2) = \cos x - 1$

4. Find all the solutions to the following equations.

   (a) $\cos 3x = \cos x$
   (b) $\cos 4x + \cos 2x = 0$
   (c) $\sin 5x + \sin 3x = 0$
   (d) $\sin 3x = \sin x$

   (e) $\sin 3x \cos x - \cos 3x \sin x = 0$
   (f) $\sin 4x \cos 2x + \cos 4x \sin 2x = 0.5$
   (g) $\cos 2x \cos x + \sin 2x \sin x = \sin 2x$
   (h) $\sin 3x \cos x - \cos 3x \sin x = \cos x$

5. Find all the solutions to the following equations.

   (a) $\sin x + \sqrt{3} \cos x - 1 = 0$
   (b) $3 \sin x + 4 \cos x = 5/2$
   (c) $\sqrt{3} \sin x = 6/5 + \cos x$
   (d) $\sqrt{2} (\sin x - \cos x) = \sqrt{3}$

   (e) $5 \sin x = 9.1 + 12 \cos x$
   (f) $7 \sin x = 24 \cos x + (25\sqrt{3}/2)$
   (g) $5 \sin x + 12 \cos x = 9.1026$
   (h) $3 \sin x = 9/2 + 4 \cos x$

6. Prove that for each real number $t \in [-1, 1]$.

   (a) $\cos (\sin^{-1} t) = \sqrt{1 - t^2}$
   (b) $\sin (\cos^{-1} t) = \sqrt{1 - t^2}$

5.11. solving triangles

In the first part of this chapter, we indicated that trigonometry had its origins in problems of geography, navigation, and astronomy. The word is derived from the Greek words for "triangle measure." Thus, one of the original uses of trigonometry was in "solving" triangles; that is, finding all the parts of a triangle when some of them are known (measure of angles, length of sides, area). In Section 5.3, we solved some right triangles—we always had a right

trigonometric functions

angle. In this section, we attack the more general question of the triangle with angles of arbitrary size.

We begin first by computing the area of a triangle, assuming that you are already aware of the derivation of the formula

(5.11.1)
$$\text{Area} = S = \frac{1}{2}b\,h$$

where b denotes the length of the base (side) of the given triangle, and h is the length of the altitude (line from the opposite vertex to the given base).

Suppose that ABC is an arbitrary triangle with vertices A, B, and C, vertex angles α, β, and γ, and opposite sides a, b, c. Further, since not every angle in a triangle can measure $90°$, suppose that α is *not* $90°$. We orient the triangle on a coordinate system so that α is in standard position with A at the origin and C lies on the positive x axis. Figure 5.11.1 illustrates the two possibilities: $\alpha < \pi/2$ and $\alpha > \pi/2$.

In either case, the y coordinate of the point (vertex) B is the altitude h of the triangle and the x coordinate of C is the length of the base b. In either case,

$$\sin \alpha = \frac{h}{c}$$

Hence,

$$h = c \sin \alpha$$

Therefore, the area is given by the formula

(5.11.2)
$$S = \frac{1}{2}b\,c \sin \alpha$$

If none of the angles of triangle ABC is a right angle; that is, if ABC is an *oblique triangle*, a similar argument also yields the formulas

$$S = \frac{1}{2}a\,b \sin \gamma$$

(5.11.3)

$$S = \frac{1}{2}c\,a \sin \beta$$

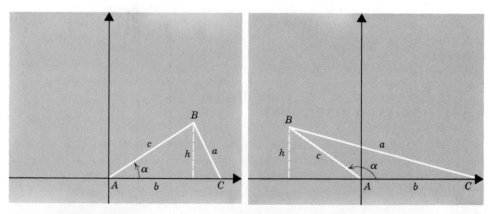

figure 5.11.1

solving triangles

If any angle were $\pi/2$, then $\sin \pi/2 = 1$, so that formula would degenerate to Formula 5.11.1. Thus we have proved the following theorem.

theorem 5.11.1

The area of any triangle is equal to half the product of the lengths of any two sides and the sine of the angle between them.

Our next formula is known as the **law of sines** and is an immediate consequence of the previous theorem. Using Equations 5.11.2 and 5.11.3 we have

$$(5.11.4) \qquad S = \frac{1}{2} b\, c \sin \alpha = \frac{1}{2} c\, a \sin \beta = \frac{1}{2} a\, b \sin \gamma$$

If we divide (5.11.4) by the nonzero real number $(1/2)\, abc$ we obtain

$$\frac{\sin \alpha}{a} = \frac{\sin \beta}{b} = \frac{\sin \gamma}{c}$$

From this we easily obtain the following theorem.

theorem 5.11.2. law of sines

In triangle ABC with sides a, b, c and angles α, β, γ, respectively, the following is true:

$$(5.11.5) \qquad \frac{a}{\sin \alpha} = \frac{b}{\sin \beta} = \frac{c}{\sin \gamma}$$

The law of sines can be used to "solve" triangles whenever two angles and any side are known, or when two sides and an angle opposite one of them is known. This last case, however, is ambiguous, as we shall see in Examples 2, 3, and 4.

example 1

Solve triangle ABC given that $\alpha = 48°$, $\beta = 71°$, and $a = 4$ in. First of all, since $\alpha + \beta + \gamma = 180°$, we readily calculate $\gamma = 61°$.
    Using the law of sines, we compute

$$b = \frac{4 \sin 71°}{\sin 48°} = \frac{4(.9455)}{(.7431)}$$

figure 5.11.2

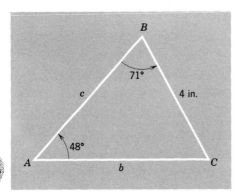

trigonometric functions

This computation is aided by logarithms. We have

$$\begin{aligned}
\log 4 &= 0.6021 \\
\log \sin 71^\circ &= \underline{0.9757 - 1} \\
& 1.5778 - 1
\end{aligned}$$

$$\begin{aligned}
-\log \sin 48^\circ &= \underline{0.8711 - 1} \\
\log b &= 0.7067 \\
b &= 5.09 \text{ in.}
\end{aligned}$$

The length of c is also obtained from the law of sines

$$c = \frac{4 \sin 61^\circ}{\sin 48^\circ} = 4.7 \text{ in.} \quad \bullet$$

We consider now the possibilities when we are given two sides and the angle opposite one of them.

example 2.   one triangle

Find angle β given triangle ABC with $\alpha = 70^\circ$, $a = 4$, and $b = 2$.

To construct the triangle, the 70° angle α is constructed and the point C is located 2 units from A. We then construct $CB = a = 4$ units as shown in Figure 5.11.3.

Then, from the law of sines,

$$\sin \beta = \frac{2 \sin 70^\circ}{4} = \frac{1}{2}(.9397) = .4699$$

Thus, $\beta = 28^\circ$ (approximately) if $\beta < 90^\circ$.

One must not forget, however, that $\sin(\pi - \beta)$ is also .4699. Thus, if $\beta' > 90^\circ$ we have

$$\pi - \beta' = 28^\circ$$

and $\beta' = 152^\circ$ might also be possible. Such is not the case, here, however, since $\alpha + \beta = 70^\circ + 152^\circ > 180^\circ$. In the next example we shall, nevertheless, see that the possibility of there being two different angles β, and $\beta' = \pi - \beta$ must always be considered. There will be but one triangle provided $\alpha + (\pi - \beta) \geq \pi$; i.e., $\alpha - \beta \geq 0$. Thus, $\alpha \geq \beta$.

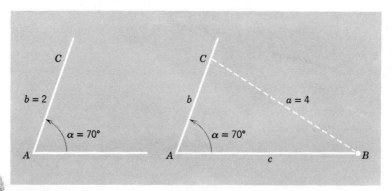

246

figure 5.11.3

solving triangles

In this triangle we have $\alpha = 70°$, $\beta = 28°$, so $\gamma = 82°$. One can also find side c by using the law of sines again.

$$c = \frac{a \sin \gamma}{\sin \alpha} = \frac{4 \sin 82°}{\sin 70°} = \frac{4(.9903)}{(.9397)}$$

$$c = 4.22 \text{ in.}$$

(The computation is most easily done using logarithms.) ●

example 3. two triangles

Suppose we know that $\alpha = 40°$, $a = 6.6$ cm, and $b = 8.4$ cm. We compute from the law of sines, aided by logarithms.

$$\sin \beta = \frac{b \sin \alpha}{a} = \frac{8.4 (\sin 40°)}{6.6} = \frac{8.4 (.6428)}{6.6} = .8183$$

Since $\sin \beta = .8183$, we have either

$$\beta \doteq 54° \ 50'$$

or

$$180° - \beta' = 54° \ 50'$$

so that

$$\beta' = 125° \ 10'$$

If $\beta = 54° \ 50' \ \alpha = 40°$, so $\gamma = 85° \ 10'$. On the other hand, if $\beta' = 125° \ 10'$ $\alpha = 40°$, so $\gamma' = 14° \ 50'$. Thus, two triangles are possible, as is illustrated in Figure 5.11.4.

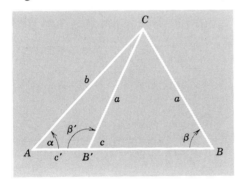

figure 5.11.4

Two triangles.

To consider the two separate cases, we set $\beta' = 14°50'$ and $\gamma' = 125°10'$ and consider the triangle $AB'C$ as well as ABC. Both sides $B'C$ and BC are $a = 6.6$ cm. We compute $c' = AB'$ and $c = AB$, again using the law of sines, obtaining $c' = 2.3$ cm and $c = 10.3$ cm. ●

example 4. no triangle

Suppose $\alpha = 70°$, $a = 2$, and $b = 4$. Solve triangle ABC.
From the law of sines we have,

$$\sin \beta = \frac{b \sin \alpha}{a}$$

$$= \frac{4}{2} \sin 70°$$

$$= 2(.9397)$$

$$= 1.8794$$

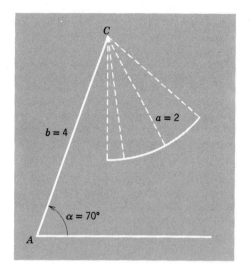

figure 5.11.5

No triangles.

But it is impossible for sin t to be larger than 1 for any real number t. Thus, there is no such angle β and no triangle ABC. This situation is depicted in Figure 5.11.5. ●

In summary, then, if two sides and the angle opposite one of them are known, one may use the law of sines to solve the triangle. However, this case is ambiguous, and there may be one, two, or no triangles. The following table summarizes these cases.

	Given a, b, and α	
One Triangle:	Solve for β,	$\alpha \geq \beta$
Two Triangles:	Solve for β,	$\alpha < \beta$
No Triangle:	sin $\beta > 1$	

Suppose now that instead of knowing the dimensions of two sides and the angle opposite one of them, or of two angles and one side, we have the dimensions of all three sides, or of two sides and an angle opposite neither of them. For these cases, the law of sines is inadequate. We derive the generalization of Pythagorus's theorem known as the *law of cosines*.

First, orient triangle ABC as was done in deriving the law of sines (see Figure 5.11.1). In Figure 5.11.6, we illustrate the two cases, where angle α is acute and where it is obtuse. We also indicate that, in either case, the coordinates of A, B, and C are, respectively, $(0, 0)$, $(c \cos \alpha, c \sin \alpha)$, and $(b, 0)$. (Why?)

solving triangles

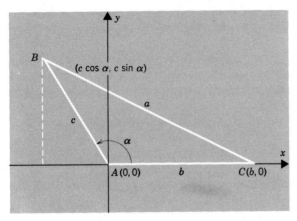

figure 5.11.6

Now the distance formula gives us the distance a from B to C to be

(5.11.6) $$a = \sqrt{(b - c \cos \alpha)^2 + (-c \sin \alpha)^2}$$

Thus,

$$a^2 = (b - c \cos \alpha)^2 + (-c \sin \alpha)^2$$
$$= b^2 - 2bc \cos \alpha + c^2 \cos^2 \alpha + c^2 \sin^2 \alpha$$
$$= b^2 + c^2 (\cos^2 \alpha + \sin^2 \alpha) - 2bc \cos \alpha$$
$$= b^2 + c^2 - 2bc \cos \alpha$$

Since a similar discussion is valid when angles β or γ are placed in standard position, we have the following theorem know as the **law of cosines**.

theorem 5.11.2.   the law of cosines

In triangle ABC with sides a, b, c and angles α, β, γ, respectively, it is true that

(5.11.7)
- (i) $a^2 = b^2 + c^2 - 2bc \cos \alpha$
- (ii) $b^2 = a^2 + c^2 - 2ac \cos \beta$
- (iii) $c^2 = a^2 + b^2 - 2ab \cos \gamma$

trigonometric functions

example 5

Given triangle ABC with $\gamma = 110°$, $a = 10$ cm and $b = 5$ cm. Find c.

$$c^2 = 10^2 + 5^2 - 2 \cdot 10 \cdot 5 \cdot \cos 110°$$
$$= 100 + 25 - 100(-0.3420)$$
$$= 125 + 34.2$$
$$= 159.2$$

so,
$$c = 12.6 \text{ cm} \qquad \bullet$$

example 6

Given the triangle ABC, with $a = 3$, $b = 4$, and $c = 7$, compute angle β. We have from (5.11.7 ii)

$$b^2 = a^2 + c^2 - 2ac \cos \beta$$

Hence,

(5.11.8) $$\cos \beta = \frac{a^2 + c^2 - b^2}{2ac}$$

Thus, in this example,

$$\cos \beta = \frac{9 + 16 - 49}{(2)(3)(7)}$$
$$= \frac{-24}{42}$$
$$= -0.5714$$

Since $\cos \beta < 0$, it must follow that $\beta \in (\pi/2, \pi)$. To use the tables, we note that

$$\cos (\pi - \beta) = -\cos \beta$$

Thus,
$$\cos (\pi - \beta) = 0.5714$$

or, since π radians $= 180°$,

$$180° - \beta = 55° \ 10' \text{ (approximately)}$$

Therefore,
$$\beta = 124° \ 50' \qquad \bullet$$

exercises

1. Solve triangle ABC given the following data.

  (a) $\beta = 60°$ $\gamma = 60°$ $b = 25$
  (b) $\alpha = 45°$ $\beta = 45°$ $a = 20$
  (c) $\beta = 30°$ $\gamma = 30°$ $c = 100$
  (d) $\alpha = 20°$ $\beta = 70°$ $c = 10$
  (e) $\alpha = 36°$ $\beta = 72°$ $a = 48$
  (f) $\beta = 44°$ $\gamma = 44°$ $a = 1$
  (g) $\alpha = 50°$ $\beta = 60°$ $b = 45$

solving triangles

2. Solve triangle ABC given the following data. (Find all solutions, if any.)

(a) $a = 5$ $b = 5$ $\alpha = 60°$
(b) $a = 5$ $b = 10$ $\beta = 60°$
(c) $b = 1$ $c = \sqrt{2}$ $\gamma = 45°$
(d) $a = 8$ $b = 8$ $\alpha = 120°$
(e) $a = 3$ $b = 2$ $\beta = 45°$
(f) $b = 3$ $c = 1$ $\gamma = 30°$
(g) $a = 3$ $b = 2$ $\beta = 30°$
(h) $b = 3$ $c = 1$ $\beta = 45°$

3. In triangle ABC given the following data, find the indicated part(s).

(a) $b = 1$ $c = \sqrt{2}$ $\alpha = 45°,$ find a, α, and γ
(b) $a = 8$ $b = 8$ $\gamma = 120°,$ find c, α, and β
(c) $a = 2$ $b = 3$ $\gamma = 40°,$ find c
(d) $b = 120$ $c = 85$ $\alpha = 54°,$ find a
(e) $b = 3.46$ $c = 5.12$ $\alpha = 108°\,30',$ find a
(f) $a = 13$ $c = 5$ $\beta = 115°\,10',$ find b
(g) $a = 2.1$ $c = 4.2$ $\beta = 72°\,20',$ find b

4. Compute formulas for $\cos \alpha$ and for $\cos \gamma$ similar to (5.11.8), and solve the following triangles.

(a) $a = 2$ $b = 2$ $c = 2$
(b) $a = 1$ $b = \sqrt{2}$ $c = 1$
(c) $a = 1$ $b = 2$ $c = \sqrt{3}$
(d) $a = 3$ $b = 4$ $c = 5$
(e) $a = 5$ $b = 12$ $c = 13$
(f) $a = 1.1$ $b = 2.2$ $c = 1.5$
(g) $a = 10.4$ $b = 6.8$ $c = 3.5$

5. The diagonals of a parallelogram are 13 cm and 19 cm in length. They intersect at an angle of 48°. Find the lengths of the sides and the angles of the parallelogram.

6. To avoid a storm, a small airplane flies for 75.8 miles in a direction 15° 10′ off course. It then turns toward its destination and reaches it after flying an additional 86.3 miles. If the average air speed of the plane was 148 mph, how much delay was caused by the storm.

7. A United Nations mediator wants to go from Cairo to Tel Aviv. The distance is 250 miles in the direction N 57° E. There is, however, no direct airline connection between the two. He must, therefore, first fly from Cairo to Athens (direction N 35° 30′ W) and then from Athens to Tel Aviv (direction S 55° W). Ignoring the curvature of the earth, determine the total distance traveled by the mediator on this trip.

8. Prove the **law of tangents**.

$$\frac{a - b}{a + b} = \frac{\tan (1/2)(\alpha - \beta)}{\tan (1/2)(\alpha + \beta)}$$

Hint: Use the law of sines; $b = a(\sin \beta/\sin \alpha)$.

9. Use the results of Exercise 8 to find the indicated parts of triangle ABC given in the following data.

(a) $a = 1$ $b = 1$ $\gamma = 50°,$ find α, β, and c
(b) $a = 1$ $b = 2$ $\gamma = 72°,$ find α and β
(c) $a = 842$ $b = 638$ $\gamma = 99°\,20',$ find α and β
(d) $b = 2.44$ $c = 8.10$ $\alpha = 46°,$ find β and γ
(e) $b = .0433$ $c = .0643$ $\alpha = 38°\,10',$ find β and γ

trigonometric functions

10. Two streets intersect at an angle of 123°40′. A lot on the corner has a frontage of 94.6 ft on one street and 93.8 ft on the other. If the lot is triangular, how long is the back fence?

11. Show that if r is the radius of a circle inscribed in triangle ABC then the area K of the triangle is $K = rs$, where $s = (1/2)(a + b + c)$ is the semiperimeter.

12. Show that with r and s as given in the previous exercise

$$r = \sqrt{\frac{(s - a)(s - b)(s - c)}{s}}$$

13. Show that, with r and s as above,

$$\tan \frac{\alpha}{2} = \frac{r}{s - a}$$

14. A proposed triangular rose garden measured 10 ft by 8 ft by 14 ft. If one rosebush is to be planted for every 9 sq ft, or fraction thereof, how many rosebushes are needed?

15. A scout walked 651 paces north from his camp and then 313 paces N 65° 20′ E to his buddy's camp. If he had walked in a straight line to the other camp, how many steps would he have saved?

16. A man on one side of a stream noted with his compass that an object on the opposite bank was in the direction N 15° 10′ E. He then walked 65.2 ft in a direction N 85° 20′ E and, from that point, noted the object was now N 45° 40′ W. How far was the object from his first location?

17. A forest fire was noted simultaneously from two lookout stations. The first recorded it in the direction N 49° 20′ E, while the second recorded it in the direction N 32° 50′ W. If the second station was 17.5 miles S 81° 30′ E from the first, how far from each station was the fire?

18. Two streets intersect at an angle of 74°, and a highway crosses them at points equidistant from the intersection. If the frontage on the highway is 210 ft, how large is a parking lot bounded by the three roads?

19. Calculations show a force of 13.8 tons applied vertically to a footing and a force of 11.3 tons applied to the footing at an angle of 76° 40′ with the horizontal. What is the resultant force on the footing, and what is its direction in relation to the horizontal?

20. A man's train pulled out just as he arrived at the station. The track was straight in the direction N 64° E to the next station, and the train was scheduled to travel 40 mph. The highway to the next station went due north 12 miles, turned, and went 21 more miles to the station. What average speed did the man need to drive to arrive simultaneously with the train at the next station?

21. A fisherman in a small boat read a bearing of N 23° 20′ E to a lighthouse and due east to a visible tower. On the map, the lighthouse and tower scaled 5390 ft apart, and the tower was S 8° 30′ E of the lighthouse. How far from the lighthouse was the boat?

22. An indicator at a lookout point showed Bald Mountain to be N 32° 10′ E and Top Peak to be N 45° 30′ W of the lookout point. On the map, Bald Mountain and Top Peak were seen to be 36.8 miles apart with Bald Mountain N 88° 20′ E of Top Peak. How far was Bald Mountain from the lookout point?

23. A triangular park measured 1250 ft by 1560 ft by 1890 ft. If the planning commission wanted one acre of park ground for every 100 people who lived in the area, how many persons would this park accommodate?

24. A thief ran from a store and dropped into the manhole of a main trunk-line storm sewer that ran in the direction N 46° 30′ E. He ran northeasterly along the sewer 248 ft then due west along a lateral line of storm drain 310 ft to another manhole, where he crawled out. How far was he when he surfaced from the spot where he entered the sewer?

the trigonometric form of a complex number

5.12. the trigonometric form of a complex number

In this section, we show how the trigonometric functions may be applied to other problems in mathematics. In particular, we show that each complex number $z = a + bi$ has a trigonometric (polar) form. This form can then be applied to solve polynomial equations of the form $x^n - a = 0$.

In previous discussions, we have indicated that a point P in the plane can be located by its distances perpendicular to the y and x axis, respectively. We have also strongly suggested that $x = r \cos \theta$ and $y = r \sin \theta$, where r is the length of the segment OP and θ is the angle in standard position that OP makes with the positive x axis (see Figure 5.12.1). We have, therefore, the relations

$$r = \sqrt{x^2 + y^2}$$

(5.12.1) $$x = r \cos \theta$$

$$y = r \sin \theta$$

Just as the line is a geometric model for the set of real numbers, the plane can serve as a model for the set of complex numbers. In Chapter 1 we developed the relationship between complex numbers and points in the plane. Thus, if z is any complex number, we discussed how to associate the unique point (x, y) in the coordinate plane with z. On the other hand, given any point $P(x, y)$ of the coordinate plane, we determine the unique complex number (x, y) associated with it. We also developed the standard form $z = x + iy$ for a complex number: $(x, y) \to x + iy$.

Pure imaginaries $yi = 0 + yi$ are, by this method, always associated with points $(0, y)$ on the y axis while real numbers $x = x + 0i$ always correspond to points $(x, 0)$ of the x axis. These axes were, therefore, called the "*real axis*" and the "*imaginary axis*," respectively.

In Figure 5.12.2 (see p. 254), we illustrate the points associated with the complex numbers $2 + i, 2 - i, -1 + (i/2), -1 - (i/2)$, respectively.

Now, by our previous comments, any complex number can represent a point (x, y) of the complex plane, and any point of the plane can be thought of as having coordinates

(5.12.2) $$(r \cos \theta, r \sin \theta),$$

for appropriate r and θ. Therefore, each complex number $z = x + iy$ has a **trigonometric representation** $z = r \cos \theta + ir \sin \theta$, or

(5.12.3) $$z = r(\cos \theta + i \sin \theta)$$

figure 5.12.1

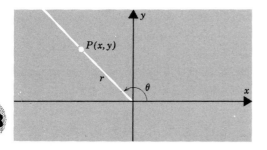

trigonometric functions

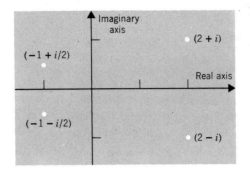

figure 5.12.2

The complex plane.

It's clear that $\bar{z} = x - iy$ has *trigonometric form*

$$\bar{z} = r(\cos \theta - i \sin \theta)$$

The trigonometric form (5.12.3) of a complex number is also called the **polar form** of the number. Frequently (5.12.3) is abbreviated

$$z = r \operatorname{cis} \theta$$

where

(5.12.4) $$r = \sqrt{x^2 + y^2}$$

(5.12.5) $$\tan \theta = \frac{y}{x}$$

The number $r = |z|$, the *absolute value* or *modulus* of z, and we have written $\tan \theta = \sin \theta / \cos \theta = y/x$. The angle θ is called the **amplitude** of z. For $z \neq 0$ there is a unique amplitude $\theta \in [0, 2\pi)$ satisfying (5.12.1), i.e., $\sin \theta = y/|z|$ and $\cos \theta = x/|z|$.

example 1

Write the complex number

$$1 + i\sqrt{3}$$

in trigonometric form. We have

$$|z| = r = \sqrt{1 + 3} = 2$$

and $\tan \theta = \sqrt{3}/1 = \sqrt{3}$. Thus, the amplitude of z is $\pi/3$. Hence, the trigonometric form of z is

$$2\left(\cos \frac{\pi}{3} + i \sin \frac{\pi}{3}\right) \quad \bullet$$

example 2

Given $|z| = 4$ and its amplitude is $5\pi/6$. Find the standard rectangular form $z = x + iy$ of z. First of all, we are given

$$z = 4\left(\cos \frac{5\pi}{6} + \sin \frac{5\pi}{6}\right)$$

the trigonometric form of a complex number

Therefore, since

$$\cos \frac{5\pi}{6} = \frac{-\sqrt{3}}{2}, \quad \sin \frac{5\pi}{6} = \frac{1}{2}$$

$$z = 4\left(-\frac{\sqrt{3}}{2} + i\frac{1}{2}\right) = -2\sqrt{3} + 2i \quad \bullet$$

example 3

Write the polar form of

$$z = \frac{4 + 12i}{1 - 2i}$$

Since z isn't even in standard form, we first multiply the denominator by its conjugate $1 + 2i$ to transform it into a real number. We must, of course, also multiply the numerator by the same number. Thus,

$$z = \frac{(4 + 12i)(1 + 2i)}{(1 - 2i)(1 + 2i)}$$

$$= \frac{-20 + 20i}{1 + 4}$$

$$= -4 + 4i$$

Then

$$r = \sqrt{(-4)^2 + 4^2} = 4\sqrt{2}$$

and

$$\tan \theta = \frac{-4}{4}$$

Thus,

$$\theta = \frac{3\pi}{4}$$

Therefore,

$$z = 4\sqrt{2}\left(\text{cis } \frac{3\pi}{4}\right) \quad \bullet$$

That the trigonometric form of a complex number is an important and useful representation, is illustrated by the following theorems.

theorem 5.12.1

Let $z_k = r_k(\cos \theta_k + i \sin \theta_k)$, $k = 1, 2, \ldots, n$ be n complex numbers. Then the product,

$$\prod_{k=1}^{n} z_k = (z_1)(z_2)(z_3) \cdots (z_n)$$

is given by

$$S(\cos \theta + i \sin \theta)$$

where

$$S = \prod_{k=1}^{n} r_k = r_1, r_2 \ldots r_n$$

and

$$\theta = \sum_{k=1}^{n} \theta_k = \theta_1 + \theta_2 + \cdots + \theta_n$$

trigonometric functions

proof. The proof is by induction on n. The theorem clearly holds for $k = 1$. Suppose

$$W = \prod_{k=1}^{n-1} z_k = r'(\cos \theta' + i \sin \theta')$$

Then

$$
\begin{aligned}
(W)(z_n) &= r'(\cos \theta' + i \sin \theta')r_n(\cos \theta_n + i \sin \theta_n) \\
&= r'r_n[\cos \theta' \cos \theta_n + i^2 \sin \theta' \sin \theta_n + i(\sin \theta' \cos \theta_n + \cos \theta' \sin \theta_n)] \\
&= r'r_n[\cos \theta' \cos \theta_n - \sin \theta' \sin \theta_n + i(\sin \theta' \cos \theta_n + \cos \theta' \sin \theta)] \\
&= r'r_n[\cos (\theta' + \theta_n) + i \sin (\theta + \theta_n)]
\end{aligned}
$$

Hence, if the theorem holds for $n - 1$, it holds for n, as well, and is therefore true for any $n \in Z$.

example 4

Multiply the following.

$$[2(\cos 30° + i \sin 30°)][\sqrt{2}(\cos 40° + i \sin 40°)][\sqrt{3}(\cos 50° + i \sin 50°)]$$

Combining them according to Theorem 5.12.1, we have

$$S = (2)(\sqrt{2})(\sqrt{3}) = 2\sqrt{6}$$

and

$$\theta = 30° + 40° + 50° = 120°$$

Thus, the product is

$$2\sqrt{6}(\cos 120° + i \sin 120°)$$

$$= 2\sqrt{6}\left(-\frac{1}{2} + \frac{i\sqrt{3}}{2}\right)$$

$$= -\sqrt{6} + 3i\sqrt{2} \qquad \bullet$$

theorem 5.12.2

The quotient

$$\frac{z_1}{z_2} = \frac{r_1(\cos \theta_1 + i \sin \theta_1)}{r_2(\cos \theta_2 + i \sin \theta_2)}$$

of two complex numbers, in trigonometric form, is

$$\frac{z_1}{z_2} = \frac{r_1}{r_2}[\cos (\theta_1 - \theta_2) + i \sin (\theta_1 - \theta_2)]$$

proof. As in Example 3, we multiply the numerator and denominator of

z_1/z_2 by $\bar{z}_2$. Now, $\bar{z}_2 = r_2[\cos \theta_2 - i \sin \theta_2] = r_2[\text{cis } (-\theta)]$ and $z_2\bar{z}_2 = |z_2|^2 = r_2^2$. (Compute this directly if you wish.) Thus, we use Theorem 5.12.1

the trigonometric form of a complex number

and obtain

$$\frac{z_1}{z_2} = \frac{r_1[\cos \theta_1 + i \sin \theta]r_2[\cos(-\theta_2) + i \sin(-\theta_2)]}{r_2^2}$$

$$= \frac{r_1 r_2}{r_2^2}[\cos(\theta_1 - \theta_2) + i \sin(\theta_1 - \theta_2)]$$

$$= \frac{r_1}{r_2}[\cos(\theta_1 - \theta_2) + i \sin(\theta_1 - \theta_2)]$$

theorem 5.12.3.   DeMoivre's theorem

If $z = r(\cos \theta + i \sin \theta)$ is a complex number and n is a positive integer, then

$$z^n = r^n(\cos n\theta + i \sin n\theta)$$

proof.   This follows from Theorem 5.12.1 when one sets $z_k = z$ for all $k = 1$, $2, \ldots, n$.

example 5

Let $z = (1 + i\sqrt{3})$. Find z^6. First of all $r = \sqrt{3 + 1} = 2$ and $\tan \theta = \sqrt{3}/1$. Thus, $\theta = \pi/3$. Therefore, $z^6 = 2^6(\cos 6\pi/3 + i \sin 6\pi/3) = 64(\cos 2\pi + i \sin 2\pi) = 64$. We see that z is a complex sixth root of 64.   ●

The proof of the final theorem is suggested in the exercises.

theorem 5.12.4.   root of a complex number

Let $z = r(\cos \theta + i \sin \theta)$ be a complex number, and let n be a positive integer. There exist exactly n distinct complex numbers $w_0, w_1, \ldots, w_{n-1}$ such that $w_k^n = z$. The numbers w_k are given by the formula

$$w_k = r^{1/n}\left(\cos \frac{\theta + 2k\pi}{n} + i \sin \frac{\theta + 2k\pi}{n}\right)$$

for $k = 0, 1, 2, \ldots, n - 1$. ($r^{1/n}$ is the principal nth root of the real number r.)

example 6

Find the three cube roots of -8. As a complex number in trigonometric form

$$-8 = -8 + 0i = 8(\cos \pi + i \sin \pi)$$

Thus, from the theorem, the cube roots are

$$w_k = 2\left(\cos \frac{\pi + 2k\pi}{3} + i \sin \frac{\pi + 2k\pi}{3}\right)$$

$k = 0, 1, 2$. We have

$$w_0 = 2\left(\cos \frac{\pi}{3} + i \sin \frac{\pi}{3}\right) = 2\left(\frac{1}{2} + \frac{\sqrt{3}}{2}i\right) = 1 + i\sqrt{3}$$

$$w_1 = 2(\cos \pi + i \sin \pi) = 2(-1 + 0i) = -2$$

$$w_2 = 2\left(\cos \frac{5\pi}{3} + i \sin \frac{5\pi}{3}\right) = 2\left(\frac{1}{2} - \frac{\sqrt{3}}{2}i\right) = 1 - i\sqrt{3}$$ ●

trigonometric functions

example 7

Find all the zeros of the polynomial $P(x) = x^6 - i$. The zeros of P will be the sixth roots of i. Thus, $i = 0 + 1i = 1(\cos \pi/2 + i \sin \pi/2)$ has sixth roots

$$w_k = \left(\cos \frac{\pi/2 + 2k\pi}{6} + i \sin \frac{\pi/2 + 2k\pi}{6} \right)$$

$k = 0, 1, \ldots, 6$. Simplifying

$$w_k = \left(\cos \frac{4k + 1}{12}\pi + i \sin \left(\frac{4k + 1}{12}\pi \right) \right)$$

Thus, the roots of P are

$$w_0 = \cos \frac{\pi}{12} + i \sin \frac{\pi}{12} = .26 + .96i$$

$$w_1 = \cos \frac{5\pi}{12} + i \sin \frac{5\pi}{12} = .96 + .26i$$

$$w_2 = \cos \frac{3\pi}{4} + i \sin \frac{3\pi}{4} = -\frac{\sqrt{2}}{2} + i\frac{\sqrt{2}}{2}$$

$$w_3 = \cos \frac{13\pi}{12} + i \sin \frac{13\pi}{12} = -.26 - .96i$$

$$w_4 = \cos \frac{17\pi}{12} + i \sin \frac{17\pi}{12} = -.96 - .26i$$

$$w_5 = \cos \frac{7\pi}{4} + i \sin \frac{7\pi}{4} = \frac{\sqrt{2}}{2} - i\frac{\sqrt{2}}{2}$$

These roots are located around the unit circle as illustrated in Figure 5.12.3. ●

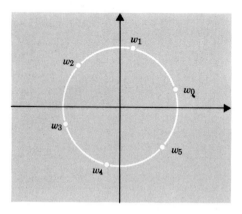

figure 5.12.3

The sixth roots of i.

exercises

1. Write each of the following complex numbers in trigonometric form.

(a) $1 + i\sqrt{3}$

(b) $1 - i\sqrt{3}$

(c) $\sqrt{3} + i$

(d) $\sqrt{3} - i$

(e) $1 + i$

(f) $1 - i$

the trigonometric form of a complex number

    (g) $2i$ (j) $-\sqrt{3} + 3i$

    (h) $-3i$ (k) -7

    (i) 8 (l) $8 - 15i$

2. Change each of the following complex numbers to the standard form $a + bi$.

    (a) $\cos 45° - i \sin 45°$ (d) $\cos \dfrac{\pi}{12} + i \sin \dfrac{\pi}{12}$

    (b) $-\cos \dfrac{\pi}{6} + i \sin \dfrac{\pi}{6}$ (e) $\cos \dfrac{4\pi}{3} + i \sin \dfrac{4\pi}{3}$

    (c) $\cos 35° - i \sin 35°$ (f) $\cos 3\pi - i \sin 3\pi$

3. Put each of the following in polar form and perform the indicated operation. Leave the answers in polar form.

    (a) $(1 + i)(1 - i)$ (e) $\dfrac{-1 - i}{1 + i}$

    (b) $(\sqrt{3} - i)(1 + i)$ (f) $\dfrac{(1 + i\sqrt{3})(-1 + i)}{-1 - i\sqrt{3}}$

    (c) $(1 + i\sqrt{3})(-1 + i)$ (g) $\dfrac{4\sqrt{3} - 4i}{(2\sqrt{3} + 2i)(\sqrt{2} - i\sqrt{2})}$

    (d) $(\sqrt{3}\,i - 1)(2 - 2i)$ (h) $\dfrac{(\sqrt{3} + i)(1 - i)}{(-1 - i)(1 - i\sqrt{3})}$

4. Use DeMoivre's Theorem to raise each complex number to the indicated power. Leave the answer in polar form.

    (a) $(1 - i)^4$ (e) $(-3 + 4i)^3$

    (b) $(-\sqrt{3} + i)^5$ (f) $(-5 + 12i)^3$

    (c) $(-1 + i)^6$ (g) $(8 + 15i)^2$

    (d) $(1 + 2i)^3$ (h) $\left(3 \text{ cis } \dfrac{\pi}{12}\right)^3$

5. Find the roots indicated.

    (a) the cube roots of $\sqrt{3} - i$ (e) the fourth roots of -16

    (b) the cube roots of 8 (f) the fifth roots of 32

    (c) the fourth roots of $16\,i$ (g) the fifth roots of $-1 - i$

    (d) the fourth roots of 81 (h) the seventh roots of $1 - i$

6. Compute the indicated roots of 1 and plot them as points in the plane.

    (a) cube   (d) sixth

    (b) fourth   (e) seventh

    (c) fifth   (f) eighth

7. Find all the zeros of the following polynomials.

    (a) $x^2 - 4$ (e) $x^6 - i$

    (b) $x^4 - 81$ (f) $x^4 - 1$

    (c) $x^5 + 32$ (g) $x^5 - 3x^4 - x + 3$

    (d) $x^3 - 27$ (h) $x^6 - x^4 + 16x^2 - 16$

8. Describe how each point P in the plane can be described in terms of two numbers (r, θ) called its *polar coordinates*, where r is the distance OP and θ is the measure of an angle.

trigonometric functions

9. Prove Theorem 5.12.4. The following steps will give a proof.

(a) Show that for integers s and t with $0 \leq s < t \leq n - 1$, the amplitudes θ_s and θ_t of the numbers w_s and w_t satisfy

$$\theta \leq \theta_s < \theta_t \leq \theta + 2\pi$$

(b) Show that $w_k^n = z$, $k = 0, 1, \ldots, n - 1$ (use Theorem 5.12.3).
(c) To complete the proof, notice that any polynomial of degree n has n roots.

quiz for review

Problems 1–5 are statements which are either (a) true or (b) false.

1. For any acute angle α, $\sin \alpha \in [0, 1]$.
2. $2 \cos \pi = \cos 2\pi$.
3. For each $t \in R$, $\cos 2t = 2 \cos^2 t - 1$
4. Cotangent is an even function.
5. If $\sin \alpha = \sin \beta$, then $\alpha = \beta$.
6. The wrapping function $w: R \to C$, where C is the unit circle, is
   (a) strictly increasing   (b) strictly decreasing   (c) injective
   (d) periodic   (e) constant
7. The projection $p_x: R^2 \to R$, given by $p_x(x, y) = x$ is
   (a) constant   (b) bijective   (c) periodic   (d) surjective
   (e) injective
8. The projection $p_y: R^2 \to R$, given by $p_y(x, y) = y$ is
   (a) surjective   (b) bijective   (c) periodic   (d) injective
   (e) constant
9. The cosine function, $\cos: R \to R$ is defined by
   (a) $\cos = w \circ p_x$ (b) $\cos = w \circ p_y$ (c) $\cos = p_x \circ w$
   (d) $\cos = p_y \circ w$ (e) $\cos (\pi/2 + t) = \sin t$, all $t \in R$
10. The sine function $\sin: R \to R$ is defined by
    (a) $\sin = w \circ p_x$ (b) $\sin = w \circ p_y$ (c) $\sin = p_x \circ w$ (d) $\sin = p_y \circ w$
    (e) $\sin (\pi - t) = \cos t$, for all $t \in R$

In problems 11–16, match the given trigonometric expression with one of the following
(a) $\cos 2t$ (b) $\sin 2t$ (c) $\sin 3t$ (d) 1 (e) none of the preceding

11. $\cos^2 t + \sin^2 t$
12. $\sec^2 t - \tan^2 t$
13. $\sin 2t \cos t + \cos 2t \sin t$
14. $1 - 2 \sin^2 t$
15. $2 \sin t \cos t$
16. $\cos^2 t - \sin^2 t$

In problems 17–21, the period of the given function is
(a) 1 (b) π (c) $\pi/2$ (d) 2π (e) 3π

17. $f(t) = \sin 2t$
18. $f(t) = 2 \sin t$
19. $f(t) = 2 \tan t/3$
20. $f(t) = [t] - t$
21. $f(t) = \pi \sin (2\pi t)$

quiz for review

Problems 22–25 refer to the following sketches.

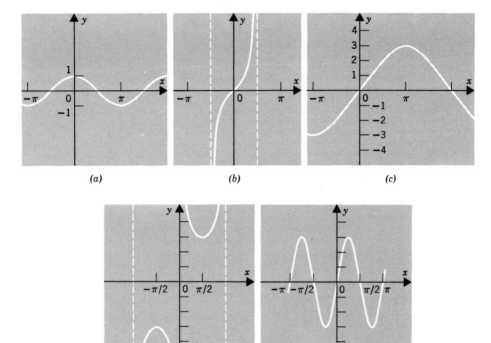

(a)   (b)   (c)

(d)   (e)

22. A sketch of the graph of $y = \tan(x + \pi)$ would look about like ———.
23. A sketch of the graph of $y = \cos x$ would look about like ———.
24. A sketch of the graph of $y = 3 \csc x$ would look about like ———.
25. A sketch of the graph of $y = 3 \sin 2x$ would look about like ———.
26. If $\cos t = \sqrt{2}/2$, then
    (a) $t = -\pi/4$ (b) $t = \pi/4$ (c) $t = \cos^{-1}(\sqrt{2}/2)$ (d) $t = 2n\pi \pm \pi/4$
    (e) $t = 2n\pi \pm \sqrt{2}/2$
27. If $\sin t = 1$, then
    (a) $t = \pi/2$ (b) $t = 0$ (c) $t = \sin^{-1} 1$ (d) $t = (4n + 1)\pi/2$
    (e) $t = n\pi + (-1)^n$
28. If $\tan t = 1$, then
    (a) $t = \tan^{-1} 1$ (b) $t = \pi/4$ (c) $t = 5\pi/4$ (d) $t = n\pi \pm 1$
    (e) $t = (4n + 1)\pi/4$
29. The function $\tan^{-1}: R \to R$
    (a) is strictly increasing (b) is strictly decreasing (c) is periodic
    (d) has image all of R (e) has domain $(-\pi/2, \pi/2)$
30. $\sin^{-1}(\sin 42° \cos 63° + \cos 42° \sin 63°) =$
    (a) $-21°$ (b) $21°$ (c) $\sin 105°$ (d) $105°$ (e) $75°$
31. $\tan(\sin^{-1} z) = (z > 0)$
    (a) $\sqrt{1 - z^2}/z$ (b) z (c) $1/z$ (d) $1/\sqrt{1 - z^2}$ (e) $z/\sqrt{1 - z^2}$
32. $\cos^{-1}(\cos 4\pi/3) =$
    (a) $-\pi/3$ (b) $2\pi/3$ (c) $\pi/3$ (d) $4\pi/3$ (e) $-2\pi/3$

trigonometric functions

33. $\sec(\cos^{-1} 1/z) = (z > 1)$

    (a) z (b) $1/z$ (c) $\sqrt{1 + z^2}$ (d) $1/\sqrt{1 + z^2}$ (e) $z/\sqrt{1 - z^2}$

Problems 34–38 refer to an arbitrary triangle ABC with sides a, b, c and angles α, β, γ, respectively (α is the angle at A, etc.).

34. $\cos \alpha$ is given by

    (a) a/c (b) $(a/b) \cos \beta$ (c) $(a^2 + b^2 - c^2)/2ab$
    (d) $(b^2 + c^2 - a^2)/2bc$ (e) b/c

35. $\sin \alpha =$

    (a) a/c (b) $(a/b) \sin \beta$ (c) $(a^2 + b^2 - c^2)/2ab$
    (d) $(b^2 + c^2 - a^2)/2bc$ (e) b/c

36. If $a = 3$ in., $b = 3$ in., $c = 4$ in., then

    (a) $\gamma = 90°$ (b) $\gamma = 60°$ (c) $\cos \gamma = 1/9$ (d) $\cos \gamma = 1/3$
    (e) $\sin \gamma = 1/3$

37. If $a = 5$ cm, $b = 5\sqrt{3}$ cm, $\alpha = 30°$, then

    (a) $\gamma = 60°$ (b) $\gamma = 90°$ (c) $\gamma = 30°$ or $90°$ (d) $\gamma = 120°$
    (e) $\gamma = 60°$ or $120°$

38. If $a = 4$ in., $\alpha = 60°$, $\beta = 45°$, then

    (a) $b = 2\sqrt{6}$ in.   (b) $\gamma = 30°$ (c) $b = 2\sqrt{2}$ in.   (d) $b = 4/3\sqrt{6}$ in.
    (e) $b = 2\sqrt{3}$ in.

39. If $\sin x - \cos x = 1$, then

    (a) $x = 0$ (b) $x = \pi/2$ (c) $x = \pi$ (d) $x = (4k + 1)\pi/2$ and
    $(2k + 1)\pi$ (e) impossible for any real number x

40. If $\sin^2 x - \cos^2 x = -\sqrt{3}/2$, then

    (a) $x = \pi/6$ (b) $x = 2n\pi \pm \pi/6$ (c) $x = n\pi \pm \pi/12$
    (d) $x = 2n\pi + \pi/12$ (e) there is no solution since $1 \neq \sqrt{3}/2$

41. The *polar form* of the real number 2 is

    (a) 2 (b) $2(\cos \pi/2 + i \sin \pi/2)$ (c) $2(\cos \pi + i \sin \pi)$
    (d) $2(\cos \theta + i \sin \theta)$ (e) imaginary

42. Which of the following is *not a fourth root* of -1?

    (a) $\cos(\pi/4) - i \sin(\pi/4)$ (b) $-\cos(\pi/4) + i \sin(\pi/4)$
    (c) $-\cos(\pi/4) - i \sin(\pi/4)$ (d) $\cos(\pi/4) + i \sin(\pi/4)$
    (e) $(\cos(\pi/4) + \sin(\pi/4)i$

43. De Moivre's Theorem tells us that $[a(\cos \theta + \sin \theta)]^n =$

    (a) $na(\cos \theta + i \sin \theta)$ (b) $a(\cos n\theta + i \sin n\theta)$ (c) $a^n(\cos \theta + i \sin \theta)$
    (d) $a^n(\cos n\theta + i \sin n\theta)$ (e) $a[\cos((\theta + 2k\pi)/n) + i \sin((\theta + 2k\pi)/n)]$

44. A cube root is i is

    (a) $1 - i$ (b) $(\sqrt{3}/2) + i(1/2)$ (c) i (d) $-i$ (e) $(1/2) + i(\sqrt{3}/2)$

45. $(1 - i)^{10}$

    (a) 32 (b) -32 (c) $\operatorname{cis}(10\pi/4)$ (d) $-32i$ (e) $32i$

systems of linear equations

6

This chapter is devoted to the study of systems of linear equations. This is a study which has traditionally been a part of college algebra courses. Frequently, problems which involve mathematics have several conditions which need to be met, giving rise to not one, but several, algebraic equations or inequalities, each of which must hold. The techniques for solving these systems when the expressions are linear is quite well developed. It, in fact, forms the basis of techniques for solving more complicated systems as well as the basis for the technique known as *linear programming* which is now widely used in the behavioral sciences.

In Chapter 7 we shall study functions of two variables

$$f(x, y)\colon D \to R$$

where $D \subseteq R^2$. We are, in this chapter, mainly concerned with such functions when f is a linear function. We are interested in linear polynomials of the form $ax + by + c$ and with their generalization to more variables. *A **linear polynomial** in n variables* has the form

$$a_1 x_1 + a_2 x_2 + \cdots + a_n x_n + k$$

A **linear equation** in n variables is an equation $p(x_1, \ldots, x_n) = 0$ which has the form

$$a_1 x_1 + a_2 x_2 + \cdots + a_n x_n - k = 0$$

or

$$a_1 x_1 + a_2 x_2 + \cdots + a_n x_n = k$$

We have already made note of the fact that, when $n = 2$, such an equation can

systems of linear equations

be considered as the equation of a line in the plane R^2, and we shall see that when $n = 3$ its graph is a plane in R^3. Geometric interpretations for more than three variables are not easy to visualize, but the algebraic formulation is straightforward.

6.1. equivalent systems

The ideas we wish to present are best illustrated by an example from elementary algebra.

example 1

A rectangle has a perimeter of 18 ft. If the length is doubled and the width is tripled, the perimeter is 40 ft. Find the dimensions of the rectangle. The algebraic formulation of the first statement is $2l + 2w = 18$ and of the second $2(2l + 3w) = 40$. Thus, we wish to find l and w so that both equations are satisfied. We write the two equations as a system

(6.1.1)
$$2l + 2w = 18$$
$$4l + 6w = 40$$

The technique for solving such a system is to replace it by an equivalent system of equations which is simpler and eventually to replace it by an equivalent system of the form

$$l = a$$
$$w = b$$

Two systems of linear equations:

(6.1.2)
$$p_1(x_1, x_2, \ldots, x_n) = 0$$
$$\vdots$$
$$p_n(x_1, x_2, \ldots, x_n) = 0$$

and

(6.1.3)
$$q_1(x_1, \ldots, x_n) = 0$$
$$\vdots$$
$$q_n(x_1, \ldots, x_n) = 0$$

are called **equivalent** *if they have the same set of solutions.* That is, the set of all n tuples $(a_1, a_2, \ldots, a_n)$ which are solutions to (6.1.3) is the same as that for (6.1.2) and vice versa. Returning to our example, we replace (6.1.1) by the following system:

(6.1.4)
$$2l + 2w = 18$$
$$2l + 3w = 20$$

264 which is obtained from (6.1.1) by replacing the second equation $4l + 6w = 40$ with the result of multiplying both its members by the constant $1/2$. You

equivalent systems

know that this operation does not change the solutions of the equation; hence (6.1.1) and (6.1.4) are equivalent. Now, the process of adding equal quantities to both sides of an equation should have no effect on the solutions, so we replace the second equation of (6.1.4) with the result of subtracting corresponding members of the first equation from the second equation. This results in the equivalent system

$$2l + 2w = 18$$
$$w = 2$$

or, replacing the first equation by half of it,

$$l + w = 9$$
$$w = 2$$

Then if we subtract the second equation from the first, we obtain the equivalent system

$$l = 7$$
$$w = 2$$

Thus, the pair (7, 2) is also a solution of (6.1.1), and our rectangle has dimensions 7 ft by 2 ft. This method of finding an equivalent system of equations which eventually leads to a solution is frequently called "*the analytic method.*"

The system (6.1.1) could easily have been solved by another method; a method frequently employed with such problems. If in (6.1.1), we alter the form of the first equation, we have

$$2l = 18 - 2w$$

or

$$l = 9 - w$$

The second equation is

$$4l + 6w = 40$$

We may substitute the value of l given by the first equation into the second equation. (Why?) This results in a single equation in one variable.

$$4(9 - w) + 6w = 40$$

or

$$36 - 4w + 6w = 40$$

This becomes

$$2w = 4$$

or

$$w = 2$$

Thus, $l = 9 - 2 = 7$. This *method of substitution*, as it is called, is also valuable in solving systems of equations when some of the equations are not linear.

We know that a linear equation in two variables has a graph which is a straight line. Thus, the system (6.1.1) has *two* straight lines as its graph. A solution of the system is a pair (l, w) which is a point on both lines; thus, their point of intersection (see Figure 6.1.1, p. 266). ●

265

systems of linear equations

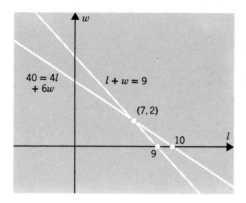

figure 6.1.1

Graphical solution of a system.

We consider two more examples, the second of which involves an equation which is not linear.

example 2

Solve the system

(6.1.5)
$$3x - y = 9$$
$$x + 4y = 8$$

We being first by sketching the two lines involved in (6.1.5). We see the point of intersection lies near (4, 1). To find the solution by the substitution method, we replace (6.1.5) by the equivalent system

$$3x - y = 9$$
$$x = 8 - 4y$$

Then, substituting the value from the second equation for x into the first, we have

$$3(8 - 4y) - y = 9$$

or

$$24 - 12y - y = 9$$
$$-13y = -15$$
$$y = \frac{15}{13}$$

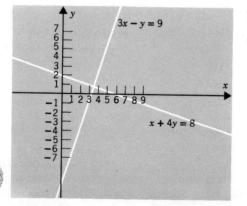

figure 6.1.2

Graphical solution.

equivalent systems

which yields the equivalent system

$$x = 8 - 4y$$

$$y = \frac{15}{13}$$

The system

$$x = \frac{44}{13}$$

$$y = \frac{15}{13}$$

is also equivalent, and yields the solution (44/13, 15/13).

The analytic method for obtaining the solution would involve replacing the second equation (6.1.5) with three times itself. This results in the system

$$3x - y = 9$$

$$3x + 12y = 24$$

which is equivalent to (6.1.5). Then, if the first equation is subtracted from the second, we have the system

$$3x - y = 9$$

$$13y = 15$$

or

$$39x - 13y = 117$$

$$13y = 15$$

Then, adding the two equations, we have

$$39x = 132$$

$$13y = 15$$

or

$$x = \frac{44}{13}$$

$$y = \frac{15}{13}$$
●

In a later section, we shall develop a systematic way of handling this analytic method, for large systems of linear equations, which will indicate its superiority over substitution. For $n = 2$, either system requires about the same amount of work.

example 3

Solve the system

$$x - y = 5$$

$$x^2 + y^2 = 17$$

267

In this example, since one of the equations is not linear, the method of substitution will be the most useful. We replace the first equation with $x = 5 + y$ and substitute this for x in the second equation. Thus, $x^2 = (5 + y)^2$, so

$$25 + 10y + y^2 + y^2 = 17$$

or

$$2y^2 + 10y + 25 = 17$$

or

$$2y^2 + 10y + 8 = 0$$

Factoring yields $(2y + 2)(y + 4) = 0$. Thus, we have two systems of linear equations

$$x = 5 + y \qquad\qquad x = 5 + y$$
$$\text{and}$$
$$2y + 2 = 0 \qquad\qquad y + 4 = 0$$

or, equivalently,

$$x = 5 + y \qquad\qquad x = 5 + y$$
$$\text{and}$$
$$y = -1 \qquad\qquad y = -4$$

These yield

$$x = 4 \qquad\qquad x = 1$$
$$\text{and}$$
$$y = -1 \qquad\qquad y = -4$$

giving the solutions as $(4, -1)$ and $(1, -4)$. ●

In summary, then, we recall that to solve a system of equations we successively replace it by equivalent systems with "simpler" equations until the solution is found. The method of substitution is, of course, based on the property of substituting "equals for equals." The following theorem indicates what sort of manipulations are allowed in using the analytic method to replace a given system by an equivalent system.

theorem 6.1.1

Given a system of equations, each of the following elementary operations on the equations of the system produces an equivalent system.

(E_1) *Interchanging the order of any two equations in the system*
(E_2) *Multiplying both sides of any equation in the system by a non-zero real number*
(E_3) *Replacing any equation in the system by its sum with a constant multiple of any other equation in the system*

The proof is omitted.

exercises

1. Without solving the system, determine which of the following systems of equations are equivalent to the system.

$$2x + 3y = 5$$
$$5x - 10y = 15$$

equivalent systems

Give reasons for your response.

(a) $2x + 3y = 5$
$x - 2y = 3$

(b) $2x + 3y = 5$
$3x + 6y = 9$

(c) $2x + 3y = 5$
$x - 9y = 4$

(d) $x + 5y = 2$
$x - 2y = 3$

(e) $x - 2y = 3$
$x + \dfrac{3}{2}y = \dfrac{5}{2}$

(f) $2x - 4y = 6$
$x + 5y = 2$

2. Solve each of the following systems.

(a) $x - y = 0$
$x + y = 1$

(b) $x + y = 2$
$2x - y = 1$

(c) $4x - y = 0$
$6x + y = 4$

(d) $2x - y = 3$
$x + 2y = 4$

(e) $4x - y = 11$
$2x + 3y = -5$

(f) $2x - y = -1$
$4x + y = 7$

(g) $x - 3y = 2$
$2x + 3y = 1$

(h) $3x - y = 1$
$4x + 2y = -2$

(i) $6x + 5y = 0$
$3x - 10y = \dfrac{5}{2}$

(j) $\sqrt{2}x + \sqrt{3}y = 5$
$\sqrt{3}x - \sqrt{2}y = 0$

3. Solve each of the systems of equations in Exercise 1.
4. Solve each of the following systems of equations.

(a) $u - 3v = 3$
$2u - 7v = 5$

(b) $6s + 5t = -5$
$-3s + 15t = 5$

(c) $11u + 14v = 1$
$3u - 5v = -2$

(d) $2a - 3b = 5$
$-4a + 6b = -10$

(e) $\dfrac{2}{3}x = 4 + \dfrac{1}{2}y$
$\dfrac{1}{3}x = -y + \dfrac{3}{4}$

(f) $a = \dfrac{1}{3}b - 2$
$-\dfrac{1}{6}a + \dfrac{1}{2}b = -1$

(g) $3x = 5 + 2(1 - y)$
$2(x + 1) = 7y + 1$

(h) $\dfrac{x + y}{2} - \dfrac{x - y}{2} = 1$
$\dfrac{2x - y}{2} + \dfrac{x - 2y}{2} = 2$

5. Sketch the graph of each of the following systems to illustrate why it has no solutions. Such systems are called *inconsistent*.

(a) $x + y = 3$
$2x + 2y = 7$

(b) $2x - y = 4$
$-6x + 3y = 12$

systems of linear equations

(c) $6x + 10y = 5$
   $9x + 6y = 9$

(d) $x - y = 1$
   $-2x + 2y = 5$

(e) $6x - 9y = 12$
   $4x - 6y = 7$

(f) $2x + 3y = 1$
   $6x + 9y = 1$

(g) $\dfrac{x}{2} + \dfrac{y}{3} = 1$

$\dfrac{x}{4} + \dfrac{y}{6} = 1$

(h) $\dfrac{x+y}{2} - \dfrac{x-y}{2} = 1$

$\dfrac{x+y}{3} - \dfrac{x-y}{3} = 2$

6. Explain why each of the following systems has infinitely many solutions. Such systems are called *dependent*. Sketch a graph for each.

(a) $x + y = 3$
   $2x + 2y = 6$

(b) $2x - y = 4$

(c) $2x - y = 4$
   $4x - 2y = 8$

(d) $6x + 10y = 5$
   $9x + 15y = 7\dfrac{1}{2}$

(e) $x - y = 1$
   $3x - 3y = 3$

(f) $6x - 9y = 12$
   $4x - 6y = 8$

(g) $\dfrac{x}{2} + \dfrac{y}{3} = 1$

$\dfrac{x}{6} + \dfrac{y}{9} = \dfrac{1}{3}$

(h) $\dfrac{x+y}{2} - \dfrac{x-y}{2} = 1$

$\dfrac{x+y}{3} + \dfrac{x-y}{3} = \dfrac{2}{3}$

7. Find two positive integers whose sum is 64 and whose difference is 20.

8. Find two positive odd integers whose sum is 20 and whose product is 99.

9. The sum of the digits of a two-digit integer is 13. The number formed by reversing the digits is 9 larger than the original number. Find the number.

10. A candy distributor wants to make a candy mix selling for $.85 per pound by combining appropriate amounts of a $.75 per pound candy with $1.15 per pound candy. How much of each should be use?

11. Solve the following system; $[x]$ is the greatest integer in x.

$$3[x] - 2[y] = -7$$
$$2[x] + 5[y] = 8$$

12. Solve the following systems

(a) $x + y + z = 0$
   $x - y + z = 2$
   $x + y + 3z = 0$

(b) $x + y + z = 3$
   $x - y + z = 2$
   $x - y - z = -1$

(c) $x + y - z = 0$
   $2x + 2y - z = 1$
   $3x + 2y + z = 3$

13. Suppose a system of linear equations contains more equations than variables; e.g., three equations in two variables. Indicate with a graph how such a system may fail to have a solution.

equivalent systems

14. Solve the following systems of equations.

(a) $x + y = 3$
$x - y = 1$
$-x + y = -1$

(b) $x^2 - y = -1$
$2x + y = 4$

(c) $y = \dfrac{6}{x^2}$
$y = 7 - x^2$

(d) $xy = 1$
$x + y = 2$

(e) $y = 3 + \log_{10} x$
$y = 5 - \log_{10}(x - 21)$

(f) $\exp_2(x + y) = 16$
$\exp_2 x = \exp_2(1 + y)$

15. Determine a and b so that the line $ax + by = 3$ intersects $x - y = 5$ at $(4, -1)$ and passes through $(1, 3)$.

6.2. solutions in echelon form

In this section, we shall expand upon the analytic method for solving a system of linear equations which we used in the previous section for simple systems of two equations in two variables. This method will also give us important information concerning the system, even if a unique solution for the system fails to exist. We shall make extensive use of Theorem 6.1.1.

Consider a system of m equations in n variables,

$$a_{11}x_1 + a_{12}x_2 + \cdots + a_{1n}x_n = b_1$$
$$a_{21}x_1 + a_{22}x_2 + \cdots + a_{2n}x_n = b_2$$
(6.2.1)
$$\vdots$$
$$a_{m1}x_1 + a_{m2}x_2 + \cdots + a_{mn}x_n = b_m$$

Note that each of the coefficients is given double subscripts, the first designating in which equation (row) of (6.2.1) it appears and the second denoting of which of the n variables in that row it is the coefficient (column). One can think of the double subscript as an address. Thus, a_{32} is the coefficient of x_2 in equation three.

Our first goal is to substitute for the system (6.2.1) an equivalent system which has a special form. *A system of equations is said to be in* **echelon form** *when the first variable having a nonzero coefficient in any equation has zero coefficient in all the equations which follow it.* Thus, we wish to replace (6.2.1) with an equivalent system in echelon form. We use Theorem 6.1.1 to accomplish this. Consider the following example.

systems of linear equations

example 1

Reduce the system which follows to an equivalent system in echelon form.

$$\begin{aligned} x_1 - 3x_2 + x_3 - x_4 &= 1 \\ 2x_1 - 9x_2 - x_3 + x_4 &= -1 \\ x_1 - 2x_2 - 3x_3 + 4x_4 &= 5 \\ 3x_1 - x_2 + 2x_3 - x_4 &= -2 \end{aligned}$$

(6.2.2)

If (6.2.2) is to be reduced to a system in echelon form, we must replace the second, third, and fourth equations with new equations having zero for the coefficient of x_1. We use Theorem 6.1.1. (E_3) to accomplish this. We replace equation two by the result of adding both sides of it to -2 times both sides of equation one; i.e., with

$$2x_1 - 9x_2 - x_3 + x_4 + (-2)(x_1 - 3x_2 + x_3 - x_4) = -1 + (-2)(1)$$

or

$$-3x_2 - 3x_3 + 3x_4 = -3$$

We replace the third equation by its sum with (-1) times the first equation; i.e., with

$$x_2 - 4x_3 + 5x_4 = 4$$

And the fourth equation with its sum with (-3) times the first equation; i.e., with $8x_2 - x_3 + 2x_4 = -5$. Thus, an equivalent system is

$$\begin{aligned} x_1 - 3x_2 + x_3 - x_4 &= 1 \\ -3x_2 - 3x_3 + 3x_4 &= -3 \\ x_2 - 4x_3 + 5x_4 &= 4 \\ 8x_2 - x_3 + 2x_4 &= -5 \end{aligned}$$

(6.2.3)

This system isn't in echelon form yet, but we have satisfied the requirement for x_1. We multiply the second equation in (6.2.3) by $-1/3$ and we get the equivalent system

$$\begin{aligned} x_1 - 3x_2 + x_3 - x_4 &= 1 \\ x_2 + x_3 - x_4 &= 1 \\ x_2 - 4x_3 + 5x_4 &= 4 \\ 8x_2 - x_3 + 2x_4 &= -5 \end{aligned}$$

(6.2.4)

We now use the second equation of (6.2.4) to take care of the x_2 coefficients below it. That is, we replace equation three with its sum with (-1) times equation two and we replace equation four with its sum with (-8) times equation two. Thus, the system which results is equivalent to (6.2.5); hence,

272

solutions in echelon form

to (6.2.2). We get

$$x_1 - 3x_2 + x_3 - x_4 = 1$$
$$x_2 + x_3 - x_4 = 1$$
(6.2.5)
$$-5x_3 + 6x_4 = 3$$
$$-9x_3 + 10x_4 = -13$$

which isn't quite in echelon form. But if we replace the fourth equation by its sum with $-9/5$, the third equation, we get the system

$$x_1 - 3x_2 + x_3 - x_4 = 1$$
$$x_2 + x_3 - x_4 = 1$$
(6.2.6)
$$-5x_3 + 6x_4 = 3$$
$$-\frac{4}{5}x_4 = -\frac{92}{5}$$

which is in echelon form. We simplify (6.2.6) even further by multiplying the fourth equation by $-5/4$ giving the system

$$x_1 - 3x_2 + x_3 - x_4 = 1$$
$$x_2 + x_3 - x_4 = 1$$
(6.2.7)
$$-5x_3 + 6x_4 = 3$$
$$x_4 = 23$$

Now, the solution to (6.2.7) and, hence, to (6.2.2) can be found by successive substitutions; in the third equation to find x_3, in the second to find x_2 and in the first to find x_1. Another method is to reduce the system even further by beginning with the fourth equation and eliminating all the coefficients of x_4 *above* it. You are invited to give the reasons why each of the following systems is equivalent to (6.2.7).

(a)
$$x_1 - 3x_2 + x_3 = 24$$
$$x_2 + x_3 = 24$$
$$-5x_3 = -135$$
$$x_4 = 23$$

(b)
$$x_1 - 3x_2 + x_3 = 24$$
$$x_2 + x_3 = 24$$
$$x_3 = 27$$
$$x_4 = 23$$

(c)
$$x_1 - 3x_2 = -3$$
$$x_2 = -3$$
$$x_3 = 27$$
$$x_4 = 23$$

(d)
$$x_1 = -12$$
$$x_2 = -3$$
$$x_3 = 27$$
$$x_4 = 23$$

273 Either way, we find that the quadruple (4 tuple), $(-12, -3, 27, 23)$ is a solution to (6.2.2). ●

systems of linear equations

example 2

Find a solution, if one exists, to the system

(6.2.8)
$$x_1 + x_2 + x_3 = 3$$
$$x_1 - x_2 - x_3 = -1$$
$$3x_1 + x_2 + x_3 = 7$$

We replace (6.2.8) by an equivalent system in echelon form by successive application of Theorem 6.1.1 in the following steps.

(a) $x_1 + x_2 + x_3 = 3$
  $-2x_2 - 2x_3 = -4$
  $-2x_2 - 2x_3 = -2$

(b) $x_1 + x_2 + x_3 = 3$
  $x_2 + x_3 = 2$
  $x_2 + x_3 = 1$

(c) $x_1 + x_2 + x_3 = 3$
  $x_2 + x_3 = 2$
  $0 = -1$

The last equation in step (c) is, of course, nonsense. Nevertheless, the system is equivalent to (6.2.8). Therefore, since (6.2.8) can have no solutions for which $0 = -1$, it is an *inconsistent* system. ●

example 3

Find a solution, if one exists, for the system

(6.2.9)
$$x_1 + x_2 + x_3 = 3$$
$$x_1 - x_2 - x_3 = -1$$
$$3x_1 + x_2 + x_3 = 5$$

We reduce (6.2.9) to echelon form with the following steps.

(a) $x_1 + x_2 + x_3 = 3$
  $-2x_2 - 2x_3 = -4$
  $-2x_2 - 2x_3 = -4$

(b) $x_1 + x_2 + x_3 = 3$
  $x_2 + x_3 = 2$
  $0 = 0$

Step (b) is a system in echelon form with three variables and essentially only two equations. Hence, (6.2.9) is a *dependent* system. If we subtract equation two of (b) from equation one, we have the following system equivalent to (6.2.9).

$$x_1 \quad\quad = 1$$
$$x_2 + x_3 = 2$$

If x_3 is any arbitrary real number k, the triple $(1, 2 - k, k)$ is a solution of (6.2.9). In particular, if $k = 1$, a solution is $(1, 1, 1)$. ●

solutions in echelon form

example 4

Find a solution, if one exists, to the system

(6.2.10)
$$2x_1 - x_2 + 3x_3 + x_4 = 0$$
$$x_1 + x_2 + x_3 = 0$$
$$3x_1 - 3x_2 + x_3 + 2x_4 = 0$$
$$-x_1 + 2x_2 - 2x_3 - x_4 = 0$$

It is easy to verify that, since the right side of each equation is zero, the system has the solution $x_1 = x_2 = x_3 = x_4 = 0$. The system (6.2.10) is called a **homogeneous* system.** The trivial solution always exists for a homogeneous system. The question is: Are there nontrivial solutions? We proceed as before. to substitute an equivalent system in echelon form in place of (6.2.10). Our work will be much easier if we interchange the first two equations and then proceed through the following sequence of steps (give reasons).

(a)
$$x_1 + x_2 + x_3 = 0$$
$$2x_1 - x_2 + 3x_3 + x_4 = 0$$
$$3x_1 - 3x_2 + x_3 + 2x_4 = 0$$
$$-x_1 + 2x_2 - 2x_3 - x_4 = 0$$

(b)
$$x_1 + x_2 + x_3 = 0$$
$$-3x_2 + x_3 + x_4 = 0$$
$$-6x_2 - 2x_3 + 2x_4 = 0$$
$$3x_2 - x_3 - x_4 = 0$$

(c)
$$x_1 + x_2 + x_3 = 0$$
$$3x_2 - x_3 - x_4 = 0$$
$$-4x_3 = 0$$
$$0 = 0$$

(d)
$$x_1 + x_2 + x_3 = 0$$
$$3x_2 - x_3 - x_4 = 0$$
$$x_3 = 0$$
$$0 = 0$$

We notice in step (b) that the second and fourth equations are dependent, resulting in the $0 = 0$ equation in step (c), which indicates that (6.2.10) was a dependent system—*only a dependent system of homogeneous equations has a nontrivial solution.*

In step (d) we note that $x_3 = 0$ in any solution. We already are allowed a free choice for x_4 so we substitute $x_4 = k$, $x_3 = 0$ in the first two equations to obtain

$$x_1 + x_2 = 0$$
$$3x_2 - k = 0$$

which is a system in two variables whose solution is seen, by substitution, to be $x_2 = (1/3)k$, $x_1 = -(1/3)k$. Thus, for any real number k, the four-tuple $(-(1/3)k, (1/3)k, 0, k)$ is a solution to (6.2.10). In particular, for $k = -3$; $x_1 = 1$, $x_2 = -1$, $x_3 = 0$, $x_4 = -3$ is a nontrivial solution. The trivial solution occurs when $k = 0$. ●

275

*An overworked term in mathematics (and elsewhere).

systems of linear equations

example 5

Find all the solutions of the homogeneous system

$$x_1 + x_2 \qquad\quad + x_4 = 0$$

$$x_2 + x_3 - x_4 = 0$$

(6.2.11)

$$x_1 \qquad\quad + x_3 - x_4 = 0$$

$$x_1 - x_2 - x_3 \qquad\quad = 0$$

We proceed to "reduce (6.2.11) to echelon form"; that is, substitute an equivalent system which is in echelon form. Justify the following sequence of steps.

(a)
$$x_1 + x_2 \qquad + x_4 = 0$$
$$x_2 + x_3 - x_4 = 0$$
$$-x_2 + x_3 - 2x_4 = 0$$
$$-2x_2 - x_3 - x_4 = 0$$

(b)
$$x_1 + x_2 \qquad + x_4 = 0$$
$$x_2 + x_3 - x_4 = 0$$
$$2x_3 - 3x_4 = 0$$
$$x_3 - 3x_4 = 0$$

(c)
$$x_1 + x_2 \qquad + x_4 = 0$$
$$x_2 + x_3 - x_4 = 0$$
$$x_3 - 3x_4 = 0$$
$$2x_3 - 3x_4 = 0$$

(d)
$$x_1 + x_2 \qquad + x_4 = 0$$
$$x_2 + x_3 - x_4 = 0$$
$$x_3 - 3x_4 = 0$$
$$3x_4 = 0$$

Step (d) is an equivalent system which is in echelon form, but is not dependent; therefore, the only solution is the trivial one. This follows, since the last equation says $x_4 = 0$. The third equation says $x_3 - 3(0) = 0$, or $x_3 = 0$. The second equation, upon substitution for x_3 and x_4 gives $x_2 = 0$, and similarly, the first equation forces $x_1 = 0$. Another way to see this is to proceed with the following steps, using Theorem 6.1.1:

(e)
$$x_1 + x_2 \qquad + x_4 = 0$$
$$x_2 + x_3 - x_4 = 0$$
$$x_3 - 3x_4 = 0$$
$$x_4 = 0$$

(f)
$$x_1 + x_2 \qquad\qquad = 0$$
$$x_2 + x_3 \qquad = 0$$
$$x_3 \qquad = 0$$
$$x_4 = 0$$

(g)
$$x_1 + x_2 \qquad = 0$$
$$x_2 \qquad = 0$$
$$x_3 \qquad = 0$$
$$x_4 = 0$$

(h)
$$x_1 \qquad\qquad = 0$$
$$x_2 \qquad = 0$$
$$x_3 \qquad = 0$$
$$x_4 = 0$$

This second procedure is used extensively in the section which follows. ●

Sometimes these methods may also be used to solve systems of equations which are not linear. This is done by making an appropriate substitution which results in a linear system. We illustrate with an example.

 276

solutions in echelon form

example 6

Solve the system

$$\frac{1}{x} + \frac{2}{y} - \frac{1}{z} = 3$$

$$\frac{2}{x} - \frac{2}{y} + \frac{1}{z} = 0$$

$$\frac{1}{x} + \frac{2}{y} + \frac{1}{z} = 1$$

We substitute $u = 1/x$, $v = 1/y$, and $w = 1/z$ into these equations to obtain the associated linear system

$$u + 2v - w = 3$$
$$2u - 2v + w = 0$$
$$u + 2v + w = 1$$

This system is equivalent to the system

$$u + 2v - w = 3$$
$$-6v + 3w = -6$$
$$2w = -2$$

or

$$u = 1$$
$$v = \frac{1}{2}$$
$$w = -1$$

Therefore, $x = 1$, $y = 2$, and $z = -1$. ●

exercises

1. Find an equivalent system in echelon form, and then solve each of the following.

(a) $2x + 4y = 0$
$3x - 2y = 8$

(d) $3x - 5y = -3$
$2x + 3y = -6$

(b) $5x - 4y = 3$
$6x - 3y = 2$

(e) $\frac{1}{3}x + y = 2$

$x - \frac{1}{4}y = -1$

(c) $x - 5y = -3$
$2x + y = -6$

(f) $\frac{x+y}{2} - \frac{x-y}{2} = 1$

$\frac{x+y}{3} + \frac{x-y}{3} = 1$

systems of linear equations

2. Find an equivalent system in echelon form and solve the following, if possible.

(a)
$$x + y + z = 3$$
$$x - 2y - z = 4$$
$$2x + y + z = 5$$

(b)
$$x + y + z = 1$$
$$2x + 3y + z = 0$$
$$x - y + 2z = 4$$

(c)
$$2x + 3y - z = -1$$
$$x - 2y - 2z = -1$$
$$x + y - z = 0$$

(d)
$$2x - 3y = 4z + 1$$
$$x + y = 5z - 1$$
$$3x - y = -1 + z$$

(e)
$$2x + 3y + 4z = 3$$
$$x + y + z = \frac{13}{12}$$
$$2x - 3y - 4z = -1$$

(f)
$$x + 3y + 2z = 0$$
$$x - 5y - z = 3$$
$$2x + 4y + 3z = 1$$

(g)
$$x + y = 2$$
$$x - y = 1$$
$$2x + 3y = 3$$

(h)
$$x + y + z = 3$$
$$x - y - z = -1$$
$$2x + 4y + 4z = 8$$

3. Find all the solutions to the following.

(a)
$$x - y + 4z = 0$$
$$2x + y - z = 0$$
$$-x - y + 2z = 0$$

(b)
$$x_1 - x_2 - x_3 = 0$$
$$3x_1 + x_2 - x_3 = 0$$
$$2x_1 + 4x_2 + x_3 = 0$$

(c)
$$x_1 + 2x_2 + x_3 = 0$$
$$2x_1 - x_2 + 2x_3 = 0$$
$$x_1 + x_2 + x_3 = 0$$

(d)
$$-x_1 + 2x_2 - 4x_3 = 0$$
$$3x_1 - 3x_2 + 2x_3 = 0$$
$$2x_1 + 3x_2 - x_3 = 0$$

(e)
$$2x - 3y + z = 0$$
$$x - 4y - z = 0$$
$$x - 9y - 4z = 0$$

(f)
$$2x_1 - x_2 - x_3 = 0$$
$$x_1 \qquad + x_3 = 0$$
$$4x_1 - x_2 + x_3 = 0$$

(g)
$$x_1 - x_2 + 2x_3 = 0$$
$$2x_1 + x_2 - x_3 = 0$$
$$-x_1 + 3x_2 - 2x_3 = 0$$

(h)
$$-x_1 + 2x_2 - 4x_3 = 0$$
$$3x_1 - 3x_2 + 2x_3 = 0$$

4. Solve each of the following systems.

(a)
$$x + y \qquad + w = -2$$
$$x \qquad + z + w = 1$$
$$x \qquad - w = -1$$
$$2x \qquad + 3z + 5w = -2$$

(b)
$$x + 3y - 2z + w = 2$$
$$2x + 4y + 2z - 3w = -1$$
$$-x + y + z + 2w = 1$$
$$2x + 5y - z - w = 1$$

(c)
$$x + y + z + w = 1$$
$$x - y - z - w = -1$$
$$x - y + z - w = -3$$
$$x + y - z + w = 3$$

(d)
$$x_1 - x_2 + 3x_3 + x_4 = 1$$
$$2x_1 + x_2 + 4x_3 + x_4 = 0$$
$$-x_1 - x_2 + x_3 + x_4 = 1$$
$$5x_1 + 2x_2 + 7x_3 - 3x_4 = 0$$

solutions in echelon form

(e)
$$\begin{aligned}
3x_1 - 4x_2 \quad + 2x_4 &= 11 \\
2x_1 - x_2 \quad + x_4 &= 5 \\
x_1 + x_2 - x_3 \quad &= 2 \\
x_2 - 3x_3 + x_4 &= 7
\end{aligned}$$

(g)
$$\begin{aligned}
2x_1 + x_2 + 6x_3 + x_4 &= 0 \\
5x_1 + 3x_2 + x_3 + 2x_4 &= 0 \\
3x_1 - x_2 + 4x_3 + 4x_4 &= 0 \\
x_1 + x_2 + 7x_3 \quad &= 0
\end{aligned}$$

(f)
$$\begin{aligned}
3x_1 - 3x_2 + x_3 + 2x_4 &= 0 \\
-x_1 + 2x_2 - 2x_3 - 3x_4 &= 0 \\
2x_1 - x_2 + 3x_3 + x_4 &= 0 \\
x_1 + x_2 + x_3 \quad &= 0
\end{aligned}$$

(h)
$$\begin{aligned}
x_1 + x_2 - x_3 - x_4 &= 0 \\
x_1 - x_2 + x_3 - x_4 &= 0 \\
x_1 - x_2 - x_3 + x_4 &= 0 \\
-x_1 + x_2 - x_3 + x_4 &= 0
\end{aligned}$$

5. Make a substitution in each of the following systems so that a linear system results and then solve; for example, let $u = 1/x$ and $v = 1/y$ in (a).

(a)
$$\begin{aligned}
\frac{3}{x} + \frac{4}{y} &= 6 \\
\frac{1}{x} - \frac{5}{y} &= 1
\end{aligned}$$

(e)
$$\begin{aligned}
\frac{4}{x} + \frac{1}{y} + \frac{1}{z} &= 5 \\
\frac{1}{x} - \frac{4}{y} + \frac{2}{z} &= 0 \\
\frac{-3}{x} + \frac{1}{y} + \frac{3}{z} &= -1
\end{aligned}$$

(b)
$$\begin{aligned}
2x^2 - 3y^2 &= -1 \\
x^2 - 2y^2 &= -3
\end{aligned}$$

(f)
$$\begin{aligned}
4 \sin x - \sin y &= 1 \\
2 \sin x + 3 \sin y &= 4
\end{aligned}$$

(c)
$$\begin{aligned}
3x^3 + y^3 &= -5 \\
16x^3 - y^3 &= 24
\end{aligned}$$

(g)
$$\begin{aligned}
2 \sin x - 2 \cos x + 3 \tan x &= 1 \\
\sin x + \sqrt{3} \cos x + 3\sqrt{3} \tan x &= 5 \\
-4 \sin x + 2 \cos x - 3 \tan x &= -2
\end{aligned}$$

(d)
$$\begin{aligned}
\frac{1}{x+y} + \frac{1}{x-y} &= 1 \\
\frac{3}{x+y} - \frac{4}{x-y} &= 2
\end{aligned}$$

(h)
$$\begin{aligned}
3(x + y) - 2(x - y) + 3xy &= -9 \\
6(x + y) + (x - y) + 2xy &= 5 \\
(x + y) - 3(x - y) + 5xy &= -18
\end{aligned}$$

6. A chemist has three solutions containing a certain acid. The first has a 10% concentration, the second 30%, and the third 50%. He wishes to use all three to mix a solution which is 40% acid and totals 25 ounces. If he uses twice as much of the 10% solution as he does of the 30% solution. How many ounces of each does he use?

7. A collection of coins—nickels, dimes, and quarters—contains 57 coins. There are three times as many nickels as dimes, and the total amount is $4.50. How many quarters are there?

8. Find the equation of the circle passing through the three points $P(-2, 6)$, $A(6, 0)$, and $R(5, 7)$.
Hint: The equation has the form $x^2 + y^2 + ax + by + c = 0$.

9. Suppose $P(x) = x^3 + ax^2 + bx + c$. If $P(1) = 4$, $P(-1) = -2$, and $P(2) = 4$, find a, b, and c.

10. For α any real number, and k_1, k_2 constants, solve the following system

$$\begin{aligned}
x_1 \cos \alpha + x_2 \sin \alpha &= k_1 \\
-x_1 \sin \alpha + x_2 \cos \alpha &= k_2
\end{aligned}$$

systems of linear equations

6.3.   a matrix method

The method for solving a system of equations by finding an equivalent system in echelon form has a simplification which greatly expands its usefulness. If you have been perceptive, you have probably already observed that it is the coefficients in each equation with which we actually work rather than the variables $x_1, x_2, \ldots$, etc. For example, suppose we wish to reduce the following system to echelon form.

$$3x_1 - x_2 + x_3 = 5$$

(6.3.1) $$x_1 + x_2 - x_3 = -1$$

$$x_1 - 2x_2 - 3x_3 = 0$$

We can, if we wish, abbreviate the left side of this system by the array of coefficients.

(6.3.2)
$$\begin{bmatrix} 3 & -1 & 1 \\ 1 & 1 & -1 \\ 1 & -2 & -3 \end{bmatrix}$$

Any rectangular array of numbers such as (6.3.2) is called a **matrix**. Our array is a square matrix called the **coefficient matrix** of the system (6.3.1). Each of the equations in the system (6.3.1) corresponds to a **row** of the matrix in (6.3.2); while the coefficients of the same variable x_i in (6.3.1) all lie in the same **column** of (6.3.2). This matrix has three rows and three columns and is called a 3×3 matrix.

The right-hand side of the system (6.3.1) can also be written as a matrix— a 3×1 matrix

(6.3.3)
$$\begin{bmatrix} 5 \\ -1 \\ 0 \end{bmatrix}$$

When the coefficient matrix of a system is combined in an array with the one column matrix of the "constant terms" of the system, we obtain a—no-longer square—matrix, called the **augmented matrix** *of the system.* Thus, the 3×4 matrix

(6.3.4)
$$\begin{bmatrix} 3 & -1 & 1 & 5 \\ 1 & 1 & -1 & -1 \\ 1 & -2 & -3 & 0 \end{bmatrix}$$

is the *augmented matrix* of the system (6.3.1).

To set up the augmented matrix of any system of equations, we need only arrange the equations of the system so that in each column we have the coefficients of the same variable. (This may necessitate using an equivalent system.) Once we have decided which column of the matrix corresponds to which

a matrix method

variable, we have a unique matrix representing that system. On the other hand, given the matrix, we can always write down a system to which it corresponds.

example 1

Write the augmented matrix for the system

$$x_1 + x_3 = x_4 - 1$$
$$x_2 - x_4 = 1 - x_3$$
$$x_1 + x_4 = 1 - x_2$$
$$x_2 - x_3 = 1 - x_1$$

We first rearrange the equations in this system so that each variable is in its own column on the left side and the constant terms are on the right side. Thus, we have the equivalent system

$$x_1 \qquad + x_3 - x_4 = -1$$
$$x_2 + x_3 - x_4 = 1$$
$$x_1 + x_2 \qquad + x_4 = 1$$
$$x_1 + x_2 - x_3 \qquad = 1$$

The augmented matrix of the system is

$$\begin{bmatrix} 1 & 0 & 1 & -1 & | & -1 \\ 0 & 1 & 1 & -1 & | & 1 \\ 1 & 1 & 0 & 1 & | & 1 \\ 1 & 1 & -1 & 0 & | & 1 \end{bmatrix} \quad \bullet$$

Now, to find a system in echelon form equivalent to the given system, we use Theorem 6.1.1. We translate the operations on equations given in that theorem to elementary operations on the rows of its augmented matrix. Thus, we have:

theorem 6.3.1

Given a matrix (the augmented matrix of a system of equations), each of the following elementary row operations on the matrix produces a matrix which is row equivalent (the augmented matrix of an equivalent system of equations) to the original matrix:

(E_1) *Interchanging any two rows of the matrix;*
(E_2) *Multiplying all the elements of the same row by the same non-zero number (called multiplying the row by the number);*
(E_3) *Replacing any row by its sum with a constant multiple of any other row.*

If we apply these operations judiciously to the matrix of (6.3.4), we can obtain a matrix in echelon form which is (row) equivalent to it.

systems of linear equations

We first interchange rows one and two, obtaining the matrix

(6.3.5)
$$\left[\begin{array}{ccc|c} 1 & 1 & -1 & -1 \\ 3 & -1 & 1 & 5 \\ 1 & -2 & -3 & 0 \end{array}\right]$$

Then, if we replace row two of (6.3.5) by its sum with -3 times row one, and row three by its sum with -1 times row one, we obtain

(6.3.6)
$$\left[\begin{array}{cccc} 1 & 1 & -1 & -1 \\ 0 & -4 & 4 & 8 \\ 0 & -3 & -2 & 1 \end{array}\right]$$

Now, multiply row two of (6.3.6) by $-1/4$ to obtain

(6.3.7)
$$\left[\begin{array}{cccc} 1 & 1 & -1 & -1 \\ 0 & 1 & -1 & -2 \\ 0 & -3 & -2 & 1 \end{array}\right]$$

Next, replace row three of (6.3.7) by its sum with 3 times row two, to obtain

(6.3.8)
$$\left[\begin{array}{cccc} 1 & 1 & -1 & -1 \\ 0 & 1 & -1 & -2 \\ 0 & 0 & -5 & -5 \end{array}\right]$$

Then, we multiply the third row by $-1/5$ to obtain

(6.3.9)
$$\left[\begin{array}{cccc} 1 & 1 & -1 & -1 \\ 0 & 1 & -1 & -2 \\ 0 & 0 & 1 & 1 \end{array}\right]$$

The matrix (6.3.9) is the augmented matrix of the system

$$x_1 + x_2 - x_3 = -1$$
$$x_2 - x_3 = -2$$
$$x_3 = 1$$

which is in echelon form and easily solved by successive substitutions. On the other hand, we can choose to perform some more elementary operations on (6.3.9) resulting in the following equivalent matrices.

$$\left[\begin{array}{cccc} 1 & 1 & 0 & 0 \\ 0 & 1 & 0 & -1 \\ 0 & 0 & 1 & 1 \end{array}\right] \qquad \text{(adding third row to rows one and two)}$$

a matrix method

$$(6.3.10) \quad \begin{bmatrix} 1 & 0 & 0 & 1 \\ 0 & 1 & 0 & -1 \\ 0 & 0 & 1 & 1 \end{bmatrix} \quad \text{(subtracting second row from the first row)}$$

The matrix (6.3.10) is in *Hermite normal form* and is the augmented matrix of the following system, which is equivalent to (6.3.1).

$$\begin{aligned} x_1 \quad & = 1 \\ x_2 \quad & = -1 \\ x_3 & = 1 \end{aligned}$$

Thus, the solution to (6.3.1) is $(1, -1, 1)$.

To find a solution to the system in Example 1, we proceed in the same way, beginning with its augmented matrix.

$$\begin{bmatrix} 1 & 0 & 1 & -1 & -1 \\ 0 & 1 & 1 & -1 & 1 \\ 1 & 1 & 0 & 1 & 1 \\ 1 & 1 & -1 & 0 & 1 \end{bmatrix}$$

You should supply the justification by means of Theorem 6.3.1 for each of the following steps which result in equivalent matrices.

(i) $\begin{bmatrix} 1 & 0 & 1 & -1 & -1 \\ 0 & 1 & 1 & -1 & 1 \\ 0 & 1 & -1 & 2 & 2 \\ 0 & 1 & -2 & 1 & 2 \end{bmatrix}$ (ii) $\begin{bmatrix} 1 & 0 & 1 & -1 & -1 \\ 0 & 1 & 1 & -1 & 1 \\ 0 & 0 & -2 & 3 & 1 \\ 0 & 0 & -3 & 2 & 1 \end{bmatrix}$

(iii) $\begin{bmatrix} 1 & 0 & 1 & -1 & -1 \\ 0 & 1 & 1 & -1 & 1 \\ 0 & 0 & 1 & -\frac{3}{2} & -\frac{1}{2} \\ 0 & 0 & -3 & 2 & 1 \end{bmatrix}$ (iv) $\begin{bmatrix} 1 & 0 & 1 & -1 & -1 \\ 0 & 1 & 1 & -1 & 1 \\ 0 & 0 & 1 & -\frac{3}{2} & -\frac{1}{2} \\ 0 & 0 & 0 & -\frac{5}{2} & -\frac{1}{2} \end{bmatrix}$

The matrix in (iv) is in echelon form. However, we proceed.

(v) $\begin{bmatrix} 1 & 0 & 1 & -1 & -1 \\ 0 & 1 & 1 & -1 & 1 \\ 0 & 0 & 1 & -\frac{3}{2} & -\frac{1}{2} \\ 0 & 0 & 0 & 1 & \frac{1}{5} \end{bmatrix}$ (vi) $\begin{bmatrix} 1 & 0 & 1 & 0 & -\frac{4}{5} \\ 0 & 1 & 1 & 0 & \frac{6}{5} \\ 0 & 0 & 1 & 0 & -\frac{1}{5} \\ 0 & 0 & 0 & 1 & \frac{1}{5} \end{bmatrix}$

systems of linear equations

(vii)

$$\begin{bmatrix} 1 & 0 & 0 & 0 & -\dfrac{3}{5} \\ 0 & 1 & 0 & 0 & \dfrac{7}{5} \\ 0 & 0 & 1 & 0 & -\dfrac{1}{5} \\ 0 & 0 & 0 & 1 & \dfrac{1}{5} \end{bmatrix}$$

In step (vii), we have a matrix in Hermite normal form which is equivalent to our original matrix. It is the augmented matrix of the system

$$x_1 = -\frac{3}{5}$$

$$x_2 = \frac{7}{5}$$

$$x_3 = -\frac{1}{5}$$

$$x_4 = \frac{1}{5}$$

which is equivalent to the original system. Thus, the four-tuple $(-3/5, 7/5, -1/5, 1/5)$ is a solution to the system.

example 2

Use the matrix method to solve the following system, if a solution exists.

$$x_1 - x_2 - x_3 = 1$$

(6.3.11)

$$x_1 + x_2 + x_3 = 3$$

$$3x_1 - x_2 - x_3 = 1$$

The augmented matrix for this system is

$$\begin{bmatrix} 1 & -1 & -1 & 1 \\ 1 & 1 & 1 & 3 \\ 3 & -1 & -1 & 1 \end{bmatrix}$$

which is row equivalent to each of the following. (Why?)

(i) $\begin{bmatrix} 1 & -1 & -1 & 1 \\ 0 & 2 & 2 & 2 \\ 0 & 2 & 2 & -2 \end{bmatrix}$ (ii) $\begin{bmatrix} 1 & -1 & -1 & 1 \\ 0 & 1 & 1 & 1 \\ 0 & 0 & 0 & -4 \end{bmatrix}$

The third row in step (ii) tells us that our system is inconsistent. (Why?) ●

a matrix method

example 3

Use the matrix method to solve the following system, if a solution exists.

(6.3.12)
$$\begin{aligned} x_1 - x_2 + x_3 &= 2 \\ 2x_1 - x_2 - x_3 &= 1 \\ 3x_1 - 2x_2 &= 3 \end{aligned}$$

The augmented matrix is

$$\begin{bmatrix} 1 & -1 & 1 & 2 \\ 2 & -1 & -1 & 1 \\ 3 & -2 & 0 & 3 \end{bmatrix}$$

which is row equivalent to each of the following. (Why?)

(i) $\begin{bmatrix} 1 & -1 & 1 & 2 \\ 0 & 1 & -3 & -3 \\ 0 & 1 & -3 & -3 \end{bmatrix}$ (ii) $\begin{bmatrix} 1 & -1 & 1 & 2 \\ 0 & 1 & -3 & -3 \\ 0 & 0 & 0 & 0 \end{bmatrix}$

The third row of the matrix in step (ii) indicates that our original system was dependent. We have replaced it with the equivalent system

$$\begin{aligned} x_1 - x_2 + x_3 &= 2 \\ x_2 - 3x_3 &= 3 \end{aligned}$$

If we continue, we can reduce to a matrix in Hermite normal form which will give us a simpler equivalent system. Thus, the following is row equivalent to our original matrix

(iii)
$$\begin{bmatrix} 1 & 0 & -2 & -1 \\ 0 & 1 & -3 & -3 \\ 0 & 0 & 0 & 0 \end{bmatrix}$$

and is simplified as much as possible. From it we obtain the system

$$\begin{aligned} x_1 \qquad 2x_3 &= -1 \\ x_2 - 3x_3 &= -3 \end{aligned}$$

or

$$\begin{aligned} x_1 &= 2x_3 - 1 \\ x_2 &= 3x_3 - 3 \end{aligned}$$

systems of linear equations

Thus, if k is any real number, the triple $(2k - 1, 3k - 3, k)$ is a solution to (6.3.12); in particular, if $k = 1$, $(1, 0, 1)$ is a solution. ●

example 4

Find all the solutions to the following homogeneous system using the matrix method

$$u + v - 2w = 0$$
$$2u - 2v + w = 0$$
$$u + 2v - 3w = 0$$

The augmented matrix is

$$\begin{bmatrix} 1 & 1 & -2 & 0 \\ 2 & -2 & 1 & 0 \\ 1 & 2 & -3 & 0 \end{bmatrix}$$

which is row equivalent to each of the following:

(i) $\begin{bmatrix} 1 & 1 & -2 & 0 \\ 0 & -4 & 5 & 0 \\ 0 & 1 & -1 & 0 \end{bmatrix}$ (ii) $\begin{bmatrix} 1 & 1 & -2 & 0 \\ 0 & 1 & -1 & 0 \\ 0 & 4 & -5 & 0 \end{bmatrix}$

(iii) $\begin{bmatrix} 1 & 1 & -2 & 0 \\ 0 & 1 & -1 & 0 \\ 0 & 0 & -1 & 0 \end{bmatrix}$ (iv) $\begin{bmatrix} 1 & 1 & -2 & 0 \\ 0 & 1 & -1 & 0 \\ 0 & 0 & 1 & 0 \end{bmatrix}$

(vi) $\begin{bmatrix} 1 & 1 & 0 & 0 \\ 0 & 1 & 0 & 0 \\ 0 & 0 & 1 & 0 \end{bmatrix}$ (v) $\begin{bmatrix} 1 & 0 & 0 & 0 \\ 0 & 1 & 0 & 0 \\ 0 & 0 & 1 & 0 \end{bmatrix}$

Step (vi) indicates that the trivial solution $u = v = w = 0$ is the only solution. You have, no doubt, already noticed that this was already apparent in step (iii) since at that point we do not show dependence of the system by a row of zeros. ●

exercises

1. Which of the following matrices are row equivalent?

(a) $\begin{bmatrix} 1 & 1 & 1 \\ 1 & 2 & 3 \\ 0 & 1 & 2 \end{bmatrix}$ (b) $\begin{bmatrix} 1 & 1 & 1 \\ 0 & 1 & 2 \\ 0 & 0 & 0 \end{bmatrix}$

a matrix method

(c) $\begin{bmatrix} 1 & 1 & 1 \\ 2 & 3 & 4 \\ 1 & 2 & 3 \end{bmatrix}$ (d) $\begin{bmatrix} 1 & 0 & -1 \\ 0 & 0 & 0 \\ 0 & 0 & 0 \end{bmatrix}$

2. Use the matrix method to solve each of the following:

(a) $x_1 + x_2 = 1$
$x_1 - x_2 = 2$

(d) $x_1 + x_2 + x_3 = 3$
$x_1 - 2x_2 - x_3 = 4$
$2x_1 - x_2 + x_3 = 5$

(b) $x_1 + 2x_2 = 3$
$2x_1 - x_2 = 1$

(e) $x_1 + x_2 - x_3 = 0$
$x_1 - x_2 - x_3 = -2$
$x_1 + 2x_2 + 3x_3 = -5$

(c) $x_1 + x_2 + x_3 = 3$
$x_1 - x_2 + x_3 = 1$
$2x_1 - x_2 - x_3 = 0$

(f) $x - y - z = 0$
$2x - 2y + z = 3$
$x - 10y + z = 1.1$

3. Use the matrix method to solve each of the following, if a solution exists.

(a) $2u + 3v - w = 1$
$u - 2v - 2w = -1$
$u + v - w = 0$

(d) $2x_1 - 3x_2 = 4$
$3x_1 - x_2 = -1$
$x_1 + x_2 = 5$

(b) $x + y - z = 2$
$2x - y + 3z = 1$
$4x + y + z = 2$

(e) $x_1 - x_2 - x_3 = 1$
$x_1 + x_2 + x_3 = -1$
$2x_1 - 4x_2 - 4x_3 = 5$

(c) $2x - y + z = 3$
$3x + y - z = 2$
$x - 3y + 3z = 3$

(f) $x_1 = x_2 - 2x_3$
$x_2 = x_1 + x_3$
$x_1 + x_2 = 2x_3$

4. Use the matrix method to find all the solutions to the following systems of equations.

(a) $x_1 - x_2 + x_3 = 0$
$2x_1 + 3x_2 - 5x_3 = 0$
$3x_1 + 2x_2 - 4x_3 = 0$

(d) $x_1 + x_2 - x_3 = 0$
$x_1 + 2x_2 + 3x_3 = 0$
$x_2 + 2x_3 = 0$

(b) $x + y + z + w = 0$
$x - y - z - w = 0$
$3x - y - z - w = 0$
$4x + y + z + w = 0$

(e) $3x_1 + x_2 + x_3 = 0$
$x_1 + 2x_2 + 3x_3 = 0$
$2x_1 - x_2 - 2x_3 = 0$

(c) $a + b + c + d = 0$
$a - c - d = 0$
$a - b + d = 0$
$b + c - d = 0$

(f) $x + y = w$
$x - y = z$
$w + x = y$
$w + z = 2x$

systems of linear equations

5. Solve the following systems of equations if possible.

   (a) $x - y = .2$
   $2x + y = 13$
   $3x - 5y = 0$
   $-x + 2y = 1$
   $-4x + 6y = -2$

   (b) $a + b + c = 9$
   $a + 2b + 3c = 22$
   $2a + 7b - 3c = 8$
   $2a - 4b + 6c = 20$
   $3a - b + 2c = 10$
   $10a + 6b - 3c = 13$

   (c) $x_1 + x_2 + x_3 + x_4 + x_5 = 0$
   $x_1 - x_2 + x_3 - x_4 + x_5 = 0$
   $2x_1 + 5x_2 - 4x_3 - 3x_4 - x_5 = 0$
   $3x_1 + 3x_2 + 2x_3 - 5x_4 + 4x_5 = 0$

   (d) $3r - s + 4t + 4u - v = 0$
   $6r - 2s + 6t + 7u + v = 0$
   $6r - 2s + 2t + 5u + 7v = 0$
   $9r - 3s + 4t + 8u + 9v = 0$

   (e) $x_1 + 2x_2 - 2x_3 + x_4 - x_5 = -1$
   $2x_1 + 3x_2 + x_3 - 4x_4 + 2x_5 = -1$
   $3x_1 + 5x_2 - x_3 + x_5 = -1$
   $x_1 + x_2 + 3x_3 - 2x_4 + 3x_5 = -1$

   (f) $x_1 + x_2 + x_3 + x_4 + x_5 + x_6 = 0$
   $x_1 - x_2 + 2x_3 + x_4 - x_5 + 2x_6 = 0$
   $x_1 + x_2 + 3x_3 + x_4 - x_5 + 3x_6 = 4$
   $x_1 - x_2 - 4x_3 - x_4 - x_5 + 4x_6 = 0$
   $2x_1 + x_2 + 3x_3 + x_4 + x_5 + x_6 = -1$
   $3x_1 + 2x_2 + 2x_3 + 3x_4 + 3x_5 + 2x_6 = -1$

6. Reduce the following matrices to Hermite normal form.

   (a) $\begin{bmatrix} 1 & -1 & 0 \\ -1 & 0 & 1 \\ 0 & 1 & -1 \end{bmatrix}$

   (c) $\begin{bmatrix} 1 & 2 & 3 & 4 & 5 \\ 2 & 3 & 4 & 5 & 6 \\ 3 & 4 & 5 & 6 & 7 \\ 4 & 5 & 6 & 7 & 8 \\ 5 & 6 & 7 & 8 & 9 \end{bmatrix}$

   (b) $\begin{bmatrix} 1 & x & -1 \\ x & 1 & -1 \\ -1 & 1 & x \end{bmatrix}$

   (d) $\begin{bmatrix} x - 1 & 1 & 2 & 3 \\ 1 & x - 1 & 1 & 0 \\ 3 & 2 & x - 2 & 1 \\ 1 & 0 & 1 & x - 2 \end{bmatrix}$

7. The general form for the equation of a parabola is

$$y = ax^2 + bx + c$$

Find the equation of the parabola which passes through the three points $(1, -2)$, $(2, 3)$, and $(-1, 3)$.

8. Find the parabola passing through $(1, -1)$, $(-1, 2)$, and $(-2, 1)$.

9. Consider the circle $x^2 + y^2 + bx + cy = d$. Find the equation of the circle which passes through $(1, 1)$, $(-1, -2)$, and $(3, 5)$.

10. Let $p(x, y, z) = ax + by + cz + d$. Find the coefficients of p, if $p(1, 1, 1) = p(1, -1, 0) = 0$ and $p(1, -1, -1) = p(-1, -1, 1) = 1$.

determinants

6.4. determinants

In the preceding section, we developed a particularly effective technique for solving a system of linear equations which involved matrix notation. We have, however, scarcely even hinted at the power available to us in the study of matrices and their properties. Separate courses in this topic are available, and research which discovers new and interesting relationships involving matrices is going forward every day. In this section, we shall describe a special function from the set of square matrices of real numbers to the real numbers. This function "*det*" is called the *determinant*. The determinant of a square matrix is a real number. We shall give an inductive definition of a determinant here, and show how its value is computed. We shall indicate also how the idea of a determinant can be useful in solving a system of equations.

For us a **square matrix** A will always be an array of *real* numbers.

(6.4.1)
$$A = \begin{bmatrix} a_{11} a_{12} \cdots a_{1n} \\ a_{21} a_{22} \cdots a_{2n} \\ \vdots \quad \vdots \qquad \vdots \\ a_{n1} a_{n2} \cdots a_{nn} \end{bmatrix}$$

is an $n \times n$ (n "by" n) matrix. The double subscripts on the entries in A denote their address (location) in the matrix. Thus, a_{34} is in row 3 column 4.

A 1×1 matrix is just a single real number.

(6.4.2)
$$A = (a)$$

A 2×2 matrix looks like

$$A = \begin{bmatrix} a_{11} & a_{12} \\ a_{21} & a_{22} \end{bmatrix}$$

A 3×3 matrix would be

$$A = \begin{bmatrix} a_{11} & a_{12} & a_{13} \\ a_{21} & a_{22} & a_{23} \\ a_{31} & a_{32} & a_{33} \end{bmatrix}$$

and so forth.

For any matrix A, the submatrix obtained from A by deleting row i and column j is called the **minor** A_{ij} of the element, a_{ij}, in row i and column j. Thus, in the matrix

(6.4.3)
$$A = \begin{bmatrix} 2 & -1 & 0 \\ 3 & 1 & 1 \\ 1 & 0 & -1 \end{bmatrix}$$

we have $A_{23} = \begin{bmatrix} 2 & -1 \\ 1 & 0 \end{bmatrix}$ is the *minor* of $a_{23} = 1$.

systems of linear equations

For a 1×1 matrix $A = (a)$, we define the determinant by

$$\det A = a$$

*We define the **determinant** of an $n \times n$ matrix A by*

(6.4.4)
$$\det A = \sum_{j=1}^{n} (-1)^{j+1} a_{1j} \det A_{1j}$$

Thus, the matrix in (6.4.3) has the following determinant. Here the straight line sides, $| \; |$, denote the determinant.

$$\begin{vmatrix} 2 & -1 & 0 \\ 3 & 1 & 1 \\ 1 & 0 & -1 \end{vmatrix} = 2 \begin{vmatrix} 1 & 1 \\ 0 & -1 \end{vmatrix} - (-1) \begin{vmatrix} 3 & 1 \\ 1 & -1 \end{vmatrix} + 0 \begin{vmatrix} 3 & 1 \\ 1 & 0 \end{vmatrix}$$

The 2×2 determinants can be evaluated by reusing (6.4.4) to obtain

$$\begin{vmatrix} 1 & 1 \\ 0 & -1 \end{vmatrix} = 1(-1) - 1(0) = -1$$

$$\begin{vmatrix} 3 & 1 \\ 1 & -1 \end{vmatrix} = 3(-1) - 1(1) = -4$$

and

$$\begin{vmatrix} 3 & 1 \\ 1 & 0 \end{vmatrix} = 3(0) - 1(1) = -1$$

Thus,

$$\begin{vmatrix} 2 & -1 & 0 \\ 3 & 1 & 1 \\ 1 & 0 & -1 \end{vmatrix} = 2(-1) + (1)(-4) + (0)(-1) = -6$$

To cut down the computations involved in direct use of the definition, for matrices with $n > 3$, we state several theorems without proof. Proofs can be given in a more careful formulation and are found in most linear algebra texts.

theorem 6.4.1

If B is a matrix which is obtained from A by multiplying a row or column of A by a constant c, then

$$\det B = c \det A$$

For example, if we have

(6.4.5)
$$B = \begin{bmatrix} 4 & -2 & 0 \\ 3 & 1 & 1 \\ 1 & 0 & -1 \end{bmatrix} = \begin{bmatrix} 2(2) & 2(-1) & 2(0) \\ 3 & 1 & 1 \\ 1 & 0 & -1 \end{bmatrix}$$

290 we see that B is obtained from the matrix A of (6.4.3) by multiplying the first row of A by 2. Therefore, $\det B = 2 \det A = -12$.

determinants

theorem 6.4.2

If B is a matrix which is obtained from the matrix A by interchanging any two rows (or columns), then

$$\det B = -\det A$$

As an example, we consider the matrix

(6.4.6)
$$B = \begin{bmatrix} 1 & 0 & -1 \\ 3 & 1 & 1 \\ 2 & -1 & 0 \end{bmatrix}$$

which is obtained from the matrix A in (6.4.3) by interchanging rows 1 and 3. Thus, Theorem 6.4.2 assures us that

$$\det B = -\det A = 6$$

theorem 6.4.3

If B is obtained from the matrix A by replacing any row (or column) of A by its sum with any nonzero multiple of any other row (column) of A, then

$$\det B = \det A$$

For example, let

(6.4.7)
$$B' = \begin{bmatrix} 0 & -1 & 0 \\ 5 & 1 & 1 \\ 1 & 0 & -1 \end{bmatrix}$$

This matrix is obtained from the matrix A of (6.4.3) by replacing the first column of A with its sum with twice the second column of A. Theorem 6.4.3 assures us that $\det B' = \det A = -6$. The value of this theorem is seen when we note that $\det B'$ is, in fact, easier to compute directly from the definition than was $\det A$. Using the minors of the elements of the first row, we obtain

$$\det B' = 0 \begin{vmatrix} 1 & 1 \\ 0 & -1 \end{vmatrix} - (-1) \begin{vmatrix} 5 & 1 \\ 1 & -1 \end{vmatrix} + 0 \begin{vmatrix} 5 & 1 \\ 1 & 0 \end{vmatrix} = 1(-5 - 1) = -6$$

It is frequently convenient to work with the columns of a matrix rather than the rows. We note that the theorems as stated apply to changes made in either columns or rows. As a partial justification for this, let $A = (a_{ji})$ be the matrix (6.4.1). We consider a different matrix, denoted by A^T, and called the **transpose** of A. The rows of A^T are the columns of A. Thus,

$$A^T = \begin{bmatrix} a_{11} & a_{21} & \cdots & a_{n1} \\ a_{12} & a_{22} & \cdots & a_{n2} \\ \vdots & \vdots & & \vdots \\ a_{1n} & a_{2n} & \cdots & a_{nn} \end{bmatrix}$$

For example, if A is the matrix of (6.4.3), we have

$$A = \begin{bmatrix} 2 & -1 & 0 \\ 3 & 1 & 1 \\ 1 & 0 & -1 \end{bmatrix} \qquad A^T = \begin{bmatrix} 2 & 3 & 1 \\ -1 & 1 & 0 \\ 0 & 1 & -1 \end{bmatrix}$$

Formally stated,

if $A = (a_{ij})$, then $A^T = (b_{ij})$, where $b_{ij} = a_{ji}$, $i, j = 1, 2 \ldots, n$

While A and A^T are obviously different matrices, and are really coefficient matrices for different systems of linear equations; nevertheless, their determinants are equal.

theorem 6.4.4

If A^T is the transpose of the matrix A, then

(6.4.8) $$\det A^T = \det A$$

If we apply the results of the above theorems, we can restate the computation of the determinant of A given in (6.4.4) in either of the two following ways.

(1) *Minors of any row*

(6.4.9) $$\det A = \sum_{j=1}^{n} (-1)^{i+j} a_{ij} \det A_{ij}, \text{ for any } i = 1, 2, \ldots, n$$

(2) *Minors of any column*

(6.4.10) $$\det A = \sum_{i=1}^{n} (-1)^{i+j} a_{ij} \det A_{ij}, \text{ for any } j = 1, 2, \ldots, n$$

To further illustrate the use of these ideas in evaluating the determinant of a matrix, consider

(6.4.11) $$C = \begin{bmatrix} 1 & 0 & 2 & 5 \\ 3 & 1 & 0 & 1 \\ -2 & 4 & 2 & 4 \\ 0 & 1 & 1 & 3 \end{bmatrix}$$

It is clear that

$$C^T = \begin{bmatrix} 1 & 3 & -2 & 0 \\ 0 & 1 & 4 & 1 \\ 2 & 0 & 2 & 1 \\ 5 & 1 & 4 & 3 \end{bmatrix}$$

292 Any column operations on C will be the same as row operations on C^T. By replacing rows two and three of C with their respective sums with appropriate

determinants

multiples (-3 and $+2$, respectively) of row one, we obtain the matrix

$$D = \begin{bmatrix} 1 & 0 & 2 & 5 \\ 0 & 1 & -6 & -14 \\ 0 & 4 & 6 & 14 \\ 0 & 1 & 1 & 3 \end{bmatrix}$$

Thus, $\det D = \det C$. By using minors of the first column of D, we obtain $\det D = (1)(\det K)$, where

$$K = \begin{bmatrix} 1 & -6 & -14 \\ 4 & 6 & 14 \\ 1 & 1 & 3 \end{bmatrix}$$

is the minor D_{11}. Then, upon replacing row two of K with its sum with row 1, we obtain

$$H = \begin{bmatrix} 1 & -6 & -14 \\ 5 & 0 & 0 \\ 1 & 1 & 3 \end{bmatrix}$$

Thus, $\det C = \det D = \det K = (-1)^{2+1} (5) \det M = -5 \det M$, where

$$M = \begin{bmatrix} -6 & -14 \\ 1 & 3 \end{bmatrix}$$

is the minor H_{21}. Now

$$\det M = -2 \begin{vmatrix} 3 & 7 \\ 1 & 3 \end{vmatrix} = (-2(9 - 7)) = -4$$

Therefore, $\det C = 20$.

Being perceptive, you will note that the properties of determinants given in the preceding theorems are related to the elementary row operations for matrices of the previous section. However, while elementary row operations on a matrix always yield an equivalent matrix, these same operations can change the value of the determinant. That is, the new matrix, though equivalent to, can have a determinant different from the original matrix.

exercises

1. Find the determinant of each of the following matrices.

   (a) $\begin{bmatrix} 1 & 2 \\ -2 & -1 \end{bmatrix}$

   (c) $\begin{bmatrix} 0 & 3 \\ -1 & 5 \end{bmatrix}$

   (b) $\begin{bmatrix} 1 & 1 \\ -2 & 2 \end{bmatrix}$

   (d) $\begin{bmatrix} -1 & 1 \\ 3 & -3 \end{bmatrix}$

293

systems of linear equations

(e) $\begin{bmatrix} 1 & -3 & -1 \\ 0 & 1 & 1 \\ -1 & 1 & 0 \end{bmatrix}$

(g) $\begin{bmatrix} 1 & 3 & -1 \\ 2 & 0 & 2 \\ 1 & -3 & 1 \end{bmatrix}$

(f) $\begin{bmatrix} 1 & 1 & -1 \\ 3 & 0 & -3 \\ 2 & 5 & -2 \end{bmatrix}$

(h) $\begin{bmatrix} \dfrac{1}{2} & 1 & \dfrac{2}{3} \\[2mm] \dfrac{1}{3} & \dfrac{3}{2} & 1 \\[2mm] \dfrac{1}{2} & \dfrac{1}{3} & 0 \end{bmatrix}$

2. Find the determinant of each of the following matrices.

(a) $\begin{bmatrix} 1 & 0 & 0 & 0 \\ 0 & 2 & 0 & 0 \\ 0 & 0 & -1 & 0 \\ 0 & 0 & 0 & 3 \end{bmatrix}$

(d) $\begin{bmatrix} 0 & 1 & 0 & 1 & 0 \\ 1 & 0 & 0 & 1 & 1 \\ 0 & 1 & 1 & 0 & 1 \\ 1 & 1 & 0 & 1 & 0 \\ 1 & 1 & 1 & 1 & 1 \end{bmatrix}$

(b) $\begin{bmatrix} 2 & 0 & 1 & 0 \\ 1 & 1 & 0 & 1 \\ -1 & 0 & -1 & 2 \\ 0 & 3 & -1 & 0 \end{bmatrix}$

(e) $\begin{bmatrix} 1 & 2 & 3 & 4 & 5 \\ 0 & 2 & 3 & 4 & 5 \\ 0 & 0 & 3 & 4 & 5 \\ 0 & 0 & 0 & 4 & 5 \\ 0 & 0 & 0 & 0 & 5 \end{bmatrix}$

(c) $\begin{bmatrix} 1 & 1 & -1 & 1 \\ -1 & 1 & 1 & -1 \\ 1 & -1 & 1 & -1 \\ -1 & 1 & -1 & 1 \end{bmatrix}$

(f) $\begin{bmatrix} 1 & 0 & 5 & 4 & 3 & 2 \\ 2 & 1 & 0 & 5 & 4 & 3 \\ 3 & 2 & 1 & 0 & 5 & 4 \\ 4 & 3 & 2 & 1 & 0 & 5 \\ 5 & 4 & 3 & 2 & 1 & 0 \\ 0 & 5 & 4 & 3 & 2 & 1 \end{bmatrix}$

3. Find the determinant of each of the following matrices Simplify the resulting expression.

(a) $\begin{bmatrix} \cos \alpha & \sin \alpha \\ -\sin \alpha & \cos \alpha \end{bmatrix}$

(d) $\begin{bmatrix} x & y \\ -y & x \end{bmatrix}$

(b) $\begin{bmatrix} 1 & \tan \alpha \\ \tan \alpha & -1 \end{bmatrix}$

(e) $\begin{bmatrix} x & 1 \\ y & 1 \end{bmatrix}$

(c) $\begin{bmatrix} 1 & \csc \alpha \\ \csc \alpha & 1 \end{bmatrix}$

(f) $\begin{bmatrix} x & y & 1 \\ 1 & 1 & 2 \\ 2 & 1 & 3 \end{bmatrix}$

determinants

(g) $\begin{bmatrix} 2 & 1 & 1 \\ 1 & 2 & 1 \\ 1 & 1 & 2 \end{bmatrix}$ (h) $\begin{bmatrix} 1 & x & x^2 \\ 1 & y & y^2 \\ 1 & z & z^2 \end{bmatrix}$

4. Write the transpose of each matrix in Problem 1. Compute the determinant of each of these transposes.

5. Compute the determinant of the matrix B in (6.4.5) directly from the definition. Compare with the results obtained using Theorem 6.4.1.

6. Same as Exercise 5 for the matrix B in (6.4.6) and Theorem 6.4.2.

7. Prove that if A is a matrix with one row of zeros then det $A = 0$.

8. Prove that if A is a matrix with two equal columns, then det $A = 0$.

9. Use (6.4.4) to prove Theorem 6.4.1 by induction.

10. Show that

$$\det \begin{bmatrix} 1 & x & x^2 & x^3 \\ 1 & y & y^2 & y^2 \\ 1 & z & z^2 & z^3 \\ 1 & w & w^2 & w^3 \end{bmatrix} = (x - y)(x - z)(x - w)(y - z)(y - w)(z - w)$$

11. Solve the equation det $A = 0$ for the given matrix A

(a) $A = \begin{bmatrix} x + 1 & 3 \\ x - 1 & 2 \end{bmatrix}$

(c) $A = \begin{bmatrix} 1 + x & 1 & 1 \\ 1 & 1 + x & 1 \\ 1 & 1 & 1 + x \end{bmatrix}$

(b) $A = \begin{bmatrix} -1 & x & x + 4 \\ 1 & 2 & 3 \\ -2 & 1 & 4 \end{bmatrix}$

(d) $A = \begin{bmatrix} x + 1 & 2 & 1 \\ 4 & x - 1 & 2 \\ 1 & -1 & x - 1 \end{bmatrix}$

12. Show that det $A_n = 0$ where A_n is the following $n \times n$ matrix.

$$A_n = \begin{bmatrix} 1 - n & 1 & 1 & \cdots & 1 \\ 1 & 1 - n & 1 & \cdots & 1 \\ 1 & 1 & 1 - n & \cdots & 1 \\ \vdots & \vdots & \vdots & & \vdots \\ 1 & 1 & 1 & \cdots & 1 - n \end{bmatrix}$$

13. A matrix is *nonsingular* if its determinant is not zero. Which of the matrices in Exercise 1 are nonsingular?

14. Reduce each of the matrices in Exercise 1 to Hermite normal form.

15. Find the determinant for each matrix of Exercise 14. Compare these values with the results of Exercise 1.

16. Reduce each matrix of Exercise 2 to Hermite normal form. Compute the determinant of the resulting matrix and compare with the results of Exercise 2.

systems of linear equations

6.5. Cramer's rule

The theory of determinants is usually attributed to some work Leibniz did in 1693 on systems of linear equations, although similar work was done about ten years earlier in Japan by Seki Kowa. In the succeeding 150 years, contributions were made by Laplace and Cauchy. While determinants still have application to the theory of solutions of systems of linear equations, the amount of work involved in large systems makes the method usually quite cumbersome. For example, a system of 25 linear equations in 25 unknowns—not unthinkably large for some applications—requires more than 26! multiplications. This number is of the order of 10^{26}, so a computer which can perform 1,000 multiplications per second would require 10^{16} years to complete them. Of necessity, other methods had to be devised.

On the other hand, determinants still have theoretical importance and are also very useful in solving small systems of equations—three or four unknowns. The method for solving a system of linear equations using determinants is known as **Cramer's rule**. We state this rule here. Consider the system of equations

$$a_{11}x_1 + a_{12}x_2 + \cdots + a_{1n}x_n = b_1$$
(6.5.1)
$$a_{21}x_1 + a_{22}x_2 + \cdots + a_{2n}x_n = b_2$$
$$a_{n1}x_1 + a_{n2}x_2 + \cdots + a_{nn}x_n = b_n$$

theorem 6.5.1

The system of Equation (6.5.1) has a unique solution if, and only if, the determinant of the coefficient matrix A is nonzero. If $\det A \neq 0$ then the solution is given by

$$x_j = \frac{\det B_j}{\det A}, \qquad j = 1, 2, \ldots, n$$

where B_j is the matrix obtained from A by replacing the jth column of A by the constant terms $b_1, b_2, \ldots, b_n$.

The proof of this theorem is a rather tedious application of the methods of the last section.

example 1

To illustrate how the theorem is used suppose we have the system

$$x_1 - 2x_2 - x_3 = 1$$
$$x_2 + x_3 = 2$$
$$x_1 + x_2 \qquad = 1$$

The following matrices are associated with this system. The coefficient matrix is

$$A = \begin{bmatrix} 1 & -2 & -1 \\ 0 & 1 & 1 \\ 1 & 1 & 0 \end{bmatrix}$$

cramer's rule

and the matrices B_i are

$$B_1 = \begin{bmatrix} 1 & -2 & -1 \\ 2 & 1 & 1 \\ 1 & 1 & 0 \end{bmatrix}, \quad B_2 = \begin{bmatrix} 1 & 1 & -1 \\ 0 & 2 & 1 \\ 1 & 1 & 0 \end{bmatrix}, \quad B_3 = \begin{bmatrix} 1 & -2 & 1 \\ 0 & 1 & 2 \\ 1 & 1 & 1 \end{bmatrix}$$

Computing each determinant one obtains $\det A = -2$, $\det B_1 = -4$, $\det B_2 = 2$, and $\det B_3 = -6$. Thus,

$$x_1 = \frac{-4}{-2} = 2, \ x_2 = \frac{2}{-2} = -1, \ x_3 = \frac{-6}{-2} = 3$$

is the unique solution. ●

example 2

Use Cramer's rule to solve the system

$$x \cos \alpha - y \sin \alpha = a$$

$$x \sin \alpha + y \cos \alpha = b$$

for x and y in terms of a and b. The coefficient matrix is

$$A = \begin{bmatrix} \cos \alpha & -\sin \alpha \\ \sin \alpha & \cos \alpha \end{bmatrix}$$

The determinant of A is 1. (Why?) The matrices B_x and B_y are

$$B_x = \begin{bmatrix} a & -\sin \alpha \\ b & \cos \alpha \end{bmatrix} \quad B_y = \begin{bmatrix} \cos \alpha & a \\ \sin \alpha & b \end{bmatrix}$$

Therefore,

$$x = \det B_x = a \cos \alpha + b \sin \alpha$$

and

$$y = \det B_y = -a \sin \alpha + b \cos \alpha \quad ●$$

If we were to have a homogeneous system of equation (in (6.5.1)) all the b's equal to zero), Theorem 6.5.1 would assure us that the system would have a unique solution when the determinant of the coefficient matrix is *not* zero. Therefore, since $(0, 0, \ldots, 0)$ is always such a solution, a homogeneous system can have *nontrivial* solutions only when the coefficient matrix has zero determinant.

theorem 6.5.2

A homogeneous system of linear equations (n equations, n variables) will have a nontrivial solution if, and only if, the determinant of the coefficient matrix is zero.

In this case, of course, Cramer's rule yields only the meaningless result $0/0$.

systems of linear equations

example 3

Find all the solutions to the system of equations

$$x + 2y + z = 0$$
$$x - y + 2z = 0$$
$$3x + 3y + z = 0$$

The coefficient matrix is

$$A = \begin{bmatrix} 1 & 2 & 1 \\ 1 & -1 & 2 \\ 3 & 3 & 1 \end{bmatrix}$$

and det $A = 9$. Therefore, since det B_x = det B_y = det B_z = 0, the only solution to this system is the trivial one

$$x = y = z = 0 \quad \bullet$$

example 4

Solve the system

$$x + y - z = 0$$
$$x - y + z = 0$$
$$x + 3y - 3z = 0$$

The coefficient matrix

$$A = \begin{bmatrix} 1 & 1 & -1 \\ 1 & -1 & 1 \\ 1 & 3 & -3 \end{bmatrix}$$

has zero determinant. Therefore, we cannot use Cramer's rule to solve this system. The augmented matrix for the system can be reduced to the matrix B by elementary row operations. (Verify this.)

$$B = \begin{bmatrix} 1 & 1 & -1 & 0 \\ 0 & 1 & -1 & 0 \\ 0 & 0 & 0 & 0 \end{bmatrix}$$

This is, in turn, equivalent to

$$C = \begin{bmatrix} 1 & 0 & 0 & 0 \\ 0 & 1 & -1 & 0 \\ 0 & 0 & 0 & 0 \end{bmatrix}$$

298 Therefore, all the solutions have the form $x = 0$, $y = z = k$, for any real number k. $\quad \bullet$

cramer's rule

example 5

Show that the system

$$x + y - z = 1$$
$$x - y + z = 3$$
$$x + 3y - 3z = 1$$

is inconsistent. The coefficient matrix A is the same as that of Example 4. We noted there that det $A = 0$. Therefore, this system fails to have a unique solution according to Theorem 6.5.1. The system is not dependent, however, since

$$B_y = \begin{bmatrix} 1 & 1 & -1 \\ 1 & 3 & 1 \\ 1 & 1 & -3 \end{bmatrix}$$

and det $B_y = -4$. You may also verify this (more easily) using the methods of Section 6.3. ●

exercises

1. Use Cramer's rule to solve the following systems.

(a) $x + y = 2$
$x - y = 1$

(b) $4x - y = 11$
$2x + 3y = -5$

(c) $6x_1 + 5x_2 = -5$
$-3x_1 + 15x_2 = 13$

(d) $u - 3v = 3$
$2u - 7v = 5$

(e) $11a + 14b = -3$
$3a - 5b = 8$

(f) $\dfrac{x}{2} + \dfrac{y}{2} = 1$

$\dfrac{x}{3} - \dfrac{y}{3} = 2$

(g) $\sqrt{2}x + \sqrt{3}y = 5$
$\sqrt{3}x - \sqrt{2}y = 0$

(h) $\dfrac{3}{2}x + \dfrac{1}{5}y = 1$

$\dfrac{3}{5}x - \dfrac{1}{2}y = 6$

2. Use Cramer's rule to solve each of the following systems of equations.

(a) $x + y + z = 3$
$x \quad\quad + z = 2$
$x - y \quad\quad = 0$

(b) $x + y + z = 3$
$x - 2y - z = 4$
$2x + y + z = 5$

(c) $x + y + z = 1$
$2x - y - z = 0$
$-y + 2z = 1$

(d) $2x - y + z = -1$
$x + 3y - z = 4$
$-3x + 5y + z = 0$

systems of linear equations

(e) $2x_1 + 3x_2 - x_3 = 4$
$-x_1 + x_2 + 2x_3 = -2$
$3x_1 - x_2 - 2x_3 = 1$

(g) $\sqrt{2}\,a + \sqrt{3}\,b - \sqrt{5}\,c = 0$
$\sqrt{8}\,a \qquad + \sqrt{5}\,c = 9$
$\sqrt{3}\,a - \sqrt{2}\,b + \qquad c = \sqrt{5}$

(f) $\dfrac{u}{2} - \dfrac{v}{2} + \dfrac{w}{2} = 0$

$\dfrac{u}{3} + \dfrac{v}{3} + \dfrac{w}{3} = 0$

$\dfrac{u}{4} + \dfrac{v}{4} - \dfrac{w}{4} = 1$

(h) $x_1 - 3x_2 + x_3 - x_4 = 1$
$2x_1 - 9x_2 - x_3 + x_4 = -1$
$x_1 - 2x_2 - 3x_3 + 4x_4 = 5$
$3x_1 - x_2 + 2x_3 - x_4 = -2$

3. Use determinants to discuss the solutions of the following systems of equations. Find all possible solutions.

(a) $11x_1 + 14x_2 = 0$
$3x_1 - 5x_2 = 0$

(d) $x_1 + 2x_2 + x_3 = 0$
$2x_1 - x_2 + 2x_3 = 0$
$x_1 + x_2 + x_3 = 0$

(b) $x_1 + x_2 + x_3 = 0$
$x_1 - x_2 + x_3 = 0$
$x_1 - x_2 - x_3 = 0$

(e) $x + y - 2z = 0$
$2x - 2y - z = 0$
$x + 2y - 3z = 0$

(c) $u - v + 4w = 0$
$2u + v - w = 0$
$-u + v + 2w = 0$

(f) $u - v - w = 1$
$u + v + w = 3$
$3u - v - w = 1$

4. Solve, where possible, the following systems.

(a) $y + 2x = 0$
$y - 5z = 1$
$z + 3x = -2$

(d) $x + iy + z = 1$
$ix + y - z = 1$
$x - y + iz = 1$

(b) $x + y = w$
$x - y = z$
$w + x = y$
$w + z = x$

(e) $(1 + i)x + 2y - z = 1$
$x + (1 + i)y - 2z = 1$
$-x + y + (1 + i)z = 1$

(c) $x + y \cos\theta = \sin\theta$
$x \cos\theta + y = 0$

(f) $x - y \tan\theta = \sec\theta$
$x \tan\theta + y = 0$

5. Show that

$$\det \begin{bmatrix} x & y & 1 \\ x_1 & y_1 & 1 \\ x_2 & y_2 & 1 \end{bmatrix} = -\det \begin{bmatrix} x - x_1 & y - y_1 \\ x_2 - x_1 & y_2 - y_1 \end{bmatrix}$$

6. Show that the discrimant of the quadratic polynomial

$$ax^2 + bx + c$$

can be written as

300

$$\Delta = \det \begin{bmatrix} b & 2c \\ 2a & b \end{bmatrix}$$

cramer's rule

7. A matrix A is called *upper triangular* if the entries below the main diagonal are all zero. ($a_{ij} = 0$ for $j < i$). Show that for an $n \times n$ upper triangular matrix A, the determinant is equal to the product of the elements on the main diagonal.

$$\det A = \prod_{i=1}^{n} a_{ii}$$

8. A different matrix function which appears repeatedly in the literature of combinatorial mathematics is the *permanent* which we define inductively as follows:

If $A = (a)$ is a 1×1 matrix, per $(A) = a$, and for the $n \times n$ matrix $A = (a_{ij})$

$$\text{per } A = \sum_{j=1}^{n} a_{ij} \text{ per } A_{ij}$$

Compute the permanent for each of the matrices in Exercise 1 of the previous section.

9. A third matrix function of importance is the *trace*. We define the trace here only for a *diagonal matrix*; that is, a matrix in which the elements off the main diagonal are all zero. Then

$$\text{trace } A = \sum_{i=1}^{n} a_{ii}$$

Compute the trace of the matrix in Exercise 2 (a) of the previous section.

10. Reduce the matrix of Exercise 2 (e) of the previous section to Hermite normal form and compute its trace.

quiz for review

PART I

Problems 1–10 refer to the following matrices.

$$A = \begin{bmatrix} 1 & 2 & 1 & 1 \\ 2 & 1 & 2 & 1 \\ 2 & 2 & 1 & 1 \end{bmatrix} \quad B = \begin{bmatrix} 1 & 2 & 1 & 1 \\ 0 & 3 & 0 & 1 \\ 0 & 0 & 3 & 1 \end{bmatrix} \quad C = \begin{bmatrix} 1 & 0 & 0 & 0 \\ 0 & 1 & 0 & \frac{1}{3} \\ 0 & 0 & 1 & \frac{1}{3} \end{bmatrix}$$

$$D = \begin{bmatrix} 1 & 2 & 1 \\ 2 & 1 & 2 \\ 2 & 2 & 1 \end{bmatrix} \quad E = \begin{bmatrix} 1 & 0 & 0 \\ 0 & 1 & 0 \\ 0 & 0 & 1 \end{bmatrix}$$

1. Which matrices are in echelon form?

   (a) all are   (b) B, C, and E (c) D and E (d) C and E
   (e) none are

2. Which matrices are in Hermite normal form?

   (a) all are   (b) B, C, and E (c) D and E (d) C and E
   (e) none are

3. Which matrices are row equivalent to A?

   (a) all are   (b) B, C, and E (c) A, B, and C (d) D and E
   (e) C and E

4. Which matrices are row equivalent to D?

   (a) all are   (b) B, C, and E (c) A, B, and C (d) D and E
   (e) C and E

systems of linear equations

5. If A is the augmented matrix of a system of linear equations in x, y, and z, then the solution of the system is:

    (a) $x = 0, y = 0, z = 0$ (b) $x = 0, y = 1, z = 1$ (c) $x = 0, y = 1/3$, $z = 1/3$ (d)   The system has infinitely many solutions.   (e)   The system has no solution; it is inconsistent.

6. det A

    (a)   doesn't exist   (b)   is zero   (c)   is nonzero   (d)   equals det D
    (e)   equals det C

7. det D

    (a)   does not exist   (b)   is zero   (c)   equals 3   (d)   equals det A
    (e)   equals 1

8. The coefficient matrix for the system of equations

$$x_1 + 2x_2 + x_3 = 1$$
$$2x_1 + x_2 + 2x_3 = 1$$
$$2x_1 + 2x_2 + x_3 = 1$$

is   (a) A (b) B (c) C (d) D (e) E

9. The augmented matrix of the system of equations

$$x_1 + 2x_2 = 1$$
$$2x_1 + x_2 = 2$$
$$2x_1 + 2x_2 = 1$$

is   (a) A (b) B (c) C (d) D (e) E

10. The system of equations in Problem 9:

    (a)   is inconsistent   (b)   is dependent   (c)   has solutions $x_1 = -1, x_2 = 1$
    (d)   has solutions $x_1 = 1/2, x_2 = 1$ (e)   cannot be a system; there are too many equations

Problems 11–29 are (a) true or (b) flase.

11. A system of equations with more unknowns than equation always has at least one solution.

12. Every homogeneous system has at least one solution.

13. Two systems of equations are *equivalent* only if they have the same augmented matrix.

14. If matrices A and B are equivalent, then their determinants are equal.

15. The system of equations

$$x + y = 3$$

has the unique solution $x = 2, y = 1$.

16. Any solution to the system of equations in Problem 15 is of the form

$$x_1 = k, x_2 = 3 - k$$

17. The systems

         $x_1 + x_2 = 0$ $x_1 + x_2 = 0$

         $x_1 - x_2 = 0$ $x_1 - x_2 = 1$

are equivalent.

quiz for review

18. The system

$$x_2 + x_3 + x_4 = 0$$
$$x_1 \qquad + x_3 + x_4 = 0$$
$$x_1 + x_2 \qquad + x_4 = 0$$
$$x_1 + x_2 + x_3 \qquad = 0$$

is homogeneous.

19. The only solution to the system of equations in Problem 18 is the trivial one $x_1 = x_2 = x_3 = x_4 = 0$.

20. The system

$$x_1 + x_2 + x_3 = 1$$
$$x_1 + x_2 - x_3 = 2$$
$$4x_1 + 4x_2 + 10x_3 = 3$$

is consistent.

PART II

Solve the following systems of equations.

(1) $x_1 + x_2 = 0$
$x_1 - x_2 = 1$

(2) $2x_1 + 3x_2 = 0$
$x_1 + 2x_2 = 1$

(3) $x_1 - 8x_2 = 3$
$2x_1 + x_2 = 1$
$4x_1 + 7x_2 = -4$

(4) $x_1 - 2x_2 = -3$
$4x_1 + 9x_2 = 11$
$7x_1 + 3x_2 = 2$

(5) $x_1 + x_2 + x_3 = 1$
$2x_1 + x_2 + x_3 = 3$
$x_1 - 2x_2 - x_3 = 0$

(6) $x_1 - 2x_2 + 5x_3 = 1$
$2x_1 - 4x_2 + 8x_3 = 2$
$-3x_1 + 6x_2 + 7x_3 = 1$

(7) $x_1 - x_2 + x_3 - x_4 = 0$
$x_1 + 2x_2 - x_3 + 2x_4 = 0$
$-2x_1 + x_2 + x_3 + x_4 = 0$
$2x_1 - 2x_2 - 3x_3 + 2x_4 = 0$

(8) $x_1 + 3x_2 - 2x_3 + x_4 = 0$
$x_1 + 2x_2 + 3x_3 - 2x_4 = 0$
$3x_1 + 8x_2 - x_3 \qquad = 0$
$2x_1 + 5x_2 + x_3 - x_4 = 0$

(9) $x_1 + x_2 + x_3 = 2$
$2x_1 + x_2 - x_3 = 1$
$3x_1 + 2x_2 \qquad = 3$
$5x_1 + 3x_2 - x_3 = 4$

(10) $x_1 - x_2 + 2x_3 + 3x_4 + x_5 = 1$
$2x_1 - x_2 - x_3 - 2x_4 + x_5 = 1$
$x_1 + x_2 - 2x_3 + x_4 + 2x_5 = 1$

303

coordinate geometry

In addition to writing poetry, the Persian Omar Khayyam (ca. 1050–1122) wrote a treatise on algebra. In this important work he wrote: "whoever thinks algebra is a trick in obtaining unknowns has thought it in vain. No attention should be paid to the fact that algebra and geometry are different in appearance."* Notwithstanding this admonition, the succeeding 500 years largely saw algebra and geometry developing in different directions. Nevertheless, in 1628 Descartes returned to this old idea and developed analytic geometry. In fact, present-day algebra and geometry are closely connected, particularly at the research level. The marriage of algebraic and geometric ideas, performed by Descartes, has indeed been a prolific union. Many problems have been solved by this couple. In the centuries since Descartes, algebra has literally increased the dimensions of geometry, while the impact of geometry on algebra has been no less significant.

Throughout this text we have attempted to explain the close relationship between geometric ideas and elementary functions. Our work in this chapter will provide additional evidence of the power of analytic geometry as we generalize and further examine some of the ideas presented in the earlier chapters. We shall stick largely to work in two- and three-dimensional Euclidean space. However, generalizations to spaces of higher dimensions are possible and are alluded to. Indeed such generalizations are necessary for the work of many physical and behavioral scientists as they *describe reality*.

*A. R. Amir-Moez, "A paper of Omar Khayyam" *Scripta Mathematica* 26(1963) p. 329.

coordinate geometry

7.1. functions of two or more variables

At various places in the previous chapters, we have indicated that functions exist with other than the set of real numbers as their domains. As an example, consider the making of a map as described in the first chapter. Chapter 6 was concerned with linear functions in several variables (unknowns). In the case of these linear equations, despite the number of unknowns, there is available a sort of geometric interpretation. We have already discussed the way in which the plane is considered as the set $R^2 = R \times R$ with points represented by ordered pairs (x_1, x_2) and distance given by the formula

$$d = \sqrt{(x_1 - x_1')^2 + (x_2 - x_2')^2}$$

This is easily generalized to the case of three-space, $R^3 = R \times R \times R$, where each point is represented by an ordered triple (x_1, x_2, x_3), and the distance between (x_1, x_2, x_3) and (x_1', x_2', x_3') is given by the formula

$$d = \sqrt{(x_1 - x_1')^2 + (x_2 - x_2')^2 + (x_3 - x_3')^2}$$

The details of this generalization is the subject of Section 7.6.

Some modern physicists, for example, cannot get along with but three dimensions; they must make relativistic corrections. For them the world is at least four dimensional with each "point" being described by a four-tpl. (x_1, x_2, x_3, x_4) or (x, y, z, t). Some behavioral scientists need an n dimensional space where points are n-tpls $(x_1, x_2, \ldots, x_n)$; e.g., "vectors of the mind." Here n is usually rather large (i.e., bigger than 4).

By **Euclidean n-space** is meant the collection $R^n = R \times R \times R \times \cdots \times R$ of ordered n-tpls $(x_1, x_2, \ldots, x_n)$ of real numbers together with a distance function generalizing those formulas given above for $n = 2$ and $n = 3$.

For the most part we shall work in Euclidean 2-space and Euclidean 3-space in this chapter. However, to give us some freedom of language, we shall define for any positive integer, a function of n variables. You have already met some examples of these in Chapter 6. A real **function of n variables** is, simply, a function

$$f: D \to R$$

whose domain D is contained in Euclidean n-space R^n.

example 1

The functions $f(x, y) = x + y$, $g(x, y) = x^2 - y^2$, and $h(x, y, z) = x^2 + y^2 + z^2$, are all functions of n variables. In the case of f and g, $n = 2$; and for h, $n = 3$. ●

example 2

The projection maps used in Chapter 5 to define the trigonometric functions: $p_x(x, y) = x$ and $p_y(x, y) = y$, are functions whose domains are the plane R^2. ●

example 3

The distance a point $(x_1, x_2, \ldots, x_n)$ lies from the origin in R^n is given by the distance formula as the function

$$d(x_1, x_2, \ldots, x_n) = \sqrt{x_1^2 + x_2^2 + \cdots + x_n^2} \quad ●$$

functions of two or more variables

example 4

Let D be the set of all points in the plane bounded by the unit circle; i.e., $D = \{(x, y)\,|\,x^2 + y^2 \leq 1\}$. Then $s(x, y) = \cos^{-1}(x^2 + y^2)$ is a function of two variables, with domain D. ●

The first five sections of this chapter are largely concerned with the sets of zeros of functions $f(x, y)$ of two variables. That is, we are interested in describing the set of all those points (x, y) in the plane for which $f(x, y) = 0$. As stated in the first chapter, this set of zeros for the function f is called *the graph of the equation*

$$f(x, y) = 0$$

The graph of an equation in n variables has an analogous definition.

example 5

Describe the set of zeros of the projection map $p_x(x, y)$; i.e., the graph of

$$p_x(x, y) = 0$$

Since $p_x(x, y) = x$, the set of zeros for this function is the set of all points in the plane for which $x = 0$; viz., the set

$$\{(0, y)\,|\,y \in R\}$$

This is, of course, the y axis. The y axis is the graph of the equation

$$x = 0 \quad ●$$

example 6

Describe the set of zeros of the function $f(x, y) = 1 - x^2 - y^2$. Setting $f(x, y) = 0$ results in the equation

$$x^2 + y^2 = 1$$

whose graph is the unit circle. ●

example 7

Consider the function

$$f(x, y) = (1 - x - y)(1 + x - y)$$

This function can be multiplied out to obtain the expression

$$f(x, y) = 1 - 2y + y^2 - x^2$$

This is the form of a polynomial in two variables. ●

In general, a *quadratic polynomial* in two variables has the form

$$p(x, y) = Ax^2 + By^2 + Cxy + Dx + Ey + F$$

The general *linear polynomial* in two variables has the form

$$p(x, y) = Ax + By + C = 0$$

coordinate geometry

A *cubic polynomial* in two variables would have the form

$$p(x, y) = a_1 x^3 + a_2 y^3 + a_3 x^2 y + a_4 xy^2 + a_5 x^2 +$$

$$a_6 y^2 + a_7 xy + a_8 x + a_9 y + a_{10}$$

We shall not attempt to give a general definition for a polynomial of degree m in n variables. This is very complicated to write down. The above examples suggest the details. In the next four sections we are interested in the set of zeros of linear and quadratic polynomials; that is, in the graphs of equations

$$p(x, y) = 0$$

where $p(x, y)$ is a linear or a quadratic polynomial.

Rational functions in two (or more) variables are quotients of polynomials in two (or more) variables.

example 8

Find the product of the two rational functions

$$\frac{2x + y}{x^2 - 2xy} \quad \text{and} \quad \frac{x^3 - 2x^2 y}{4x^2 - y^2}$$

We have

$$\frac{2x + y}{x^2 - 2xy} \cdot \frac{x^3 - 2x^2 y}{4x^2 - y^2} = \frac{(2x + y)}{x(x - 2y)} \cdot \frac{x^2(x - 2y)}{(2x - y)(2x + y)}$$

When common factors are reduced from the numerator and denominator this reduces to

$$\frac{x}{2x - y} \bullet$$

example 9

Divide the two rational functions

$$\frac{x^2 - 2xy + y^2}{x^2 - y^2} \quad \text{and} \quad \frac{y(x^2 + y^2)}{(x + y)(y - x)}$$

This quotient becomes

$$\frac{(x - y)(x - y)/(x - y)(x + y)}{y(x^2 + y^2)/(x + y)(y - x)} = \left(\frac{(x - y)(x - y)}{(x - y)(x + y)} \right) \left(\frac{-(x - y)(x + y)}{y(x^2 + y^2)} \right)$$

$$= \frac{-(x - y)^2}{y(x^2 + y^2)} = \frac{-x^2 + 2xy - y^2}{x^2 y + y^3} \bullet$$

exercises

1. Find the sum of each of the following pairs of polynomials.

(a) $(1 - 2x + y) + (7 - 4y - 3x)$
(b) $(-5x + 2y - 3) + (4x + y + 6)$
(c) $(x^2 + y^2 - xy) + (2x^2 - 2y^2)$

functions of two or more variables

    (d) $(2x^2 - 3y^2 + 8) + (x - 4y - 3)$
    (e) $(x + y - z) + (x^2 + xy + z)$
    (f) $(x^3 + x^2 + x + y) + (y^2 - xy - y)$
    (g) $(x^3 + y^3 + 1) + (-3xy)$
    (h) $(x^3 - y^3 - x^2y) + (x^2y + y - x)$
    (i) $(56x^2 - 9y^2 + 37x) + (8y^2 - 12x^2 + 1)$
    (j) $(27x^3 - 9y^2 + 3) + (3y^2 - 9x^2 + xy - 4)$

2. Multiply each of the following.

    (a) $(x - y)(2x^2y)$
    (b) $(2x + 3y)(3x + 2y)$
    (c) $(3x - 5y)(2x + y)$
    (d) $(3x^2 - 2y^3)(2x^2 + 5y^3)$
    (e) $(x + y + 3)(x + y - 3)$

    (f) $(f + 2x + y)(4 - 2x - y)$
    (g) $(3x + y - 2)(3x - y + 2)$
    (h) $(5x - 3y + 2)(5x + 3y - 2)$
    (i) $(3x + x^2y - y^2)(x^2 - xy + y^2)$
    (j) $(2x^2 + xy^2 + y)(x^2 - xy + y^2)$

3. Perform the indicated operations on the given rational function.

    (a) $\dfrac{28x^3}{5xy^2} \cdot \dfrac{10xy^3}{21x^6}$

    (b) $\left(\dfrac{x^3 + y^3}{x^2 - y^2}\right) \cdot \left(\dfrac{x^2 + 2xy - 3y^2}{x^2 - xy + y^2}\right)$

    (c) $\dfrac{\dfrac{x^3}{3y^3}}{3x^3}$

    (d) $\left(\dfrac{y^2 - 4}{2x - xy}\right) \cdot \left(\dfrac{2x + xy}{y^2 + 4y + 4}\right)$

    (e) $\dfrac{\dfrac{x^2y - 9y^3}{x^2 + 2xy - 3y^2}}{\dfrac{3y - x}{y - x}}$

    (f) $\dfrac{\dfrac{1}{x} - \dfrac{1}{y}}{\dfrac{1}{x^2} - \dfrac{1}{y^2}}$

    (g) $\dfrac{\dfrac{1}{2} - \dfrac{1}{y} + \dfrac{2}{y^2}}{\dfrac{1}{4} + \dfrac{2}{y^3}}$

4. Find the distance between each of the following pairs of "points."

    (a) $(2, 1)$ and $(-1, 3)$
    (b) $(-1, -1)$ and $(1, 4)$
    (c) $(0, 1/2)$ and $(1/4, 0)$
    (d) $(1, 1, 1)$ and $(1, 0, 1)$
    (e) $(2, 1, -3)$ and $(1, 3, -2)$

    (f) $(1/2, 1/2, 0)$ and $(0, 1/2, 1/2)$
    (g) $(-2/3, 1, 1/2)$ and $(3, -1/2, 4)$
    (h) $(1, 1, 1, 1)$ and $(1, 0, 1, 0)$
    (i) $(-1, 1, -1, 1)$ and $(2, 1, 3, 1)$
    (j) $(-1, -1, 1, 1)$ and $(4, 2, 3, -1)$

5. Describe the set of zeros for each of the following functions as a subset of the plane.

    (a) $f(x, y) = xy$
    (b) $p_y(x, y) = y$
    (c) $f(x, y) = x + y$
    (d) $f(x, y) = x - y$

    (e) $f(x, y) = (x - 1)^2 + (y - 2)^2$
    (f) $f(x, y) = x^2 + y^2$
    (g) $f(x, y) = x^2 - y^2$
    (h) $f(x, y) = x + y - 2$

6. Write an equation or an inequality to describe each of the following subsets of the plane.

    (a) D is the y axis
    (b) D is both axes
    (c) D is the unit circle with center $(0, 0)$
    (d) D is the interior of the unit circle with center $(0, 0)$
    (e) D is the empty set

coordinate geometry

7. Write an equation in x and y which expresses the fact that certain points (x, y) satisfy the given condition (see Chapter 1 if necessary).

(a) The distance between (x, y) and $(0, 0)$ is z.
(b) The distance between (x, y) and $(1, 1)$ is z.
((c) The distance between (x, y) and $(-1, 4)$ is 3.
(d) The distance between (x, y) and $(1, 1)$ is the same as the distance from (x, y) to the line $y = 2$.
(e) The point (x, y) lies on the perpendicular bisector of the line segment joining $(1, 0)$ and $(0, 1)$.
(f) The distance from (x, y) to $(1, 0)$ is the same as the distance from (x, y) to $(-1, 1)$.

7.2. lines in the plane

It has already been pointed out in previous chapters that an equation of the form

(7.2.1) $$Ax + By - C = 0$$

has a graph which is a straight line in the plane. This fact will be established in what follows. We shall see that every line in the plane can be considered as the set of zeros for a linear polynomial in two variables

(7.2.2) $$p(x, y) = Ax + By - C$$

and, furthermore, that every set of solutions to Equation (7.2.1) is a line in the plane. We base our demonstration on the axiom of high school plane geometry which states that given any line L there is a uniquely determined line which passes through a given point and is parallel to L. We also use the analytic geometry techniques available to us.

Consider a line L in the Euclidean plane R^2. There is a unique line L_o which passes through the origin $(0, 0)$ and is parallel to L. In Figure 7.2.1, a possible circumstance is sketched. In every case, the line L_o makes an angle α with the positive x axis. Of the two possible choices of angles there is always one, α such that

$$0 \leq \alpha < \pi$$

This angle α is called the **inclination** of L_o and also of the given line L. The well-known properties of parallel lines assure us that L itself will intersect the x axis, or any line parallel to it, at an angle whose measure is the same as α.

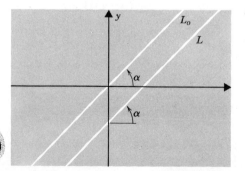

figure 7.2.1

The inclination of L.

lines in the plane

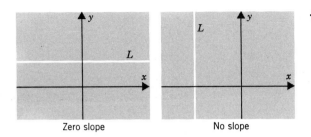

Zero slope   No slope

figure 7.2.2

The number

$$m = \tan \alpha$$

is called the **slope** of the line L (and of L_o). Notice that if L, hence L_o, is horizontal then $\alpha = 0$ and $\tan \alpha = 0$. On the other hand if L, hence L_o, is vertical $\alpha = \pi/2$ and $\tan \pi/2$ fails to exist. Horizontal lines have *zero slope*, while vertical lines have *no slope* (see Figure 7.2.2).

The slope of a line is an invariant of the line. That is, given any two distance points $P_1(x_1, y_1)$ and $P_2(x_2, y_2)$ on a given line L, the slope can be computed, and will be always the same no matter which two points of L are chosen. This is what you ought to expect of any quantity proposing to describe a line, since you recall that any two points of a line determine the line. Therefore, let L, P_1, and P_2, be given. Let L_o be the line through $O(0, 0)$ parallel to L, and let $P(x, y)$ be a point of L such that OP and $P_1 P_2$ are of the same length (see Figure 7.2.3). Now α is angle POX. If Q denotes the intersection of the perpendicular drawn from P to the x axis, its coordinates will be $(x, 0)$. Then, let M be the point $(x_1, 0)$ where the perpendicular from P_1 meets the x axis, and let $N(x_2, 0)$ be the corresponding point for P_2. By using similar triangles, it is apparent that

$$OQ = MN$$

Thus,

$$x = MN = x_2 - x_1$$

a corresponding discussion with perpendiculars drawn to the y axis shows that

$$y = y_2 - y_1$$

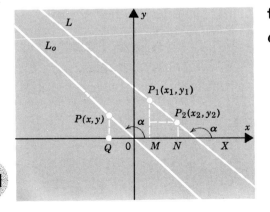

figure 7.2.3

Computing the slope of L.

311

coordinate geometry

Therefore,

$$\tan \alpha = \frac{y}{x} = \frac{y_2 - y_1}{x_2 - x_1}$$

The essential points of this argument do not depend on the way in which the figure is drawn. Therefore, this proves that the slope of any line L can be obtained from the coordinates of any two points on it using the formula

(7.2.3)
$$m = \frac{y_2 - y_1}{x_2 - x_1}$$

example 1

Determine the slope of the line passing through the points $(-1, 3)$ and $(2, -5)$.

$$m = \frac{-5 - 3}{2 - (-1)} = \frac{-8}{3}$$

Notice that had we interchanged points P_1 and P_2, we would arrive at

$$m = \frac{3 - (-5)}{-1 - 2} = \frac{8}{-3} = -\frac{8}{3}$$

the same value. ●

To continue the determination of the equation of a given line let (x_3, y_3) be any other point, different from (x_1, y_1) and (x_2, y_2) of line L. A glance at Figure 7.2.4 and the similar triangles drawn there should convince you that

$$m = \frac{y_2 - y_3}{x_2 - x_3} = \frac{y_1 - y_3}{x_1 - x_3} = \frac{y_2 - y_1}{x_2 - x_1} \quad \text{etc.}$$

Thus, the slope m is an invariant for a given line L. Therefore, given line L and slope m, let $P_1(x_1, y_1)$ and $P_2(x_2, y_2)$ be two distinct points of L. Then, for any point $P(x, y)$ of L, it is true that (if $P \neq P_1$)

$$m = \frac{y_2 - y_1}{x_2 - x_1} = \frac{y - y_1}{x - x_1}$$

or

(7.2.4)
$$y - y_1 = \frac{y_2 - y_1}{x_2 - x_1}(x - x_1)$$

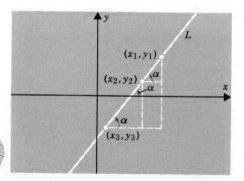

figure 7.2.4

The slope is invariant.

lines in the plane

This equation is also satisfied by the points P_1 and P_2. Hence, this is the equation of the line L. This form is called the **two point form** of the equation of L. Upon substituting (7.2.3) into (7.2.4), we have a simpler form, the **point-slope form**,

(7.2.5)
$$y - y_1 = m(x - x_1)$$

To obtain the form (7.2.1) known as the **standard form**, set $m = -A/B$. Then (7.2.5) becomes

$$y - y_1 = \frac{-A}{B}x + \frac{Ax_1}{B}$$

Multiply by B to obtain

$$By - By_1 = -Ax + Ax_1$$

Rearranging terms

$$Ax + By - (Ax_1 + By_1) = 0$$

Since (x_1, y_1) is known, set $Ax_1 + By_1 = C$. Therefore, every nonvertical line has an equation of the form (7.2.1). Vertical lines also have equations of the form (7.2.1) with $B = 0$.

The converse that the equation (7.2.1) has a graph which is a straight line was discussed in Chapter 3.

example 2

Write the equation of the line of Example 1 in standard form. In Example 1 we showed that $m = -8/3$. If P_1 is $(-1, 3)$ and $P(x, y)$ is any point of the line, the slope intercept form of the equation is

$$y - 3 = -8/3 \left(x - (-1) \right)$$

or

$$3y - 9 = -8(x + 1)$$

Thus,

$$8x + 3y - 1 = 0$$

is the desired equation. ●

example 3

Write the equation of the line determined by the points $(1/2, 3)$ and $(-1, -2/3)$. Using the two-point form we have

$$(y - 3) = \frac{-(2/3) - 3}{-1 - (1/2)} \left(x - \frac{1}{2} \right)$$

This can be simplified as follows:

$$y - 3 = \frac{-2 - 9}{(3[(-2 - 1)/2])} \left(x - \frac{1}{2} \right)$$

$$= \frac{22}{9} \left(x - \frac{1}{2} \right)$$

Hence, $9y - 27 = 22x - 11$, or $-22x + 9y - 16 = 0$, or $22x - 9y + 16 = 0$.

●

coordinate geometry

example 4

Write the equation of the line through the point $(-1, -3)$ and parallel with the line whose equation is $3x - 2y = 6$. Sketch a graph.

The equation of the given line can be written in the form

$$-2y = -3x + 6$$

or

$$y = \frac{3}{2}x - 3$$

Hence it passes through $(0, -3)$ and has slope $3/2$. Any line parallel to it also has slope $3/2$. Hence, the desired line has equation

$$y - (-3) = \frac{3}{2}(x - (-1))$$

or

$$y + 3 = \frac{3}{2}(x + 1)$$

Putting this in standard form gives

$$3x - 2y - 3 = 0$$

for the desired equation. These lines are sketched in Figure 7.2.5 using the given slope and the given points. ●

Let L_1 and L_2 be two perpendicular lines, neither of which is vertical. Let L_1 have slope $m_1 = \tan \alpha_1$, and let L_2 have slope $m_2 = \tan \alpha_2$. Since $L_1 \perp L_2$ it is true that $\alpha_1 = (\pi/2) + \alpha_2$ (or vice versa). Now, from trigonometry,

$$\tan \left(\frac{\pi}{2} + \alpha_2\right) = -\cot \alpha_2 = -\frac{1}{\tan \alpha_2}$$

is an identity. Therefore,

$$m_1 = \tan \alpha_1 = -\frac{1}{\tan \alpha_2} = -\frac{1}{m_2}$$

Thus, if two nonvertical lines are perpendicular, their slopes are negative reciprocals of each other.

Conversely, suppose $m_1 m_2 = -1$ for two lines L_1 and L_2. Neither could

figure 7.2.5

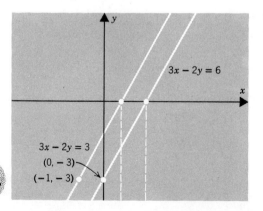

$3x - 2y = 6$

$3x - 2y = 3$
$(0, -3)$
$(-1, -3)$

314

lines in the plane

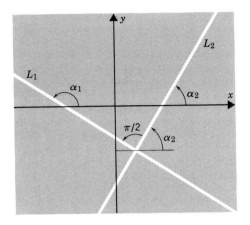

figure 7.2.6

Perpendicular lines.

be vertical since their slopes exist. Let Θ be the angle from L_2 to L_1 (see Figure 7.2.6). We have

$$m_1 m_2 = -1$$

$$\tan \alpha_1 = -\frac{1}{\tan \alpha_2} = -\cot \alpha_2$$

$$\tan(\alpha_2 + \Theta) = -\cot \alpha_2 = \tan\left(\alpha_2 + \frac{\pi}{2}\right)$$

Recall that $0 \leq \alpha_1 \leq \pi$ and $0 \leq \alpha_2 \leq \pi$. Therefore,

$$\alpha_2 + \Theta = \alpha_2 + \frac{\pi}{2}$$

or

$$\Theta = \frac{\pi}{2}$$

and L_1 and L_2 are perpendicular. We state this result as a theorem.

theorem 7.2.1

> Let L_1 and L_2 be two nonvertical lines with slopes m_1 and m_2, respectively. Then L_1 and L_2 are perpendicular, if and only if,
>
> $$m_1 m_2 = -1$$

example 5

Write the equation of the line passing through $(1/2, 2/3)$ and perpendicular to $4x + 3y = 9$. Writing the given line in the form

$$y = -\frac{4}{3}x + 3$$

we see that it has slope $-4/3$. Hence, the slope of any line perpendicular to it is $3/4$. The desired line, then, has equation

$$y - \frac{2}{3} = \frac{3}{4}\left(x - \frac{1}{2}\right)$$

315 Simplifying this, it becomes $18x - 24y + 7 = 0$. This is the equation of the perpendicular. ●

coordinate geometry

exercises

1. Find the slope of the line passing through each of the following pairs of points.

(a) $(0, 0), (1, 1)$

(f) $(2, 0), \left(0, -\dfrac{1}{2}\right)$

(b) $(0, 0), (-1, 1)$

(g) $(1, 1), (-1, 1)$

(c) $(0, 0), \left(\dfrac{1}{2}, -\dfrac{1}{2}\right)$

(h) $(4, -3), (-5, -3)$

(d) $(3, 0), (0, 2)$

(i) $(1, 2), (-1, 3)$

(e) $(-1, 0), (0, 5)$

(j) $\left(\dfrac{1}{2}, \dfrac{1}{3}\right), \left(-\dfrac{3}{5}, \dfrac{1}{7}\right)$

2. Write the equation, in standard form, for each of the lines in Exercise 1.

3. Sketch the graph for each of the following equations.

(a) $y = 0$

(b) $x = 0$

(c) $x + 1 = 0$

(d) $y - 1 = 0$

(e) $x + y = 0$

(f) $y = x$

(g) $3x - y = 4$

(h) $2x + 5y = 6$

(i) $y - 4x = 8$

(j) $7y - 12x + 11 = 0$

4. Sketch the graph of each line in Exercise 1.

5. Show that the line passing through the points $(a, 0)$ and $(0, b)$ has equation

$$\frac{x}{a} + \frac{y}{b} = 1$$

This is called the **intercepts form** of the equation of a line.

6. Use the results of Exercise 5 to write the equation of each of the lines with the following given intercepts. Sketch the graph of each line.

(a) $(1, 0)$ and $(0, 1)$

(f) $\left(\dfrac{1}{2}, 0\right)$ and $\left(0, \dfrac{1}{2}\right)$

(b) $(2, 0)$ and $(0, -2)$

(g) $\left(\dfrac{1}{2}, 0\right)$ and $\left(0, -\dfrac{1}{2}\right)$

(c) $(3, 0)$ and $(0, -1)$

(h) $\left(\dfrac{1}{3}, 0\right)$ and $\left(0, -\dfrac{4}{3}\right)$

(d) $(-1, 0)$ and $(0, 2)$

(i) $\left(\dfrac{5}{9}, 0\right)$ and $\left(0, \dfrac{7}{3}\right)$

(e) $(-1, 0)$ and $(0, -2)$

(j) $(1.34, 0)$ and $(0, 0.01)$

7. Write the equation of the *vertical* line through each of the following points.

(a) $(1, 1)$

(f) $\left(-\dfrac{1}{2}, \dfrac{2}{3}\right)$

(b) $(3, 0)$

(g) $\left(-\dfrac{3}{5}, 7\right)$

(c) $(0, 3)$

(d) $(-3, 0)$

(e) $(5, 7)$

(h) (π, e)

(i) (e, π)

(j) $(\sqrt{2}, 4)$

lines in the plane

8. Write the equation of each line L which satisfies the following:

   (a) L has slope $1/2$ and passes through $(-1, 3)$.
   (b) L has slope $-4/3$ and passes through $(6, -4)$.
   (c) L has slope -2 and passes through $(1/2, -1/2)$.
   (d) L has slope 0 and passes through (π, e).
   (e) L is parallel to $3x + y = 2$ and passes through $(1, -1)$.
   (f) L is parallel to $x - 3y = 4$ and passes through $(-2, 1)$.
   (g) L is parallel to $2x + 3y = 5$ and passes through $(-3, -1)$.
   (h) L is parallel to $3x - 5y = 8$ and passes through $(-5, 8)$.
   (i) L is parallel to $x = 5$ and passes through $(4, 7)$.
   (j) L is parallel to $y = 7$ and passes through (π, e).

9. In each of the following, write the equation in standard form of the line L which is perpendicular to the given line and passes through $(1, 1)$.

   (a) $x = 0$ (g) $3x - 4y = 2$
   (b) $y = 0$ (h) $7x - 16y = 32$
   (c) $x + y = 0$ (i) $12x - 8y = 6$
   (d) $x - y = 0$ (j) $\dfrac{x}{2} + \dfrac{y}{3} = 1$
   (e) $2x + 3y = 1$ (k) $\dfrac{x}{2} + \dfrac{y}{5} = 2$
   (f) $x + y = 1$ (l) $\dfrac{x}{4} + \dfrac{y}{6} = 3$

10. Find the point of intersection of each of the following pairs of lines, sketch their graphs.

   (a) $x + y = 1$, $x - y = 1$ (e) $x + y + 4 = 0$, $x - 5 = 0$
   (b) $2x + y = 4$, $x - y = 1$ (f) $3x - y + 2 = 0$, $6x - 2y - 7 = 0$
   (c) $x - y = 4$, $x + 2y = 6$ (g) $6 - 2x + 3y = 0$, $4x + y + 5 = 0$
   (d) $2x - y = 1$, $3x + 4y = 2$ (h) $\pi x + ey = 1$, $\pi x - ey = 0$

11. The following table lists the results of an experiment. Carefully plot the points and notice that they approximate a straight line. Write an equation for the line by determining its slope.

x	1	2	3	4	5	6
y	4.5	4.0	3.4	3.0	2.6	2.0

12. Use the line of Exercise 11 to determine y when $x = 0$ and also y when $x = 5.5$.

13. Given the equation $y = mx + b$ for a line L, suppose that m is fixed. Then, as b takes on all real values, we obtain the set of *all* lines with slope m. This set is called a *family of lines*. Similarly for fixed b, as m takes on all real values, we obtain the *family* of *all* lines with y intercept b. Sketch about seven members of each of the following families.

   (a) $y = 2x + b$ (f) $y = mx + 1$
   (b) $y = \dfrac{1}{2}x + b$ (g) $y = mx - 1$
   (c) $y = -2x + b$ (h) $y = mx + 3$
   (d) $y = -\dfrac{1}{2}x + b$ (i) $y = mx - 4$
   (e) $y = b$ (j) $x = a$

coordinate geometry

14. Write an equation for the family of lines passing through the point (1, 2).

15. Write the equation of the family of all lines with inclination $\alpha = 60°$.

16. Given the points $A(4, -2)$ and $B(-1, 3)$, write the equation of the perpendicular bisector of AB.

17. Show that the points (1, 3), (−2, 5), and (3, 6) are the vertices of a right triangle. Find the other two angles.

18. Show that the perpendicular distance from the point (a, b) to the line $Ax + By - C = 0$ is given by the formula

$$d = \frac{|Aa + Bb - C|}{\sqrt{A^2 + B^2}}$$

19. Use the formula in Exercise 18 to find the distance from the given point to the given line:

(a) $4x - y = 6$, (0, 0)   (e) $2x + 4y = -5$, $(1, -1)$
(b) $4x - y = 6$, $(1, -3)$ (f) $4x - y + 15 = 0$, $(-1, 1)$
(c) $4x - y = 6$, $(-3, -7)$ (g) $x - y - 8 = 0$, (3, 2)
(d) $x + 2y = 4$, $(-1, -7)$ (h) $\pi x + ey = 0$, (0, 0)

20. Show that the equation of a line L passing through the points (x_1, y_1) and (x_2, y_2) can be written as the determinant equation

$$\det A = 0$$

where

$$A = \begin{bmatrix} x & y & 1 \\ x_1 & y_1 & 1 \\ x_2 & y_2 & 1 \end{bmatrix}$$

Hint: Eliminate the ones in column three, rows one and two; expand and consider the two-point form of the equation of the line.

21. Show that the area of the triangle with vertices (x_o, y_o), (x_1, y_1) and (x_2, y_2) is 1/2 $|\det A|$ where A is the matrix

$$A = \begin{bmatrix} x_o & y_o & 1 \\ x_1 & y_1 & 1 \\ x_2 & y_2 & 1 \end{bmatrix}$$

22. Show that if L_1 and L_2 are lines with slopes m_1 and m_2 respectively, the angle Θ from L_2 to L_1 satisfies

$$\tan \Theta = \frac{m_1 - m_2}{1 + m_1 m_2}$$

7.3. conic sections—circles & parabolas

Suppose that we have a right circular cone with two nappes. When you buy an ice-cream cone, you only get half a cone, but then the ice-cream would fall out the bottom, anyway. Both of the curves discussed in this section result when such a cone is cut through by a plane. This is also true of the two curves discussed in Section 7.4. For this reason, *circles, parabolas, ellipses, and hyperbolas* are called **conic sections** (or **conics**).

conic sections—circles & parabolas

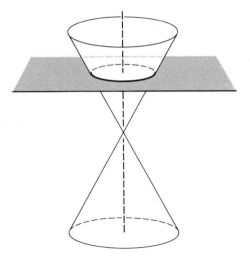

figure 7.3.1

The circle as a conic section.

In Figure 7.3.1, we illustrate the fact that when the plane cuts entirely across one nappe (your ice-cream cone will do most of the time) and is perpendicular to the axis of the cone the curve which is the intersection of the two surfaces is a **circle**.

When the intersecting plane is tipped, that is, it is not perpendicular to the axis of the cone, the curve of intersection is an **ellipse**. This is illustrated in Figure 7.3.2.

Actually, considered as a conic section, a circle is a special case of the ellipse. Another special case occurs if the plane passes through the point V, the vertex of the cone. This gives a "point ellipse."

If the plane cuts only one nappe of the cone, but does not cut all the way across it, the plane makes the same angle with the axis of the cone as does the line which generated the cone. As illustrated in Figure 7.3.3, (p. 320) the curve of intersection in this case is a **parabola**.

Finally, if the plane intersects both nappes of the cone as illustrated in Figure 7.3.4, (p. 320) the curve of intersection is a **hyperbola,** one branch coming from each nappe.

figure 7.3.2

The ellipse as a conic section.

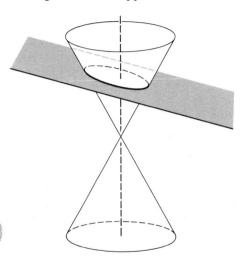

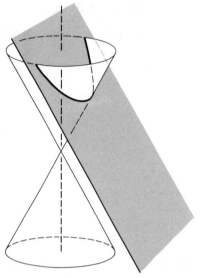

figure 7.3.3

The parabola as a conic.

Actually, as with the case of the ellipse, certain **degenerate cases** occur with parabolas and hyperbolas as well. If, in the parabolic case, the cutting plane contains the generator of the cone, a *straight line* results. Similarly, in the hyperbolic case, if the axis of the cone lies in the cutting plane, the curve of intersection is *two straight lines*.

We have been deliberately sketchy here. We have not really defined cone, axis, generator, etc. This approach to conic sections goes back to the ancient Greeks who studied these curves in great detail, but without the benefit of analytic geometry. They did, however, determine many of the properties of these curves; particularly relating them to certain fixed points (*foci*) and fixed

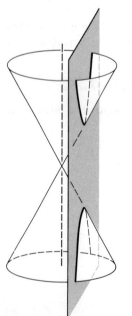

figure 7.3.4

The hyperbola as a conic.

conic sections—circles & parabolas

lines (*directrices*). These properties are the things we shall exploit in discussing conics from a present-day point of view. The fact that the curves, which we shall analytically define here, are the same as the intersections of a plane and a cone, as anciently studied, can be rigorously proved. However, we omit the proof.

Even though the conic sections have an ancient history, they are still important in applications. From planetary orbits to paths of a projectile, to the design of headlights, radar antenna, and auditoriums, these curves, because of their remarkable properties, are applied by present-day scientists, engineers, and draftsmen.

In this section, we discuss circles and parabolas, both of which have been mentioned in earlier chapters. In Section 7.4, we shall consider ellipses and hyperbolas. In Section 7.5, we shall consider all conics as the set of zeros for the general quadratic polynomial

$$P(x, y) = ax^2 + bxy + cy^2 + dx + ey + f$$

circles

*A **circle** is defined to be the set of all those points (x, y) in the plane which lie a given fixed distance r from a given point (h, k) called the **center** of the circle. The number r is called the **radius** of the circle.*

example 1

Write the equation of the circle whose radius is 2 and whose center is at $(1, -2)$.

According to the definition, a point (x, y) of R^2 will lie on the circle provided that its distance from $(1, -2)$ is 2. Therefore, the distance formula tells us

$$2 = \sqrt{(x - 1)^2 + (y + 2)^2}$$

or, by squaring both sides we obtain

$$4 = (x - 1)^2 + (y + 2)^2$$

This is the equation of the desired circle (see Figure 7.3.5).

In general, the distance formula implies that the set (*locus*) of all points

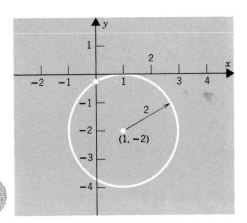

figure 7.3.5

The circle $(x - 1)^2 + (y + 2)^2 = 4$.

coordinate geometry

(x, y) a fixed distance r from the point (h, k) must satisfy

$$r = \sqrt{(x - h)^2 + (y - k)^2}$$

when squared, the resulting equation

(7.3.1) $$(x - h)^2 + (y - k)^2 = r^2$$

is the *equation of the circle of radius r and center (k, k)*.

example 2

Show that the set of all points (x, y) in the plane which satisfy the equation

$$x^2 + y^2 - 4x + 6y - 8 = 0$$

is a circle.

To do this we "complete the square" as follows:

$$(x^2 - 4x \qquad) + (y^2 + 6y \qquad) = 8$$

This becomes

$$(x^2 - 4x + 4) + (y^2 + 6y + 9) = 8 + 4 + 9$$

or

$$(x - 2)^2 + (y + 3)^2 = 21$$

This is the equation of a circle with center at $(2, -3)$ and radius $\sqrt{21}$. ●

example 3

A circle can be determined by any three points lying on it. Write the equation of the circle passing though $(1, 1)$, $(1, -1)$, and $(-2, 3)$. Determine the center and the radius.

If this is to be a circle, its equation will have the form

$$(x - h)^2 + (y - k)^2 = r^2$$

or

$$x^2 - 2hx + h^2 + y^2 - 2ky + k^2 = r^2$$

If the given three points are to lie on the circle, they must each satisfy this equation. Substituting the coordinates of each point for x, and y gives the following three equations in h, k, and r.

$(1, 1)$: $\qquad 1 - 2h + h^2 + 1 - 2k + k^2 = r^2$

$(1, -1)$: $\qquad 1 - 2h + h^2 + 1 + 2k + k^2 = r^2$

$(-2, 3)$: $\qquad 4 + 4h + h^2 + 9 - 6k + k^2 = r^2$

These three equations simplify to

$$h^2 - 2h + k^2 - 2k - r^2 = -2$$
$$h^2 - 2h + k^2 + 2k - r^2 = -2$$
$$h^2 + 4h + k^2 - 6k - r^2 = -13$$

To solve this system, which is not a linear system, we adapt the techniques

conic sections—circles & parabolas

given in Section 6.1. An equivalent system is obtained by substituting for the second equation the result of subtracting the first equation from it and re-arranging.

$$4k = 0$$
$$h^2 - 2h + k^2 - 2k - r^2 = -2$$
$$h^2 + 4h + k^2 - 6k - r^2 = -13$$

Since $k = 0$, a simpler equivalent system is

$$k = 0$$
$$h^2 - 2h - r^2 = -2$$
$$h^2 + 4h - r^2 = -13$$

Subtracting the second equation from the third, we have the following equivalent system.

$$k = 0$$
$$h^2 - 2h - r^2 = -2$$
$$6h = -11$$

or

$$k = 0$$
$$h = -\frac{11}{6}$$
$$h^2 - 2h - r^2 = -2$$

From this we determine that $r = (5/6)\sqrt{13}$. Therefore, the equation of the circle is,

$$\left(x + \frac{11}{6}\right)^2 + y^2 = \frac{325}{36} \quad \bullet$$

parabolas

*A **parabola** is defined to be the set (locus) of all points (x, y) in the plane whose distance from a given fixed point is the same as its distance from a given fixed line. The point is called the **focus** and the line is called the **directrix** of the parabola.*

The definitions of circles and parabolas are geometric in the sense that a coordinatization is not specified. To write an equation for a parabola, we choose a set of axes oriented in such a way that the focus and directrix are conveniently located. Suppose we set up our coordinate system so that the focus of the parabola is located at $F(h, k + p/2)$ and the directrix L is the line $y = k - p/2$, as shown in Figure 7.3.6 on the following page. Thus, the distance from the focus to the directrix is denoted by p, $p > 0$.

The point V of the parabola, which lies along the line FM drawn through F perpendicular to the directrix L, is called the **vertex** of the parabola and has coordinates (h, k), in this case. The line through FM is called the **axis** of the parabola.

coordinate geometry

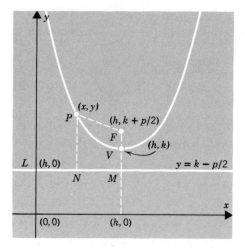

figure 7.3.6

The parabola $(x - h)^2 = 2p(y - k)$.

The equation comes via the distance formula. If $P(x, y)$ is any point on the parabola then the length of FP is

$$(FP) = \sqrt{(x - h)^2 + \left(y - \left(k + \frac{p}{2}\right)\right)^2}$$

The distance P lies from the directrix is $|PN| = |y - [k - (p/2)]|$. Therefore, the definition of the parabola states that $|PN| = |FP|$, or

$$\left|y - k + \frac{p}{2}\right| = \sqrt{(x - h)^2 + \left(y - k - \frac{p}{2}\right)^2}$$

Square both sides to obtain

$$\left(y - k + \frac{p}{2}\right)^2 = (x - h)^2 + \left(y - k - \frac{p}{2}\right)^2$$

or

$$(y - k)^2 + p(y - k) + \frac{p^2}{4} = (x - h)^2 + (y - k)^2 - p(y - k) + \frac{p^2}{4}$$

This reduces to

(7.3.2) $$2p(y - k) = (x - h)^2$$

as the *equation of the parabola* described. In particular, if the vertex is located at the origin $(0, 0)$ this becomes

$$x^2 = 2py$$

If the focus and directrix are oriented in other positions with respect to the coordinate axes, the equation will, of course, be different. There are essentially four *standard forms* for the equation of a parabola with vertex (h, k). These are illustrated in Figures 7.3.7, 7.3.8, and 7.3.9 in addition to 7.3.6. In each case the distance from the focus to the directrix is denoted by p, and p is positive in each case. When the directrix L is p units above the focus, F has coordinates $(h, k - p/2)$, and L has equation $y = k + p/2$; the parabola has equation

(7.3.3) $$(x - h)^2 = -2p(y - k)$$

conic sections—circles & parabolas

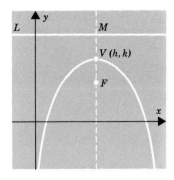

figure 7.3.7

The parabola $(x - h)^2 = -2p(y - k)$.

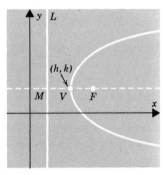

figure 7.3.8

The parabola $(y - k)^2 = 2p(x - h)$.

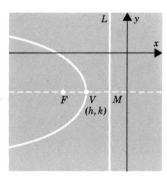

figure 7.3.9

The parabola $(y - k)^2 = -2p(x - h)$.

When the axis of the parabola is parallel to the x axis, the forms are

(7.3.4) $$(y - k)^2 = 2p(x - h)$$

and

(7.3.5) $$(y - k)^2 = -2p(x - h)$$

according as the directrix is to the left or right, respectively, of the focus.

    You will be given the opportunity to verify these equations in the exercises. Which standard form is used depends upon the relative positions of the focus and directrix.

example 4

Write the equation of the parabola whose focus is $F(1, 1)$ and whose directrix L has equation $x = -2$. Since, in this case, the axis of the parabola is the line $y = 1$ and the focus is to the right of the directrix, we use the form typified

325

coordinate geometry

in Figure 7.3.8; i.e., $(y - k)^2 = 2p(x - h)$. The value of p is the distance of F from 1. That is,

$$p = 3$$

Hence, the vertex V is located at $(3/2, 1)$. The equation of the parabola is, therefore,

$$(y - 1)^2 = 6\left(x - \frac{3}{2}\right)$$

or

$$y^2 - 2y - 6x + 10 = 0 \quad \bullet$$

example 5

Sketch the graph of the parabola whose equation is

$$y^2 = -8x$$

Find its focus, directrix, and vertex. We compare this equation with Figure 7.3.9 and note that $h = 0$, $k = 0$, so the origin is the vertex. Since $-2p = -8$, $p = 4$. The equation of the directrix L, which is located p units from the focus F and $p/2$ units from the vertex, is

$$x = 2$$

The focus is located at $(h - p/2, k)$ or $(-2, 0)$. The graph is sketched in Figure 7.3.10. The axis of the parabola is the x axis. $\quad \bullet$

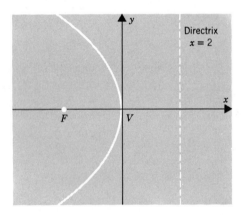

Directrix
$x = 2$

figure 7.3.10

$$y^2 = -8x$$

example 6

Determine the focus, vertex, directrix, and axis and sketch the graph of the parabola whose equation is

$$x^2 - 2x + 6y - 5 = 0$$

To put this equation into one of the standard forms, separate the x's and y's as follows:

326

$$x^2 - 2x \quad\quad = -6y + 5$$

conic sections—circles & parabolas

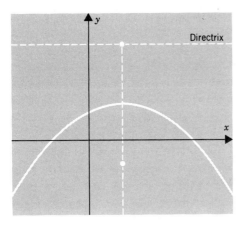

figure 7.3.11

$$x^2 - 2x + 6y - 5 = 0$$

The space on the left is to enable us to complete the square. By doing so, the equation becomes

$$x^2 - 2x + 1 = -6y + 5 + 1$$

or

$$(x - 1)^2 = -6y + 6$$

Thus, the equation in standard form becomes

$$(x - 1)^2 = -6(y - 1)$$

This equation indicates that the vertex of the parabola is at (1, 1), the axis of the parabola is parallel to the y axis (the line $x = 1$). The value of p is 3. Therefore, the focus has coordinates $(1, -1/2)$, and the equation of the directric $y = k + p/2$ is $y = 5/2$. The graph is sketched in Figure 7.3.11. ●

exercises

1. Write the equation of the circle in each of the following cases given the center C and radius r.

(a) $C(-1, 1)$ $r = 1$ (d) $C(2, 3)$ $r = 2$
(b) $C(0, 0)$ $r = 1$ (e) $C(-3, 4)$ $r = 3$
(c) $C(0, 0)$ $r = 5$ (f) $C(2, -1)$ $r = 2$

2. Write the equation of the circle which satisfies the given conditions:

(a) A diameter is the segment from $(-2, 3)$ to $(4, 3)$.
(b) A diameter is the segment from $(4, 2)$ to $(8, 6)$.
(c) The center is at $(-3, -5)$, and it is tangent to the x axis.
(d) The center is at $(-2, 6)$, and it is tangent to the y axis.
(e) The center is $C(-1, -2)$, and it passes through $(-2, 2)$.
(f) The center is $C(2, 3)$, and it is tangent to the line $3x + 4y + 2 = 0$.

3. Given each of the following equations, determine that the graph of each is a circle by completing the square. Identify the center and radius of each.

(a) $x^2 + y^2 + 6x - 8y = 0$ (d) $2x^2 + 2y^2 + 6x - 10y = 1$
(b) $x^2 + y^2 + 2x - 4y = 11$ (e) $2x^2 + 2y^2 + 3x + 5y + 2 = 0$
(c) $x^2 + y^2 + 6x - 4y = 3$ (f) $2x^2 + 2y^2 - 5x + 7y + 9 = 0$

coordinate geometry

4. Sketch the graph of each circle in Problem 3.

5. Show that the graph of the equation

$$ax^2 + ay^2 + cx + dy + f = 0$$

is

(a) a circle if $c^2 + d^2 > 4fa$
(b) a point if $c^2 + d^2 = 4fa$
(c) empty if $c^2 + d^2 < 4fa$

Hint: Complete the square as in Problem 3 and consider the radius.

6. Determine the coordinates of the focus, the equation of the directrix, and sketch the graph of each of the following parabolas.

(a) $y = x^2$
(b) $x^2 = 6y$

(c) $y^2 = -12x$

(d) $x^2 + 16y = 0$
(e) $x^2 = -12y$

(f) $y = \dfrac{1}{10}x^2$

7. Find the equation of each parabola whose vertex is at the origin which satisfies the following additional condition.

(a) focus at $(-3, 0)$
(b) focus at $(0, 4)$
(c) directrix $x = -3$

(d) directrix $y = \dfrac{5}{3}$

(e) axis along x axis and $(3, 2)$ is on the graph
(f) axis along y axis and $(3, 2)$ is on the graph

8. Determine the focus, vertex, directrix, and axis of the parabolas with the following equations. Sketch the graph of each.

(a) $y = x^2 - 2x + 3$
(b) $y = -x^2 + 6x + 4$
(c) $x = y^2 + 2y - 4$

(d) $x = 8 - y^2 - 6y$
(e) $x^2 + 2y - 3x + 5 = 0$
(f) $y^2 + 2x - 4y + 7 = 0$

9. Determine the equation of the parabola which satisfies the given conditions in each of the following.

(a) directrix $x = 0$, focus $(4, 0)$
(b) directrix $x = -2$, focus $(1, -3)$
(c) directrix $y = 0$, focus $(0, -4)$
(d) directrix $y = -2$, focus $(-2, 3)$
(e) vertex $(1, 5)$, focus $(-2, 5)$
(f) vertex $(-1, -3)$, focus $(-1, 2)$
(g) vertex $(-2, -1)$, focus $(4, -1)$
(h) focus $(0, 2)$ and passes through $(-1, 4)$ and $(1, 4)$

10. Write the equation of the circle which is determined by the given points in each of the following.

328

(a) $(1, 0), (-1, 0), (0, 1)$
(b) $(1, 1), (1, 2), (-1, -1)$
(c) $(0, 0), (0, 3), (-1, -1)$

(d) $(\sqrt{2}, \sqrt{2}), (1, \sqrt{3}), (0, 2)$
(e) $(1, 1), (-1, 2), (3, -1)$
(f) $(1, 3), (2, 4), (3, 2)$

conic sections—ellipses & hyperbolas

11. Show that any parabola whose axis is parallel to the y axis has an equation which may be written in the form

$$y = ax^2 + bx + c$$

Hint: Consider equations (7.3.2) and (7.3.3).

12. Show that any parabola whose axis is parallel to the x axis has an equation which may be written in the form

$$x = ay^2 + by + c$$

13. Use the results of Exercise 11 to determine the equation of each parabola whose axis is vertical which is determined by the points in Exercise 10.

14. Find the vertex, the focus, and the directrix for each of the parabolas of Exercise 13.

15. Use the results of Exercise 12 to determine the equation of each parabola whose axis is horizontal which is determined by the points in Exercise 10.

16. Find the vertex, the focus, and the directrix for each of the parabolas of Exercise 15.

17. Verify equation (7.3.3).

18. Verify equation (7.3.4).

19. Verify equation (7.3.5).

20. A radar antenna is constructed so that a cross section through its axis is a parabola with the receiver located at the focus. Find the focus if the antenna is 5 ft across at the opening and 1.5 ft maximum depth.

7.4. conic sections—ellipses & hyperbolas

In Section 7.3 the general idea of a conic was introduced and circles and parabolas were discussed. In this section, equations will be derived for ellipses and hyperbolas. In Section 7.3 as well as in this section, these curves are oriented on a coordinate system in such a way that the axes of the curves are parallel to the coordinate axes. In Section 7.5, we shall examine the circumstance where the coordinate axes make an angle ϕ with the axes of the curve.

ellipse

An **ellipse** *is the collection (locus) of all points (x, y) in the plane, the sum of whose distances from two fixed points, called the* **foci,** *is a constant.*

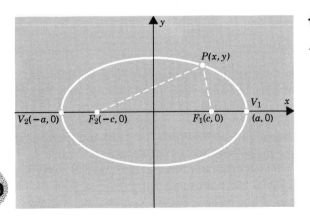

figure 7.4.1

An ellipse

coordinate geometry

To simplify the computation involved, let us choose our coordinate system so that the two fixed points, foci, are the points $F_1 : (c, 0)$ and $F_2 : (-c, 0)$, for $c > 0$. The line $F_1 F_2$, therefore, lies along the x axis. The mid-point of the segment $F_1 F_2$ is called the **center** of the ellipse. In our case, the center is located at the origin. The intersection of the ellipse with its *focal axis*, the line through $F_1 F_2$, determines two points, V_1 and V_2 the **vertices** of the ellipse. In our situation, we let these have coordinates $(a, 0)$ and $(-a, 0)$, $a > 0$. To determine the equation of this ellipse, let $P : (x, y)$ be any point on it. The distance formula and the definition of an ellipse tell us that

$$|F_1 P| + |F_2 P| = \sqrt{(x - c)^2 + y^2} + \sqrt{(x + c)^2 + y^2}$$

The vertex V_1 is also on the ellipse and

$$|F_1 V_1| + |F_2 V_1| = (a - c) + (a + c) = 2a$$

Therefore,

$$\sqrt{(x - c)^2 + y^2} + \sqrt{(x + c)^2 + y^2} = 2a$$

or

$$\sqrt{(x - c)^2 + y^2} = 2a - \sqrt{(x + c)^2 + y^2}$$

Squaring both sides, this equation becomes

$$(x - c)^2 + y^2 = 4a^2 - 4a\sqrt{(x + c)^2 + y^2} + (x - c)^2 + y^2$$

which simplifies to

$$a\sqrt{(x + c)^2 + y^2} = a^2 + cx$$

Square this again to obtain

$$a^2[(x^2 + 2cx + c^2) + y^2] = a^4 + 2a^2cx + c^2x^2$$

which can be written in the form

$$x^2(a^2 - c^2) + a^2 y^2 = a^2(a^2 - c^2)$$

Now let $b > 0$ be such that $a^2 - c^2 = b^2$. The above equation is

$$x^2 b^2 + a^2 y^2 = a^2 b^2$$

Upon dividing both sides by $a^2 b^2 > 0$ we have

(7.4.1)
$$\frac{x^2}{a^2} + \frac{y^2}{b^2} = 1$$

as the equation of this ellipse. This is a **standard form** for the equation.

The segment between the two vertices of an ellipse is called the **major axis** of the ellipse. In the above equation, we see that the points $(0, b)$ and $(0, -b)$ lie on the ellipse. The segment between these two points is called the **minor axis** of the ellipse. Since $a > c$, we have $a^2 - c^2 = b^2$, so $a > b$ and the major axis is longer than the minor axis.

If the ellipse is oriented on a coordinate system in such a way that its foci are along the y axis, say at the points $(0, c)$ and $(0, -c)$, then the major axis will lie along this axis. If the vertices are at $(0, a)$ and $(0, -a)$, the same sort

conic sections—ellipses & hyperbolas

of derivation given above will result in the equation

(7.4.2)
$$\frac{y^2}{a^2} + \frac{x^2}{b^2} = 1$$

where, as before, $b^2 = c^2 - a^2$, so $b < a$, and the points $(b, 0)$ and $(-b, 0)$ are on the minor axis which lies along the x axis.

example 1

Discuss and sketch the graph of the equation

$$4x^2 + 25y^2 = 100$$

Divide this equation by 100 to obtain

$$\frac{x^2}{25} + \frac{y^2}{4} = 1$$

which is the standard form for an ellipse whose major axis lies along the x axis; $a^2 = 25$ so $a = 5$, $b^2 = 4$, so $b = 2$. Therefore, $c^2 = a^2 - b^2 = 25 - 4 = 21$ and the foci are located at $(\sqrt{21}, 0)$ and $(-\sqrt{21}, 0)$. The major axis has length 10 with the vertices being at $(5, 0)$ and $(-5, 0)$. The graph is sketched in Figure 7.4.2. ●

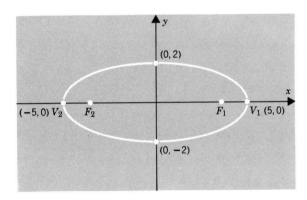

figure 7.4.2

$4x^2 + 25y^2 = 100$

example 2

Sketch the graph of the equation

$$9x^2 + 4y^2 = 36$$

Upon dividing both sides of this equation by 36, we obtain a standard form resembling equation 7.4.2 for the ellipse

$$\frac{x^2}{4} + \frac{y^2}{9} = 1$$

Since $9 > 4$, we write this in the form

331

$$\frac{y^2}{9} + \frac{x^2}{4} = 1$$

coordinate geometry

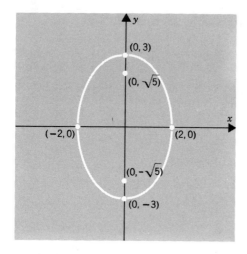

figure 7.4.3

$$9x^2 + 4y^2 = 36$$

This ellipse has its major axis along the y axis with vertices $(0, 3)$ and $(0, -3)$. Since $c^2 = a^2 - b^2$, $c = \sqrt{5}$, and the foci are located at $(0, \sqrt{5})$ and $(0, -\sqrt{5})$. The graph is sketched in Figure 7.4.3. ●

In the event that the ellipse is oriented on the coordinate system in such a way that its major axis is *parallel* with the x axis but the center is located at (h, k) with foci $(h \pm c, k)$ and vertices $(h \pm a, k)$, the equation of the ellipse would be

(7.4.3)
$$\frac{(x - h)^2}{a^2} + \frac{(y - k)^2}{b^2} = 1$$

In a similar situation, but with the major axis along the y axis, the center at (h, k), the foci at $(h, k \pm c)$, and the vertices at $(h, k \pm a)$, the equation is

(7.4.4)
$$\frac{(y^2 - k)^2}{a^2} + \frac{(x - h)^2}{b^2} = 1$$

example 3

Write the equation of the ellipse whose foci are located at $F_1 : (1, 3)$ and $F_2 : (1, 7)$ and whose major axis has length 10.

Note first that since $2a = 10$, $a = 5$. The center of the ellipse is the midpoint of the segment $F_1 F_2$ which is $(1, 5)$. Thus, (h, k) is $(1, 5)$, and $c = 2$. Compute, $b^2 = 25 - 4 = 21$. Therefore, the equation, using 7.4.3, is

$$\frac{(x - 1)^2}{25} + \frac{(y - 5)^2}{21} = 1$$

or

$$21x^2 + 25y^2 - 42x - 250y + 121 = 0 \quad ●$$

hyperbolas

A **hyperbola** *is the set (locus) of all points in the plane the difference of whose distances from two fixed points* **(foci)** *is a constant.*

If, as in the case of the ellipse, we let the foci, the two given fixed points, be located on the x axis at the points $F_1(c, 0)$ and $F_2(-c, 0)$, $c > 0$, and let

conic sections—ellipses & hyperbolas

the constant be $2a$, $a > 0$, then, for any point $P(x, y)$ on the hyperbola,

$$2a = \sqrt{(x - c)^2 + y^2} - \sqrt{(x + c)^2 + y^2}$$

Proceeding as before, this equation simplifies to

$$\frac{x^2}{a^2} - \frac{y^2}{c^2 - a^2} = 1$$

Note that in the case of a hyperbola, in contrast to the ellipse, $a < c$. The **vertices** are the points where the curve intersects the x axis and are located at $V_1(a, 0)$ and $V_2(-a, 0)$.

In any hyperbola, the line determined by the foci is the **axis**. The vertices V_1 and V_2 are the intersections of the graph with this axis and the segment $V_1 V_2$ is called the **transverse axis** of the hyperbola.

If we set $b^2 = c^2 - a^2$, with $b > 0$, the standard form for the equation of the hyperbola, whose transverse axis is on the x axis and lies between $(a, 0)$ and $(-a, 0)$, is

(7.4.5)
$$\frac{x^2}{a^2} - \frac{y^2}{b^2} = 1$$

The graph of this curve does not intersect the y axis since x cannot be zero in this equation. In fact, you should verify that there are no points (x, y) which satisfy this equation when $|x| < a$.

As was the case with many rational functions, hyperbolas have asymptotes. To discover these we consider the position of the point $P(x, y)$ on the hyperbola when x is large (in absolute value). When $x \geq a$ we may rewrite (7.4.5) in the form (solve for y)

$$y = \pm \frac{b}{a} \sqrt{x^2 - a^2}$$

$$= \pm \frac{b}{a} x \sqrt{1 - \frac{a^2}{x^2}}$$

Now, if $|x|$ becomes very large, that is, x is in some neighborhood of $+\infty$ or of $-\infty$, you see that a^2/x^2 is very small (it is in a neighborhood of 0). Thus,

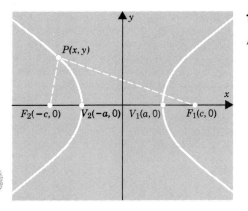

figure 7.4.4

Hyperbola.

coordinate geometry

y is quite near the numbers $\pm(b/a)x$. Therefore, the point P is located quite close to one of the lines

(7.4.6)
$$y = \pm \frac{b}{a} x$$

These are the **asymptotes** of the hyperbola.

You will note that, in contrast with the ellipse, the points $W_1(0, b)$ and $W_2(0, -b)$ *do not* lie on the graph of the hyperbola, However, the number b is involved in the equation of the asymptotes. The line segment between W_1 and W_2 is called the **conjugate axis** of the hyperbola. If one sketches a rectangle which has its sides passing through the vertices, V_1, and V_2 and through the ends of the conjugate axis, W_1, and W_2, and, respectively, perpendicular to these axes, the diagonals of this rectangle are segments of the lines $y = \pm(b/a)x$, these are the asymptotes of the hyperbola.

example 4

Sketch the graph of the equation

$$4x^2 - 9y^2 = 36$$

First of all we write this equation, by dividing both sides by 36, in the form

$$\frac{x^2}{9} - \frac{y^2}{4} = 1$$

Thus, the graph is a hyperbola with $a = 3$ and $b = 2$. We plot the points $V_1(3, 0)$, $V_2(-3, 0)$, $W_1(2, 0)$, and $W_2(-2, 0)$. Draw the rectangle they determine (see Figure 7.4.5). The asymptotes are then drawn in, and finally the graph is sketched. The asymptotes of this hyperbola are the lines

$$y = \pm \frac{2}{3} x$$

The foci of the hyperbola are determined from the fact that $b^2 = c^2 - a^2$, so, $c^2 = a^2 + b^2 = 9 + 4 = 13$. Hence $F_1:(\sqrt{13}, 0)$ and $F_2:(-\sqrt{13}, 0)$ are the foci.

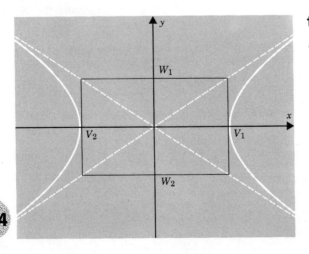

figure 7.4.5

$4x^2 - 9y^2 = 36$

conic sections—ellipses & hyperbolas

As was the case with the ellipse, if the hyperbola has its foci, and vertices on the y axis, the equation has a different although similar form. If $F_1(0, c)$ and $F_2(0, -c)$ are the foci and $V_1(0, a)$ and $V_2(0, -a)$ are the vertices, then the equation of the hyperbola is

(7.4.7)
$$\frac{y^2}{a^2} - \frac{x^2}{b^2} = 1$$

example 5

Sketch the graph of the equation

$$4x^2 - 16y^2 = -25$$

Divide both sides by -25 to obtain

$$\frac{4x^2}{-25} + \frac{16y^2}{25} = 1$$

which can be written

$$\frac{y^2}{25/16} - \frac{x^2}{25/4} = 1$$

This is the standard form for a hyperbola with transverse axis along the y axis. The values of a and b are

$$a = \frac{5}{4}$$

$$b = \frac{5}{2}$$

The asymptotes in this case will have equation of the form

(7.4.8)
$$y = \pm\frac{a}{b}x$$

as you can verify (Exercise 10). For this example, then, the asymptotes have equation

$$y = \pm\frac{1}{2}x$$

The foci are the points $F_1(0, (5/2)\sqrt{5})$ and $F_2(0, -(5/2)\sqrt{5})$. The graph is sketched in Figure 7.4.6 at the top of the following page.

This example illustrates that, again in contrast to an ellipse, it is not always true for a hyperbola that $a > b$. In fact we can have $a > b$, $a < b$, or $a = b$ for hyperbolas.

example 6

Write the equation of the hyperbola whose foci are located at $(\pm 5, 0)$ and which passes through the point $(3\sqrt{2}, 4)$. To do this, we recall that $b^2 = c^2 - a^2$, and $c = 5$. Hence the equation has the form

335

$$\frac{x^2}{a^2} - \frac{y^2}{25 - a^2} = 1$$

coordinate geometry

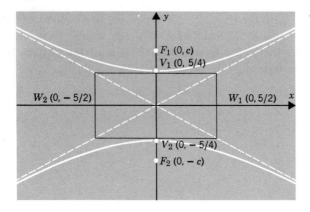

figure 7.4.6

or

$$(25 - a^2)x^2 - a^2y^2 = 25a^2 - a^4$$

Since $(3\sqrt{2}, 4)$ lies on the graph, it must satisfy this equation. Substituting $x = 3\sqrt{2}$, $y = 4$, we get

$$(25 - a^2)(18) - 16a^2 = 25a^2 - a^4$$

or

$$a^4 - 59a^2 + 450 = 0$$

$$(a^2 - 9)(a^2 - 50) = 0$$

Thus, $a^2 = 9$ or $a^2 = 50$, but since $c > a$, and $c^2 = 25$, we reject $a^2 = 50$. We compute $b^2 = c^2 - a^2 = 25 - 9 = 16$. Therefore, the desired equation is

$$\frac{x^2}{9} - \frac{y^2}{16} = 1 \qquad \bullet$$

As was the case with ellipses, in the event that the hyperbola is oriented on the coordinate system in such a way that its transverse axis is *parallel* with, but does not coincide with, a coordinate axis the standard form of the equations is somewhat different. If $C:(h, k)$ is the mid-point of the transverse axis, the **center** of the hyperbola, then the two standard forms are

(7.4.9)
$$\frac{(x - h)^2}{a^2} - \frac{(y - k)^2}{b^2} = 1$$

and

(7.4.10)
$$\frac{(y - k)^2}{a^2} - \frac{(x - h)^2}{b^2} = 1$$

depending upon whether the transverse axis is horizontal, or vertical, respectively.

example 7

Write the equation, find the foci and the asymptotes of the hyperbola whose vertices are the points $V_1(-1, 2)$ and $V_2(-1, -4)$, which passes through the point $P_1(2, 5)$. Sketch the graph. It is clear that the transverse axis is vertical along the line $x = -1$. The center $C:(h, k)$ is located at $(-1, -1)$. Then the

conic sections—ellipses & hyperbolas

vertices are $V_1(h, k + a)$ and $V_2(h, k - a)$, so that $a = 3$. Therefore, the equation has the form

$$\frac{(y + 1)^2}{9} - \frac{(x + 1)^2}{b^2} = 1$$

Substituting the point $(2, 5)$ in this equation, we obtain

$$\frac{36}{9} - \frac{9}{b^2} = 1$$

This is solved for b, as follows:

$$\frac{36}{9} - 1 = \frac{9}{b^2}$$

$$\frac{27}{9}b^2 = 9$$

$$b^2 = 3$$

Hence, the desired equation is

$$\frac{(y + 1)^2}{9} - \frac{(x + 1)^2}{3} = 1$$

The conjugate axis is horizontal and has length $2b = 2\sqrt{3}$. This means that the points W_1 and W_2 are located at $(h \pm b, k)$, or $W_1(-1 + \sqrt{3}, -1)$ and $W_2(-1 - \sqrt{3}, -1)$. The foci are found by computing

$$c^2 = a^2 + b^2 = 9 + 3 = 12$$

$$c = 2\sqrt{3}$$

Thus, they are at $(h, k \pm c)$, or $F_1(-1, -1 + 2\sqrt{3})$ and $F_2(-1, -1 - 2\sqrt{3})$. The asymptotes are drawn by constructing the rectangle as before. You may

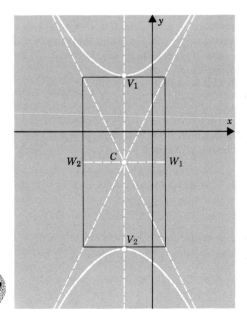

figure 7.4.7

coordinate geometry

verify in Exercise 16 that they are the lines whose equations are

(7.4.11) $$y - k = \pm \frac{a}{b}(x - h)$$

Hence, they are, in this example, the lines

$$y - 1 = \pm \sqrt{3}\,(x - 1)$$

The graph is found in Figure 7.4.7. ●

exercises

1. In each of the following, find the lengths of the major and minor axes, the coordinates of the vertices, and the coordinates of the foci. Sketch the graph of each.

(a) $16x^2 + 25y^2 = 400$

(b) $9x^2 + 16y^2 = 144$

(c) $25x^2 + 169y^2 = 4225$

(d) $4x^2 + 18y^2 = 36$

(e) $5x^2 + 2y^2 = 10$

(f) $x^2 + 4y^2 = 9$

(g) $\frac{1}{2}x^2 + 2y^2 = 8$

(h) $x^2 + 10y^2 = 5$

2. In each of the following, find the equation of the ellipse which satisfies the given conditions.

(a) vertices at $(\pm 5, 0)$, foci at $(\pm 4, 0)$

(b) vertices at $(\pm 8, 0)$, foci at $(\pm 5, 0)$

(c) vertices at $(0, \pm 5)$, foci at $(0, \pm 3)$

(d) vertices at $(0, \pm 5)$, length of minor axis 3

(e) foci at $(\pm 3, 0)$, length of minor axis 2

(f) vertices at $(0, \pm 6)$, passing through $(3, 2)$

(g) vertices at $(0, \pm 5)$, passing through $(3, 3)$

3. Determine the equation of each of the following ellipses satisfying the conditions given.

(a) center at $(2, 3)$, major axis along $y = 3$ of length 6, and minor axis of length 4

(b) center at $(5, -3)$, major axis along $x = 5$ of length 12, and minor axis of length 8

(c) vertices $(2, 5)$ and $(2, -3)$ foci at $(2, 4)$ and $(2, -2)$

(d) vertices $(6, 3)$ and $(-4, 3)$ foci at $(5, 3)$ and $(-3, 3)$

(e) vertices $(-1, 8)$ and $(-1, -4)$ passing through $(2, 4)$

(f) foci at $(-2, -1)$ and $(-5, -1)$ length of minor axis 2

(g) foci $(13, -2)$ and $(-17, -2)$ major axis of length 34

4. In each of the following, find the lengths of transverse and conjugate axes, the coordinates of the vertices, and the coordinates of the foci. Sketch the graph of each.

(a) $9x^2 - 16y^2 = 144$

(b) $9x^2 - 4y^2 = 36$

(c) $25y^2 - 144x^2 = 3600$

(d) $5x^2 - 2y^2 = 10$

(e) $x^2 - y^2 = 1$

(f) $x^2 - 4y^2 = 9$

(g) $2x^2 - 3y^2 = -1$

(h) $x^2 - 2y^2 + 8 = 0$

5. Find the equations of the asymptotes for each hyperbola in Exercise 4.

6. In each of the following, find the equation of the hyperbola which satisfies the given conditions.

(a) vertices $(\pm 3, 0)$, foci $(\pm 5, 0)$

(b) vertices $(0, \pm 2)$, foci $(0, \pm 3)$

      (c)   foci $(\pm 3, 0)$ conjugate axis of length 10

      (d)   foci $(0, \pm 5)$ conjugate axis of length 8

      (e)   vertices $(\pm 4, 0)$ passing through $(8, 2)$

      (f)   vertices $(\pm 10, 0)$ asymptotes $y = \pm 3x$

      (g)   vertices $(0, \pm 3)$ asymptotes $y = \pm (1/2)x$

7. Determine the equation of each of the following hyperbolas satisfying the given conditions.

      (a)   center the origin, axes on the coordinate axes, and passing through the point $(0, -4)$ and $(3, 5)$

      (b)   vertices $(4, 2)$ and $(-2, 2)$, foci $(6, 2)$ and $(-4, 2)$

      (c)   vertices $(-1, 1)$ and $(-1, -3)$, foci $(-1, 2)$ and $(-1, -4)$

      (d)   foci $(2, 2)$ and $(2, -8)$, conjugate axis of length 8

      (e)   vertices $(3, 3)$ and $(-5, 3)$, passing through $(7, 5)$

      (f)   transverse axis $y = 3$ of length 3, center $(2, 3)$, conjugate axis of length 8

8. Derive Equation (7.4.2) in detail.

9. Derive Equation (7.4.7) in detail.

10. Derive Equation (7.4.8) for the asymptotes of the hyperbola of the form (7.4.7).

11. Derive Equation (7.4.3).

12. Derive Equation (7.4.4).

13. Derive Equation (7.4.9).

14. Derive Equation (7.4.10).

15. By constructing a rectangle as in Figure 7.4.8, verify that a hyperbola whose equation is the form

$$\frac{(x - h)^2}{a^2} - \frac{(y - k)^2}{b^2} = 1$$

has asymptotes whose equations are

(7.4.12) $$(y - k) = \pm \frac{b}{a}(x - h)$$

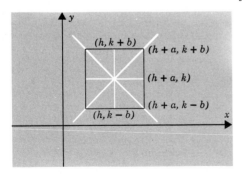

figure 7.4.8

16. Use the method of Exercise 15 to derive Equation (7.4.11).

17. The **eccentricity** e of an ellipse or a hyperbola is defined as the number

$$e = \frac{c}{a}$$

      (a)   Show that $0 < e < 1$ for an ellipse.

      (b)   Discuss how if a is kept fixed the number e, depending on the size of c, determines how far the ellipse is from a circle. That is, what if e is near 0? What if it is near 1?

      (c)   Show that $e > 1$ for a hyperbola.

18. Calculate the eccentricity of each of the ellipses in Exercise 1.

19. Calculate the eccentricity of each of the hyperbolas in Exercise 4.

20. The Mormon Tabernacle in Salt Lake City has an elliptical floor plan. The building is 250 ft long and 150 ft wide. Locate the foci from the front and rear of the building.

7.5. translation & rotation of axes

So far in this chapter, we have shown that the set of zeros of the general linear polynomial in two variables

$$p(x, y) = Ax + By + C$$

is a straight line in the plane. We promised also to demonstrate that the set of zeros for the general quadratic polynomial

$$p(x, y) = Ax^2 + By^2 + Cxy + Dx + Ey + F$$

is a conic section. We shall do just that in this section. To do so, we shall consider two cases: $C = 0$ and $C \neq 0$. In the first case, we shall demonstrate that translating the coordinate axes to a new origin located at (h, k) will reduce the equation to a standard form for one of the conics. In the second case, $C \neq 0$, we shall perform a rotation of our coordinate system which will result in a new equation with $C'' = 0$.

The general question of translating the coordinate axes to a new origin proceeds as follows. Let $Q(h, k)$ be a point in a given xy coordinate system. We introduce a new $x'y'$ coordinate system with origin O' located at Q such that the x' axis is parallel to the given x axis and the y' axis is parallel to the given y axis. We have already alluded to this procedure in the previous sections when we derived equations for conics whose center (vertex for the parabola) was at (h, k) rather than at $O(0, 0)$.

A typical situation is illustrated in Figure 7.5.1 where Q is in the first quadrant so that, in this case, h and k are both positive. In our new coordinate system, we use primes to denote the coordinates. Thus (h, k) becomes $(0', 0')$ and the point P whose coordinates are (x, y) will have coordinates (x', y') in the new system. Given such a point P, project it on the various x, y, x', and y'

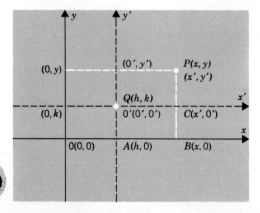

figure 7.5.1

Translated axes.

translation & rotation of axes

axes. The following relationships hold between the old and new coordinates.

(7.5.1)
$$x = x' + h$$
$$y = y' + k'$$

These follow since, for example,

$$x = |OB| = |OA| + |AB| = |OA| + |QC| = h + x'$$

The other equation is similarly verified. Equations (7.5.1) may be written in the equivalent form

(7.5.2)
$$x' = x - h$$
$$y' = y - k$$

These formulas enable one to change from the primed to the unprimed co-ordinate system or vice versa.

example 1

Equation (7.4.3) for an ellipse whose center is at (h, k) and whose major axis lies along the line $y = k$ parallel to the x axis is

$$\frac{(x - h)^2}{a^2} + \frac{(y - k)^2}{b^2} = 1$$

Using Equations (7.5.2) this equation becomes

$$\frac{(x')^2}{a^2} + \frac{(y')^2}{b^2} = 1$$

which describes the same ellipse, but with its center located at the primed origin $O'(0', 0')$. ●

example 2

Show that the equation

$$x^2 + y^2 - 4x - 6y + 12 = 0$$

has a graph which is a unit circle. The method of completing the square described in Section 7.3 will accomplish this. Thus,

$$x^2 - 4x \quad + y^2 - 6y \quad = -12$$

Adding 4 and 9 in the appropriate spaces, we have

$$(x^2 - 4x + 4) + (y^2 - 6y + 9) = -12 + 4 + 9$$

or

$$(x - 2)^2 + (y - 3)^2 = 1$$

Then with $h = 2$ and $k = 3$, Equations (7.5.2) transform this equation to the form

$$(x')^2 + (y')^2 = 1$$

a unit circle with center at the $x'y'$ origin. The graph is sketched in Figure 7.5.2.

coordinate geometry

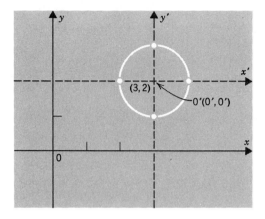

figure 7.5.2

$$x^2 + y^2 - 4x - 6y + 12 = 0$$

The procedure outline in this example illustrates the general method used to identify the set of zeros for the general quadratic in two variables

$$Ax^2 + By^2 + Dx + Ey + F$$

when the coefficient C of the term xy is zero. That is, translate the axes by use of Equations (7.5.1) and eliminate the terms of first degree, if possible. In the new $x'y'$ coordinate system, we'll have the standard form of a conic. In the event that $A = 0$ or $B = 0$, the graph will be a parabola, so we only eliminate the first degree term in x if $B = 0$ or in y if $A = 0$. We have otherwise

$$Ax^2 + Dx \qquad + By^2 + Ey \qquad = -F$$

or

$$A\left(x^2 + \frac{D}{A}x + \frac{D^2}{4A^2}\right) + B\left(y^2 + \frac{E}{B}y + \frac{E^2}{4B^2}\right) = -F + \frac{D^2}{4A} + \frac{E^2}{4B}$$

This is

$$A\left(x + \frac{D}{2A}\right)^2 + B\left(y + \frac{E}{2B}\right)^2 = \frac{D^2}{4A} + \frac{E^2}{4B} - F$$

Then set $h = -D/2A$ and $k = -E/2B$. Equations (7.5.1) give

$$A(x')^2 + B(y')^2 = Ah^2 + Bk^2 - F$$

which will be a conic or one of the degenerate cases depending on the nature of the number $Ah^2 + Bk^2 - F$. Which conic, assuming $Ah^2 + Bk^2 - F > 0$, will depend on the following:

    (i)   *if $A = B$, a circle*
    (ii) *if $A \neq B$, but A and B have the same sign, an ellipse*
    (iii) *if $A \neq B$ and A and B are opposite in sign, a hyperbola*

If the number $Ah^2 + Bk^2 - F$ is not positive then, of course, the equation is either one of the degenerate cases; two straight lines, or a point, when $Ah^2 + Bk^2 = F$, or an empty graph, when $Ah^2 + Bk^2 - F$ is negative.

As we mentioned previously, if one of A or B is zero, the curve is a parabola, and of course, the above calculations cannot be carried out as outlined. The next example illustrates this eventuality.

342

translation & rotation of axes

example 3

Discuss and sketch· the graph of the equation

$$y^2 - 8x + 2y - 15 = 0$$

Note that in this case the coefficient A of x^2 is 0, so we suspect that this is the equation of a parabola. Rewrite the equation in the form

$$y^2 + 2y \qquad = 8x + 15$$

completing the square, it becomes

$$(y^2 + 2y + 1) = 8x + 15 + 1$$

or

$$(y + 1)^2 = 8(x + 2)$$

Set $h = -2, k = -1$, and Equations (7.5.2) give us

$$(y')^2 = 8x'$$

which is the equation of a parabola. It has its vertex at the translated origin $O'(0', 0')$, which is $(-2, -1)$ in the old (unprimed) system. Since $2p = 8, p = 4$, and the *primed* coordinates of the focus F are $(2, 0)$. In the primed system the directrix is the line whose equation is

$$x' = -2$$

The graph is sketched in Figure 7.5.3.

In the *unprimed* system, the focus is located at $(2, -1)$ and the directrix has equation $x = -4$.

In discussing the set of zeros for the quadratic polynomial

$$Ax^2 + By^2 + Cxy + Dx + Ey + F = 0$$

we have seen that, when $C = 0$ and the "xy term" is missing, a translation of the axes generally will change the equation to the standard form of one of the conics. In a *translation*, the new (primed) axes are *parallel* to the original

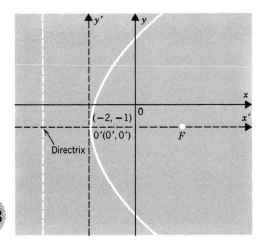

figure 7.5.3

$$y^2 - 8x + 2y - 15 = 0$$

343

coordinate geometry

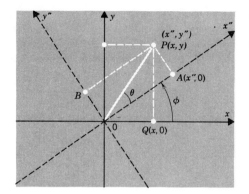

figure 7.5.4

Rotated axes.

coordinate axes, but the *origin is moved* to a new point. We turn now to a transformation which leaves the *origin fixed*, but which rotates the axes about it to new positions which we call the x'' and y'' axes. This *rotation of axes* will be done so as to eliminate the $x''y''$ term in the general quadratic.

Suppose we perform a rotation, as depicted in Figure 7.5.4, in which the positive x axis is rotated through an angle ϕ to the positive x'' axis. Denote the new coordinates of a point $P(x, y)$ by (x'', y'') when referred to the rotated system. If the line OP from the origin to P makes an angle θ with the x'' axis, and if p is the length of OP, we see that

$$x'' = |OA| = p \cos \theta$$

and

$$y'' = |OB| = |AP| = p \sin \theta$$

It is also true that

$$x = |OQ| = p \cos (\theta + \phi)$$

and

$$y = |PQ| = p \sin (\theta + \phi)$$

The addition formulas of trigonometry then give

$$x = p \cos \theta \cos \phi - p \sin \theta \sin \phi$$

$$y = p \sin \theta \cos \phi + p \cos \theta \sin \phi$$

Substitute the values of x'' and y'' already derived in these equations and obtain the transformation equations.

(7.5.4)
$$x = x'' \cos \phi - y'' \sin \phi$$

$$y = x'' \sin \phi + y'' \cos \phi$$

If these equations are solved for x'' and y'' using Cramer's rule, we have

$$x'' = \frac{\begin{vmatrix} x & -\sin \phi \\ y & \cos \phi \end{vmatrix}}{\begin{vmatrix} \cos \phi & -\sin \phi \\ \sin \phi & \cos \phi \end{vmatrix}} \qquad y'' = \frac{\begin{vmatrix} \cos \phi & x \\ \sin \phi & y \end{vmatrix}}{\begin{vmatrix} \cos \phi & -\sin \phi \\ \sin \phi & \cos \phi \end{vmatrix}}$$

The denominator is $\cos^2 \phi + \sin^2 \phi = 1$, so

344 (7.5.5)
$$x'' = x \cos \phi + y \sin \phi$$

$$y'' = -x \sin \phi + y \cos \phi$$

translation & rotation of axes

example 4

By rotating the coordinate axes through an angle $\phi = \pi/4$, show that the graph of the equation

$$y = \frac{1}{x}$$

is a hyperbola. First, write the equation in the form $xy = 1$. Then use Equations (7.5.4) to change to a primed equation. Thus,

$$xy = \left(x'' \cos \frac{\pi}{4} - y'' \sin \frac{\pi}{4}\right)\left(x'' \sin \frac{\pi}{4} + y'' \cos \frac{\pi}{4}\right) = 1$$

$$= \left(\frac{\sqrt{2}}{2}x'' - \frac{\sqrt{2}}{2}y''\right)\left(\frac{\sqrt{2}}{2}x'' + \frac{\sqrt{2}}{2}y''\right) = 1$$

$$= \left(\frac{\sqrt{2}}{2}\right)(x'' - y'')\left(\frac{\sqrt{2}}{2}\right)(x'' + y'') = 1$$

$$= \frac{1}{2}((x'')^2 - (y'')^2) = 1$$

Therefore,

$$\frac{(x'')^2}{2} - \frac{(y'')^2}{2} = 1$$

This is the standard form for the equation of a hyperbola whose transverse axis lies along the x'' axis (at a $45°$ angle with the x axis), whose center is at the origin, and whose asymptotes have equations

$$y'' = \pm x''$$

These asymptotes are the x axis and the y axis of the unprimed system. The graph is sketched in Figure 7.5.5.

The method used in this example suggests a general procedure for eliminating the xy term from the general quadratic polynomial and, thereby, reducing

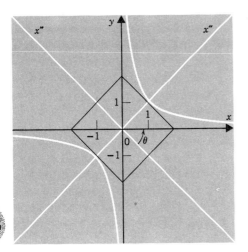

figure 7.5.5

$xy = 1$

the question of describing its set of zeros to a question we have already handled. We proceed as follows:

If

(7.5.6) $$Ax^2 + By^2 + Cxy + Dx + Ey + F = 0$$

use Equations (7.5.4) to change to a (double-) primed coordinate system making an angle ϕ with the original system. Thus,

$$A(x'' \cos \phi - y'' \sin \phi)^2 + B(x'' \sin \phi + y'' \cos \phi)^2$$

(7.5.7) $$+ C(x'' \cos \phi - y'' \sin \phi)(x'' \sin \phi + y'' \cos \phi)$$

$$+ D(x'' \cos \phi - y'' \sin \phi) + E(x'' \sin \phi + y'' \cos \phi) + F = 0$$

Upon expanding (7.5.7) and collecting terms, it has the form

$$A''(x'')^2 + B''(y'')^2 + C''(x''y'') + Dx'' + Ey'' + F = 0$$

where

$$C'' = 2(B - A) \sin \phi \cos \phi + C(\cos^2 \phi - \sin^2 \phi)$$

$$A'' = (A \cos^2 \phi + C \sin \phi \cos \phi + B \sin^2 \phi)$$

(7.5.8) $$B'' = (A \sin^2 \phi - C \sin \phi \cos \phi + B \cos^2 \phi)$$

$$D'' = (D \cos \phi + E \sin \phi)$$

$$E'' = (-D \sin \phi + E \cos \phi)$$

Since our plan is to eliminate the $x''y''$ term, set $C'' = 0$. By using the *double-angle* formulas in the first of Equation (7.5.8), we obtain

$$C'' = (B - A) \sin 2\phi + C \cos 2\phi = 0$$

Hence

$$\cot 2\phi = \frac{A - B}{C}$$

provided $C \neq 0$. Therefore,

(7.5.9) $$\phi = \frac{1}{2} \cot^{-1} \left(\frac{A - B}{C} \right)$$

is an angle through which the axes may be rotated to eliminate the xy term from our original equation. Of course, if $C = 0$, there was no point to the entire procedure. In Example 4 we had $A = B = 0$, and $C = 1$. Therefore, the angle ϕ suggested was

$$\phi = \frac{1}{2} \cot^{-1} 0 = \frac{1}{2} \cdot \frac{\pi}{2} = \frac{\pi}{4}$$

This was the angle we used.

example 5

Discuss the graph of the equation

$$6x^2 + xy = 7y^2 - 1 = 0$$

This equation has the form (7.5.6), with $A = 6$, $B = 7$, $C = 1$, and $D = E =$

translation & rotation of axes

$F = 0$. Using the formula (7.5.9) we have

$$\phi = \frac{1}{2} \cot^{-1} \left(\frac{-1}{1} \right)$$

Since $\cot^{-1}(-1) = 3\pi/4$, $\phi = 67\frac{1}{2}°$. We could use Table 6 to find $\sin \phi$ and $\cos \phi$. Instead, we note that if $\cot 2\phi = -1$, then $\tan 2\phi = -1$, and,

$$\sin 2\phi = \frac{\sqrt{2}}{2}, \text{ while } \cos 2\phi = \frac{-\sqrt{2}}{2}.$$

Why? Therefore, the *half-angle formulas* give us

$$\sin \phi = \frac{\sqrt{1 + (\sqrt{2}/2)}}{2} = \frac{\sqrt{2 + \sqrt{2}}}{2}$$

and

$$\cos \phi = \frac{\sqrt{2 - \sqrt{2}}}{2}$$

Substituting these values in (7.5.8), we obtain

$$C'' = 0$$

$$A'' = \frac{1}{2}(13 + \sqrt{2})$$

$$B'' = \frac{1}{2}(13 - \sqrt{2})$$

$$D'' = E'' = 0$$

Thus, Equation (7.5.6) becomes

$$\frac{1}{2}(13 + \sqrt{2})(x'')^2 + \frac{1}{2}(13 - \sqrt{2})(y'')^2 = 1$$

which is the equation of an ellipse whose center is at the origin and whose major axis lies along the y axis which makes an angle of $67\frac{1}{2}°$ with the given y axis. In standard form, this equation would read

$$\frac{(y'')^2}{(2/(13 - \sqrt{2}))} + \frac{(x'')^2}{(2/(13 + \sqrt{2}))} = 1$$

Using rational approximations for the number $\sqrt{2}$ this has the form

$$\frac{(y'')^2}{(0.041)^2} + \frac{(x'')}{(0.037)^2} = 1$$

The graph is sketched in Figure 7.5.6 at the top of the following page.

example 6

Consider the equation

$$x^2 + 2xy + y^2 - x + y = 0$$

347

Equation (7.5.9) suggests a rotation through an angle $\phi = 45°$. Making the

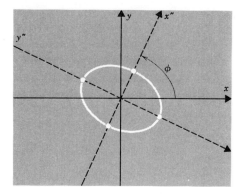

figure 7.5.6

$$6x^2 + xy + 7y^2 - 1 = 0$$

substitutions results in the transformed equation

$$\sqrt{2}\,(x'')^2 + y'' = 0$$

You should carry out the details. This equation can be written

$$(x'')^2 = -\frac{\sqrt{2}}{2}\,y''$$

which is the standard form for a parabola whose vertex is at the origin, and which opens on the negative side of the y'' axis. The new coordinates of the focus are $(0,\ -\sqrt{2}/8)$ and the directrix has equation

$$y'' = \frac{\sqrt{2}}{8}$$

The unprimed equation of the directrix is $4x - 4y + 1 = 0$, obtained from Equations (7.5.5). The focus is located at $(1/4,\ -1/4)$ in the original system. This is obtained using Equations (7.5.4). The graph is sketched in Figure 7.5.7.

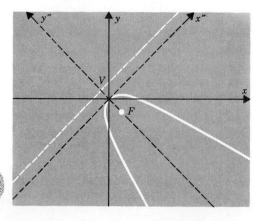

figure 7.5.7

$$x^2 + 2xy + y^2 - x + y = 0$$

translation & rotation of axes

exercises

1. Find primed coordinates for each of the following points P when the axes are translated so that the origin is at the point O' whose unprimed coordinates are as given.

(a) $P:(1, 4)\ O':(0, 1)$

(b) $P:(2, 3)\ O':(0, 1)$

(c) $P:(-1, 1)\ O':(1, 0)$

(d) $P:(-3, 1)\ O':(1, 0)$

(e) $P:(5, 0)\ O':(1, 1)$

(f) $P:(-2, -5)\ O':(1, 1)$

(g) $P:(\sqrt{2}, 1)\ O':(-1, 2)$

(h) $P:(2, 7)\ O':(-1, 2)$

(i) $P:(1, 4)\ O':(-2, -3)$

(j) $P:(\sqrt{2}, -\sqrt{3})\ O':(-2, -3)$

2. Find new coordinates (double primed) for each point P in Exercise 1 if the origin remains fixed but the axes are rotated through an angle $\phi = 30°$.

3. Work Problem 2 for $\phi = 60°$.

4. Translate the axes to $(-2, 4)$ and give new equations relative to these new axes for each of the following.

(a) $x + y = 1$

(b) $xy = 1$

(c) $x - y = 2$

(d) $4x + 2y = 1$

(e) $2(x + 2) - 3(y - 4) = 5$

(f) $x^2 + y^2 + 4x - 8y + 19 = 0$

(g) $4x^2 + y^2 + 16x - 8y + 28 = 0$

(h) $4x^2 - y^2 + 16x + 8y - 1 = 0$

(i) $x^2 + 4x - 4y + 20 = 0$

(j) $y^2 - 8x - 8y = 32$

5. Find the new center, translate the axes, and describe each of the following curves.

(a) $9x^2 + 16y^2 + 54x - 32y - 47 = 0$

(b) $4x^2 + 9y^2 + 24x + 18y + 9 = 0$

(c) $4x^2 + 25y^2 - 16x - 250y = -541$

(d) $x^2 + 4y^2 = 2x$

(e) $4y^2 - 32x + 4y = 49$

(f) $4x^2 - y - 16x + 13 = 0$

(g) $4x^2 - y^2 + 40x - 4y + 60 = 0$

(h) $25y^2 - 9x^2 - 54x - 100y = -10$

(i) $9x^2 - y^2 - 36x + 12y = 36$

(j) $4y^2 - 8x - x^2 + 32y + 49 = 0$

6. Sketch the graph of each of the curves in Exercise 5.

7. Eliminate the xy term by performing a suitable rotation of the axes for each of the following.

(a) $xy = 2$

(b) $xy = -4$

(c) $53x^2 - 72xy + 32y^2 = 80$

(d) $7y^2 - 48xy - 7x^2 = 225$

(e) $x^2 + 10\sqrt{3}xy + 11y^2 = 4$

(f) $x^2 - 3xy - 3y^2 = 5$

(g) $16x^2 - 24xy + 9y^2 - 60x - 80y + 100 = 0$

(h) $41x^2 - 24xy + 34y^2 - 25 = 0$

(i) $64x^2 - 240xy + 225y^2 + 1020x - 544y = 0$

(j) $5y^2 + 2xy + 5x^2 - 12y - 12x = 6$

8. Sketch the graphs of each of the curves in Exercise 7.

9. Show that the slope of a line does not change when axes are translated.

10. Prove that the equation of a circle with center at the origin will always have the xy term missing.

11. Show that the *sum* of the coefficients of x^2 and y^2 in the general quadratic does not change when the axes are rotated.

12. Show that the discriminant $\Delta = C^2 - 4AB$ of the general quadratic does not change under either a rotation or a translation of axes.

13. Let A, B, C, D, E, F be as in (7.5.6). Show that

$$R = \frac{1}{2} \det \begin{bmatrix} 2A & C & D \\ C & 2B & E \\ D & E & 2F \end{bmatrix}$$

is invariant under rotations and translations.

14. Show that if $R = 0$, R in Exercise 13, the conic is a point or one or two lines.

15. Show that Δ, Exercise 12 is negative, zero, or positive when the conic is an ellipse, parabola, or hyperbola, respectively.

7.6. three-dimensional rectangular coordinates

As was indicated in Section 7.1, a "point" in R^3 can be described by an *ordered triple* (a, b, c) of real numbers. To relate this algebraic set with points in three-dimensional space, we introduce a coordinate system in space analogous with the system used in the plane. To do this, first choose a point of origin and consider three mutually perpendicular lines intersecting at this point. We have sketched an approximation to this in Figure 7.6.1. Of course, representing three-dimensional space on a flat piece of paper is always only suggestive at best.

The coordinate system in Figure 7.6.1 has the y and z axes in the plane of the page while the x axis is perpendicular to the page, projecting out toward you. The order in which these axes are named is slightly arbitrary. Our system is often called a *right-handed system*. If the fingers of your right hand are bent perpendicular to your wrist and are made to correspond to the positive x axis while your wrist corresponds to the positive y axis, your thumb will point up in the direction of the positive z axis. What would a left-handed system look like?

The coordinate plane determined by the x and the y axes is called the *xy plane*. The *xz plane* and the *yz plane* are defined similarly. The plane of the page in our right-handed system, is therefore, the *yz plane*.

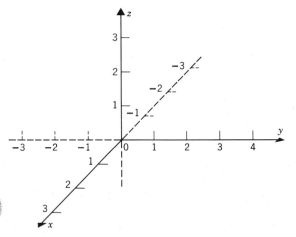

figure 7.6.1

A rectangular coordinate system

three-dimensional rectangular coordinates

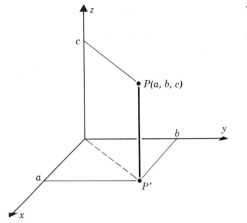

figure 7.6.2

Locating (a, b, c).

Let P be a point in three-space. The projection of P on the x axis deter-mines a real number a which is called the x coordinate of P. Similarly the y and z *coordinates*, b and c, respectively, of P are determined by its projections on the y axis and z axis, respectively. The x coordinate may also be thought of as the directed distance from the yz plane to P. The y coordinate can also be considered as the directed distance from the xz plane to P and the z coordinate as the directed distance from the xy plane to P. The coordinates of P form, then, an ordered triple (a, b, c) and correspond to an element of R^3. We shall use the notation $P:(a, b, c)$ for the point. Conversely, to each element, ordered triple (a, b, c) of R^3, there exists a point P in three-space whose coordinates this triple gives. The point P is located by first locating a point P' in one of the coordinate planes, say $P':(a, b)$ in the xy plane, in the usual way, and then measuring c units on a line perpendicular to the plane through P' (see Figure 7.6.2). Thus, there is the desired one-to-one correspondence between points in three-space and elements of R^3.

The concept of plotting points in three-space is, thus, analogous with the process in the plane described in Chapter 1. To aid in this, it is fre-quently convenient to construct a rectangular parallelepiped as illustrated in Figure 7.6.3 for the point $Q(2, -1, -2)$.

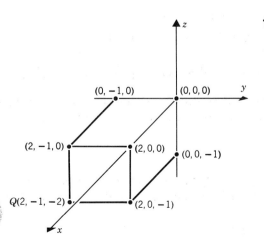

figure 7.6.3

Plotting points.

coordinate geometry

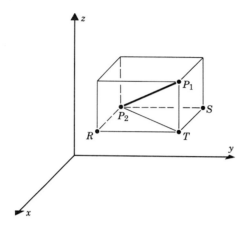

figure 7.6.4

Distance in three-space.

The three coordinate planes divide three-space up into eight parts. These are called *octants*. The *first octant* is the set of all points $P(a, b, c)$ for which a, b, and c are positive.

The concept of distance in three-space is formalized in the same way in which it was done in the plane. We derive the formula for the distance between two points $P_1(x_1, y_1, z_1)$ and $P_2(x_2, y_2, z_2)$ as follows: Draw a rectangular parallelepiped with P_1 and P_2 at opposite corners as in Figure 7.6.4 by constructing planes containing P_1 and P_2 which are parallel to the coordinate planes.

The coordinates of the point T, which lies in the same plane as P_2 parallel to the xy plane and in the same planes as P_1 parallel to the yz and xz planes, are (x_1, y_1, z_2). Since P_2 and T lie in a plane, their distance apart in that plane is given by the distance formula of Chapter 1 (ignore their common z coordinate) as

$$|P_2 T|^2 = (x_2 - x_1)^2 + (y_2 - y_1)^2$$

The distance from P_1 to T is, clearly, along a line parallel to the z axis, hence is

$$|P_1 T| = |z_2 - z_1|$$

Then, the Pythogorean theorem tells us that

$$|P_1 P_2|^2 = |P_2 T|^2 + |P_1 T|^2$$

or

$$|P_1 P_2|^2 = [(x_2 - x_1)^2 + (y_2 - y_1)^2] + (z_2 - z_1)^2$$

Therefore, the undirected distance between P_1 and P_2 is given by the square root of this equation. We state this result formally as a theorem.

theorem 7.6.1

Let $P_1:(x_1, y_1, z_1)$ and $P_2:(x_2, y_2, z_2)$ be two points in three-space. The undirected distance $|P_1 P_2|$ between P_1 and P_2 is given by the formula

(7.6.1) $\qquad |P_1 P_2| = \sqrt{(x_2 - x_1)^2 + (y_2 - y_1)^2 + (z_2 - z_1)^2}$

three-dimensional rectangular coordinates

example 1

Find the distance between $A(1, 2, 3)$ and $B(-2, -1, 4)$. The distance formula (7.6.1) gives

$$|AB| = \sqrt{(-2 - 1)^2 + (-1 - 2)^2 + (4 - 3)^2}$$
$$= \sqrt{(-3)^2 + (-3)^2 + (1)^2}$$
$$= \sqrt{19} \qquad \bullet$$

example 2

Determine a point $C(4, y, z)$ so that given $A(-2, 4, -3)$ and $B(-3, -2, 4)$, triangle ABC is equilateral. To be equilateral we must have $|AB| = |AC| = |BC|$. Using Formula (7.6.1) we have

$$\sqrt{1^2 + 6^2 + (-7)^2} = \sqrt{(6)^2 + (y - 4)^2 + (z + 3)^2} = \sqrt{7^2 + (y + 2)^2 + (z - 4)^2}$$

or

$$\sqrt{86} = \sqrt{36 + (y - 4)^2 + (z + 3)^2} = \sqrt{49 + (y + 2)^2 + (z - 4)^2}$$

This simplifies to the system of equations

$$6y - 7z = -4$$
$$y^2 + z^2 - 8y + 6z = 25$$
$$y^2 + z^2 + 4y - 8z = 17$$

which is equivalent to the system

$$y^2 + z^2 - 8y + 6z = 25$$
$$-12y + 14z = 8$$

This is solved by substituting $z = (6/7)y + 4/7$ in the first equation to obtain the quadratic equation

$$85y^2 - 92y - 1041 = 0$$

which has roots $y = -3$ and $y = 347/85$. Thus, the point C can be either the point $(4, -3, -2)$ or $(4, 347/85, 346/85)$. $\quad\bullet$

Analogous with the mid-point formula in the plane, one may prove the following **mid-point formula** for three-space.

theorem 7.6.2

The mid-point of the line segment between $P_1:(x_1, y_1, z_1)$ and $P_2:(x_2, y_2, z_2)$ is the point

(7.6.2)
$$M:\left(\frac{x_1 + x_2}{2}, \frac{y_1 + y_2}{2}, \frac{z_1 + z_2}{2} \right)$$

example 3

353

Find the mid-point M of the segment joining the points $D(-1, 2, -2)$ and $E(-5, 0, -6)$, and show that the triangle formed by M, D, and the origin O is isosceles.

coordinate geometry

First determine the coordinates of M, using (7.6.2) to be

$$M\left(\frac{-1-5}{2}, \frac{2+0}{2}, \frac{-2-6}{2}\right)$$

so that $M:(-3, 1 -4)$ is the mid-point of DE. Now (7.6.1) gives

$$|OD| = \sqrt{1^2 + 2^2 + 2^2} = 3$$

$$|OM| = \sqrt{3^2 + 1^2 + 4^2} = \sqrt{26}$$

$$|DM| = \sqrt{2^2 + 1^2 + 2^2} = 3$$

Since $|OD| = |DM|$, triangle ODM is isosceles. ●

example 4

A **sphere** is the collection of all points in three-space whose distance from the center is a fixed number r, the *radius*. Write the equation of a sphere whose radius is r and whose center is located at $C:(h, k, j)$. To do this, let $P(x, y, z)$ be any point of the sphere. Then the distance formula (7.6.1) gives

$$r = \sqrt{(x - h)^2 + (y - k)^2 + (z - j)^2}$$

Upon squaring both sides, we obtain the *standard form* for the equation of a sphere.

(7.6.3) $$(x - h)^2 + (y - k)^2 + (z - j)^2 = r^2$$ ●

exercises

1. Sketch a right-handed coordinate system for three-space and locate the following points.

(a) $(1, 0, 0)$

(b) $(0, 1, 0)$

(c) $(0, 0, 1)$

(d) $(1, 1, 1)$

(e) $(-1, -1, -1)$

(f) $(1, -1, 1)$

(g) $(2, -1, -3)$

(h) $(-3, 5, -1)$

(i) $\left(\frac{1}{2}, -\frac{1}{2}, \frac{2}{3}\right)$

(j) $\left(-\frac{3}{4}, 1, -\frac{2}{3}\right)$

2. Given the two points A and B, find the length of the segment AB.

(a) $A:(1, 2, 5)\ B:(2, 5, 8)$

(b) $A:(2, 4, -5)\ B:(4, -2, 3)$

(c) $A:(1, -2, 7)\ B:(2, 4, -2)$

(d) $A:(0, 0, 0)\ B:(1, 1, 1)$

(e) $A:(-4, 0, 1)\ B:(3, -2, 1)$

(f) $A:(1, 0, 0)\ B:(0, 0, 0)$

(g) $A:(0, 5, -4)\ B:(1, 1, 0)$

(h) $A:(-1, 0, \sqrt{2})\ B:(1, \sqrt{2}, 0)$

3. Find the mid-point for each of the segments AB in Exercise 2.

4. Find the perimeter of the triangle ABC where A and B are in the corresponding parts of Exercise 2 and C is as follows:

(a) $C:(0, 0, 0)$

(b) $C:(1, 1, 1)$

(c) $C:(2, 1, 2)$

(d) $C:(-1, 0, 0)$

(e) $C:(1, 1, 0)$

(f) $C:(0, 0, 1)$

(g) $C:(1, 0, 1)$

(h) $C:(\sqrt{2}, 0, \sqrt{2})$

5. Show that $A:(1, 2, 0)$, $B:(2, 0, 1)$, and $C:(3, 1, 2)$ are the vertices of a right triangle and find its area.

6. Show that $A:(4, -3, 2)$, $B:(7, -6, 5)$, and $C:(6, -2, 1)$ are the vertices of a right triangle and find its area.

7. Find the point $C:(k, k, k)$ so that triangle ABC is isosceles given $A:(1, 0, 2)$ and $B:(2, 0, 1)$.

8. Let S be a sphere with given center C and radius r. Write the equation of S with C and r as follows:

(a) $C:(0, 0, 0)$, $r = 2$ (e) $C:(3, -1, 2)$, $r = 3$

(b) $C:(1, -1, 0)$, $r = 2$ (f) $C:(4, -5, 1)$, $r = 5$

(c) $C:(0, 1, -1)$, $r = 2$ (g) $C:(-5, 0, 1)$, $r = \dfrac{1}{2}$

(d) $C:(-1, -1, -1)$, $r = 1$ (h) $C:(0, -3, -6)$, $r = \sqrt{3}$

9. Write an equation for the set of all points P such that their distance from $A:(2, -1, 3)$ is the same as their distance from $B:(-1, 5, 1)$.

10. From the points P of Exercise 9, select those such that triangle APB is equilateral.

11. Describe the collection of all points $P(x, y, z)$ for which

(a) $x = 0$ (e) $x = 1/2$

(b) $x = 1$ (f) $y = -2$

(c) $y = 1$ (g) $z = -3$

(d) $z = 1$ (h) $x = y$

12. Describe the graph of the equation $xyz = 0$.

7.7. lines & planes in three-space

Suppose that we are given a rectangular coordinate system for three-space. Let us consider the set of zeros for the linear polynomial

$$p(x, y, z) = Ax + By + Cz + D$$

We saw that, in the two-dimensional case, the graph of $Ax + By + C = 0$ was always a straight line in the plane. Such will *not* be the case in three-space, as can be easily seen from the following example.

example 1

Show that the set S of zeros for the linear polynomial $p(x, y, z) = 3x + 3y + 2z - 6$ intersects each coordinate plane in a straight line. Sketch the graph of each such line and suggest what S might be.

The intersection of S with the yz plane will occur when $x = 0$. That is, it is the graph of the equation

$$3y + 2z = 6$$

which passes through $(0, 0, 3)$ and $(0, 2, 0)$. Similarly, the intersection of S with the xy plane will be the line $3x + 3y = 6$, and the intersection of S with the xz plane is the line $3x + 2z = 6$. These lines are sketched in Figure 7.7.1 by determining the points where they cross the coordinate axes. It is quite clear that no single line in three-space will result in the triangle of this figure. On

coordinate geometry

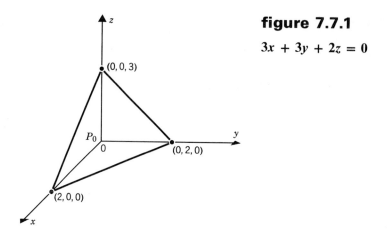

figure 7.7.1

$$3x + 3y + 2z = 0$$

the other hand, any point $P(x, y, z)$ lying in the plane determined by that triangle will be a zero of the given polynomial. ●

You will have the opportunity in Exercise 12 to show that the set of zeros for a linear polynomial of three variables always determines a plane. However, before discussing planes in three-space, let us see how to describe algebraically a *line* in three-space.

lines

As was the case in R^2, if L is a line in R^3, there is a unique line L_0, in the plane determined by L and the origin, which passes through the origin and is parallel to L. In R^2 we described the inclination of a line in terms of the angle α which L_0 made with the x axis. Since here our line L_0 does not necessarily lie in a coordinate plane, we use *three* angles to describe it, as shown in Figure 7.7.2. Thus, α *is the angle from the positive x axis to L_0, β is the angle L_0 makes with the positive y axis, and γ is the angle L_0 makes with the positive z axis.* These angles, whose radian measures satisfy

$$0 \le \alpha \le \pi, \qquad 0 \le \beta \le \pi, \qquad 0 \le \gamma \le \pi$$

are called the **direction angles** *of L_0 and of L.*

Rather than trying to define a single number as the "slope," it is more

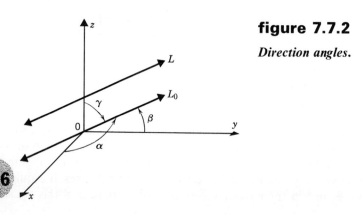

figure 7.7.2

Direction angles.

356

lines & planes in three-space

convenient, in three-space, to consider the cosines of the three direction angles of a line. *The numbers*

$$\cos \alpha, \qquad \cos \beta, \qquad \text{and } \cos \gamma$$

are called the **direction cosines** *of* L_0, *and of* L. Let $P_1:(x_1, y_1, z_1)$ and $P_2:(x_2, y_2, z_2)$ be two points in R^3. These two points determine a line L. To compute the direction cosines for L, we construct a rectangular parallelepiped in which the segment $P_1 P_2$ is a diagonal. A typical situation is shown in Figure 7.7.3.

Let the length of the segment $P_1 P_2$ be r. Since the lines $P_1 C$, $P_1 A$, and $P_1 B$ are parallel to the coordinate axes, by considering the respective right triangles $P_1 B P_2$, $P_1 A P_2$, and $P_1 C P_2$, we see that

$$(7.7.1) \quad \cos \alpha = \frac{x_2 - x_1}{r}, \qquad \cos \beta = \frac{y_2 - y_1}{r}, \qquad \cos \gamma = \frac{z_2 - z_1}{r}$$

The numbers

$$x_2 - x_1 = r \cos \alpha, \qquad y_2 - y_1 = r \cos \beta, \qquad z_2 - z_1 = r \cos \gamma$$

are called *direction numbers* for the line L. In fact, *three numbers a, b, and c are* **direction numbers** *for a line L if they are proportional to the direction cosines of L*; that is, if there exists a positive real number λ such that

$$a = \lambda \cos \alpha, \qquad b = \lambda \cos \beta, \qquad c = \lambda \cos \gamma$$

The three direction angles for a given line L are not completely independent. The fact that they are related to a single line forces them to satisfy the relation

$$(7.7.2) \qquad \cos^2 \alpha + \cos^2 \beta + \cos^2 \gamma = 1$$

This follows from the preceding derivation, since the distance formula tells us that $r = |P_1 P_2|$ is given by

$$(7.7.3) \qquad r = \sqrt{(x_2 - x_1)^2 + (y_2 - y_1)^2 + (z_2 - z_1)^2}$$

Therefore, using (7.7.1) and (7.7.3), we have

$$(x_2 - x_1)^2 = r^2 \cos^2 \alpha, \qquad (y_2 - y_1)^2 = r^2 \cos^2 \beta, \qquad (z_2 - z_1)^2 = r^2 \cos^2 \gamma$$

and

$$r = \sqrt{r^2 \cos^2 \alpha + r^2 \cos^2 \beta + r^2 \cos^2 \gamma} = \sqrt{r^2(\cos^2 \alpha + \cos^2 \beta + \cos^2 \gamma)}$$

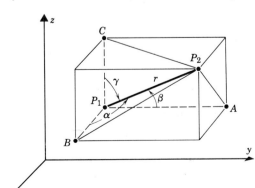

figure 7.7.3

Direction cosines for $P_1 P_2$.

coordinate geometry

example 2

Determine the direction cosines for the line L passing through $P_1(1, -1, -3)$ and $P_2(2, 1, -2)$. Using the coordinates of P_1 and P_2, we have

$$a = x_2 - x_1 = 2 - 1 = 1, \qquad b = y_2 - y_1 = 1 - (-1) = 2,$$

$$c = z_2 - z_1 = (-2 - (-3)) = 1$$

These are a set of direction numbers for L. They are not direction cosines since $r^2 = a^2 + b^2 + c^2 = 6 \neq 1$. Therefore, dividing by r results in

$$\cos \alpha = \frac{1}{\sqrt{6}}, \qquad \cos \beta = \frac{2}{\sqrt{6}}, \qquad \cos \gamma = \frac{1}{\sqrt{6}} \qquad \bullet$$

Rather than being described by a single equation, a line L in three-space is described by *three* equations (of course, only two of these are independent). Equations (7.7.1) for the direction cosines for L are one such description. There are several alternate formulations which we give here.

Let L have direction angles α, β, γ, and let L pass through a point (x_1, y_1, z_1). Then if $P:(x, y, z)$ is any other point of L, (7.7.1) tells us that

$$\cos \alpha = \frac{x - x_1}{r}, \qquad \cos \beta = \frac{y - y_1}{r}, \qquad \cos \gamma = \frac{z - z_1}{r}$$

By solving each of these for r and equating the results, we obtain the **symmetric form** of a line in three-space.

$$(7.7.4) \qquad \frac{x - x_1}{\cos \alpha} = \frac{y - y_1}{\cos \beta} = \frac{z - z_1}{\cos \gamma}$$

Since (7.7.4) would still hold if all the denominators were multiplied by a positive constant λ, resulting in them being direction numbers a, b, and c for L, we have the **point direction** form

$$(7.7.5) \qquad \frac{x - x_1}{a} = \frac{y - y_1}{b} = \frac{z - z_1}{c}$$

The direction numbers can be calculated directly from the coordinates of two points on a line. Therefore, if we have two points (x_1, y_1, z_1) and (x_2, y_2, z_2) of L, the **two-point forms** of the equations of L are,

$$\frac{x - x_1}{x_2 - x_1} = \frac{y - y_1}{y_2 - y_1} = \frac{z - z_1}{z_2 - z_1}$$

Finally, if we let $t = r/\sqrt{a^2 + b^2 + c^2}$, where a, b, and c are direction numbers for L, and r is the distance between two given points, (7.7.1) will yield

$$x - x_1 = r \cos \alpha = at$$

$$y - y_1 = r \cos \beta = bt$$

$$z - z_1 = r \cos \gamma = ct$$

These are rewritten in the **parametric form**

$$x = x_1 + at$$

$(7.7.7)$ $$y = y_1 + bt$$

$$z = z_1 + ct$$

lines & planes in three-space

Conversely, if t is any real number, the line L contains the point (x, y, z) provided that Equations (7.7.7) are true, where a, b, and c are direction numbers for L and (x_1, y_1, z_1) is a point of L. You may verify this as Exercise 26.

example 3

Write the parametric form for the equation of the line L determined by $P:(1, -1, 3)$ and $Q:(2, 1, 5)$. First compute the direction numbers as follows:

$$a = (2 - 1) = 1 \qquad b = \left(1 - (-1)\right) = 2 \qquad c = (5 - 3) = 2$$

Using P, then, the equations for L are

$$x = 1 + t$$
$$y = -1 + 2t$$
$$z = 3 + 2t \qquad \bullet$$

example 4

Sketch the graph of the line in three-space whose equations are

$$x = 1 - t$$
$$y = 1 + t$$
$$z = 3t$$

By comparing these equations with Equations (7.7.7), we see that the line passes through the point $P:(1, 1, 0)$ and has direction numbers -1, 1, and 3. Thus, another point $Q:(x_2, y_2, z_2)$ of the line can be computed from the equations

$$x_2 - 1 = -1, \qquad y_2 - 1 = 1 \qquad z_2 - 0 = 3$$

Hence, $Q:(0, 2, 3)$ is also a point on the line. These points determine the line. We sketch them, and the line in Figure 7.7.4.

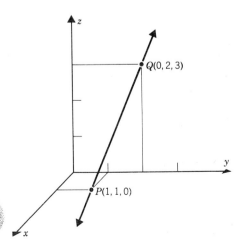

figure 7.7.4

The line of Example 4.

coordinate geometry

the angle between two lines in R^3

Let us now see how the direction cosines help us determine the angle between two lines in R^3. Let L_1 have direction angles α_1, β_1, and γ_1, and let L_2 have direction angles α_2, β_2, and γ_2. Further, let us suppose that L_1 and L_2 intersect at the origin. We can do this without loss of generality since the angle between these two lines will be the same as if they intersect elsewhere. (Why?) Select P on L_1 and Q on L_2 so that $|OP| = |OQ| = 1$. Triangle OPQ is isosceles and contains θ (see Figure 7.7.5).

The law of cosines then gives

$$|PQ|^2 = |OP|^2 + |OQ|^2 - 2|OP|\,|OQ| \cos \theta$$

But, by our choice of the points P and Q, this becomes

$$|PQ|^2 = 2 - 2 \cos \theta$$

Furthermore, the coordinates of P are $(\cos \alpha_1, \cos \beta_1, \cos \gamma_1)$, and those of Q are $(\cos \alpha_2, \cos \beta_2, \cos \gamma_2)$. (Why?) Therefore, from the distance formula,

$$|PQ|^2 = (\cos \alpha_2 - \cos \alpha_1)^2 + (\cos \beta_2 - \cos \beta_1)^2 + (\cos \gamma_2 - \cos \gamma_1)^2$$
$$= 2 - 2(\cos \alpha_1 \cos \alpha_2 + \cos \beta_1 \cos \beta_2 + \cos \gamma_1 \cos \gamma_2)$$

You should verify this simplification.

Equate the two values of $|PQ|^2$ to obtain the relationship

(7.7.8) $\cos \theta = \cos \alpha_1 \cos \alpha_2 + \cos \beta_1 \cos \beta_2 + \cos \gamma_1 \cos \gamma_2$

for the angle θ between L_1 and L_2.

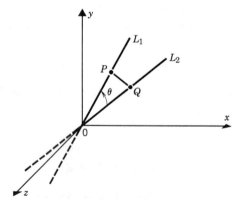

figure 7.7.5

The angle between two lines.

example 5

Find the angle between the two lines L_1 with direction numbers 1, 0, and 1, and L_2 with direction numbers 1, 3, and -1. The direction cosines for L_1 are computed from the direction numbers as

$$\cos \alpha_1 = \frac{1}{\sqrt{2}}, \qquad \cos \beta_1 = 0, \qquad \cos \gamma_1 = \frac{1}{\sqrt{2}}$$

Those for L_2 are similarly computed to be

$$\cos \alpha_2 = \frac{1}{\sqrt{11}}, \qquad \cos \beta_2 = \frac{3}{\sqrt{11}}, \qquad \cos \gamma_2 = \frac{-1}{\sqrt{11}}$$

lines & planes in three-space

Therefore, (7.7.8) becomes, in this case,

$$\cos\theta = \left(\frac{1}{\sqrt{2}}\right)\left(\frac{1}{\sqrt{11}}\right) + (0)\left(\frac{3}{\sqrt{11}}\right) + \left(\frac{1}{\sqrt{2}}\right)\left(-\frac{1}{\sqrt{11}}\right) = 0$$

Hence $\theta = \pi/2$: (Why?) L_1 and L_2 are perpendicular or **orthogonal**. ●

planes

As we stated earlier in this section, the set of zeros for the linear polynomial in three variables

$$p(x, y, z) = Ax + By = Cz + D$$

is not a line, but rather a **plane** in three-space. Conversely, every plane has an equation, $p(x, y, z) = 0$; e.g., the xy plane has the equation

$$z = 0$$

Suppose that Γ is a plane in R^3. In high school geometry you learned that there is a unique line L through the origin which is perpendicular to Γ (see Figure 7.7.6). If this line is considered as directed from the origin to the plane, it is known as the **normal** L to Γ. Let α, β, and γ be the direction angles of L and let ρ be the distance along L between the origin O and the intersection M of L and the plane Γ. Let $P:(x, y, z)$ be any point of Γ. Since L is perpendicular to Γ, it is perpendicular to every line in Γ which passes through M. (We suppose $P \neq M$). Therefore, L is perpendicular to the line MP. Let θ be the angle MOP, and set $r = |OP|$. Then $r \cos\theta = \rho$. Let the line OP have direction angles α', β', and γ'. From (7.7.8) we have

$$\cos\theta = \cos\alpha\cos\alpha' + \cos\beta\cos\beta' + \cos\gamma\cos\gamma'$$

Hence, $\rho = r\cos\theta = r[\cos\alpha\cos\alpha' + \cos\beta\cos\beta' + \cos\gamma\cos\gamma']$. Now, the coordinates of P are (x, y, z), therefore,

$$x = r\cos\alpha', \qquad y = r\cos\beta', \qquad y = r\cos\gamma'$$

(Why?) Thus, substituting these, we obtain

$$\rho = \cos\theta = x\cos\alpha + y\cos\beta + z\cos\gamma$$

Therefore, $P:(x, y, z)$ is on the plane Γ, whose normal L has direction cosines α, β, and γ, if and only if,

(7.7.9) $$x\cos\alpha + y\cos\beta + z\cos\gamma - \rho = 0$$

This is an equation for the plane Γ.

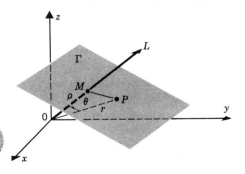

figure 7.7.6

The plane Γ and its normal L.

coordinate geometry

example 6

Find the equation of the plane whose normal OM passes through the point $(2, -2, 1)$, and such that $P:(1, 1, 3)$ lies in the plane. First of all, some direction numbers for the normal are 2, -2, and 1; hence, its direction cosines are

$$\cos \alpha = \frac{2}{3}, \qquad \cos \beta = -\frac{2}{3}, \qquad \cos \gamma = \frac{1}{3}$$

(Why?) Therefore, a point (x, y, z) is on the plane if, and only if,

$$\frac{2}{3}x - \frac{2}{3}y + \frac{1}{3}z - \rho = 0$$

To compute ρ, we note that the point $P:(1, 1, 3)$ is on the plane; hence, its coordinates satisfy this equation; i.e.,

$$\frac{2}{3}(1) - \frac{2}{3}(1) + \left(\frac{1}{3}\right)(3) - \rho = 0$$

Thus, $\rho = 1$, and the equation of the desired plane is

$$\frac{2}{3}x - \frac{2}{3}y + \frac{1}{3}z - 1 = 0$$

Equivalently, the plane has equation

$$2x - 2y + z - 3 = 0 \quad \bullet$$

example 7

Describe the plane whose equation is

$$2x + 3y + 4z = 12$$

If we apply the above, we see that direction numbers for the normal to this plane are 2, 3, and 4. Hence, the direction cosines are

$$\cos \alpha = \frac{2}{\sqrt{29}}, \qquad \cos \beta = \frac{3}{\sqrt{29}}, \qquad \cos \gamma = \frac{4}{\sqrt{29}}$$

Therefore, $\rho = 12/\sqrt{29}$. A sketch of the plane can be obtained by plotting the normal OM, given this information.

An alternative, and often easier, method is that suggested in Example 1 at the beginning of this section. It is often convenient to sketch the **traces** of the graph in the coordinate planes; that is, the lines in which the given plane intersects the coordinate planes. To find the trace in the xy plane, for example, we solve the equation

$$z = 0$$

for the xy plane, and the equation

$$2x + 3y + 4z = 12$$

simultaneously. This is most easily done by substituting $z = 0$ in the second equation. Thus, we obtain the equation

$$2x + 3y = 12$$

lines & planes in three-space

of a line in the xy plane as the *trace* there. Similarly, the trace in the xz plane is the line

$$2x + 4z = 12$$

and the trace in the yz plane is the line

$$3y + 4z = 12$$

These lines are most easily sketched from their intercepts, the points $(0, 4, 0)$, $(0, 0, 3)$, and $(6, 0, 0)$. These points satisfy the equations involved. These points, the traces, and then the given plane are sketched in Figure 7.7.7. ●

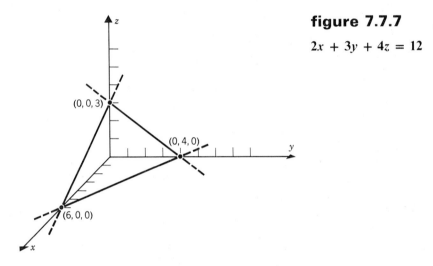

figure 7.7.7

$2x + 3y + 4z = 12$

example 8

Find an equation of the unique plane determined by the three points $P:(1, 2, -1)$, $Q:(3, 0, 1)$, and $S:(2, -1, 1)$. Since a plane has equation of the form

$$Ax + By + Cz + D = 0$$

the coordinates of P, Q, and S must satisfy this equation. Substituting each gives the following system of equations.

$$A + 2B - C = -D$$
$$3A \qquad + C = -D$$
$$2A - B + C = -D$$

The augmented matrix for this system is

$$\begin{bmatrix} 1 & 2 & -1 & -D \\ 3 & 0 & 1 & -D \\ 2 & -1 & 1 & -D \end{bmatrix}$$

363

coordinate geometry

which is equivalent to

$$\begin{bmatrix} 1 & 2 & -1 & -D \\ 0 & 1 & -1 & -D \\ 0 & 0 & 1 & 2D \end{bmatrix}$$

This has Hermite normal form

$$\begin{bmatrix} 1 & 0 & 0 & -D \\ 0 & 1 & 0 & D \\ 0 & 0 & 1 & 2D \end{bmatrix}$$

Thus, for any D, the equation

$$-Dx + Dy + 2Dz + D = 0$$

is an equation for the plane. In particular if $D = -1$, an equation is

$$x - y - 2z - 1 = 0 \quad \bullet$$

exercises

1. Find the direction cosines for the lines determined by each of the following pairs of points.

(a) $(1, 0, 0)$ and $(0, 0, 0)$
(b) $(0, 0, 0)$ and $(1, 0, 1)$
(c) $(0, 0, 0)$ and $(1, 1, 1)$
(d) $(1, 1, 1)$ and $(1, 0, 1)$
(e) $(3, -1, 9)$ and $(-1, 2, 5)$

(f) $(3, -2, 4)$ and $(-7, -1, 11)$
(g) $(-13, 0, 5)$ and $(1, 12, 0)$
(h) $(12, -4, 3)$ and $(5, 1, -1)$
(i) $(0, -3, -2)$ and $(6, 15, 0)$
(j) $(\sqrt{2}, 0, -\sqrt{2})$ and $(0, \sqrt{2}, \sqrt{2})$

2. Write the two-point form for the equation of each line in Exercise 1.

3. Write the symmetric form of the equation of each line in Exercise 1.

4. Write the parametric form of each line in Exercise 1.

5. Let L_1 and L_2 have direction numbers as follows. In each case find the cosine of the angle between L_1 and L_2.

(a) $L_1: 1, 0,$ and 0; $L_2: 0, 1,$ and 0
(b) $L_1: 1, 0,$ and 0; $L_2: 0, 1,$ and 1
(c) $L_1: 3, 2,$ and 0; $L_2: 0, 0,$ and 1
(d) $L_1: 1, 0,$ and 1; $L_2: 3, 2,$ and -1
(e) $L_1: 4, -1,$ and 1; $L_2: -2, 3,$ and 0
(f) $L_1: 8, 7,$ and 2; $L_2: 0, 2,$ and -4
(g) $L_1: 8, 9,$ and 1; $L_2: 3, 8,$ and 1
(h) $L_1: 2, 8,$ and 3; $L_2: 1, 4,$ and 3

6. Find the angles in the triangle whose vertices are the points $(0, 0, 0)$, $(1, 0, 0)$, and $(0, 0, 1)$. Sketch its graph.

7. Work Exercise 6 for the triangle whose vertices are $(1, -3, 2)$, $(2, -1, 3)$, and $(4, 1, 0)$.

8. Find the parametric form for the equations of the line through $(1, 1, 1)$ which is perpendicular both to L_1, whose direction numbers are 5, 2, and 0; and to L_2, whose direction numbers are -2, 1, and 3.

lines & planes in three-space

9. Work Exercises 8 when the line passes through (8, 3, 2) and is perpendicular to L_1: 4, 3, and 1 and L_2: 0, 1, and 3.

10. Sketch the graphs of the lines in Exercise 1.

11. Sketch the graphs of each of the following planes.

(a) $y = 3$
(b) $x = -2$
(c) $z = 4$
(d) $x = 5$
(e) $y = -1$
(f) $z = -3/2$

(g) $2x + y - 6 = 0$
(h) $3y - 2z = 24$
(i) $4x - 2z = 15$
(j) $3y + 7z + 21 = 0$
(k) $x + y + z = 1$
(l) $x + y + z = 0$

12. Use Equation (7.7.9) to prove that the set of zeros for every linear polynomial in three variables $Ax + By + Cz + D$ is a plane in three-space.

13. Find an equation for the plane determined by each of the following sets of three points.

(a) $(0,0,0), (1,0,0), (0,1,0)$
(b) $(0,0,0), (0,1,0), (0,0,1)$
(c) $(1,0,0), (0,1,0), (0,0,1)$
(d) $(1,1,1), (1,0,0), (0,0,1)$
(e) $(1,1,0), (0,1,1), (1,0,1)$
(f) $(1,2,3), (-1,-1,2), (1,0,1)$

(g) $(1,5,7), (2,2,-1), (-3,2,4)$
(h) $(-1,2,-1), (0,1,1), (2,2,2)$
(i) $(5,0,1), (6,-1,-1), (4,2,-1)$
(j) $(4,-4,1), (5,3,2), (0,-2,3)$
(k) $(8,3,2), (4,3,1), (0,1,3)$
(l) $(6,1,-1), (2,-1,2), (1,-3,1)$

14. Show that two linear equations

$$a_1 x + b_1 y + c_1 z = d_1$$
$$a_2 x + b_2 y + c_2 z = d_2$$

represent the same plane Γ in R^3 if, and only if, their corresponding coefficients are all in the same ratio.

15. Show that, for any two planes Γ_1 and Γ_2, with equations

$$a_1 x + b_1 y + c_1 z = d_1$$

and

$$a_2 x + b_2 y + c_2 z = d_2$$

respectively, Γ_1 is parallel to Γ_2 if, and only if, there exists a nonzero real number λ such that

$$a_1 = \lambda a_2, \qquad b_1 = \lambda b_2, \qquad c_1 = \lambda c_2$$

16. Show that the planes Γ_1 and Γ_2 of Exercise 15 are perpendicular (orthogonal) if, and only if,

$$a_1 a_2 + b_1 b_2 + c_1 c_2 = 0$$

17. Show that, for three nonzero real numbers a, b, and c, the equation of the plane Γ which is determined by the points $(a, 0, 0)$, $(0, b, 0)$, and $(0, 0, c)$ has an "intercept form"

$$\frac{x}{a} + \frac{y}{b} + \frac{z}{c} = 1$$

18. Show that the two planes whose equations are

$$x + y + z = 1$$

and

$$x + 2y - z = 2$$

intersect in a line.

coordinate geometry

19. Find the equation of the plane which is determined by the line of intersection of Exercise 18 and the origin.

20. Show that the three planes whose equations are

$$x + y + z = 3, \qquad x + 2y + 3z = 6, \qquad x - y + 2z = 8$$

intersect in a point.

21. Show that the equation of the plane determined by the three points (x_1, y_1, z_1), (x_2, y_2, z_2), and (x_3, y_3, z_3) can be written in the form det $A = 0$, where A is the matrix

$$\begin{bmatrix} x & y & z & 1 \\ x_1 & y_1 & z_1 & 1 \\ x_2 & y_2 & z_2 & 1 \\ x_3 & y_3 & z_3 & 1 \end{bmatrix}$$

22. Show that the distance between two parallel planes, using the notation of Exercise 15, is $|d_2 - \lambda d_1|$.

23. Find the equation of the plane through $(2, 5, -9)$ which is parallel to $3x - 2y + 9z = 4$.

24. Show that the distance from the plane

$$x \cos \alpha + y \cos \beta + z \cos \gamma = \rho$$

to the point $P:(x_1, y_1, z_1)$ is given by the formula

$$d = |x_1 \cos \alpha + y_1 \cos \beta + z_1 \cos \gamma - \rho|$$

Hint: Draw a parallel plane containing P and write its equation in the form (7.7.9). Find the distance between the two planes.

25. Find the distance from the given plane to the given point.

(a) $x + y + z = 0$ to $(1,1,4)$
(b) $x + y + z = 0$ to $(1,-6,2)$
(c) $4x - 2y + z + 2 = 0$ to $(-1,-3,-2)$
(d) $4x - 2y + z + 2 = 0$ to $(1,-6,2)$
(e) $3y - x - 4z = -12$ to $(1,-3,0)$
(f) $3y - x - 4z + 12 = 0$ to $(1, -6,2)$
(g) $4x - 3y - 2z = 1$ to $(1,1,1)$
(h) $3x - 15y - 2z = 4$ to $(-1,-3,0)$
(i) $3x - 15y - 2z - 4$ to $(1,-6,2)$
(j) $x + 2y - 5z = -7$ to $(-8,3,-14)$

26. Verify that a point $P(x, y, z)$ is on a line L if it satisfies Equations (7.7.7).

7.8. polar coordinates & functions

In Chapter 1 we stated that the graph of the unit circle was *not* the graph of a function. This is surely true in the sense that there is no function

$$f: X \to Y$$

whose graph in the usual *Cartesian-coordinate system* gives the unit circle.

On the other hand, there is a method of assigning coordinates to points in **366** the plane which does make the graph of the unit circle, the graph of a certain function. This is the method of **polar coordinates**. We again use ordered pairs

polar coordinates & functions

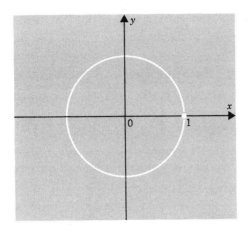

figure 7.8.1

The unit circle.

of real numbers (r, θ), but this time the second real number θ is the measure of a plane angle. The system is based upon a fixed point O called the origin or **pole** and a directed ray, called the **polar axis**, with endpoint O. We usually orient the polar axis as if it were the positive x axis, although this is not essential (see Figure 7.8.2).

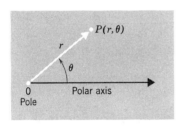

figure 7.8.2

Polar coordinates.

If P is any point in the plane, draw OP. The length of OP is the real number r, and the angle which OP makes with the polar axis is θ. As we discussed in Chapter 5, θ is positive if it is generated by a counterclockwise rotation of the line OP about the pole from the polar axis. If the rotation is clockwise, then θ is negative. A glance at Figure 7.8.3 suggests that, unlike rectangular coordinates, the polar coordinates of a point are *not unique*. For example, θ and the opposite rotation $-(2\pi - \theta)$ will put OP in the same position relative to the polar axis. We also allow r to be negative. In this event, instead of measuring $|r|$ units along the terminal side of θ, measure $|r|$ units along the opposite ray beginning at O. In Figure 7.8.3, we have suggested a number of different pairs (r, θ) for the same point P.

example 1

Sketch the point whose polar coordinates are $(2, 3\pi/4)$, $(2, 11\pi/4)$, $(2, -5\pi/4)$, $(-2, -\pi/4)$, and $(-2, 7\pi/4)$. These are all different coordinatizations for the same point as sketched in the series of diagrams in Figure 7.8.3. ●

Despite this lack of uniqueness, polar coordinates are often very useful. If we agree that the pole O has polar coordinates $(0, \theta)$ *for any* θ, we have an assignment of ordered pairs (r, θ) to every point in the plane. On the other hand, if a pair (r, θ) is given in $R \times R$ they are the polar coordinates of a *unique* point P in the plane. Thus, there is uniqueness in one direction.

coordinate geometry

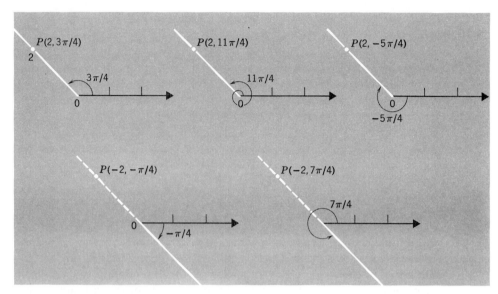

figure 7.8.3

Polar coordinates are not unique.

If you have been perceptive, you will have noticed that the process of polar coordinatizing the plane is a function

$$\text{pol}: R \times \Theta \to \Gamma$$

where R and Θ are sets of real numbers, and Γ is the plane. Is pol an injective function? Is it surjective?

We stated at the beginning that the unit circle could be considered as a function. In a manner analogous with the connection between equations involving x and y in rectangular coordinates, **polar equations** are equations in r and θ. A pair of real numbers (a, b) is a *solution* to such an equation provided the equation is true when a is substituted for r and b for θ. We can also consider polar functions from the reals to the reals given by equations (rules of correspondence) of the form

$$r = f(\theta) \qquad \text{or} \qquad \theta = f(r)$$

With this sort of interpretation, the unit circle is easily seen to be the constant function given by the equation

$$r = 1$$

example 2

Sketch the graph of the polar function given by

$$r = 2 \sin \theta$$

To do this, we plot points, as before, and connect them carefully with a smooth curve. Real care must be taken to connect the points in increasing order with θ,

polar coordinates & functions

as we did before with x. This equation has solutions (points which satisfy it) which include those in the following table.

θ	0	$\frac{\pi}{6}$	$\frac{\pi}{4}$	$\frac{\pi}{3}$	$\frac{\pi}{2}$	$\frac{2\pi}{3}$	$\frac{3\pi}{4}$	$\frac{5\pi}{6}$	π	$\frac{7\pi}{6}$	$\frac{5\pi}{4}$
r	0	1	$\sqrt{2}$	$\sqrt{3}$	2	$\sqrt{3}$	$\sqrt{2}$	1	0	-1	$-\sqrt{2}$

In rectangular coordinates the graph of $y = 2 \sin x$ would be a sine wave with amplitude 2 and period 2π. In this case, however, the *polar graph* is a unit circle which passes through the pole. This is seen by plotting points as in Figure 7.8.4. Notice that for $\theta \in (\pi, 2\pi)$, the negative values of r give the same points over again that we got for $\theta \in [0, \pi]$. It this function injective? As θ runs through all the real numbers, the same points are traversed over and over again. This is, of course, both because the sine is a periodic function and also because of the interpretation given for negative values of r. ●

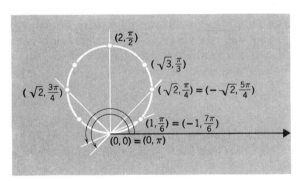

figure 7.8.4

Graph of $r = 2 \sin \theta$.

example 3

Sketch the graph of $r = 3 - 4 \cos \theta$. The values of cosine θ decrease from 1 to -1 as θ increases from 0 to π. Therefore, in this equation r will increase from -1 to 7 on the θ interval $[0, \pi]$. If the points are tabulated for various values of θ we obtain the following:

θ	0	$\frac{\pi}{6}$	$\frac{\pi}{4}$	$\frac{\pi}{3}$	$\frac{\pi}{2}$	$\frac{2\pi}{3}$	$\frac{3\pi}{4}$	$\frac{5\pi}{6}$	π
r	-1	$3-2\sqrt{3}$	$3-2\sqrt{2}$	1	3	5	$3+2\sqrt{2}$	$3+2\sqrt{3}$	7

If θ runs through the interval $[\pi, 2\pi]$ cos θ increases from -1 to 1, and r decreases from 7 to -1. This gives a reflection in the line along the polar axis. Plotting points and connecting them gives the graph of Figure 7.8.5. Because of its heart shape, this curve is called a **cardioid**. ●

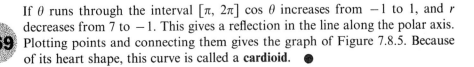

369

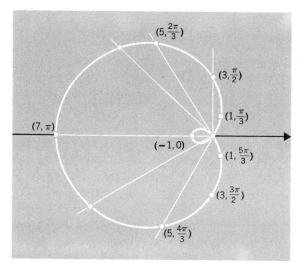

figure 7.8.5

$r = 3 - 4 \cos \theta$

Let $P:(r, \theta)$ be a point in the polar coordinate plane. Let us determine rectangular coordinates (x, y) for P when the xy origin coincides with the pole and the polar axis is the positive x axis. If the x and y axes are positioned as described, then the angle θ of the point $P:(r, \theta)$ is in *standard position,* and P is a point on its terminal side. See the sketch of a particular situation in Figure 7.8.6. Therefore, if r is positive, the definitions of the trigonometric functions immediately yield the relationships

(7.8.1) $\qquad\qquad x = r \cos \theta \qquad\qquad y = r \sin \theta$

On the other hand, if r is negative, then $-r > 0$; and the angle $\pi - \theta$ is in standard position also. You should verify this fact not only for the case $\theta > \pi$ as depicted in Figure 7.8.6, but also for $\theta \le \pi$. In this event, the point P also has polar coordinates $(-r, \pi - \theta)$, and the definitions of the sine and cosine yield

$$x = (-r) \cos (\pi - \theta), \qquad\qquad y = (-r)(\sin \pi - \theta)$$

which reduce to (7.7.1). If $r = 0$, then P is at the pole, and Equations (7.8.1) still hold.

It is easily seen that the following formulas follow directly from Equations (7.8.1).

(7.8.2) $\qquad\qquad \tan \theta = \dfrac{y}{x} \qquad$ and $\qquad r^2 = x^2 + y^2$

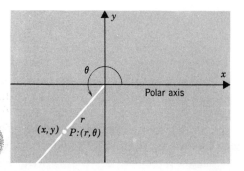

figure 7.8.6

Polar and rectangular coordinates.

polar coordinates & functions

example 4

Write the polar equation for the set of all points whose rectangular coordinates satisfy the equation $xy = 4$. We transform this equation using Formulas (7.8.1) as follows:

$$(r \cos \theta)(r \sin \theta) = 4$$

Thus

$$r^2(\sin \theta \cos \theta) = 4$$

or

$$r^2 \sin 2\theta = 8 \quad \bullet$$

example 5

Find the polynomial equation in rectangular coordinates, x and y, whose graph is the same as the graph of the polar equation

$$r = \frac{1}{1 - \cos \theta}$$

We first write

$$r - r \cos \theta = 1 \qquad \text{or} \qquad r = 1 + r \cos \theta$$

Then Equations (7.8.1) and (7.8.2) yield

$$\pm\sqrt{x^2 + y^2} = 1 + x$$

Squaring both sides gives the equation

$$x^2 + y^2 = 1 + 2x + x^2$$

or

$$y^2 = 2x + 1$$

In squaring both sides, however, we included in our result the graph of the equation $r = -(1 + r \cos \theta)$. This is, however, the same as the graph of the equation

$$-r = \frac{1}{1 + \cos \theta}$$

But this is the same as the equation

$$-r = \frac{1}{1 - \cos (\pi + \theta)}$$

because $\cos (\pi + \theta) = -\cos \theta$. However, the points (r, θ) and $(-r, \pi + \theta)$ are the same point, as we remarked earlier. Thus, the extraneous graph is the same as the original, so that the graph of the equation $r = 1/(1 - \cos \theta)$ is the same as that for the parabola given by the rectangular equation $y^2 = 2x + 1$. We sketch this graph in Figure 7.8.7. You may wish to plot polar points on it also. $\bullet$

coordinate geometry

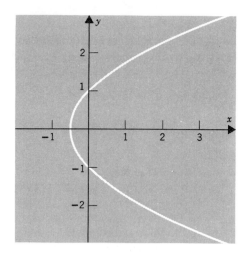

figure 7.8.7
$$y^2 = 2x + 1$$

exercises

1. Sketch the graphs of the following polar equations.

(a) $r = 4$

(b) $r = -1$

(c) $\theta = \pi/4$

(d) $\theta = -\pi/3$

(e) $r = 4 \sin \theta$

(f) $r = 2 \cos \theta$

(g) $r = -4 \cos \theta$

(h) $r = 4(1 - \sin \theta)$

2. Sketch the graphs of the following polar equations.

(a) $r = 1 + 2 \cos \theta$ (limaçon)

(b) $r = 2 \cos 3\theta$ (three-leaved rose)

(c) $r = 2 \sin 4\theta$ (eight-leaved rose)

(d) $r = 3 \sin^2 \theta/2$ (cardioid)

(e) $r^2 = 4 \cos 2\theta$ (lemniscate)

(f) $r^2\theta = 4, r > 0$, (lituus)

(g) $r = 2 + 2 \sec \theta$ (conchoid)

(h) $r = 3^\theta, \theta \geq 0$ (spiral)

(i) $r\theta = 1, \theta > 0$ (spiral)

3. In each of the following, find a polar equation whose graph will be the same as the given rectangular equation.

(a) $x = 3$

(b) $y = -2$

(c) $x - y = 0$

(d) $xy = 3$

(e) $x^2 + y^2 = 25$

(f) $x^2 = 8y$

(g) $y^2 = 2x$

(h) $x^2 - y^2 = 16$

(i) $9x^2 + 4y^2 = 36$

(j) $x^2 + y^2 + 2x + 6y = 0$

4. In each of the following, find an equation in rectangular coordinates whose graph contains the graph of the given polar equation. Discuss possible extraneous solutions.

(a) $r = 6$

(b) $\theta = \pi/4$

(c) $r = 2 \cos \theta$

(d) $r = 5 \sin \theta$

(e) $r \cos \theta = 6$

(f) $r(\cos \theta - \pi/4) = 2$

(g) $r = 2(1 + \cos \theta)$

(h) $r = 4 \sec \theta$

(i) $r^2 = \cos 2\theta$

(j) $r = 2 \sec \theta \tan \theta$

5. Derive a distance formula in polar coordinates. *Hint*: Use the law of cosines.

372

6. A circle has center at $(5, \pi/3)$ with radius 2. Find its equation (polar).

polar coordinates & functions

7. Show that the polar form of the equation of a straight line passing through $N(p, \alpha)$ and perpendicular to ON is

$$r \cos (\theta - \alpha) = p$$

8. Use Exercise 7 to derive the **normal form** of the equation of the straight line L in the Cartesian (rectangular) plane as

$$x \cos \alpha + y \sin \alpha = p$$

Here α is the angle of a line drawn from the origin perpendicular to L, and p is the distance L is from the origin along this perpendicular.

quiz for review

PART I. True or False

1. Each point in the plane, except for the origin $(0, 0)$, has a unique pair of polar coordinates.

2. The polar equation of *the* unit circle is $r = 1$.

3. The graph of the equation $xy = 0$ is a hyperbola.

4. A parabola has two foci.

5. The graph of the equation $x^2 + x + y = 0$ is a parabola.

6. The graph of the equation $3x + 4y = 1$ is a plane in three-space.

7. If two foci of an ellipse coincide, then the ellipse is either a circle, a point, or the empty set.

8. The curve whose polar equation is $\theta = c$, c a constant, is a circle.

9. The zeros of a linear polynomial $p(x_1, x_2, \ldots, x_n)$ in n variables will always have a straight line in n space for a graph.

10. The distance formula for three-space is $\sqrt{(x_2 - x_1) + (y_2 - y_1) + (z_2 - z_1)}$

PART II. Multiple Choice

In problems 1–10, name the curve whose equation is given as a

    (a) hyperbola   (b) parabola   (c) circle   (d) ellipse   (e) line(s)

1. $x = 25 - y^2$
2. $x^2 = 25 - y^2$
3. $x^2 = 25 + y^2$
4. $xy = 4$
5. $x^2 - y^2 = 0$

6. $x^2 + y^2 + 2x + 3y = 7$
7. $2x^2 - y^2 + 4x + 4y = 8$
8. $4x^2 + 9y^2 + 8x + 6y = 12$
9. $3x^2 + 3y^2 + 2xy - 1 = 0$
10. $2x + 3y = 7$

11. The equation $5x - 4y + 2z = 1$ has a graph which is

    (a) is a line in the plane   (b) is a line in three-space   (c) a plane
    (d) a conic section

12. The point $(-1, -\pi/4)$ in polar coordinates lies in which quadrant?

    (a) first   (b) second   (c) third   (d) fourth   (e) both first and third

13. If $c > 0$ is a real number, the point whose rectangular coordinates are (c, c) has polar coordinates

    (a) (c, c) (b) $(c\sqrt{2}, \pi/4)$ (c) $(c\sqrt{2}, -\pi/4)$ (d) $(-c\sqrt{2}, \pi/4)$
    (e) $(c, \pi/4)$

14. Through what angle must the axes be rotated to eliminate the xy term from the equation $x^2 + y^2 + xy - 1 = 0$?

    (a) $\pi/4$ (b) $\pi/6$ (c) $\pi/3$ (d) $\tan^{-1}(-1)$ (e) $\tan^{-1} 0$

coordinate geometry

15. The ellipse whose equation is $(x^2/16) + (y^2/25) = 1$ has major axis of length
    (a) 4   (b) 5   (c) 8   (d) 10   (e) 16

16. The hyperbola whose equation is $(x^2/16) - (y^2/25) = 1$ has transverse axis of length
    (a) 4   (b) 5   (c) 8   (d) 10   (e) 25

17. The asymptotes of the hyperbola in Problem 16 have equations
    (a) $y = \pm\dfrac{5}{4}x$ (b) $y = \pm\dfrac{4}{5}x$ (c) $y = \pm\dfrac{25}{16}x$ (d) $y = \pm\dfrac{16}{25}x$
    (e) $25x = 16y$

18. The parabola $x^2 = -4y$ opens
    (a) upward   (b) downward   (c) to the right   (d) to the left
    (e) this isn't a parabola

19. The graph of $[(x - 1)^2/16] - [(y + 2)^2/9] = 0$ is
    (a) a hyperbola whose transverse axis is parallel to the x axis
    (b) a hyperbola whose transverse axis is parallel to the y axis
    (c) an ellipse whose major axis is parallel to the y axis
    (d) an ellipse whose major axis is parallel to the x axis
    (e) two lines

20. The ellipse $[(x - 1)^2/25] + [(y + 3)^2/9] = 1$ has its center at
    (a) $(1, 3)$ (b) $(1, -3)$ (c) $(4, 3)$ (d) $(16, 9)$ (e) $(-1, 3)$

21. A focus of the ellipse in Problem 20 is located at
    (a) $(1, 3)$ (b) $(1, 1)$ (c) $(4, 3)$ (d) $(5, -3)$ (e) $(-5, 3)$

22. The equation of the line passing through $P:(1, 1, 1)$ and $Q:(0, 2, 0)$ is
    (a) $x + y + z = 3$ (b) $x + y = 1$ (c) $x - 1 = 1 - y = z - 1$
    (d) $x = y/2 = z$ (e) $x - y - z = 1$

23. The distance from P to Q, in the previous problem, is
    (a) 1   (b) $\sqrt{3}$ (c) 3   (d) $\sqrt{5}$ (e) 5

24. The direction cosines of the normal to the plane $x + y + z = 5$ are
    (a) 1, 1, and 1   (b) 1/5, 1/5, and 1/5   (c) $1/\sqrt{3}, 1/\sqrt{3}$, and $1/\sqrt{3}$
    (d) α, β, and γ (e) $-5/\sqrt{3}, -5/\sqrt{3}$, and $-5/\sqrt{3}$

25. The graph of the polar equation $r = 3 \sin 4\theta$ is
    (a) a circle   (b) a cardioid   (c) a rose   (d) a line   (e) a spiral

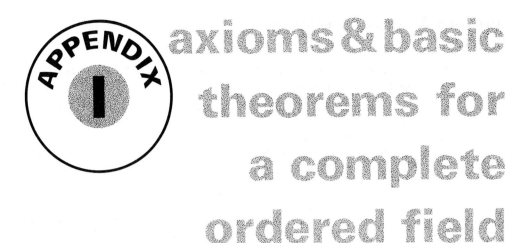

axioms & basic theorems for a complete ordered field

There are several ways to characterize the real numbers in rigorous mathematical terms. A rather natural method is to follow the line of their cultural development by beginning with the natural (counting) numbers. In this method, the natural numbers are taken as primitive concepts. Some axioms are stated, theorems proved, and the other number sets built up. A detailed account of this procedure can be found in the now classic book, *Foundations of Analysis*, by E. Landau (New York, Chelsea Publishing Company, 1951).

In this section, we shall adopt the approach of most present-day intermediate algebra texts and characterize the real numbers by axioms for a complete ordered field. We shall omit most of the details and discussion. One can pursue this further in the references listed at the end.

A.1. the field axioms

We begin with the set R of at least two elements called *real numbers* on which two binary operations (functions from $R \times R$ to R) are defined. These operations, **addition** and **multiplication**, are denoted by $+$ and juxtaposition or $\cdot$, respectively. That is, for any pair of elements a, b in R, $a + b$ and $ab = a \cdot b$, both belong to R. Additionally, these operations are assumed to have the following properties for each a, b, and c in R.

1. *Commutative laws:*

$$a + b = b + a$$
$$ab = ba$$

2. *Associative laws:*

$$a + (b + c) = (a + b) + c$$
$$a(bc) = (ab)c$$

3. *Distributive law:*

$$a(b + c) = ab + ac$$

axioms & basic theorems

4. *Identity elements: There exist real numbers 0, and 1, 1 ≠ 0, in R such that, for each a in R,*

$$a + 0 = a, \text{ and}$$

$$1a = a$$

5. *Inverse elements: For each a in R, there exists an element −a* (called the **additive inverse** of a) *such that*

$$a + (-a) = 0$$

Furthermore, for each nonzero real number a ∈ R there exists a real number a^{-1} (called the **multiplicative inverse** of a) *such that*

$$aa^{-1} = 1$$

The identity elements of Axiom 4 are called the **additive** and **multiplicative identities**, respectively.

Frequently, the statement that addition and multiplication are defined on R is formally stated as a *closure axiom*. The system $(R, +, \cdot)$ which satisfies these axioms is called a **field**.

We turn now to examine some easy implications of these axioms. These are properties possessed by every field. We are primarily concerned when R is the real numbers, or when it is the set of complex numbers.

A.2. basic theorem

The following basic theorems are a consequence of the axioms given for a field. Many of the proofs are left as exercises.

theorem A.1

For each $a, b, c \in R$, the right distributive law $(a + b)c = ac + bc$ also holds.

This theorem easily follows from Axioms 1 and 3.

The next two theorems follow mainly from Axiom 1.

theorem A.2

For each $a \in R$, $0 + a = a$, and $(-a) + a = 0$.

theorem A.3

For each $a \in R$, $a1 = a$, and $a^{-1}a = 1$.

theorem A.4

The additive and multiplicative identity elements of R are unique.

proof. We prove 0 is unique. Suppose there are *two* additive identities 0, and $0'$. Then for each $a \in R$ $a + 0 = 0 + a = a$ (Why?) and $a + 0' + 0' + a = a$. This is also true when $a = 0'$ in the first statement, and when $a = 0$ in the second; i.e., $0' + 0 = 0 + 0' = 0'$, and $0 + 0' = 0' + 0 = 0$. Thus, $0 = 0'$.

 The proof that $1 = 1'$ is left for you as an exercise.
These unique identities are generally called **zero** and **one**.

basic theorem

theorem A.5

For each $a \in R$ and each $b \neq 0$ in $R(-a)$ and b^{-1} are unique.

proof. Exercise.

theorem A.6 additive cancellation

If $a, b, c \in R$ and $a + b = c + b$, then $a = c$.

proof. Given $a + b = c + b$. Justify each of the following steps.

$$(a + b) + (-b) = (c + b) + (-b)$$
$$a + (b + (-b)) = c + (b + (-b))$$
$$a + 0 = c + 0$$
$$a = c$$

theorem A.7

For each $a \in R$, $0 \cdot a = a \cdot 0 = 0$.

proof. Since $0 + 0 = 0$, we have $(0 + 0)a = 0 \cdot a$, or $0 \cdot a + 0 \cdot a = 0 \cdot a$. Thus, $0 \cdot a + 0 \cdot a = 0 \cdot a + 0$, or, by using Theorem A.6, $0 \cdot a = 0$.

theorem A.8

For each $a, b \in R$, $a(-b) = -ab$.

proof. You should justify each of the following steps.

$$a(-b) + ab = a(-b + b)$$
$$= a \cdot 0$$
$$= 0$$

Thus, $a(-b)$ is the unique additive inverse of ab; viz., $-ab$.

corollary

For each $a, b \in R$, $-ab = (-a)b = a(-b)$.

theorem A.9

For each $a \in R$, $-(-a) = a$, and if $a \neq 0$, $(a^{-1})^{-1} = a$.

proof. Exercise.

theorem A.10 multiplicative cancellation

If $a, b, c \in R$, with $c \neq 0$, and if $ac = bc$, then $a = b$.

proof. Exercise.

We now define **subtraction** and **division** in R by

$$a - b = a + (-b), \text{ and}$$

$$\frac{a}{b} = a \cdot b^{-1} \text{ (when } b \neq 0)$$

axioms & basic theorems

theorem A.11

For each a, b and $c \in R$, $a(b - c) = ab - ac$.

proof. Exercise.

theorem A.12

If a, $b \in R$ with $b \neq 0$, then

(i) $b\left(\dfrac{a}{b}\right) = a$

(ii) $\dfrac{b}{b} = 1$

(iii) $\dfrac{0}{b} = 0$

(iv) $\dfrac{a}{1} = a$

proof. Exercise.

theorem A.13

For each a, $b \in R$,

(i) $-(a + b) = -a - b$
(ii) $-(a - b) = -a + b$
(iii) $(-1)b = -b$
(iv) $(-a)(-b) = ab$

proof. Exercise.

theorem A.14

For each a, $b \in R$, with $b \neq 0$, $\dfrac{a}{-b} = -\dfrac{a}{b} = \dfrac{-a}{b}$.

proof. Exercise.

theorem A.15

For each a, b, c, $d \in R$, with $a \neq 0$ and $b \neq 0$,

(i) $\dfrac{1}{a} \cdot \dfrac{1}{b} = \dfrac{1}{ab}$

(ii) $\dfrac{c}{a} \cdot \dfrac{d}{b} = \dfrac{cd}{ab}$

(iii) $\dfrac{ac}{ab} = \dfrac{c}{b}$

378

proof. Exercise.

ordered fields

theorem A.16

For $a, b, c, d \in R$, with $b \neq 0$ and $d \neq 0$,

(i) $\dfrac{a}{b} + \dfrac{c}{d} = \dfrac{ad + bc}{bd}$

(ii) $\dfrac{a/b}{c/d} = \dfrac{a}{b} \cdot \dfrac{d}{c}$, provided $c \neq 0$

proof. Exercise.

theorem A.17

For each $a, b \in R$,

(i) $(a + b)^2 = a^2 + 2ab + b^2$
(ii) $(a + b)(a - b) = a^2 - b^2$
(iii) $(a - b)^2 = a^2 - 2ab + b^2$
(iv) $(a - b)(a^2 + ab + b^2) = a^3 - b^3$
(v) $(a + b)(a^2 - ab + b^2) = a^3 + b^3$

proof. Exercise. Use the distributive laws and the previous theorems.

theorem A.18

For $a, b \in R$, we have $ab = 0$ if and only if $a = 0$ or $b = 0$.

proof. If either a or b is zero, then $ab = 0$, by Theorem A.7. Conversely, suppose $ab = 0$ and $b \neq 0$. Then $ab = 0 \cdot b = 0$. Hence by Theorem A.10, $a = 0$.

A.3. ordered fields

We now define an order relation on the field R of real numbers. To do this, we first distinguish a subset $R_p \subset R$ called the set of **positive elements** of R. Our set R_p must have the following properties *(Axioms)*:

6. If $a, b \in R_p$, then $a + b$ and ab belong to R_p *(closure under addition and multiplication)*.

7. *The tricotomy law:* For each $a \in R_p$ exactly one of the following is true.

$$a \in R_p, \qquad a = 0, \qquad \text{or} \ -a \in R_p$$

The **order relation** $<$ *is then defined as follows on our field* R: For $a, b \in R$,

$$a < b \text{ provided } b - a \in R_p$$

That is, a is *less than* b, when $b - a$ is a positive real number.

theorem A.19

If $a, b \in R$ then exactly one of the following is true.

(i) $a < b$
(ii) $a = b$
(iii) $b < a$

axioms & basic theorems

proof. Use the tricotomy law.

We define an associated **relation,** $>$, on R by

$$a > b \text{ if, and only if, } b < a$$

corollary

Exactly one of the following is true, for each

  (i) $b > a$
  (ii) $b = a$
  (iii) $b < a$

We now show how these relations combine with the operations of addition and multiplication in R.

theorem A.20

For $a, b, c \in R$, the following are true.

  (i)   *If $a < b$, then $a + c < b + c$.*
  (ii) *If $a < b$, and $0 < c$, then $ac < bc$.*
  (iii) *If $a < b$, and $c < 0$, then $ac > bc$.*
  (iv) *If $a < b$ and $b < c$, then $a < c$.*

proof of (iii).   Given that $a < b$ and $c < 0$, we have $b - a \in R_p$ and $0 - c = -c \in R_p$. Thus $(b - a)(-c) \in R_p$, but $(b - a)(-c) = -bc + ac$, since $ac - bc \in R_p$ $bc < ac$, or $ac > bc$.

The set of real numbers satisfies the properties of an ordered field as does also the set of rational numbers. You are invited to verify that the set of complex numbers does not. It fails to satisfy the following theorem.

theorem A.21

For each nonzero $a \in R$, $a^2 > 0$.

proof. If $a > 0$ then $a \in R_p$ so $a^2 = (a \cdot a) \in R_p$. If $a < 0$, then $-a \in R_p$; hence $(-a)(-a) \in R_p$, but $(-a)(-a) = a \cdot a = a^2$ by Theorem A.13 (iv).

The real numbers, as distinguished from the ordered field Q of rational numbers, have an additional property called **completeness.** We shall state the *completeness axiom* formally after we introduce the concept of upper and lower bounds for a set. Informally, it states that there are no holes in the real line. For example, the sequence of rational numbers 1, 1.4, 1.41, 1.414 . . . actually is approaching a number; viz., $\sqrt{2}$, but $\sqrt{2}$ is not rational.

A subset S of R is said to be **bounded above** *if there exists $b \in R$ such that $s \leq b$ for all $s \in S$. A set S is* **bounded below** *if there exists $c \in R$ such that $c \leq s$ for all $s \in S$. The number b is called an* **upper bound** *for S, and c is called a* **lower bound** *for S.*

The set of non-negative integers has no upper bound, but 0, -1, -2, etc., are all lower bounds. The set of all positive integers whose square is less than 2 has 3 as an upper bound, also 1.5, etc. The number 0 is a lower bound for this set.

ordered fields

A number u is called the **least upper bound** *(l.u.b.) for a set $S \subset R$ provided (1) u is an upper bound and (2) for any upper bound b of S, $u \leq b$. A* **greatest lower bound** *for S is defined similarly.*

The set Q of rational numbers is an ordered field yet the set of those positive numbers whose square is less than 2 has upper bounds which are rational, but *no rational least upper bound.* Considered as real numbers, however, the number $\sqrt{2}$ is the least upper bound for this set.

The real numbers form a complete ordered field in that in addition to the axioms for an ordered field they also satisfy:

8. **The completeness axiom:** *Every nonempty subset S or R which has an upper bound, has a least upper bound.*

exercises

1. Prove each of the following theorems.

(a)	Theorem A.1	(f)	Theorem A.10	(k)	Theorem A.15
(b)	Theorem A.2	(g)	Theorem A.11	(l)	Theorem A.16
(c)	Theorem A.3	(h)	Theorem A.12	(m)	Theorem A.17
(d)	Theorem A.5	(i)	Theorem A.13	(n)	Theorem A.19
(e)	Theorem A.9	(j)	Theorem A.14	(o)	Theorem A.20

2. Verify that the rational numbers Q form an ordered field.

3. Prove that the complex numbers cannot be ordered so as to form an ordered field. (Consider *i*.)

4. Give some examples of unbounded sets of real numbers.

5. Give some examples of bounded sets of real numbers and find the l.u.b. for each.

6. Define the greatest lower bound (g.l.b.) for a set S of real numbers.

7. Show that every nonempty set of real numbers which has a lower bound has a g.l.b.

8. Give some examples of bounded sets of real numbers, and find the g.l.b. for each.

references

Willerding, M. F. and Hoffman, S., *College Algebra*, New York: John Wiley & Sons, Inc., 1971.

McCoy, Neal H., *Introduction to Modern Algebra*, Boston: Allyn & Bacon, 1972.

Peterson, John M., *Foundations of Algebra & Number Theory*, Chicago: Markham, 1971.

greek alphabet

A	α	alpha	N	ν	nu
B	β	beta	Ξ	ξ	xi
Γ	γ	gamma	O	o	omicron
Δ	δ	delta	Π	π	pi
E	ε	epsilon	P	ρ	rho
Z	ζ	zeta	Σ	σ	sigma
H	η	eta	T	τ	tau
Θ	θ	theta	Υ	υ	upsilon
I	ι	iota	Φ	ϕ	phi
K	κ	kappa	X	χ	chi
Λ	λ	lambda	Ψ	ψ	psi
M	μ	mu	Ω	ω	omega

table 1
Power and Roots

Number	Square	Square Root	Cube	Cube Root	Number	Square	Square Root	Cube	Cube Root
1	1	1.000	1	1.000	51	2,601	7.141	132,651	3.708
2	4	1.414	8	1.260	52	2.704	7.211	140,608	3.733
3	9	1.732	27	1.442	53	2,809	7.280	148,877	3.756
4	16	2.000	64	1.587	54	2,916	7.348	157,464	3.780
5	25	2.236	125	1.710	55	3.025	7.416	166,375	3.803
6	36	2.449	216	1.817	56	3,136	7.483	175,616	3.826
7	49	2.464	343	1.913	57	3,249	7.550	185,193	3.849
8	64	2.828	512	2.000	58	3,364	7.616	195,112	3.871
9	81	3.000	729	2.080	59	3,481	7.681	205,379	3.893
10	100	3.162	1,000	2.154	60	3,600	7.746	216,000	3.915
11	121	3.317	1,331	2.224	61	3,721	7.810	226,981	3.936
12	144	3.464	1,728	2.289	62	3,844	7.874	238,328	3.958
13	169	3.606	2,197	2.351	63	3,969	7.937	250,047	3.979
14	196	3.742	2,744	2.410	64	4,096	8.000	262,144	4.000
15	225	3.873	3,375	2.466	65	4,225	8.062	274,625	4.021
16	256	4.000	4,096	2.520	66	4,356	8.124	287,496	4.041
17	289	4.123	4,913	2.571	67	4,489	8.185	300,763	4.062
18	324	4.243	5,832	2.621	68	4,624	8.246	314,432	4.082
19	361	4.359	6,859	2.668	69	4,761	8.307	328,509	4.102
20	400	4.472	8,000	2.714	70	4,900	8.367	343,000	4.121
21	441	4.583	9,261	2.759	71	5,041	8.426	357,911	4.141
22	484	4.690	10,648	2.802	72	5,184	8.485	373,248	4.160
23	529	4.796	12,167	2.844	73	5,329	8.544	389,017	4.179
24	576	4.899	13,824	2.884	74	5,476	8.602	405,224	4.198
25	625	5.000	15,625	2.924	75	5,265	8.660	421,875	4.217
26	676	5.099	17,576	2.962	76	5,776	8.718	438,976	4.236
27	729	5.196	19,638	3.000	77	5,929	8.775	456,533	4.254
28	784	5.292	21,592	3.037	78	6,084	8.832	474,552	4.273
29	841	5.385	24,389	3.072	79	6,241	8.888	493,039	4.291
30	900	5.477	27,000	3.107	80	6,400	8.944	512,000	4.309
31	961	5.568	29,791	3.141	81	6,561	9.000	531,441	4.327
32	1,024	5.657	32,768	3.175	82	6,724	9.055	551,368	4.344
33	1,089	5.745	35,937	3.208	83	6,889	9.110	571,787	4.362
34	1,156	5.831	39,304	3.240	84	7,056	9.165	592,704	4.380
35	1,225	5.916	42,875	3.271	85	7,225	9.220	614,125	4.397
36	1,296	6.000	46,656	3.302	86	7,396	9.274	636,056	4.414
37	1,369	6.083	50,653	3.332	87	7,569	9.327	658,503	4.431
38	1,444	6.164	54,872	3.362	88	7,744	9.381	681,472	4.448
39	1,521	6.245	59,319	3.391	89	7,921	9.434	704,969	4.465
40	1,600	6.325	64,000	3.420	90	8,100	9.487	729,000	4.481
41	1,681	6.403	68,921	3.448	91	8,281	9.539	753,571	4.498
42	1,764	6.481	74,088	3.476	92	8,464	9.592	778,688	4.514
43	1,849	6.557	79,507	3.503	93	8,649	9.644	804,357	4.531
44	1,936	6.633	85,184	3.530	94	8,836	9.695	830,584	4.547
45	2,025	6.708	91,125	3.557	95	9,025	9.747	857,375	4.563
46	2,116	6.782	97,336	3.583	96	9,216	9.798	884,736	4.579
47	2,209	6.856	103,823	3.609	97	9,409	9.849	912,673	4.595
48	2,304	6.928	110,592	3.634	98	9,604	9.899	941,192	4.610
49	2,401	7.000	117,649	3.659	99	9,801	9.950	970,299	4.626
50	2,500	7.071	125,000	3.684	100	10,000	10.000	1,000,000	4.642

table 2

Exponential Functions e^t and e^{-t} ($t = 0.00$ through 10.00)

t	e^t	e^{-t}	t	e^t	e^{-t}
0.00	1.0000	1.0000	2.50	12.182	0.0821
0.05	1.0513	0.9512	2.55	12.807	0.0781
0.10	1.1052	0.9048	2.60	13.464	0.0743
0.15	1.1618	0.8607	2.65	14.154	0.0707
0.20	1.2214	0.8187	2.70	14.880	0.0672
0.25	1.2840	0.7788	2.75	15.643	0.0639
0.30	1.3499	0.7408	2.80	16.445	0.0608
0.35	1.4191	0.7047	2.85	17.288	0.0578
0.40	1.4918	0.6703	2.90	18.174	0.0550
0.45	1.5683	0.6376	2.95	19.106	0.0523
0.50	1.6487	0.6065			
0.55	1.7333	0.5769	3.00	20.086	0.0498
0.60	1.8221	0.5488	3.05	21.115	0.0474
0.65	1.9155	0.5220	3.10	22.198	0.0450
0.70	2.0138	0.4966	3.15	23.336	0.0429
0.75	2.1170	0.4724	3.20	24.533	0.0408
0.80	2.2255	0.4493	3.25	25.790	0.0388
0.85	2.3396	0.4274	3.30	27.113	0.0369
0.90	2.4596	0.4066	3.35	28.503	0.0351
0.95	2.5857	0.3867	3.40	29.964	0.0334
			3.45	31.500	0.0317
1.00	2.7183	0.3679	3.50	33.115	0.0302
1.05	2.8577	0.3499	3.55	34.813	0.0287
1.10	3.0042	0.3329	3.60	36.598	0.0273
1.15	3.1582	0.3166	3.65	38.475	0.0260
1.20	3.3201	0.3012	3.70	40.447	0.0247
1.25	3.4903	0.2865	3.75	42.521	0.0235
1.30	3.6693	0.2725	3.80	44.701	0.0224
1.35	3.8574	0.2592	3.85	46.993	0.0213
1.40	4.0552	0.2466	3.90	49.402	0.0202
1.45	4.2631	0.2346	3.95	51.935	0.0193
1.50	4.4817	0.2231			
1.55	4.7115	0.2122	4.00	54.598	0.0183
1.60	4.9530	0.2019	4.10	60.340	0.0166
1.65	5.2070	0.1920	4.20	66.686	0.0150
1.70	5.4739	0.1827	4.30	73.700	0.0136
1.75	5.7546	0.1738	4.40	81.451	0.0123
1.80	6.0496	0.1653	4.50	90.017	0.0111
1.85	6.3598	0.1572	4.60	99.484	0.0101
1.90	6.6859	0.1496	4.70	109.95	0.0091
1.95	7.0287	0.1423	4.80	121.51	0.0082
			4.90	134.29	0.0074
2.00	7.3891	0.1353	5.00	148.41	0.0067
2.05	7.7679	0.1287	5.20	181.27	0.0055
2.10	8.1662	0.1225	5.40	221.41	0.0045
2.15	8.5849	0.1165	5.60	270.43	0.0037
2.20	9.0250	0.1108	5.80	330.30	0.0030
2.25	9.4877	0.1054	6.00	403.43	0.0025
2.30	9.9742	0.1003	7.00	1096.6	0.0009
2.35	10.486	0.0954	8.00	2981.0	0.0003
2.40	11.023	0.0907	9.00	8103.1	0.0001
2.45	11.588	0.0863	10.00	22026.0	0.00005

table 3

Four-Place Logarithms of Numbers from 1 to 10

$P(t) = A e^{kt}$

n	0	1	2	3	4	5	6	7	8	9
1.0	+0.0000	0043	0086	0128	0170	0212	0253	0294	0334	0374
1.1	.0414	0453	0492	0531	0569	0607	0645	0682	0719	0755
1.2	.0792	0828	0864	0899	0934	0969	1004	1038	1072	1106
1.3	.1139	1173	1206	1239	1271	1303	1335	1367	1399	1430
1.4	.1461	1492	1523	1553	1584	1614	1644	1673	1703	1732
1.5	.1761	1790	1818	1847	1875	1903	1931	1959	1987	2014
1.6	.2041	2068	2095	2122	2148	2175	2201	2227	2253	2279
1.7	.2034	2330	2355	2380	2405	2430	2455	2480	2504	2529
1.8	.2553	2577	2601	2625	2648	2672	2695	2718	2742	2765
1.9	.2788	2810	2833	2856	2878	2900	2923	2945	2967	2989
2.0	.3010	3032	3054	3075	3096	3118	3139	3160	3181	3201
2.1	.3222	3243	3263	3284	3304	3324	3345	3365	3385	3404
2.2	.3424	3444	3464	3483	3502	3522	3541	3560	3579	3598
2.3	.3617	3636	3655	3674	3692	3711	3729	3747	3766	3784
2.4	.3802	3820	3838	3856	3874	3892	3909	3927	3945	3962
2.5	.3979	3997	4014	4031	4048	4065	4082	4099	4116	4133
2.6	.4150	4166	4183	4200	4216	4232	4249	4265	4281	4298
2.7	.4314	4330	4346	4362	4378	4393	4409	4425	4440	4456
2.8	.4472	4487	4502	4518	4533	4548	4564	4579	4594	4609
2.9	.4624	4639	4654	4669	4983	4698	4713	4728	4742	4757
3.0	.4771	4786	4800	4814	4829	4843	4857	4871	4886	4900
3.1	.4914	4928	4942	4955	4969	4983	4997	5011	5024	5038
3.2	.5051	5065	5079	5092	5105	5119	5132	5145	5159	5172
3.3	.5185	5198	5211	5224	5237	5250	5263	5276	5289	5302
3.4	.5315	5328	5340	5353	5366	5378	5391	5403	5416	5428
3.5	.5441	5453	5465	5478	5490	5502	5514	5527	5539	5551
3.6	.5563	5575	5587	5599	5611	5623	5635	5647	5658	5670
3.7	.5682	5694	5705	5717	5729	5740	5752	5763	5775	5786
3.8	.5798	5809	5821	5832	5843	5855	5866	5877	5888	5899
3.9	.5911	5922	5933	5944	5955	5966	5977	5988	5999	6010

4

$1950 - 30,000$

$1960 - 70,000$

$A = 30,000$

$70,000 = 30,000\, e^{(k)(10)}$

$7/3 = e^{10k}$

$\ln 2.33 = 10k$

$.8459 = 10k$

$k = .08459$

$1970 = 30,000\, e^{(.08459)(20)}$

$1980 = 30,000\, e^{(.08459)(30)}$

$12,000 - 1964$

$24,000 - 2044$

$12,000 = (12,000)\, e^{.09}$

$24,000 = (12,000)\, e^{14}$

$2 = e^9 \quad \ln 2 = .6981$

$9 = .6981$

$V_t = (N_0)\, e^{qt}$

$\dfrac{20}{80}\, \dot{} \dfrac{.25}{gm}$

$N_{dyn} = (12,000)\, e^{(.25)(.6981)}$

$= (12,000)(1.189)\, 14700$

table 3 (continued)

n	0	1	2	3	4	5	6	7	8	9
4.0	.6021	6031	6042	6053	6064	6075	6085	6096	6107	6117
4.1	.6128	6138	6149	6160	6170	6180	6191	6201	6212	6222
4.2	.6232	6243	6253	6263	6274	6284	6294	6304	6314	6325
4.3	.6335	6345	6355	6365	6375	6385	6395	6405	6415	6425
4.4	.6435	6444	6454	6464	6474	6484	6493	6503	6513	6522
4.5	.6532	6542	6551	6561	6571	6580	6590	6599	6609	6618
4.6	.6628	6637	6646	6656	6665	6675	6684	6693	6702	6712
4.7	.6721	6730	6739	6749	6758	6767	6776	6785	6794	6803
4.8	.6812	6821	6830	6839	6848	6857	6866	6875	6884	6893
4.9	.6902	6911	6920	6928	6937	6946	6955	6964	6972	6981
5.0	+.6990	6998	7007	7016	7024	7033	7042	7050	7059	7067
5.1	.7076	7084	7093	7101	7110	7118	7126	7135	7143	7152
5.2	.7160	7168	7177	7185	7193	7202	7210	7218	7226	7235
5.3	.7243	7251	7259	7267	7275	7284	7292	7300	7308	7316
5.4	.7324	7332	7340	7348	7356	7364	7372	7380	7388	7396
5.5	.7404	7412	7419	7427	7435	7443	7451	7459	7466	7474
5.6	.7482	7490	7497	7505	7513	7520	7528	7536	7543	7551
5.7	.7559	7566	7574	7582	7589	7597	7604	7612	7619	7627
5.8	.7634	7642	7649	7657	7664	7672	7679	7686	7694	7701
5.9	.7709	7716	7723	7731	7738	7745	7752	7760	7767	7774
6.0	.7782	7789	7796	7803	7810	7818	7825	7832	7839	7846
6.1	.7853	7860	7868	7875	7882	7889	7896	7903	7910	7917
6.2	.7924	7931	7938	7945	7952	7959	7966	7973	7980	7987
6.3	.7993	8000	8007	8014	8021	8028	8035	8041	8048	8055
6.4	.8062	8069	8075	8082	8089	8096	8102	8109	8116	8122
6.5	.8129	8136	8142	8149	8156	8162	8169	8176	8182	8189
6.6	.8195	8202	8209	8215	8222	8228	8235	8241	8248	8254
6.7	.8261	8267	8274	8280	8287	2893	8299	8306	8312	8319
6.8	.8325	8331	8338	8344	8351	8357	8363	8370	8376	8382
6.9	.8388	8395	8401	8407	8414	8420	8426	8432	8439	8445

$\log_2 (x+2) + \log_2 (x) = 3$

$\log_2 [(x+2)(x)] = 3$

$2^3 = (x+2)(x)$

$2^3 = x^2 + 2x$

$8 = x^2 + 2x$

$x^2 + 2x - 8 = 0$

$x = -4$

$x = 2$

table 3 (*continued*)

n	0	1	2	3	4	5	6	7	8	9
7.0	.8451	8457	8463	8470	8476	8482	8488	8494	8500	8506
7.1	.8513	8519	8525	8531	8537	8543	8549	8555	8561	8567
7.2	.8573	8579	8585	8591	8597	8603	8609	8615	8621	8627
7.3	.8633	8639	8645	8651	8657	8663	8669	8675	8681	8686
7.4	.8692	8698	8704	8710	8716	8722	8727	8733	8739	8745
7.5	.8751	8756	8762	8768	8774	8779	8785	8791	8797	8802
7.6	.8808	8814	8820	8825	8861	8837	8842	8848	8854	8859
7.7	.8865	8871	8876	8882	8887	8893	8899	8904	8910	8915
7.8	.8921	8927	8932	8938	8943	8949	8954	8960	8965	8971
7.9	.8976	8982	8987	8993	8998	9004	9009	9015	9020	9025
8.0	.9031	9036	9042	9047	9053	9058	9063	9069	9074	9079
8.1	.9085	9090	9096	9101	9106	9112	9117	9122	9128	9133
8.2	.9138	9143	9149	9154	9159	9165	9170	9175	9180	9186
8.3	.9191	9196	9201	9206	9212	9217	9222	9227	9232	9238
8.4	.9243	9248	9253	9258	9263	9269	9274	9279	9284	9289
8.5	.9294	9299	9304	9309	9315	9320	9325	9330	9335	9340
8.6	.9345	9350	9355	9360	9365	9370	9375	9380	9385	9390
8.7	.9395	9400	9405	9410	9415	9420	9425	9430	9435	9440
8.8	.9445	9450	9455	9460	9465	9469	9474	9479	9484	9489
8.9	.9494	9499	9504	9509	9513	9518	9523	9528	9533	9538
9.0	.9542	9547	9552	9557	9562	9566	9571	9576	9581	9586
9.1	.9590	9595	9600	9605	9609	9614	9619	9624	9628	9633
9.2	.9638	9643	9647	9652	9657	9661	9666	9671	9675	9680
9.3	.9685	9689	9694	9699	9703	9708	9713	9717	9722	9727
9.4	.9731	9736	9741	9745	9750	9754	9759	9763	9768	9773
9.5	.9777	9782	9786	9791	9795	9800	9805	9809	9814	9818
9.6	.9823	9827	9832	9836	9841	9845	9850	9854	9859	9863
9.7	.9868	9872	9877	9881	9886	9890	9894	9899	9903	9908
9.8	.9912	9917	9921	9926	9930	9934	9939	9943	9948	9952
9.9	.9956	9961	9965	9969	9974	9978	9983	9987	9991	9996

table 4
Natural Logarithms $\log_l t$*

t	0.00	0.01	0.02	0.03	0.04	0.05	0.06	0.07	0.08	0.09
1.0	0.0000	0.0100	0.0198	0.0296	0.0392	0.0488	0.0583	0.0677	0.0770	0.0862
1.1	0.0953	0.1044	0.1133	0.1222	0.1310	0.1398	0.1484	0.1570	0.1655	0.1740
1.2	0.1823	0.1906	0.1989	0.2070	0.2151	0.2231	0.2311	0.2390	0.2469	0.2546
1.3	0.2624	0.2700	0.2776	0.2852	0.2927	0.3001	0.3075	0.3148	0.3221	0.3293
1.4	0.3365	0.3436	0.3507	0.3577	0.3646	0.3716	0.3784	0.3853	0.3920	0.3988
1.5	0.4055	0.4121	0.4187	0.4253	0.4318	0.4383	0.4447	0.4511	0.4574	0.4637
1.6	0.4700	0.4762	0.4824	0.4886	0.4947	0.5008	0.5068	0.5128	0.5188	0.5247
1.7	0.5306	0.5365	0.5423	0.5481	0.5539	0.5596	0.5653	0.5710	0.5766	0.5822
1.8	0.5878	0.5933	0.5988	0.6043	0.6098	0.6152	0.6206	0.6259	0.6313	0.6366
1.9	0.6419	0.6471	0.6523	0.6575	0.6627	0.6678	0.6729	0.6780	0.6831	0.6881
2.0	0.6931	0.6981	0.7031	0.7080	0.7130	0.7178	0.7227	0.7275	0.7324	0.7372
2.1	0.7419	0.7467	0.7514	0.7561	0.7608	0.7655	0.7701	0.7747	0.7793	0.7839
2.2	0.7885	0.7930	0.7975	0.8020	0.8065	0.8109	0.8154	0.8198	0.8242	0.8286
2.3	0.8329	0.8372	0.8416	0.8459	0.8502	0.8544	0.8587	0.8629	0.8671	0.8713
2.4	0.8755	0.8796	0.8838	0.8879	0.8920	0.8961	0.9002	0.9042	0.9083	0.9123
2.5	0.9163	0.9203	0.9243	0.9282	0.9322	0.9361	0.9400	0.9439	0.9478	0.9517
2.6	0.9555	0.9594	0.9632	0.9670	0.9708	0.9746	0.9783	0.9821	0.9858	0.9895
2.7	0.9933	0.9969	1.0006	1.0043	1.0080	1.0116	1.0152	1.0188	1.0225	1.0260
2.8	1.0296	1.0332	1.0367	1.0403	1.0438	1.0473	1.0508	1.0543	1.0578	1.0613
2.9	1.0647	1.0682	1.0716	1.0750	1.0784	1.0818	1.0852	1.0886	1.0919	1.0953
3.0	1.0986	1.1019	1.1053	1.1086	1.1119	1.1151	1.1184	1.1217	1.1249	1.1282
3.1	1.1314	1.1346	1.1378	1.1410	1.1442	1.1474	1.1506	1.1537	1.1569	1.1600
3.2	1.1632	1.1663	1.1694	1.1725	1.1756	1.1787	1.1817	1.1848	1.1878	1.1909
3.3	1.1939	1.1970	1.2000	1.2030	1.2060	1.2090	1.2119	1.2149	1.2179	1.2208
3.4	1.2238	1.2267	1.2296	1.2326	1.2355	1.2384	1.2413	1.2442	1.2470	1.2499
3.5	1.2528	1.2556	1.2585	1.2613	1.2641	1.2669	1.2698	1.2726	1.2754	1.2782
3.6	1.2809	1.2837	1.2865	1.2892	1.2920	1.2947	1.2975	1.3002	1.3029	1.3056
3.7	1.3083	1.3110	1.3137	1.3164	1.3191	1.3218	1.3244	1.3271	1.3297	1.3324
3.8	1.3350	1.3376	1.3403	1.3429	1.3455	1.3481	1.3507	1.3533	1.3558	1.3584
3.9	1.3610	1.3635	1.3661	1.3686	1.3712	1.3737	1.3762	1.3788	1.3813	1.3838
4.0	1.3863	1.3888	1.3913	1.3938	1.3962	1.3987	1.4012	1.4036	1.4061	1.4085
4.1	1.4110	1.4134	1.4159	1.4183	1.4207	1.4231	1.4255	1.4279	1.4303	1.4327
4.2	1.4351	1.4375	1.4398	1.4422	1.4446	1.4469	1.4493	1.4516	1.4540	1.4563
4.3	1.4586	1.4609	1.4633	1.5656	1.4679	1.4702	1.4725	1.4748	1.4770	1.4793
4.4	1.4816	1.4839	1.4861	1.4884	1.4907	1.4929	1.4952	1.4974	1.4996	1.5019
4.5	1.5041	1.5063	1.5085	1.5107	1.5129	1.5151	1.5173	1.5195	1.5217	1.5239
4.6	1.5261	1.5282	1.5304	1.5326	1.5347	1.5369	1.5390	1.5412	1.5433	1.5454
4.7	1.5476	1.5497	1.5518	1.5539	1.5560	1.5581	1.5602	1.5623	1.5644	1.5665
4.8	1.5686	1.5707	1.5728	1.5748	1.5769	1.5790	1.5810	1.5831	1.5851	1.5872
4.9	1.5892	1.5913	1.5933	1.5953	1.5974	1.5994	1.6014	1.6034	1.6054	1.6074
5.0	1.6094	1.6114	1.6134	1.6154	1.6174	1.6194	1.6214	1.6233	1.6253	1.6273
5.1	1.6292	1.6312	1.6332	1.6351	1.6371	1.6390	1.6409	1.6429	1.6448	1.6467
5.2	1.6487	1.6506	1.6525	1.6544	1.6563	1.6582	1.6601	1.6620	1.6639	1.6658
5.3	1.6677	1.6696	1.6715	1.6734	1.6752	1.6771	1.6790	1.6808	1.6827	1.6845
5.4	1.6864	1.6882	1.6901	1.6919	1.6938	1.6956	1.6974	1.6993	1.7011	1.7029

table 4 (continued)

t	0.00	0.01	0.02	0.03	0.04	0.05	0.06	0.07	0.08	0.09
5.5	1.7047	1.7066	1.7084	1.7102	1.7120	1.7138	1.7156	1.7174	1.7192	1.7210
5.6	1.7228	1.7246	1.7263	1.7281	1.7299	1.7317	1.7334	1.7352	1.7370	1.7387
5.7	1.7405	1.7422	1.7440	1.7457	1.7475	1.7492	1.7509	1.7527	1.7544	1.7561
5.8	1.7579	1.7596	1.7613	1.7630	1.7647	1.7664	1.7682	1.7699	1.7716	1.7733
5.9	1.7750	1.7766	1.7783	1.7800	1.7817	1.7834	1.7851	1.7867	1.7884	1.7901
6.0	1.7918	1.7934	1.7951	1.7967	1.7984	1.8001	1.8017	1.8034	1.8050	1.8066
6.1	1.8083	1.8099	1.8116	1.8132	1.8148	1.8165	1.8181	1.8197	1.8213	1.8229
6.2	1.8245	1.8262	1.8278	1.8294	1.8310	1.8326	1.8342	1.8358	1.8374	1.8390
6.3	1.8406	1.8421	1.8437	1.8453	1.8469	1.8485	1.8500	1.8516	1.8532	1.8547
6.4	1.8563	1.8579	1.8594	1.8610	1.8625	1.8641	1.8656	1.8672	1.8687	1.8703
6.5	1.8718	1.8733	1.8749	1.8764	1.8779	1.8795	1.8810	1.8825	1.8840	1.8856
6.6	1.8871	1.8886	1.8901	1.8916	1.8931	1.8946	1.8961	1.8976	1.8991	1.9006
6.7	1.9021	1.9036	1.9051	1.9066	1.9081	1.9095	1.9110	1.9125	1.9140	1.9155
6.8	1.9169	1.9184	1.9199	1.9213	1.9228	1.9242	1.9257	1.9272	1.9286	1.9301
6.9	1.9315	1.9330	1.9344	1.9359	1.9373	1.9387	1.9402	1.9416	1.9430	1.9445
7.0	1.9459	1.9473	1.9488	1.9502	1.9516	1.9530	1.9544	1.9559	1.9573	1.9587
7.1	1.9601	1.9615	1.9629	1.9643	1.9657	1.9671	1.9685	1.9699	1.9713	1.9727
7.2	1.9741	1.9755	1.9769	1.9782	1.9796	1.9810	1.9824	1.9838	1.9851	1.9865
7.3	1.9879	1.9892	1.9906	1.9920	1.9933	1.9947	1.9961	1.9974	1.9988	2.0001
7.4	2.0015	2.0028	2.0042	2.0055	2.0069	2.0082	2.0096	2.0109	2.0122	2.0136
7.5	2.0149	2.0162	2.0176	2.0189	2.0202	2.0215	2.0229	2.0242	2.0255	2.0268
7.6	2.0282	2.0295	2.0308	2.0321	2.0334	2.0347	2.0360	2.0373	2.0386	2.0399
7.7	2.0412	2.0425	2.0438	2.0451	2.0464	2.0477	2.0490	2.0503	2.0516	2.0528
7.8	2.0541	2.0554	2.0567	2.0580	2.0592	2.0605	2.0618	2.0631	2.0643	2.0665
7.9	2.0669	2.0681	2.0694	2.0707	2.0719	2.0732	2.0744	2.0757	2.0769	2.0782
8.0	2.0794	2.0807	2.0819	2.0832	2.0844	2.0857	2.0869	2.0882	2.0894	2.0906
8.1	2.0919	2.0931	2.0943	2.0956	2.0968	2.0980	2.0992	2.1005	2.1017	2.1029
8.2	2.1041	2.1054	2.1066	2.1078	2.1090	2.1102	2.1114	2.1126	2.1138	2.1150
8.3	2.1163	2.1175	2.1187	2.1199	2.1211	2.1223	2.1235	2.1247	2.1258	2.1270
8.4	2.1282	2.1294	2.1306	2.1318	2.1330	2.1342	2.1353	2.1365	2.1377	2.1389
8.5	2.1401	2.1412	2.1424	2.1436	2.1448	2.1459	2.1471	2.1483	2.1494	2.1506
8.6	2.1518	2.1529	2.1541	2.1552	2.1564	2.1576	2.1587	2.1599	2.1610	2.1622
8.7	2.1633	2.1645	2.1656	2.1668	2.1679	2.1691	2.1702	2.1713	2.1725	2.1736
8.8	2.1748	2.1759	2.1770	2.1782	2.1793	2.1804	2.1815	2.1827	2.1838	2.1849
8.9	2.1861	2.1872	2.1883	2.1894	2.1905	2.1917	2.1928	2.1939	2.1950	2.1961
9.0	2.1972	2.1983	2.1994	2.2006	2.2017	2.2028	2.2039	2.2050	2.2061	2.2072
9.1	2.2083	2.2094	2.2105	2.2116	2.2127	2.2138	2.2148	2.2159	2.2170	2.2181
9.2	2.2192	2.2203	2.2214	2.2225	2.2235	2.2246	2.2257	2.2268	2.2279	2.2289
9.3	2.2300	2.2311	2.2322	2.2332	2.2343	2.2354	2.2364	2.2375	2.2386	2.2396
9.4	2.2407	2.2418	2.2428	2.2439	2.2450	2.2460	2.2471	2.2481	2.2492	2.2502
9.5	2.2513	2.2523	2.2534	2.2544	2.2555	2.2565	2.2576	2.2586	2.2597	2.2607
9.6	2.2618	2.2628	2.2638	2.2649	2.2659	2.2670	2.2680	2.2690	2.2701	2.2711
9.7	2.2721	2.2732	2.2742	2.2752	2.2762	2.2773	2.2783	2.2793	2.2803	2.2814
9.8	2.2824	2.2834	2.2844	2.2854	2.2865	2.2875	2.2885	2.2895	2.2905	2.2915
9.9	2.2925	2.2935	2.2946	2.2956	2.2966	2.2976	2.2986	2.2996	2.3006	2.3016

$*\log_{10} e \doteq 0.4342944819 \ldots$

table 5

Four-Place Values of Trigonometric Functions, Real Numbers θ or Angles α, in Radians and Degrees

Real Number θ or α Radians	α Degrees	sin α or sin θ	csc α or csc θ	tan α or tan θ	cot α or cot θ	sec α or sec θ	cos α or cos θ
0.00	0° 00′	0.0000	No value	0.0000	No value	1.000	1.000
.01	0° 34′	.0100	100.0	.0100	100.0	1.000	1.000
.02	1° 09′	.0200	50.00	.0200	49.99	1.000	0.9998
.03	1° 43′	.0300	33.34	.0300	33.32	1.000	0.9996
.04	2° 18′	.0400	25.01	.0400	24.99	1.001	0.9992
0.05	2° 52′	0.0500	20.01	0.0500	19.98	1.001	0.9988
.06	3° 26′	.0600	16.68	.0601	16.65	1.002	.9982
.07	4° 01′	.0699	14.30	.0701	14.26	1.002	.9976
.08	4° 35′	.0799	12.51	.0802	12.47	1.003	.9968
.09	5° 09′	.0899	11.13	.0902	11.08	1.004	.9960
0.10	5° 44′	0.0998	10.02	0.1003	9.967	1.005	0.9950
.11	6° 18′	.1098	9.109	.1104	9.054	1.006	.9940
.12	6° 53′	.1197	8.353	.1206	8.293	1.007	.9928
.13	7° 27′	.1296	7.714	.1307	7.649	1.009	.9916
.14	8° 01′	.1395	7.166	.1409	7.096	1.010	.9902
0.15	8° 36′	0.1494	6.692	0.1511	6.617	1.011	0.9888
.16	9° 10′	.1593	6.277	.1614	6.197	1.013	.9872
.17	9° 44′	.1692	5.911	.1717	5.826	1.015	.9856
.18	10° 19′	.1790	5.586	.1820	5.495	1.016	.9838
.19	10° 53′	.1889	5.295	.1923	5.200	1.018	.9820
0.20	11° 28′	0.1987	5.033	0.2027	4.933	1.020	0.9801
.21	12° 02′	.2085	4.797	.2131	4.692	1.022	.9780
.22	12° 36′	.2182	4.582	.2236	4.472	1.025	.9759
.23	13° 11′	.2280	4.386	.2341	4.271	1.027	.9737
.24	13° 45′	.2377	4.207	.2447	4.086	1.030	.9713
Real Number θ or α Radians	α Degrees	sin α or sin θ	csc α or csc θ	tan α or tan θ	cot α or cot θ	sec α or sec θ	cos α or cos θ

table 5 (*continued*)

Real Number θ or α Radians	α Degrees	sin α or sin θ	csc α or csc θ	tan α or tan θ	cot α or cot θ	sec α or sec θ	cos α or cos θ
0.25	14° 19'	0.2474	4.042	0.2553	3.916	1.032	0.9689
.26	14° 54'	.2571	3.890	.2660	3.759	1.035	.9664
..27	15° 28'	.2667	3.749	.2768	3.613	1.038	.9638
.28	16° 03'	.2764	3.619	.2876	3.478	1.041	.9611
.29	16° 37'	.2860	3.497	.2984	3.351	1.044	.9582
0.30	17° 11'	0.2955	3.384	0.3093	3.233	1.047	0.9553
.31	17° 46'	.3051	3.278	.3203	3.122	1.050	.9523
.32	18° 20'	.3146	3.179	.3314	3.018	1.053	.9492
.33	18° 54'	.3240	3.086	.3425	2.920	1.057	.9460
.34	19° 29'	.3335	2.999	.3537	2.827	1.061	.9428
0.35	20° 03'	0.3429	2.916	0.3650	2.740	1.065	0.9394
.36	20° 38'	.3523	2.839	.3764	2.657	1.068	.9359
.37	21° 12'	.3616	2.765	.3879	2.578	1.073	.9323
.38	21° 46'	.3709	2.696	.3994	2.504	1.077	.9287
.39	22° 21'	.3802	2.630	.4111	2.433	1.081	.9249
0.40	22° 55'	0.3894	2.568	0.4228	2.365	1.086	0.9211
.41	23° 29'	.3986	2.509	.4346	2.301	1.090	.9171
.42	24° 04'	.4078	2.452	.4466	2.239	1.095	.9131
.43	20° 38'	.4169	2.399	.4586	2.180	1.100	.9090
.44	25° 13'	.4259	2.348	.4708	2.124	1.105	.9048
0.45	25° 47'	0.4350	2.299	0.4831	2.070	1.111	0.9004
.46	26° 21'	.4439	2.253	.4954	2.018	1.116	.8961
.47	26° 56'	.4529	2.208	.5080	1.969	1.122	.8916
.48	27° 30'	.4618	2.166	.5206	1.921	1.127	.8870
.49	28° 04'	.4706	2.125	.5334	1.875	1.133	.8823
Real Number θ or α Radians	α Degrees	sin α or sin θ	csc α or csc θ	tan α or tan θ	cot α or cot θ	sec α or sec θ	cos α or cos θ

table 5 (*continued*)

Real Number θ or α Radians	α Degrees	sin α or sin θ	csc α or csc θ	tan α or tan θ	cot α or cot θ	sec α or sec θ	cos α or cos θ
0.50	28° 39'	0.4794	2.086	0.5463	1.830	1.139	0.8776
.51	29° 13'	.4882	2.048	.5594	1.788	1.146	.8727
.52	29° 48'	.4969	2.013	.5726	1.747	1.152	.8678
.53	30° 22'	.5055	1.978	.5859	1.707	1.159	.8628
.54	30° 56'	.5141	1.945	.5994	1.668	1.166	.8577
0.55	31° 31'	0.5227	1.913	0.6131	1.631	1.173	0.8525
.56	32° 05'	.5312	1.883	.6269	1.595	1.180	.8473
.57	32° 40'	.5396	1.853	.6410	1.560	1.188	.8419
.58	33° 14'	.5480	1.825	.6552	1.526	1.196	.8365
.59	33° 48'	.5564	1.797	.6696	1.494	1.203	.8309
0.60	34° 23'	0.5646	1.771	0.6841	1.462	1.212	0.8253
.61	34° 57'	.5729	1.746	.6989	1.431	1.220	.8196
.62	35° 31'	.5810	1.721	.7139	1.401	1.229	.8139
.63	36° 06'	.5891	1.697	.7291	1.372	1.238	.8080
.64	36° 40'	.5972	1.674	.7445	1.343	1.247	.8021
0.65	37° 15'	0.6052	1.652	0.7602	1.315	1.256	0.7961
.66	37° 49'	.6131	1.631	.7761	1.288	1.266	.7900
.67	38° 23'	.6210	1.610	.7923	1.262	1.276	.7838
.68	38° 58'	.6288	1.590	.8087	1.237	1.286	.7776
.69	39° 32'	.6365	1.571	.8253	1.212	1.297	.7712
0.70	40° 06'	0.6442	1.552	0.8423	1.187	1.307	0.7648
.71	40° 41'	.6518	1.534	.8595	1.163	1.319	.7584
.72	41° 15'	.6594	1.517	.8771	1.140	1.330	.7518
.73	41° 50'	.6669	1.500	.8949	1.117	1.342	.7452
.74	42° 24'	.6743	1.483	.9131	1.095	1.354	.7385
Real Number θ or α Radians	α Degrees	sin α or sin θ	csc α or csc θ	tan α or tan θ	cot α or cot θ	sec α or sec θ	cos α or cos θ

table 5 (*continued*)

Real Number θ or α Radians	α Degrees	sin α or sin θ	csc α or csc θ	tan α or tan θ	cot α or cot θ	sec α or sec θ	cos α or cos θ
0.75	42° 58′	0.6816	1.467	0.9316	1.073	1.367	0.7317
.76	43° 33′	.6889	1.452	.9505	1.052	1.380	.7248
.77	44° 07′	.6961	1.436	.9697	1.031	1.393	.7179
.78	44° 41′	.7033	1.422	.9893	1.011	1.407	.7109
.79	45° 16′	.7104	1.408	1.009	.9908	1.421	.7038
0.80	45° 50′	0.7174	1.394	1.030	0.9712	1.435	0.6967
.81	46° 25′	.7243	1.381	1.050	.9520	1.450	.6895
.82	46° 59′	.7311	1.368	1.072	.9331	1.466	.6822
.83	47° 33′	.7379	1.355	1.093	.9146	1.482	.6749
.84	48° 08′	.7446	1.343	1.116	.8964	1.498	.6675
0.85	48° 42′	0.7513	1.331	1.138	0.8785	1.515	0.6600
.86	49° 16′	.7578	1.320	1.162	.8609	1.533	.6524
.87	49° 51′	.7643	1.308	1.185	.8437	1.551	.6448
.88	50° 25′	.7707	1.297	1.210	.8267	1.569	.6372
.89	51° 00′	.7771	1.287	1.235	.8100	1.589	.6294
0.90	51° 34′	0.7833	1.277	1.260	0.7936	1.609	0.6216
.91	52° 08′	.7895	1.267	1.286	.7774	1.629	.6137
.92	52° 43′	.7956	1.257	1.313	.7615	1.651	.6058
.93	53° 17′	.8016	1.247	1.341	.7458	1.673	.5978
.94	53° 51′	.8076	1.238	1.369	.7303	1.696	.5898
0.95	54° 26′	0.8134	1.229	1.398	0.7151	1.719	0.5817
.96	55° 00′	.8192	1.221	1.428	.7001	1.744	.5735
.97	55° 35′	.8249	1.212	1.459	.6853	1.769	.5653
.98	56° 09′	.8305	1.204	1.491	.6707	1.795	.5570
.99	56° 43′	.8360	1.196	1.524	.6563	1.823	.5487
Real Number θ or α Radians	α Degrees	sin α or sin θ	csc α or csc θ	tan α or tan θ	cot α or cot θ	sec α or sec θ	cos α or cos θ

table 5 (*continued*)

Real Number θ or α Radians	α Degrees	sin α or sin θ	csc α or csc θ	tan α or tan θ	cot α or cot θ	sec α or sec θ	cos α or cos θ
1.00	57° 18′	0.8415	1.188	1.557	0.6421	1.851	0.5403
1.01	57° 52′	.8468	1.181	1.592	.6281	1.880	.5319
1.02	58° 27′	.8521	1.174	1.628	.6142	1.911	.5234
1.03	59° 01′	.8573	1.166	1.665	.6005	1.942	.5148
1.04	59° 35′	.8624	1.160	1.704	.5870	1.975	.5062
1.05	60° 10′	0.8674	1.153	1.743	0.5736	2.010	0.4976
1.06	66° 44′	.8724	1.146	1.784	.5604	2.046	.4889
1.07	61° 18′	.8772	1.140	1.827	.5473	2.083	.4801
1.08	61° 53′	.8820	1.134	1.871	.5344	2.122	.4713
1.09	62° 27′	.8866	1.128	1.917	.5216	2.162	.4625
1.10	63° 02′	0.8912	1.122	1.965	0.5090	2.205	0.4536
1.11	63°36′	.8957	1.116	2.014	.4964	2.249	.4447
1.12	64° 10′	.9001	1.111	2.066	.4840	2.295	.4357
1.13	64° 45′	.9044	1.106	2.120	.4718	2.344	.4267
1.14	65° 19′	.9086	1.101	2.176	.4596	2.395	.4176
1.15	65° 53′	0.9128	1.096	2.234	0.4475	2.448	0.4085
1.16	66° 28′	.9168	1.091	2.296	.4356	2.504	.3993
1.17	67° 02′	.9208	1.086	2.360	.4237	2.563	.3902
1.18	67° 37′	.9246	1.082	2.427	.4120	2.625	.3809
1.19	68° 11′	.9284	1.077	2.498	.4003	2.691	.3717
1.20	68° 45′	0.9320	1.073	2.572	0.3888	2.760	0.3624
1.21	69° 20′	.9356	1.069	2.650	.3773	2.833	.3530
1.22	69° 54′	.9391	1.065	2.733	.3659	2.910	.3436
1.23	70° 28′	.9425	1.061	2.820	.3546	2.992	.3342
1.24	71° 03′	.9458	1.057	2.912	.3434	3.079	.3248
Real Number θ or α Radians	α Degrees	sin α or sin θ	csc α or csc θ	tan α or tan θ	cot α or cot θ	sec α or sec θ	cos α or cos θ

table 5 (*continued*)

Real Number θ or α Radians	α Degrees	sin α or sin θ	csc α or csc θ	tan α or tan θ	cot α or cot θ	sec α or sec θ	cos α or cos θ
1.25	71° 37′	0.9490	1.054	3.010	0.3323	3.171	0.3153
1.26	72° 12′	.9521	1.050	3.113	.3212	3.270	.3058
1.27	72° 46′	.9551	1.047	3.224	.3102	3.375	.2963
1.28	73° 20′	.9580	1.044	3.341	.2993	3.488	.2867
1.29	73° 55′	.9608	1.041	3.467	.2884	3.609	.2771
1.30	74° 29′	0.9636	1.038	3.602	0.2776	3.738	0.2675
1.31	75° 03′	.9662	1.035	3.747	.2669	3.878	.2579
1.32	75° 38′	.9687	1.032	3.903	.2562	4.029	.2482
1.33	76° 12′	.9711	1.030	4.072	.2456	4.193	.2385
1.34	76° 47′	.9735	1.027	4.256	.2350	4.372	.2288
1.35	77° 21′	0.9757	1.025	4.455	0.2245	4.566	0.2190
1.36	77° 55′	.9779	1.023	4.673	.2140	4.779	.2092
1.37	78° 30′	.9799	1.021	4.913	.2035	5.014	.1994
1.38	79° 04′	.9819	1.018	5.177	.1931	5.273	.1896
1.39	79° 38′	.9837	1.017	5.471	.1828	5.561	.1798
1.40	80° 13′	0.9854	1.015	5.798	0.1725	5.883	0.1700
Real Number θ or α Radians	α Degrees	sin α or sin θ	csc α or csc θ	tan α or tan θ	cot α or cot θ	sec α or sec θ	cos α or cos θ

table 6

Four-Place Values of Trigonometric Functions, Angle α in Degrees and Radians

Angle α									
Degrees	Radians	sin α	csc α	tan α	cot α	sec α	cos α		
0° 00′	.0000	.0000	No value	.0000	No value	1.000	1.0000	1.5708	90° 00′
10	029	029	343.8	029	343.8	000	000	679	50
20	058	058	171.9	058	171.9	000	000	650	40
30	.0087	.0087	114.6	.0087	114.6	1.000	1.0000	1.5621	30
40	116	116	85.95	116	85.94	000	.9999	592	20
50	145	145	68.76	145	68.75	000	999	563	10
1° 00′	.0175	.0175	57.30	.0175	57.29	1.000	.9998	1.5533	89° 00′
10	204	204	49.11	204	49.10	000	998	504	50
20	233	233	42.98	233	42.96	000	997	475	40
30	.0262	.0262	38.20	.0262	38.19	1.000	.9997	1.5446	30
40	291	291	34.38	291	34.37	000	996	417	20
50	320	320	31.26	320	31.24	001	995	388	10
2° 00′	.0349	.0349	28.65	.0349	28.64	1.001	.9994	1.5359	88° 00′
10	378	378	26.45	378	26.43	001	993	330	50
20	407	407	24.56	407	24.54	001	992	301	40
30	.0436	.0436	22.93	.0437	22.90	1.001	.9990	1.5272	30
40	465	465	21.49	466	21.47	001	989	243	20
50	495	494	20.23	495	20.21	001	988	213	10
3° 00′	.0524	.0523	19.11	0.524	19.08	1.001	.9986	1.5184	87° 00′
10	553	552	18.10	553	18.07	002	985	155	50
20	582	581	17.20	582	17.17	002	983	126	40
30	.0611	.0610	16.38	.0612	16.35	1.002	.9981	1.5097	30
40	640	640	15.64	641	15.60	002	980	068	20
50	669	669	14.96	670	14.92	002	978	039	10
4° 00′	.0698	.0698	14.34	.0699	14.30	1.002	.9976	1.5010	86° 00′
10	727	727	13.76	729	13.73	003	974	981	50
20	756	756	13.23	758	13.20	003	971	952	40
30	.0785	.0785	12.75	.0787	12.71	1.003	.9969	1.4923	30
40	814	814	12.29	816	12.25	003	967	893	20
50	844	843	11.87	846	11.83	004	964	864	10
5° 00′	.0873	.0872	11.47	.0875	11.43	1.004	.9962	1.4835	85° 00′
		cos α	sec α	cot α	tan α	csc α	sin α	Radians	Degrees
								Angle α	

table 6 (*continued*)

Angle α									
Degrees	Radians	sin α	csc α	tan α	cot α	sec α	cos α		
5° 00′	.0873	.0872	11.47	.0875	11.43	1.004	.9962	1.4835	85° 00′
10	902	901	11.10	904	11.06	004	959	806	50
20	931	929	10.76	934	10.71	004	957	777	40
30	.0960	.0958	10.43	.0963	10.39	1.005	.9954	1.4748	30
40	989	987	10.13	992	10.08	005	951	719	20
50	.1018	.1016	9.839	.1022	9.788	005	948	690	10
6° 00′	.1047	.1045	9.567	.1051	9.514	1.006	.9945	1.4661	84° 00′
10	076	074	9.309	080	9.255	006	942	632	50
20	105	103	9.065	110	9.010	006	939	603	40
30	.1134	.1132	8.834	.1139	8.777	1.006	.9936	1.4573	30
40	164	161	8.614	169	8.556	007	932	544	20
50	193	190	8.405	198	8.345	007	929	515	10
7° 00′	.1222	.1219	8.206	.1228	8.144	1.008	.9925	1.4486	83° 00′
10	251	248	8.016	257	7.953	008	922	457	50
20	280	276	7.834	287	7.770	008	918	428	40
30	.1309	.1305	7.661	.1317	7.596	1.009	.9914	1.4399	30
40	338	334	7.496	346	7.429	009	911	370	20
50	367	363	7.337	376	7.269	009	907	341	10
8° 00′	.1396	.1392	7.185	.1405	7.115	1.010	.9903	1.4312	82° 00′
10	425	421	7.040	435	6.968	010	899	283	50
20	454	449	6.900	465	6.827	011	894	254	40
30	.1484	.1478	6.765	.1495	6.691	1.011	.9890	1.4224	30
40	513	507	6.636	524	6.561	012	886	195	20
50	542	536	6.512	554	6.435	012	881	166	10
9° 00′	.1571	.1564	6.392	.1584	6.314	1.012	.9877	1.4137	81° 00′
10	600	593	277	614	197	013	872	108	50
20	629	622	166	644	084	013	868	079	40
30	.1658	.1650	6.059	.1673	5.976	1.014	.9863	1.4050	30
40	687	679	5.955	703	871	014	858	1.4021	20
50	716	708	855	733	769	015	853	992	10
10° 00′	.1745	.1736	5.759	.1763	5.671	1.015	.9848	1.3963	80° 00′
		cos α	sec α	cot α	tan α	csc α	sin α	Radians	Degrees
									Angle α

table 6 (*continued*)

Degrees	Radians	sin α	csc α	tan α	cot α	sec α	cos α		
10° 00′	.1745	.1736	5.759	.1763	5.671	1.015	.9848	1.3963	80° 00′
10	774	765	665	793	576	016	843	934	50
20	804	794	575	823	485	016	838	904	40
30	.1833	.1822	5.487	.1853	5.396	1.017	.9833	1.3875	30
40	862	851	403	883	309	018	827	846	20
50	891	880	320	914	226	018	822	817	10
11° 00′	.1920	.1908	5.241	.1944	5.145	1.019	.9816	1.3788	79° 00′
10	949	937	164	974	066	019	811	759	50
20	978	965	089	.2004	4.989	020	805	730	40
30	.2007	.1994	5.016	.2035	4.915	1.020	.9799	1.3701	30
40	036	.2022	4.945	065	843	021	793	672	20
50	065	051	876	095	773	022	787	643	10
12° 00′	.2094	.2079	4.810	.2126	4.705	1.022	.9781	1.3614	78° 00′
10	123	108	745	156	638	023	775	584	50
20	153	136	682	186	574	024	769	555	40
30	.2182	.2164	4.620	.2217	4.511	1.024	.9763	1.3526	30
40	211	193	560	247	449	025	757	497	20
50	240	221	502	278	390	026	750	468	10
13° 00	.2269	.2250	4.445	.2309	4.331	1.026	.9744	1.3439	77° 00′
10	298	278	390	339	275	027	737	410	50
20	327	306	336	370	219	028	730	381	40
30	.2356	.2334	4.284	.2401	4.165	1.028	.9724	1.3352	30
40	386	363	232	432	113	029	717	323	20
50	414	391	182	462	061	030	710	294	10
14° 00′	.2443	.2419	4.134	.2493	4.011	1.031	.9703	1.3265	76° 00′
10	473	447	086	524	3.962	031	696	235	50
20	502	476	039	555	914	032	689	206	40
30	.2531	.2504	3.994	.2586	3.867	1.033	.9681	1.3177	30
40	560	532	950	617	821	034	674	148	20
50	589	560	906	648	776	034	667	119	10
15° 00′	.2618	.2588	3.864	.2679	3.732	1.035	.9659	1.3090	75° 00′
		cos α	sec α	cot α	tan α	csc α	sin α	Radians	Degrees
								Angle α	

table 6 (*continued*)

Angle α									
Degrees	Radians	sin α	csc α	tan α	cot α	sec α	cos α		
15° 00′	.2618	.2588	3.864	.2679	3.732	1.035	.9659	1.3090	75° 00′
10	647	616	822	711	689	036	652	061	50
20	676	644	782	742	647	037	644	032	40
30	.2705	.2672	3.742	.2773	3.606	1.038	.9639	1.3003	30
40	734	700	703	805	566	039	628	974	20
50	763	728	665	836	526	039	621	945	10
16° 00′	.2793	.2756	3.628	.2867	3.487	1.040	.9613	1.2915	74° 00′
10	822	784	592	899	450	041	605	886	50
20	851	812	556	931	412	042	596	857	40
30	.2880	.2840	3.521	.2962	3.376	1.043	.9588	1.2828	30
40	909	868	487	994	340	044	580	799	20
50	938	896	453	.3026	305	045	572	770	10
17° 00′	.2967	.2924	3.420	.3057	3.271	1.046	.9563	1.2741	73° 00′
10	996	952	388	089	237	047	555	712	50
20	.3025	979	357	121	204	048	546	683	40
30	.3054	.3007	3.326	.3153	3.172	1.048	.9537	1.2654	30
40	083	035	295	185	140	049	528	625	20
50	113	062	265	217	108	050	520	595	10
18° 00′	.3142	.3090	3.236	.3249	3.078	1.051	.9511	1.2566	72° 00′
10	171	118	207	281	047	052	502	537	50
20	200	145	179	314	018	053	492	508	40
30	.3229	.3173	3.152	.3346	2.989	1.054	.9483	1.2479	30
40	258	201	124	378	960	056	474	450	20
50	287	228	098	411	932	057	465	421	10
19° 00′	.3316	.3256	3.072	.3443	2.904	1.058	.9455	1.2392	71° 00′
10	345	283	046	476	877	059	446	363	50
20	374	311	021	508	850	060	436	334	40
30	.3403	.3338	2.996	.3541	2.824	1.061	.9426	1.2305	30
40	432	365	971	574	798	062	417	275	20
50	462	393	947	607	773	063	407	246	10
20° 00′	.3491	.3420	2.924	.3640	2.747	1.064	.9397	1.2217	70° 00′
		cos α	sec α	cot α	tan α	csc α	sin α	Radians	Degrees
									Angle α

table 6 (*continued*)

Angle α Degrees	Angle α Radians	sin α	csc α	tan α	cot α	sec α	cos α		
20° 00′	.3491	.3420	2.924	.3640	2.747	1.064	.9397	1.2217	70° 00′
10	520	448	901	673	723	065	387	188	50
20	549	475	878	706	699	066	377	159	40
30	.3578	.3502	2.855	.3739	2.675	1.068	.9367	1.2130	30
40	607	529	833	772	651	069	356	101	20
50	636	557	812	805	628	070	346	072	10
21° 00′	.3665	.3584	2.790	.3839	2.605	1.071	.9336	1.2043	69° 00′
10	694	611	769	872	583	072	325	1.2014	50
20	723	638	749	906	560	074	315	985	40
30	.3752	.3665	2.729	.3939	2.539	1.075	.9304	1.1956	30
40	782	692	709	973	517	076	293	926	20
50	811	719	689	.4006	496	077	283	897	10
22° 00′	.3840	.3746	2.669	.4040	2.475	1.079	.9272	1.1868	68° 00′
10	869	773	650	074	455	080	261	839	50
20	898	800	632	108	434	081	250	810	40
30	.3927	.3827	2.613	.4142	2.414	1.082	.9239	1.1781	30
40	956	854	595	176	394	084	228	752	20
50	985	881	577	210	375	085	216	723	10
23° 00′	.4014	.3907	2.559	.4245	2.356	1.086	.9205	1.1694	67° 00′
10	043	934	542	279	337	088	194	665	50
20	072	961	525	314	318	089	182	636	40
30	.4102	.3987	2.508	.4348	2.300	1.090	.9171	1.1606	30
40	131	.4014	491	383	282	092	159	577	20
50	160	041	475	417	264	093	147	548	10
24° 00′	.4189	.4067	2.459	.4452	2.246	1.095	.9135	1.1519	66° 00′
10	218	094	443	487	229	096	124	490	50
20	247	120	427	522	211	097	112	461	40
30	.4276	.4147	2.411	.4557	2.194	1.099	.9100	1.1432	30
40	305	173	396	592	177	100	088	403	20
50	334	200	381	628	161	102	075	374	10
25° 00′	.4363	.4226	2.366	.4663	2.145	1.103	.9063	1.1345	65° 00′
		cos α	sec α	cot α	tan α	csc α	sin α	Radians	Degrees
								Angle α	

table 6 (*continued*)

Angle α									
Degrees	Radians	sin α	csc α	tan α	cot α	sec α	cos α		
25° 00′	.4363	.4226	2.366	.4663	2.145	1.103	.9063	1.1345	65° 00′
10	392	253	352	699	128	105	051	316	50
20	422	279	337	734	112	106	038	286	40
30	.4451	.4305	2.323	.4770	2.097	1.108	.9026	1.1257	30
40	480	331	309	806	081	109	013	228	20
50	509	358	295	841	066	111	001	199	10
26° 00′	.4538	.4384	2.281	.4877	2.050	1.113	.8988	1.1170	64° 00′
10	567	410	268	913	035	114	975	141	50
20	596	436	254	950	020	116	962	112	40
30	.4625	.4462	2.241	.4986	2.006	1.117	.8949	1.1083	30
40	654	488	228	.5022	1.991	119	936	054	20
50	683	514	215	059	977	121	923	1.1025	10
27° 00′	.4712	.4540	2.203	.5095	1.963	1.122	.8910	1.0996	63° 00′
10	741	566	190	132	949	124	897	966	50
20	771	592	178	169	935	126	884	937	40
30	.4800	.4617	2.166	.5206	1.921	1.127	.8870	1.0908	30
40	829	643	154	243	907	129	857	879	20
50	858	669	142	280	894	131	843	850	10
28° 00′	.4887	.4695	2.130	.5317	1.881	1.133	.8829	1.0821	62° 00′
10	916	720	118	354	868	134	816	792	50
20	945	746	107	392	855	136	802	763	40
30	.4974	.4772	2.096	.5430	1.842	1.138	.8788	1.0734	30
40	.5003	797	085	467	829	140	774	705	20
50	032	823	074	505	816	142	760	676	10
29° 00′	.5061	.4848	2.063	.5543	1.804	1.143	.8746	1.0647	61° 00′
10	091	874	052	581	792	145	732	617	50
20	120	899	041	619	780	147	718	588	40
30	.5149	.4924	2.031	.5658	1.767	1.149	.8704	1.0559	30
40	178	950	020	696	756	151	689	530	20
50	207	975	010	735	744	153	675	501	10
30° 00′	.5236	.5000	2.000	.5774	1.732	1.155	.8660	1.0472	60° 00′
		cos α	sec α	cot α	tan α	csc α	sin α	Radians	Degrees
								Angle α	

table 6 (*continued*)

Angle α Degrees	Radians	sin α	csc α	tan α	cot α	sec α	cos α		
30° 00′	.5236	.5000	2.000	.5774	1.732	1.155	.8660	1.0472	60° 00′
10	265	025	1.990	812	720	157	646	443	50
20	294	050	980	851	709	159	631	414	40
30	.5323	.5075	1.970	.5890	1.698	1.161	.8616	1.0385	30
40	352	100	961	930	686	163	601	356	20
50	381	125	951	969	675	165	587	327	10
31° 00′	.5411	.5150	1.942	.6009	1.664	1.167	.8572	1.0297	59° 00′
10	440	175	932	048	653	169	557	268	50
20	469	200	923	088	643	171	542	239	40
30	.5498	.5225	1.914	.6128	1.632	1.173	.8526	1.0210	30
40	527	250	905	168	621	175	511	181	20
50	556	275	896	208	611	177	496	152	10
32° 00′	.5585	.5299	1.887	.6249	1.600	1.179	.8480	1.0123	58° 00′
10	614	324	878	289	590	181	465	094	50
20	643	348	870	330	580	184	450	065	40
30	.5672	.5373	1.861	.6371	1.570	1.186	.8434	1.0036	30
40	701	398	853	412	560	188	418	1.007	20
50	730	422	844	453	550	190	403	977	10
33° 00′	.5760	.5446	1.836	.6494	1.540	1.192	.8387	.9948	57° 00′
10	789	471	828	536	530	195	371	919	50
20	818	495	820	577	520	197	355	890	40
30	.5847	.5519	1.812	.6619	1.511	1.199	.8339	.9861	30
40	876	544	804	661	501	202	323	832	20
50	905	568	796	703	1.492	204	307	803	10
34° 00′	.5934	.5592	1.788	.6745	1.483	1.206	.8290	.9774	56° 00′
10	963	616	781	787	473	209	274	745	50
20	992	640	773	830	464	211	258	716	40
30	.6021	.5664	1.766	.6873	1.455	1.213	.8241	.9687	30
40	050	688	758	916	446	216	225	657	20
50	080	712	751	959	437	218	208	628	10
35° 00′	.6109	.5736	1.743	.7002	1.428	1.221	.8192	.9599	55° 00′
		cos α	sec α	cot α	tan α	csc α	sin α	Radians	Degrees
								Angle α	

table 6 (*continued*)

Angle α Degrees	Radians	sin α	csc α	tan α	cot α	sec α	cos α		
35° 00′	.6109	.5736	1.743	.7002	1.428	1.221	.8192	.9599	55° 00′
10	138	760	736	046	419	223	175	570	50
20	167	783	729	089	411	226	158	541	40
30	.6196	.5807	1.722	.7133	1.402	1.228	.8141	.9512	30
40	225	831	715	177	393	231	124	483	20
50	254	854	708	221	385	233	107	454	10
36° 00′	.6283	.5878	1.701	.7265	1.376	1.236	.8090	.9425	54° 00′
10	312	901	695	310	368	239	073	396	50
20	341	925	688	355	360	241	056	367	40
30	.6370	.5948	1.681	.7400	1.351	1.244	.8039	.9338	30
40	400	972	675	445	343	247	021	308	20
50	429	995	668	490	335	249	004	279	10
37° 00′	.6458	.6018	1.662	.7536	1.327	1.252	.7986	.9250	53° 00′
10	487	041	655	581	319	255	969	221	50
20	516	065	649	627	311	258	951	192	40
30	.6545	.6088	1.643	.7673	1.303	1.260	.7934	.9163	30
40	574	111	636	720	295	263	916	134	20
50	603	134	630	766	288	266	898	105	10
38° 00′	.6632	.6157	1.624	.7813	1.280	1.269	.7880	.9076	52° 00′
10	661	180	618	860	272	272	862	047	50
20	690	202	612	907	265	275	844	.9018	40
30	.6720	.6225	1.606	.7954	1.257	1.278	.7826	.8988	30
40	749	248	601	.8002	250	281	808	959	20
50	778	271	595	050	242	284	790	930	10
39° 00′	.6807	.6293	1.589	.8098	1.235	1.287	.7771	.8901	51° 00′
10	836	316	583	146	228	290	753	872	50
20	865	338	578	195	220	293	735	843	40
30	.6894	.6361	1.572	.8243	1.213	1.296	.7716	.8814	30
40	923	383	567	292	206	299	698	785	20
50	952	406	561	342	199	302	679	756	10
40° 00′	.6981	.6428	1.556	.8391	1.192	1.305	.7660	.8727	50° 00′
		cos α	sec α	cot α	tan α	csc α	sin α	Radians	Degrees
								Angle α	

table 6 (continued)

Degrees	Radians	sin α	csc α	tan α	cot α	sec α	cos α		
40° 00′	.6981	.6428	1.556	.8391	1.192	1.305	.7660	.8727	50° 00′
10	.7010	450	550	441	185	309	642	698	50
20	039	472	545	491	178	312	623	668	40
30	.7069	.6494	1.540	.8541	1.171	1.315	.7604	.8639	30
40	098	517	535	591	164	318	585	610	20
50	127	539	529	642	157	322	566	581	10
41° 00′	.7156	.6561	1.524	.8693	1.150	1.325	.7547	.8552	49° 00′
10	185	583	519	744	144	328	528	523	50
20	214	604	514	796	137	332	509	494	40
30	.7243	.6626	1.509	.8847	1.130	1.335	.7490	.8465	30
40	272	648	504	899	124	339	470	436	20
50	301	670	499	952	117	342	451	407	10
42° 00′	.7330	.6691	1.494	.9004	1.111	1.346	.7431	.8378	48° 00′
10	359	713	490	057	104	349	412	348	60
20	389	734	485	110	098	353	392	319	40
30	.7418	.6756	1.480	.9163	1.091	1.356	.7373	.8290	30
40	447	777	476	217	085	360	353	261	20
50	476	799	471	271	079	364	333	232	10
43° 00′	.7505	.6820	1.466	.9325	1.072	1.367	.7314	.8203	47° 00′
10	534	841	462	380	066	371	294	174	50
20	563	862	457	435	060	375	274	145	40
30	.7592	.6884	1.453	.9490	1.054	1.379	.7254	.8116	30
40	621	905	448	545	048	382	234	087	20
50	650	926	444	601	042	386	214	058	10
44° 00′	.7679	.6947	1.440	.9657	1.036	1.390	.7193	.8029	46° 00′
10	709	967	435	713	030	394	173	.7999	50
20	738	988	431	770	024	398	153	970	40
30	.7767	.7009	1.427	.9827	1.018	1.402	.7133	.7941	30
40	796	030	423	884	012	406	112	912	20
50	825	050	418	942	006	410	092	883	10
45° 00′	.7854	.7071	1.414	1.000	1.000	1.414	.7071	.7854	45° 00′
		cos α	sec α	cot α	tan α	csc α	sin α	Radians	Degrees
								Angle α	

answers to selected exercises

chapter 1.

section 1.1 (pp. 5–7).

1. (b) $\{-3, -2, -1, 0, 1, 2, 3\}$; (c) $\{1, 2\}$; (d) $\{1\}$.

2. (a) B; (b) A; (c) C; (d) D; (e) A; (f) $\{(1,1), (1,2), (1,3), (2,1), (2,2) (2,3)\}$; (i) B; (j) D; (e) $\{2\}$; (o) $\{\varnothing, C, \{2\}, \{3\}\}$; (p) $\{\varnothing, D\}$.

3. (b) $\{$F. D. Roosevelt$\}$; (d) $\{$Grover Cleveland$\}$; (e) $\varnothing$.

4. Correct are (a), (c), (d), (e), (g), and (h).

7. (a) $\varnothing, \{x, y, z\}, \{x\}, \{y\}, \{z\}, \{x, y\}, \{x, z\}, \{y, z\}$; (b) 2^n.

9. (a) $A \subseteq B$, so if $x \in A$, then $x \in B$. But $B \subseteq C$, so $x \in C$. Therefore, if $x \in A$, then $x \in C$. So $A \subseteq C$.
   (c) $x \in B \backslash A$ if and only if $x \in B$ and $x \notin A$. But $x \notin A$ means $x \in \tilde{A}$. Thus, $x \in B$ and $x \in \tilde{A}$; i.e., if and only if $x \in \tilde{A} \cap B$. Thus, both $B \backslash A \subseteq \tilde{A} \cap B$ and $\tilde{A} \cap B \subseteq B \backslash A$.

11. (a) yes; (b) no; (c) not unless $A = \varnothing$; (d) 0; (e) 1; (f) no; (g) yes; (h) $\varnothing$; (i) the universe; (j) A; (k) A; (l) A.

12. (a) Let $x \in \widetilde{(A \cap B)}$, then $x \notin A \cap B$; hence, it is false that x belongs to both A and to B. Thus, either $x \in \tilde{A}$ or $x \in \tilde{B}$; i.e., $x \in \tilde{A} \cup \tilde{B}$.
   (b) If $A \subset B$, then $x \in A$ implies $x \in B$. Thus, if $x \in A \cup B$, x is either in A, hence in B, or else in B. In either case $x \in B$, so $A \cup B \subset B$. But $B \subset A \cup B$.

section 1.2 (pp. 11–13).

1.

N	Z	Q	R	C
2	-7	$22/7$	π	$\sqrt{-5}$
4	-6	3.1416	$\sqrt{5}$	$3 - 7i$
$\sqrt[3]{27}$	$\sqrt[3]{-8}$		π^2	$\sqrt{-16}$
1	0		$1 - \sqrt{\pi}$	
			$1 + \sqrt{2}$	

2. (a) real numbers
   (b) rational numbers
   (c) integers
   (d) positive integers
   (e) primes

5. 2, 3, 5, 7, 11, 13, 17, 19, 23, 29, 31, 37, 41, 47.

6. (a) $2^2 3$; (c) $2^2 3^2$; (e) $5^2 2$; (g) $-7 \cdot 2$; (i) $-11 \cdot 3$.

7. (a) $x = 3$ or -3; (c) $x < 0$; (e) $x < 1$; (g) $x = -5$ or -1; (i) $x = 0$ or 2; (j) $x = -4/5$ ($-8/3$ is extraneous).

8. (a) $2/3, 2$; (b) none ($-1/4$ and -4 are both extraneous); (c) $5/27, 3/29$; (e) 1; (g) 1; (i) 1.

9. $-16 < -\pi < 1 - \pi < -2 < -\sqrt{3} < -\sqrt{2} < 0 < 1/2 < 1 < \pi/3 < \sqrt{2} < \pi/2 < 7/3 < 3 < \pi$.

11. (a) $99a = 118$, $a = 118/99$; (c) $999a = 202$, $a = 202/999$.

13. $b - a = (b + c) - (a + c)$, so $(b + c) - (a + c) \in R^+$ whenever $b - a$ does.

answers to selected exercises

15. $b - a \in R^+$ and $-c \in R^+$ imply that $(b - a)(-c) = ac - bc \in R^+$.

17. $b - a \in R^+$ and $d - c \in R^+$ imply $(b - a) + (d - c) = (b + d) - (a + c) \in R^+$.

section 1.3 (pp. 18–19).

1. (a) $x < -2$; (c) $x > -2$; (e) $x \le 4$; (g) $x > -6/5$; (i) $0 < x < 2$.

2. (a) $-5 < x < 5$; (c) $x \ge 4$ or $x \le -4$; (d) no real number; (e) $1 \le x \le 3$;
    (g) $-2/3 \le x \le 2$; (i) $x \le -1/5$ or $x \ge 3$.

3. (a) $0 < x < 1$; (d) $x < -2$ or $x > 2$; (e) $x > 0$; (f) $x < 0$.

4. (a) $x = 2$ or -2; (c) $x = 0$ or $5/8$; (e) $x \ge 5$ or $x \le -5$; (g) $-3 \le x \le 3$;
    (i) $x = 0$; (j) no x.

5. (a) $-3 \le x \le 4$; (c) $x < -7/3$ or $x > 14/3$; (e) $-4 \le x \le 12$; (g) all x;
    (h) no x; (i) $x < 1/2$.

6. (a) Use Theorem 1.3.1(v) with $c = -1$.
    (c) $r > 0$, so $r \cdot r < r \cdot s$, $s > 0$, so $r \cdot s < s \cdot s$. Thus, $r^2 < rs < s^2$.
    (e) False; let $r = 2$, $s = 3$. (g) False; let $r = 2$, $s = 3$.

7. $a \le a$ iff $a - a \in R^+$ or $a - a = 0$. But $a - a = 0$ for all real numbers a.

9. (a) If $x = 0$, then $0 = -x = |-x| = |x| = 0$. If $x > 0$, $-x < 0$, so $|-x| = -(-x) = x$
    and $x = |x|$. If $x < 0$, then $-x > 0$; thus, $-x = |x| = |-x|$.
    (c) $|x - y|^2 = (x - y)^2 = x^2 - 2xy + y^2 = (y - x)^2 = |y - x|^2$.
    (e) $(a + b)^2 = a^2 + 2ab + b^2$, but if $a, b > 0$, then $2ab > 0$.
    (f) Use the fact that $a^2 + b^2 > 2ab$ and $ab > 0$. Divide both sides by ab.
    (g) $\dfrac{a^2 + 2ab + b^2}{2} > \dfrac{a^2 + b^2}{2} > \dfrac{2ab}{2} > ab$, so $\dfrac{(a + b)}{2} > \sqrt{ab}$.
    (i) Consider each of the four cases (1) $x \ge 0$, $y \ge 0$; (2) $x \ge 0$, $y < 0$;
    (3) $x < 0$, $y \ge 0$; (4) $x < 0$, $y < 0$. Case 1 follows because $x + y \ge 0$, and
    $|x + y| = x + y = |x| + |y|$. In Case 4, $x + y < 0$, so $|x + y| = -(x + y)$
    and $-x + (-y) = |x| + |y|$. Cases 2 and 3 are essentially the same; we do
    Case 2. $|x| = x$, $|y| = -y$. If $x + y \ge 0$ then $|x + y| = x + y \le x + (-y) = x - y$.
    If $x + y < 0$, then $|x + y| = -(x + y) = -x - y \le -x + y \le x - y$. In either
    case $|x + y| \le |x| + |y|$.
    (j) $|x| = |y - (y - x)| = |y + (-(y - x))| \le |y| + |x - y|$. Therefore,
    $|x| - |y| \le |x - y|$.

11. $\left| x - \dfrac{(x + y)}{2} \right| = \left| \dfrac{2x - x - y}{2} \right| = \left| \dfrac{x - y}{2} \right| = \left| \dfrac{y - x}{2} \right| = \left| y - \dfrac{(x + y)}{2} \right|$. Then

    apply Exercise 6(b).

12. $3.141 < \pi < 3.1416$ 13. (a) $1.570 < \pi/2 < 1.5708$; (c) $0.785 < \pi/4 < 0.7854$;
    (e) $2.094 < 2\pi/3 < 2.0944$; (g) $1.732 < \sqrt{3} < 1.7321$;
    (j) $1.3660 < (1 + \sqrt{3})/2 < 1.3661$.

15. If $x^2 + 1 = 0$, then $x^2 = -1$ contradicting Exercise 14.

16. $|2x - 6| = |2(x - 3)| = |2| \, |x - 3| < 2(0.001) = 0.002$.

section 1.4 (pp. 25–26).

1. $-1, 0, 1, 2, 3$; $-2, -1, 0, 1, 2, 3, 4$; there are infinitely many.

2. 6; 0; there are none. 3. $(-1/2, 3/2)$. 4. $(3, \infty) = \{x | x > 3\}$.

5. $1, 0, 1/2, \pi/4, \sqrt{2}/2$. 7. $(-3, 2)$

8. (a) $(-3,3)$; (c) $(-2.001, -1.999)$; (e) $(-\infty, -4.01] \cup [-3.99, \infty)$; (g) $[-2,2]$;

9. (a) $(0,3)$; (c) $[1,3)$; (d) $\emptyset$; (e) $\{(x, y) | 0 < x < 1, 0 < y < 1\}$
    (g) $\{(x, y) | 0 < x < 1, 0 < y < 3\}$

answers to selected exercises

10. $[2, 97]$ 11. (a) $\varnothing$; (b) $(0, 1)$ 12. There is none; $[0, \sqrt{2}]$.

13. $I \times I = \{(x, y) \mid 0 \leq x \leq 1, 0 \leq y \leq 1\}$

18. (a) $(1, 1)$; (c) $(-1, 3/4)$; (e) $(3\pi/4, 1)$.

19. (a) $2\sqrt{2}$; (c) $1/2$; (e) $\frac{1}{2}(\sqrt{64 + \pi^2})$.

section 1.5 (pp. 31–32).

2. (a) i; (b) 5; (c) $1 + i$; (e) 0; (f) $-6i$; (h) $1/2 + 3i/5$; (k) π; (l) $\pi + i$;
   (m) $1/2 - i\sqrt{3}/2$; (n) $1 - i\sqrt{2}$.

4. (a) 1; (b) 5; (d) $\sqrt{2}$; (e) 0; (f) 6; (h) $\sqrt{61}/10$; (j) $\sqrt{74}$;
   (l) $\sqrt{1 + \pi^2}$; (m) $\sqrt{1 - \sqrt{3}/2}$; (o) $\sqrt{5}$.

5. (a) $5 + 6i$; (c) $1/6 - 3i/4$; (e) $7/5$; (g) $8 + i$; (i) $(1 + \sqrt{6}) + (\sqrt{3} - \sqrt{2})i$.

6. (a) $77/53 - 22i/53$; (c) $-5/13 - 12i/13$; (d) i; (f) $1/10 + 17i/10$;
   (h) $1/2 - i/2$; (k) i.

7. (a) -1; (c) $-i$; (e) i; (g) 1; (i) 1; (k) 1.

9. (a) Use the fact that the reals are closed under addition and multiplication.
   (c) Use the appropriate properties of the reals and calculate the result.

11. Let $z = x + iy$, $w = s + it$. Then $z + w = (x + s) + i(y + t)$, and
    $\bar{z} = x - iy$, $\bar{w} = s - it$, so $\bar{z} + \bar{w} = (x + s) - i(y + t) = \overline{z + w}$.

13. For $z = x + iy$ $|z| = \sqrt{x^2 + y^2}$. Then $\bar{z} = x - iy$ and
    $|\bar{z}| = \sqrt{x^2 + (-y)^2} = \sqrt{x^2 + y^2} = |z|$.

14. (a) $\bar{z} = x - iy$, so $(\bar{\bar{z}}) = x - (-iy) = x + iy = z$.
    (c) $|0| = 0$. If $\sqrt{x^2 + y^2} = 0$, then $x^2 + y^2 = 0$. But $x^2 \geq 0$ and $y^2 \geq 0$ for every
       real number x and y. We must, therefore, have $x^2 = y^2 = 0$, or $x = y = 0$.
    (e) $z + \bar{z} = x + iy + x - iy = 2x \in R$.

15. (a) $x = 5$, $y = -3$; (c) $x = -8$, $y = \sqrt{2}/6$; (e) $x = -10$, $y = 5$;
    (f) $x = -4$, $y = -5/2$.

16. (a) $|zw|^2 = (zw)(\overline{zw}) = (zw)(\bar{z}\bar{w}) = (z\bar{z})(w\bar{w}) = |z|^2|w|^2$.
    (c) $|z + w|^2 = (z + w)(\overline{z + w}) = (z + w)(\bar{z} + \bar{w}) = z\bar{z} + w\bar{w} + z\bar{w} + w\bar{z}$
       $= |z|^2 + |w|^2 + z\bar{w} + w\bar{z}$.
    If $z = x + iy$ and $w = s + it$, then $z\bar{w} = xs + yt + i(ys - xt)$ and
    $w\bar{z} = xs + yt + i(xt - ys)$. Thus, $z\bar{w} + w\bar{z} = 2(xs + yt)$. Hence,
    $|z + w|^2 = |z|^2 + |w|^2 + 2(xs + yt) \leq |z|^2 + |w|^2 + 2|z| |w| = (|z| + |w|)^2$.

17. Note that $|z| = r$ imples that $x^2 + y^2 = r^2$. Sketch this circle in the plane.

18. If $z = x + 0i$ then $|z| = \sqrt{x^2 + 0^2} = \sqrt{x^2} = |x|$.

19. If $z = (a + ib)$ then $iz = -b + ai$. If z is the point (a, b) then iz is the point $(-b, a)$.
    These points are as desired.

section 1.6 (pp. 39–40).

1. (a) yes; (b) no, 2 has two images; (c) yes; (d) no, 1 has two images;
   (e) no, 3 has no image; (f) yes.

3. No, two images; yes.

4. No. Some presidents were vice-presidents and took office on the death of the previous president. They, therefore, had no vice-president serving at that time.

5. $f(0) = -9$, $f(1) = -12$, $f(\sqrt{2}) = -3 - 6\sqrt{2}$, $f(3) = 0$, $f(-3) = 63$,
   $f(h) = 3h^2 - 6h - 9$, $f(x + h) = 3(x + h)^2 - 6(x + h) - 9$,
   $f(x + h) - f(x) = 6xh - 6h + 3h^2$.

answers to selected exercises

6. (a)
$$
\begin{array}{cccccc}
2 & 4 & 6 & 8 & 10 & 12 \\
\downarrow & \downarrow & \downarrow & \downarrow & \downarrow & \downarrow \\
0 & 0 & 1 & 0 & 0 & 1
\end{array}
$$

7. Yes. It sends states to two element subsets in the set of all subsets of the senate.

9. (a)

n	-3	-2	-1	0	1	2	3
$s(n)$	-2	-1	0	1	2	3	4

(c)

α	a	b	c	$\ldots$	x	y	z
$s(\alpha)$	b	c	d	$\ldots$	y	z	a

10. Let r denote the current rate (in cents), and x the weight in ounces, then
$$p(x) = r([x] + 1).$$

11.

	(a)	(c)	(e)	(g)	(h)	(i)
$f(0) =$	0	1	1	1	0	0
$f(1) =$	1	3	0	2	2	2
$f(-1) =$	1	-1	2	0	-2	-2
$f(2) =$	4	5	1	3	4	4
$f(-2) =$	4	-3	3	-1	-4	-4
$f(\pi) =$	π^2	$2\pi + 1$	$\pi - 1$	4	6	6
$f(3/2) =$	$9/4$	4	$1/2$	2	2	3
$f(\sqrt{2}) =$	2	$2\sqrt{2}+1$	$\sqrt{2}-1$	2	2	2
$f(1.6) =$	2.56	4.2	0.6	2	2	3

12. (a) f is a function from R to R: $f(0) = 1, f(1) = 3, f(-1) = 1, f(2) = 7,$
$f(\pi) = \pi^2 + \pi + 1, f(1/2) = 7/4, f(\sqrt{2}) = 3 + \sqrt{2}.$
(c) f is a function from R to R: $f(0) = 0, f(1) = 1, f(-1) = -1, f(2) = 4,$
$f(\pi) = \pi^2, f(1/2) = 1/4, f(\sqrt{2}) = 2.$
(e) 1 is not in the domain of this f. (g) -1 is not in the domain of this f.

13. $A = 50w - w^2$.

14. $T_C = \frac{5}{9}(T_F - 32)$.

section 1.7 (pp. 47–50).

1. (h) is a surjection (onto); (d), (e), (f), and (i) are bijections (both).
3. (a) $X = \{0,1,2\}, Y = \{1,2,3\}$; (c) $X = R = Y$; (d) $X = R, Y = \{2\}$;
(f) $X = R\backslash\{0\} = Y$; (b), (e), (g), and (h) are not functions.
5. (a) and (d).
6. (b) and (c).
9. Only if its range consists of a single point.
10. 7(c) only.
11. The functions are (a), (b), (c), (f), and (g). (b) and (g) are injections.
12. No. Set $f(x) = 0$ the constant function.
13. (a) No. $abs\ z = abs\ \bar{z}$. (b) Yes.
14. (a) 26; (b) 37; (c) 100; (d) 435; (e) 9.
15. (a), (b), (d), (e), and (f).

section 1.8 (pp. 56–57).

1. (a) $(f + g)(x) = 2x + 1$; (c) $(f + g)(x) = (x + 1)^2$;
(e) $(f + g)(x) = x^2 + \sqrt{1 - x}, x \leq 1$; (g) $(f + g)(x) = 2x^2 - x - 1$;
(i) $(f + g)(x) = \sqrt{x} + x + 1, x \geq 0$; (k) $(f + g)(x) = 3x/2 - [x] - 1$;
(m) $(f + g)(x) = x^2 + x - 20$; (o) $(f + g)(x) = x^3 + x^2 - x + 2$.
2. (b) $fg(x) = 1 - x^2$; (d) $fg(x) = -x^4 - 2x^3 + 2x + 1$;
(f) $fg(x) = x^{7/2}, x \geq 0$; (h) $fg(x) = x^4 - x^3 - x + 1$; (j) $fg(x) = 2x[x] + x$;
(l) $fg(x) = x^3 + x^2 - x - 1$; (n) $fg(x) = 27 + 9x - 3x^2 - x^3$.

answers to selected exercises

3. (a) $\dfrac{f}{g}(x) = \dfrac{2}{2x - 1}$, $x \neq 1/2$; (c) $\dfrac{f}{g}(x) = \dfrac{x^2}{2x + 1}$, $x \neq -1/2$;

(e) $\dfrac{f}{g}(x) = \dfrac{\sqrt{1 - x}}{x^2}$, $x < 0$ or $0 < x \leq 1$; (g) $\dfrac{f}{g}(x) = \dfrac{x + 3}{x - 2}$, $x \neq 1, 2$;

(i) $\dfrac{f}{g}(x) = \dfrac{\sqrt{x}}{x + 1}$, $x \geq 0$; (k) $\dfrac{f}{g}(x) = \dfrac{3x - 2}{-2[x]}$, $x < 0$ or $x \geq 1$;

(m) $\dfrac{f}{g}(x) = x + 4$, $x \neq 4$; (o) $\dfrac{f}{g}(x) = x + 1$, all x.

4. (a) $f(g(x)) = 2$, $g(f(x)) = 3$; (c) $f(g(x)) = (2x + 1)^2$, $g(f(x)) = 2x^2 + 1$;
(e) $f(g(x)) = \sqrt{1 - x^2}$, $|x| \leq 1$; $g(f(x)) = 1 - x$, $x \leq 1$;
(g) $f(g(x)) = x^4 - 6x^3 + 13x^2 - 10x + 1$, $g(f(x)) = x^4 + 4x^3 - 2x^2 - 15x + 11$;
(i) $f(g(x)) = \sqrt{|x + 1|}$, $g(f(x)) = \sqrt{x} + 1$, $x \geq 0$;
(k) $f(g(x)) = -\frac{3}{2}[x] - 1$, $g(f(x)) = -[3x/2 - 1]$;
(m) $f(g(x)) = x^2 - 8x$, $g(f(x)) = x^2 - 20$;
(o) $f(g(x)) = x^6 - 3x^5 + 6x^4 - 6x^3 + 4x^2 - 3x + 2$,
$g(f(x)) = x^6 + 2x^3 - x + 2$.

5. For $x \in X$, there are two cases: (1) $x \in E$ and (2) $x \in E'$. For case (1), x in E implies
$(f_E + f_{E'})(x) = f(x) + 0 = f(x)$; and $f_E f_{E'}(x) = f(x) \cdot 0 = 0$. For case (2) we have
$x \in E'$, so $(f_E + f_{E'})(x) = 0 + f(x) = f(x)$ and $f_E f_{E'}(x) = 0f(x) = 0$.

7. $f(i(x)) = f(x)$ and $i(f(x)) = f(x)$.

8. Compare Exercises 2 and 4. There are many examples there.

9. $\text{conj} \circ \text{abs}(z) = \text{conj}(\text{abs}(z)) = \sqrt{x^2 + y^2} = \text{abs}(z)$, and $\text{abs}(\text{conj}(z)) = \text{abs}(z)$.

11. (a) $2x - 3x^2 + \frac{1}{2}x^3$; (c) $4x - 7/2$; (e) $2/x - 3\sqrt{1 - x} + x^2/2 + 1/2$.

13. $f^2(2) = 1$, $f^3(2) = -1$, $f^4(0) = 1$, $[f(2)]^{-1} = 1$, $[f(3)]^{-2} = 1$.

14. (b) If $h(a) = h(b)$, $a, b \in X$, then $g(f(a)) = g(f(b))$. Since g is one-to-one, this
implies that $f(a) = f(b)$. But f is one-to-one, so $a = b$. Thus, h is one-to-one.
(c) Combine the results of parts (a) and (b).

15. No. Contrast the results of Exercise 6(a) with Theorem 18 of Appendix I.

section 1.9 (pp. 65–66).

2. f and g are increasing on $[0,2]$ as are the functions in parts (a), (b), (c), (d), (e), (f),
(g), (i), (k), (l), and (m); (h) and (j) are decreasing.

4. (a) $f^{-1}(x) = (x - 1)/2$; (c) $f^{-1}(x) = \sqrt[3]{x}$; (d) $f^{-1}(x) = x - 3$;
(e) $f^{-1}(x) = x^2$; (f) $f^{-1}(x) = \sqrt[5]{x}$; (h) $f^{-1}(x) = \sqrt[3]{x - 1}$; (j) $f^{-1}(x) = x$;
(m) $f^{-1}(x) = x^3$; (p) $f^{-1}(x) = f(x)$. The function in (b), (g), (i), (k), (l), (n), and
(o) do not have inverses which are functions.

5. (a) $f^{-1}(x) = x + 1$; (b) $g^{-1}(a) = \sqrt[3]{a - 1}$; (c) $H^{-1}(x) = x/2$;
(d) $Q^{-1}(x) = -\sqrt[4]{x}$, $x \geq 0$; (e) $C^{-1}(x) = \sqrt{1 - x}$, $0 \leq x \leq 1$.

6. (a) Take $x \geq 0$ for $f(x)$, then $f^{-1}(x) = \sqrt{1 - x}$, $0 \leq x \leq 1$.
(c) Take $x \geq -1$ for $f(x)$, then $f^{-1}(x) = \sqrt{x} - 1$, $x \geq 0$.
(e) Take $x \geq 0$ for $f(x)$, then $f^{-1}(x) = \sqrt[4]{x}$, $x \geq 0$.
(g) Take $x \geq 1$, for $f(x)$, then $f^{-1}(x) = \sqrt{\sqrt{x} + 1}$, $x \geq 0$.

9. Domain S Image for Each Function S to S.

	(1)	(2)	(3)	(4)	(5)	(6)
$x_1 \longrightarrow$	x_1	x_1	x_3	x_2	x_2	x_3
$x_2 \longrightarrow$	x_2	x_3	x_2	x_1	x_3	x_1
$x_3 \longrightarrow$	x_3	x_2	x_1	x_3	x_1	x_2

answers to selected exercises

10. No, Yes, Yes, Yes.   11. If $a < b$ implies $f(a) < f(b)$, suppose $f(a) = f(b)$. Then $a \neq b$ leads to a contradiction.

12. (a) $|x^3|$;   (c) $|1 + x|$;   (e) $\sqrt[3]{[x]}$.

13. Let $x \in X$, $f^{-1}(f(x)) = x$, since f is one-to-one.
Let $y \in Y$, since f is onto, $f^{-1}(y) \in X$ for each y in Y, and $f(f^{-1}(y)) = y$.

15. $f + g$: no, take $f(x) = -g(x)$, then $f + g$ is the zero function.
fg: no, take $f(x) = g(x) = x$, x^2 is not injective.
$f - g$: no, take $g(x) = f(x)$, then $f - g$ is the zero function.
$f \circ g$: yes, see Exercise 14 of Section 8.

17. No, see Exercise 15.

chapter 2.

section 2.1 (pp. 75–77).

1. $1 = 1(2)/2$. If $1 + 2 + \cdots + k = k(k + 1)/2$, then
$$(1 + 2 + \cdots + k) + k + 1 = k(k + 1)/2 + (k + 1) = [k(k + 1) + 2(k + 1)]/2$$
$$= (k + 1)(k + 2)/2 \text{ as desired.}$$

3. $1^3 = 1^2(2)^2/4$. If $1^3 + 2^3 + \cdots + k^3 = k^2(k + 1)^2/4$, then
$$(1^3 + 2^3 + \cdots + k^3) + (k + 1)^3 = k^2(k + 1)^2/4 + (k + 1)^3$$
$$= [k^2(k + 1)^2 + 4(k + 1)^3]/4$$
$$= \frac{(k + 1)^2[k^2 + 4k + 1]}{4}$$
$$= \frac{(k + 1)^2(k + 2)^2}{4}, \text{ as desired.}$$

5. $2 = 1(2)$. If $2 + 4 + 6 + \cdots + 2k = k(k + 1)$, then
$$(2 + 4 + 6 + \cdots + 2k) + 2(k + 1) = k(k + 1) + 2(k + 1) = (k + 1)(k + 2).$$

7. $\dfrac{1}{1 \cdot 2} = \dfrac{1}{2}$. If $\dfrac{1}{1 \cdot 2} + \dfrac{1}{2 \cdot 3} + \cdots + \dfrac{1}{k(k + 1)} = \dfrac{k}{k + 1}$, then
$$\frac{1}{1 \cdot 2} + \frac{1}{2 \cdot 3} + \cdots + \frac{1}{k(k + 1)} + \frac{1}{(k + 1)(k + 2)} = \frac{k}{k + 1} + \frac{1}{(k + 1)(k + 2)}$$
$$= \frac{k(k + 2) + 1}{(k + 1)(k + 2)}$$
$$= \frac{k^2 + 2k + 1}{(k + 1)(k + 2)}$$
$$= \frac{k + 1}{k + 2}, \text{ as desired.}$$

9. For $n = 1$ we have a triangle the sum of whose angles is $180°$. Suppose that the proposition is true when there are k sides. Consider a polygon with $k + 1$ sides. Let the vertices be $A_1, A_2, \ldots, A_{k+1}$ (given clockwise) so that the sides are $A_1 A_2$, $A_2 A_3, \ldots, A_k A_{k+1}$. Denote the corresponding angles by $\alpha_1, \alpha_2, \ldots, \alpha_k$. Now connect point A_1 with point A_3 so that $A_1 A_2 A_3$ is a triangle. Let $\angle A_2 A_1 A_3$ be called β_2 and call $\angle A_2 A_3 A_1$, β_3. Thus, $\alpha_2 + \beta_2 + \beta_3 = 180°$. Denote $\angle A_{k+1} A_1 A_3$

answers to selected exercises

by β_1 and $\angle A_1 A_3 A_4$ by β_4. We have chosen notation so that $\alpha_1 = \beta_1 + \beta_2$ and $\alpha_3 = \beta_3 + \beta_4$. The polygon $A_1 A_3 A_4 \ldots A_k A_{k+1}$ has k sides so that

$$\beta_1 + \beta_4 + \alpha_4 + \alpha_5 + \cdots + \alpha_{k+1} = k(180°).$$

Hence, $\alpha_1 + \alpha_2 + \alpha_3 + \alpha_4 + \cdots + \alpha_{k+1} = \beta_1 + \beta_2 + \beta_3 + \beta_4 + \alpha_2 + \alpha_4 + \cdots + \alpha_{k+1}$
$= (\beta_2 + \beta_3 + \alpha_2) + (\beta_1 + \beta_3 + \alpha_4 + \cdots + \alpha_{k+1}) = 180° + k(180°)' = (k+1)(180°).$

11. $16 < 24$, so $2^4 < 4!$ Suppose $2^{k+3} < (k+3)!$. Then $2^{k+1+3} = 2^{k+4}$ and
$2^{k+4} = 2 \cdot 2^{k+3} < 2(k+3)! < (k+4)(k+3)!$, since $2 < k+4$. Thus,
$2^{k+1+3} < (k+1+3)!$

13. $1^5/5 + 1^3/3 + 7(1)/15 = 15/15 = 1$. Suppose that $k^5/5 + k^3/3 + 7k/15 = m$ is an integer. Then consider $(k+1)^5/5 + (k+1)^3/3 + 7(k+1)/15$. From the assumption that m is an integer we conclude that $3k^5 + 5k^3 + 7k = 15m$. Therefore $\frac{1}{15}[3(k+1)^5 + 5(k+1)^3 + 7k] = \frac{1}{15}[3k^5 + 1 + 3A + 5k^3 + 1 + 5B + 7k + 7]$, for appropriate A and B. Thus the left member equals $m + 1 + (3A + 5B)/15$. Now A is the rest of the expansion of $(k+1)^5$ and B is the rest of the expansion of $(k+1)^3$, and since 5 and 3 are prime each of the coefficients of $3A$ and $5B$ is divisible by 15. Thus, $(3A + 5B)/15$ is an integer. Therefore, the left member of the equation above (for $k+1$) is divisible by 15 and the desired induction holds.

15. $7^2 + 16 - 1 = 64$ is divisible by 64. Suppose that $7^{2k} + 16k - 1$ is divisible by 64, and consider $7^{2(k+1)} + 16(k+1) - 1$. This becomes

$$7^{2k+2} + 16(k+1) - 1 = 7^{2k+2} + 7^2 - 7^2 + 7^2(16k) - 7^2(16k) + 16k + 16 - 1$$
$$= 7^2(7^{2k} + 16k - 1) - 16k(7^2 - 1) + 7^2 + 16 - 1$$
$$= 7^2(7^{2k} + 16k - 1) - k(16)(48) + 64$$

Now each of the terms in this last expression is divisible by 64, the first by our induction assumption; hence, the sum is divisible by 64 as desired.

17. Let $z = a + bi$, then $\bar{z} = a - bi$. Then $(\bar{z})^1 = a - bi = \overline{(z^1)}$. Now suppose $\overline{(z^k)} = (\bar{z})^k$, then $\overline{(z^{k+1})} = \overline{(z \cdot z^k)} = (\bar{z})\overline{(z^k)}$ because the conjugate of the product of two complex numbers is the product of their conjugates (Theorem 1.5.1(ii)). Thus, $\overline{(z^{k+1})} = (\bar{z})\overline{(z^k)} = (\bar{z})(\bar{z})^k = (\bar{z})^{k+1}$, as desired.

19. $P(1)$ states $3 = 3(1)(2)/2 + 0$ is true. $P(2)$, however, is false. $P(2)$ states that $3 + 6 = 3(2)(3)/2 + 1$.

21. When $n = 1$ we have $a = (a + ar)/(1 - r) = a(1 - r)/(1 - r)$ is true. Suppose for $n = k$ the statement is true, and consider the sum of $k + 1$ terms

$$(a + ar + ar^2 + \cdots + ar^{k-1}) + ar^k = (a - ar^k)/(1 - r) + ar^k$$
$$\frac{a - ar^k + (1 - r)ar^k}{1 - r}$$
$$= \frac{a - ar^k + ar^k - ar^{k+1}}{1 - r}$$
$$(a - ar^{k+1})/(1 - r).$$

23. Try to do the proof for $k = 2$. It will fail because if g_1 has red hair and g_2 does not, then dropping g_1 leaves the set $\{g_2\}$ which is not a set of girls having at least one with red hair. Yet this is tascitly assumed in the "proof" given.

25. From Exercise 24 we see that the induction principle is equivalent to the well ordering principle. Assume (a') and (b'). Let M be the set of positive integers n for which $P(n)$ is false. If $M \neq \emptyset$ it has a least element m. Now $P(1)$ is true so $1 \neq m$. For any $k < m$, since m is the least element of M, we must have $k \notin M$, so $P(k)$ must be true. But then $P(m)$ is true, contradicting $m \in M$. Therefore, $M = \emptyset$ and $P(n)$ is true for

answers to selected exercises

every positive integer n. Hence, the original induction principle implies well ordering implies this new formulation of induction.

27. The case $n = 1$ is trivial. Assume that $\sum_{i=1}^{n-1} ka_i = k \sum_{i=1}^{n-1} a_i$. Then

$$\sum_{i=1}^{n} ka_i = \sum_{i=1}^{n-1} ka_i + ka_n = k\left(\sum_{i=1}^{n-1} a_i\right) + ka_n$$

$$= k\left(\sum_{i=1}^{n-1} a_i + a_n\right) = k \sum_{i=1}^{n} a_i$$

29. 1. $\displaystyle\sum_{i=1}^{n} i = \frac{n(n+1)}{2}$;   2. $\displaystyle\sum_{i=1}^{n} i^2 = \frac{n(n+1)(2n+1)}{6}$

3. $\displaystyle\sum_{i=1}^{n} i^3 = \frac{n^2(n+1)^2}{4} = \left(\sum_{i=1}^{n} i\right)^2$;   4. $\displaystyle\sum_{i=1}^{n} \frac{i}{2} = \frac{n(n+1)}{4}$

29. 5. $\displaystyle\sum_{i=1}^{n} 2i = 2 \sum_{i=1}^{n} i = n(n+1)$ 6. $\displaystyle\sum_{i=1}^{n} i \cdot i! = (n+1)! - 1$

7. $\displaystyle\sum_{i=1}^{n} \frac{1}{i(i+1)} = \frac{n}{n+1}$

section 2.2 (pp. 82–83).

1. (a) 3; (b) 15; (c) 21; (d) (40)(31)(29); (e) 550; (f) 30; (g) 6; (h) 60;
   (i) undefined $1 + 1 + 1 > 2$.
3. $(3)(4)^3 = 192$.   5. 17,576,000.   7. $5! = 120$ 9. 2^5.
11. 5^{20}.   13. (b) 14.   15. 6.
17. $\displaystyle\binom{p}{2} = \frac{p!}{2!(p-2)!} = \frac{p(p-1)}{2}$, and $p - 1$ is even.
19. $\displaystyle\binom{n}{k}$ counts the number of k-element subsets of a set with n elements. Thus $\displaystyle\sum_{k=0}^{n} \binom{n}{k}$

counts *all* the subsets of an n-element set, hence, equals 2^n.

section 2.3 (pp. 85–86).

1. (a) $a^7 + 7a^6b + 21a^5b^2 + 35a^4b^3 + 35a^3b^4 + 21a^2b^5 + 7ab^6 + b^7$.
   (b) $1 + 6b + 15b^2 + 20b^3 + 15b^4 + 6b^5 + b^6$.
   (c) $1 - 5b + 10b^2 - 10b^3 + 5b^4 - b^5$.
   (d) $a^8 - 8a^7b + 28a^6b^2 - 56a^5b^3 + 70a^4b^4 - 56a^3b^5 + 28a^2b^6 - 8ab^7 + b^8$.
   (e) $p^7 - 7p^6q + 21p^5q^2 - 35p^4q^3 + 35p^3q^4 - 21p^2q^5 + 7pq^6 - q^7$.
   (f) $81c^4 - 108c^3d + 54c^2d^2 - 12cd^3 + d^4$.   (g) $8x^3 - 36x^4y + 54x^2y^2 - 27y^3$.
   (h) 0.941192   (i) 1.0030301
3. $2^n = (1+1)^n = \displaystyle\sum_{r=0}^{n} \binom{n}{r}(1)^{n-r}(1)^r = \sum_{r=0}^{n} \binom{n}{r}$.
5. Add the results of exercises 3 and 4 together, where n is even, to obtain

$$2^n = 2\binom{n}{0} + 2\binom{n}{2} + 2\binom{n}{4} + \cdots + 2\binom{n}{n}$$

Now divide both sides by two to obtain the desired result.

7. (a) $15x^2y$; (b) $4x^2$; (c) $2160x^2$; (d) $495x^2$.
9. (a) 22; (b) 41; (c) 44; (d) 210.

answers to selected exercises

section 2.4 (pp. 93–94).

1. (a) 60; (b) 88; (c) -87; (d) 172; (e) 57/6; (f) -81; (g) -39;
   (h) -79; (i) $29b - 28a$; (j) $29v + 30a$.

3. (a) 240; (b) 330; (c) -315; (d) 600; (e) 65/3; (f) -225; (g) -145;
   (h) -165; (i) $105b - 90a$; (j) $120a + 105v$.

5. 1950.

7. (a) 16; (b) 81; (c) 81; (d) 8/27; (e) 4/9; (f) 1/16; (g) 1/81; (h) $9\sqrt{2}$;
   (i) 0.00004; (j) b^4/a^3.

9. (a) 31; (b) 121; (c) 61; (d) 121/27; (e) 209/36; (f) 31/64; (g) 121/81;
   (h) $13\sqrt{2} + 4\sqrt{6}$; (i) 0.44444.

11. $a_1 = 8$, $a_2 = 12$.

13. (a) 4/9; (b) 2/9; (c) 39/99; (d) 2/5; (e) 4/99; (f) 40/9; (g) 2; (h) 22/7;

15. 6/7   17. \$1378.00.

chapter 3.

section 3.1 (pp. 102–103).

1. (a) deg 2; (c) not a polynomial, x^{-1} is not allowed; (e) deg 0;

   (g) not a polynomial, $\dfrac{1}{x}$ is not allowed; (i) deg 1;

   (j) the zero polynomial has no degree; (l) deg 7.

2. (a) $2x$, deg 1; (c) $x^5 - x^4 + x^3 - 10x - 1$, deg 5; (e) $x^5 + 2x^2 - 2x$, deg 5.

3. (a) $x^2 - 1$, deg 2; (b) $x^5 + x^4 + x^3 + 2x^2 - 1$, deg 5;
   (c) $-x^9 + x^8 + x^7 - 7x^6 + 3x^5 - 2x^4 - 4x^3 + 20x^2 - 7x$, deg 9;
   (d) $-x^{12} + x^{10} - x^8 + 2x^6 - x^4 + x^2 - 1$, deg 12;
   (e) $x^9 - x^8 + x^7 - 3x^6 + 3x^5 + 3x^4 - 2x^3 - 3x^2 - 4x - 1$, deg 9.

5. (a) $-1,0,-3,-2$; (b) $-1,1,-17,-5$; (c) $0,-6,26,6$; (d) $1,0,-51,0$;
   (e) $1,3,-45,-1$.

7. (a) $x^2 + 1$, deg 2; (b) $x^3 + (1 - i)x^2 + (1 - i)x - i$, deg 3;
   (c) $x^4 + (1 - i)x^3 - (1 - 2i)x^2 + 4ix - 3i$, deg 4; (d) $x^2 - 2x + 2$, deg 2;
   (e) $x^2 - \sqrt{3}x + 1$, deg 2; (f) $x^4 - i\sqrt{2}x^3 - ix^2 - (\sqrt{2} - i)x + \sqrt{2}$, deg 3;
   (g) $-3x^4 - 1$, deg 4; (h) $x^3 - x^2 - (2 + i)x + (1 + i)$, deg 3.

9. (a) $p(x) = q(x) = x^3$; (b) $p(x) = x^3 + x^2$, $q(x) = -x^3$;
   (c) $p(x) = x^3 + x$, $q(x) = -x^3$; (d) $p(x) = x^3 + 1$, $q(x) = -x^3$.

11. Suppose that deg $p(x) = m$ and deg $q(x) = n$. Then deg $p(x)q(x) = m + n$. But if $p(x)q(x) = $ the zero polynomial, it has no degree. This is a contradiction if both m and n are positive integers. Hence, either $p(x)$ or $q(x)$ must fail to have a positive degree; i.e., must be the zero polynomial.

13. Let $p(x) = a_n x^n + a_{n-1} x^{n-1} + \cdots + a_1 x + a_0$, and let $q(x) = b_m x^m + \cdots + b_1 x + b_0$, with deg $p(x) = n$ and deg $q(x) = m$. Thus, $a_n \neq 0$ and $b_m \neq 0$. Then the highest degree term of $p(x)q(x)$ is $a_n b_m x^{n+m}$, and $a_n b_m \neq 0$. Thus, deg $p(x)q(x) = m + n = $ deg $p(x) + $ deg $q(x)$. If either $p(x)$ or $q(x)$ is the zero polynomial, then both sides of the equation are vacuously equal.

15. $c = 0$.

17. $b = 2$, $m = 1$.

answers to selected exercises

section 3.2 (pp. 111–112).

3. (a) See 2(c); (b) See 2(a); (c) $(0,-3)$, $(\frac{3}{2},0)$; (d) $(0,-2)$, $(4,0)$;
   (e) $(0,-2)$, $(3,0)$; (f) $(0,3)$, $(2,0)$; (g) $(0,-1)$, $(2,0)$; (h) $(0,-6)$, $(3,0)$;
   (i) $(0,\frac{9}{7})$, $(\frac{9}{5},0)$; (j) $(0,-\frac{3}{2})$, $(4,0)$; (k) $(0,3)$, $(2,0)$; (l) $(0,-2)$, $(3,0)$.
4. (a) $(0,1)$, $(1,0)$; (c) $(0,-9)$, $(3,0)$, $(-3,0)$; (e) $(0,-2)$, $(1,0)$, $(-2,0)$;
   (f) $(0,6)$ no x-intercepts; (g) $(0,16)$, $(\frac{4}{3},0)$, $(-\frac{4}{3},0)$; (i) $(0,-1)$, $(\frac{1}{2},0)$, $(-1,0)$;
   (k) $(0,-2)$, $(\frac{2}{3},0)$, $(-\frac{1}{2},0)$.
5. (a) $(1,0)$ min.; (c) $(0,-9)$ min; (d) $(0,9)$ max.; (f) $(\frac{1}{2},\frac{23}{4})$ min.;
   (h) $(0,8)$ max.; (k) $(\frac{1}{12},-\frac{49}{24})$ min.

section 3.3 (pp. 115–116).

1. (a) $(x-3)(x+3)$; (c) $(x-1)(x^2+x+1)$;
   (e) $(x+1)(x-1)(x^2+1)(x^2-\sqrt{2}x+1)(x^2+\sqrt{2}x+1)$;
   (g) $(x-4)(x-1)$; (i) $(x-1)(x-2)(x+2)$; (j) is prime.
2. (a) $x^2-9=(x+3)(x-3)+0$; (c) $x^3-x^2+4=(x^2-4x)(x+3)+(12x+4)$;
   (e) $3x^3-7x^2+2x+1=(x^2+x+1)(3x-10)+(9x+11)$;
   (g) $x^3+1=(x^2-x+1)(x+1)+0$; (i) $6x+9=(9x^2+6)(0)+(6x+9)$.
3. (a) $x^2+7x+38$, 194; (c) x^2-3x+6, -8; (e) $4x^2-8x+7$, -10;
   (g) $x^2-11x+1$, $99x-13$; (i) x^2-1, 0.
4. If $f(x)=g(x)h(x)$ in $F[x]$, then $\deg f = \deg g + \deg h$. But $\deg f = 1$; hence, $\deg g$ or $\deg h$ must be zero. That is, one or the other of g or h is a constant polynomial. Thus, f is irreducible.

section 3.4 (pp. 119–120).

1. (a) -5; (b) 2; (c) 7; (d) 5; (e) $-1+3i$; (f) $2+6i$; (g) $-3i$.
3. (a) x^2+4x-7; (b) x^3+2x^2-x+1; (c) $x^2-(b+c)x+bc$;
   (d) $0.2x^2-0.04x-0.022$, $r(x)=0.0194$;
   (e) $x^2-2ix+(-4+2i)$, $r(x)=-3-6i$;
   (f) $x-(7-i)$, $r(x)=18-10i$.
5. $c = \pm\dfrac{\sqrt{2}}{2}$.
7. For a real number a, $a^4 \geq 0$, and $a^2 \geq 0$, so $a^4+2a^2+1 \geq 1$. Therefore, a cannot be a root, so $x-a$ cannot be a factor. Since $p(x)=(x^2+1)^2$, it is not prime.
9. By induction: set $n=2m$. For $m=1$, the statement is $x^2-y^2=(x-y)(x+y)$. Suppose the statement is true for $m=k$ and consider it for $m=k+1$. Then,
$$x^{2k+2}-y^{2k+2}=x^{2k+2}-x^{2k}y^2+x^{2k}y^2-y^{2k+2}=x^{2k}(x^2-y^2)+y^2(x^{2k}-y^{2k}).$$
Both terms of the right member are divisible by $x+y$, so we have the desired result.

section 3.5 (p. 127).

1. (a) x^4-1; (c) x^4-5x^2+4; (e) $x^4-6ix^3-11x^2+6ix$;
   (g) $x^4-(3+i)x^3+(3+3i)x^2-(1+3i)x+i$; (i) x^4+x^2.
2. (a) -1 (mult. 3), 3, -2; (c) -1 (mult. 2), i, $-i$;
   (e) -5 (mult. 3), 5 (mult. 3); (g) 0 (mult. 3), -1 (mult. 2), i (mult. 3), $-i$ (mult. 3);
   (i) $-\dfrac{1}{2}-i\dfrac{\sqrt{3}}{2}$ (mult. 2), $-\dfrac{1}{2}+i\dfrac{\sqrt{3}}{2}$ (mult. 2), $\dfrac{1+\sqrt{5}}{2}$ (mult. 3), $\dfrac{1-\sqrt{5}}{2}$ (mult. 3).
3. $(x-3)^2(x-2i)(x+2i)$.

answers to selected exercises

5. If $p(x)$ in $C[x]$ has degree $n > 0$ and m distinct roots, $r_1, r_2, \ldots, r_m$, then $(x - r_1)(x - r_2) \cdots (x - r_m) = t(x)$ is a factor of $p(x)$. Clearly deg $t(x) = m$. If $m > n$, then deg $t(x) >$ deg $p(x)$, which is impossible.

6. $2x^3 - 4x^2 - 10x + 12$.   7. $2x^3 + 4x^2 - 2x - 4$.

9. Suppose that $p(x) = (x - r)q(x) + p(r)$, $p(r) \geq 0$, and coefficients of $q(x)$ non-negative. Let $s > r$. Then $p(s) = (s - r)q(s) + p(r) > 0$, since $q(s)$ and $(s - r)$ are both positive. Hence $p(s) \neq 0$, so s cannot be a root.

11. -2.624.

12. (a) $1, -2$; (c) $0, -1$; (e) $6, -2$; (g) $2, -2$.

section 3.6 (pp. 132–133).

1. (a) ± 1; (c) $\pm 1, \pm 3, \pm \frac{1}{2}, \pm \frac{3}{2}$; (e) $\pm 1, \pm \frac{1}{3}$; (g) $\pm 1, \pm 2$;
   (i) $\pm 1, \pm 3, \pm 5, \pm 15, \pm 25, \pm 75, \pm \frac{1}{2}, \pm \frac{3}{2}, \pm \frac{5}{2}, \pm \frac{15}{2}, \pm \frac{25}{2}, \pm \frac{1}{3}, \pm \frac{5}{3}, \pm \frac{25}{3}$;
   (k) $\pm 1, \pm \frac{1}{3}, \pm \frac{1}{2}, \pm \frac{1}{4}, \pm \frac{1}{6}, \pm \frac{1}{12}$.

2. (a) $-\frac{1}{4}$; (c) $\frac{1}{2}, -\frac{2}{3}$; (e) $\frac{5}{2}, \frac{2}{3}$; (g) $\frac{3}{2}, \frac{1}{3}, 4$; (i) $-\frac{2}{3}$; (k) $\frac{7}{5}$.

3. (a) $4x + 1$; (c) $x^2 + x + 1$; (e) $(x - 1)(6x^2 + 11x + 4)$;
   (g) $(2x - 11)(x^2 - 2x - 2)$; (i) $(x - 1)(x^3 - 3x^2 - 4)$; (k) $(x + 1)^3(x - 2)^2$.

4. (a) $\frac{2}{3}$; (c) $1, \frac{1}{2} \pm i\dfrac{\sqrt{7}}{2}$; (e) $\frac{1}{2}, 2 \pm \sqrt{2}$; (g) $\frac{3}{2}, 1 \pm \sqrt{5}$; (i) $2, -2, -\frac{1}{2} \pm i\dfrac{\sqrt{7}}{2}$.

5. (a) $1, 1$; (b) $-1, -1$; (c) $\frac{1}{2} \pm i\dfrac{\sqrt{5}}{2}$; (d) $1, \frac{1}{3}$.

6. (b) $\frac{5}{3}$; (d) $\frac{5}{2}, -\frac{1}{3}$; (f) $\frac{1}{3} \pm i\dfrac{\sqrt{15}}{2}$; (h) $\dfrac{-2}{5}, \dfrac{-3 \pm \sqrt{5}}{2}$; (j) $1, \frac{1}{3}, -2, -\frac{1}{2}$;
   (l) $0, \frac{1}{2}, -2, 3, -\frac{2}{5}$.

7. The only possible rational roots are ± 1 and ± 2. None of these is a root as can readily be tested by synthetic division.

9. (a) $-6.00, -1.73, 1.73$; (c) $-4.00, 1.62, -0.62$; (e) $0.67, 1.32, -1.32$.

11. 1 inch.   13. Rational roots $\dfrac{p}{q}$ must be such that q divides $a_n = 1$, so $q = \pm 1$.

14. (a) $0, 2, 2$; (c) $0, 3, 3, -2, -2$; (e) $0, 3, 1, \frac{1}{2}, -\frac{1}{2}$.

section 3.7 (pp. 137–138).

1. (a) $x^2 + 9$; (c) $3x^2 + 27$; (e) $2x^2 + 10$; (g) $x^2 - 4x + 13$;
   (i) $2x^2 + 8$; (k) $2x^2 - 2\sqrt{2}x + 2$.

2. (a) $x^3 - 3x^2 + x - 3$; (c) $x^3 - 2x - 4$; (e) $x^3 - 3x^2 - 5x + 39$;
   (g) $x^3 - x^2 + 5x - 5$; (i) $2x^3 + (3 - 4\sqrt{3})x^2 + (8 - 6\sqrt{3})x + 12$;
   (k) $x^3 - \sqrt{3}x^2 + 3x - 3\sqrt{3}$.

3. (a) $\pm i, -2 \pm \sqrt{2}$; (c) $1 \pm i, \frac{1}{2}$; (e) $1 \pm \sqrt{3}, \frac{1}{3}$; (g) $\pm i\sqrt{2}, -1$;
   (i) $1, 1, 1 \pm 2i$; (k) $-1 \pm i\sqrt{3}$, each of multiplicity 2, and 2.

4. $b = 4, c = -12$.   5. (c) complete the square.

6. (a) $1, -\frac{1}{2} \pm i\dfrac{\sqrt{3}}{2}$; (c) $\dfrac{\sqrt{2}}{2} \pm i\dfrac{\sqrt{2}}{2}$; (e) $\pm 1, \pm i$.

7. $i^4 = (i^2)^2 = (-1)^2 = 1$; $i^k = i^{k-4} \cdot i^4 = i^{k-4} \cdot 1$.

9. (a) $2(x - 2)(2x - 5)(x + 3)$; (c) $(3x + 5)(x - 1)(x + 2)$;
   (e) $(4x - 1)(3x + 2)(x - 1)$; (g) $(x - 2)^2(x^2 + x + 1)$.

10. (a) $(x - 1)(3x + 2)$; (c) $2x^2 - 5x - 12$; (e) $(x^2 + 1)(x + 1)^2$;
   (g) $(x^2 - 3)(2x - 1)(x - 2)$.

answers to selected exercises

11. Use induction on the multiplicity k of z. The case $k = 1$ is Theorem 3.7.1. Suppose that the statement is true for $k = n$. Now consider $k = n + 1$; viz., z is a root of multiplicity $n + 1$ of the polynomial $p(x)$ in $R[x]$. Since z is a root, so is $\bar{z}$ (Theorem 3.7.1). Thus, $p(x) = (x - z)(x - \bar{z})q(x)$. Now, $(x - z)(x - \bar{z})$ has real coefficients, so $q(x)$ has real coefficients. Furthermore, z is a root of $q(x)$ of multiplicity n; hence, by our assumption $\bar{z}$ is a root of $q(x)$ of multiplicity n. Therefore $p(x)$ has $\bar{z}$ as a root of multiplicity $n + 1$ as desired.

13. (a) 1; (c) 1; (e) 2; (g) $2x - 4$; (i) $10x^4 + 6x$.

14. If r is a double root of $p(x)$, then $p(x) = (x - r)^2 s(x)$. Now set $g(x) = (x - r)s(x)$, so $g(r) = 0$. Since $p(x) = (x - r)g(x)$, it follows that $p'(x) = (x - r)g'(x) + g(x)$. Therefore, $p'(r) = (r - r)g'(r) + g(r) = 0$.

section 3.8 (p. 145)

1. (a) $x = 0$; (c) $x = \frac{1}{2}$; (e) $x = 1, x = -1$; (g) $x = -1$;

   (i) $x = \dfrac{\sqrt{2}}{2}, x = -\dfrac{\sqrt{2}}{2}$.

2. (g) $y = 1$; (i) $y = \frac{1}{2}$; (j) $y = 1$.

4. (a) $x = 1, y = 1$; (c) $x = -2, x = 1, y = 0$; (e) no vertical, $y = 1$;
   (g) $x = 1, x = 2, x = 3, y = 1$; (i) $x = 5, x = 3, y = 1$.

6. Yes, see 1(h), 4(e), 4(i), etc.

7. $A = \frac{1}{4}, B = -\frac{3}{8}, C = \frac{9}{8}$.

chapter 4.

section 4.1 (pp. 156–158).

1. (a) 3^3; (c) $5^0 = 1$; (e) $(5)(2^7)$; (g) $\dfrac{2^2}{3}$; (i) $\dfrac{3^6}{5^2 2^{10}}$;

   (k) $7(10^3) + 3(10^2) + 9(10) + 2$.

2. (a) 4; (c) 2; (e) $\frac{1}{12}$; (g) $\frac{1}{25}$; (i) 3.375.

3. (a) 7; (c) -4; (e) -2; (g) $\frac{1}{2}$; (i) $-\frac{1}{4}$; (k) $\frac{1}{2}$.

4. (a) $a^{7/6}$; (c) u^4; (e) 72.

5. (a) a^2; (c) $ab^{8/3}$; (e) $b^{13/3}$; (g) a^5; (i) $a^2 - \dfrac{1}{a^2}$; (k) $2 + b + \dfrac{2}{b} + \dfrac{3}{b^2}$.

6. (a) $a^{1/4}$; (c) $a^{1/2}$; (e) $3^{1/2}a^{1/2}$; (g) 1; (i) $2^{5/2}b^5$; (k) $\dfrac{3^{2/3}a^{2/3}}{2^{1/2}b^{1/6}}$.

7. (a) $3^{3/4}$; (c) $a^{1/6}$; (e) $\dfrac{1}{a^{23/12}b^{97/60}}$; (g) $\dfrac{a^2 + ab^2}{b^2 + ba^2}$; (i) $\dfrac{1}{a^3 + b^3}$; (k) $\dfrac{ab}{2b - 3a}$.

9. (a) Use induction. For $n = 1$ the second factor equals 1. Assume the statement is true for $n = k$ and consider
$$b^{k+1} - 1 = b^{k+1} - b^k + b^k - 1 = b^k(b - 1) + b^k - 1$$
$$= b^k(b - 1) + (b - 1)(b^{k-1} + \cdots + b + 1)$$
$$= (b - 1)[b^k + (b^{k-1} + b^{k-2} + \cdots + 1)]$$

   (b) Set $b^{n-1} + b^{n-2} + \cdots + b + 1 = p(b)$. Note that $p(b)$ is a polynomial with integral coefficients which are all positive; therefore, for $r \geq 0, p(r) > 0$. (i) For $b = 1, b^n - 1 = (1 - 1)p(1) = 0$, so $b^n = 1$. (ii) If $b > 1$, then $b - 1 > 0$, and $p(b) > 0$. Hence, $b^n - 1 > 0$, and $b^n > 1$. (iii) If $0 < b < 1$, then $b - 1 < 0$ and $p(b) > 0$. Hence, $b^n - 1 < 0$, and $b^n < 1$.

answers to selected exercises

11. (a) $b^r = b^{m/n} = (b^{1/n})^m$, but $b > 1$ implies that $s = b^{1/n} > 1$. Since $s^m > 1$ we have $b^r > 1$.

    (b) For $b^r = (b^m)^{1/n}$ we have, if $b = 1$, $b^m = 1$; and $1^{1/n} = 1$ by a proof similar to 10(b).

    (c) For b^r as in part (b), if $0 < b < 1$, then $0 < b^m < 1$ by 9. Hence, $0 < (b^m)^{1/n} < 1$ by 10.

14. If $a, b \geq 0$ and $\sqrt{a} \leq \sqrt{b}$, then from Chapter 1 $(\sqrt{a})^2 \leq (\sqrt{a})(\sqrt{b}) \leq (\sqrt{b})^2$, or $a \leq b$, so $0 \leq a \leq b$. Conversely, if $\sqrt{b} < \sqrt{a}$, then, since both are positive, $(\sqrt{b})^2 < (\sqrt{b})(\sqrt{a}) < (\sqrt{a})^2$, or $b < a$, and it is false that $0 \leq a \leq b$.

section 4.2 (pp. 164–165).

5. (a) 3.004; (c) 0.298; (e) 0.1003; (g) 0.0497; (i) 0.0043; (j) 22033.4.

6. (a) 2.718; (c) 8.166; (e) 0.896.

9. (a), (c), (f), (g), and (h) are even functions.

11. (a) $b^{x+y} = b^x \cdot b^y$; (c) $b^{xy} = (b^x)^y = (b^y)^x$.

14. (a) 6; (c) -3; (e) -5; (g) 1; (i) $-3, 2$; (k) 0.3.

section 4.3 (pp. 170–172).

2. (a) 5; (c) 4; (e) 5; (g) $\frac{4}{3}$; (i) 3; (k) -1.7.

3. (a) 5; (c) 3; (e) 25; (g) 27; (i) $\frac{1}{2}$.

4. (a) 0.9030; (c) $-2 + 0.9030 = -1.0970$; (e) 0.4313; (g) 0.7781; (i) -2.2219.

5. (a) 2.8; (c) -0.4; (e) 0.4; (g) 0.3; (i) 2.1.

6. (a) 0; (c) $\pm\sqrt{1 - e^{-1}}$; (e) $3, -2$; (g) $-4, 2$; (i) $-5, 2$.

7. Set $m = \log_a b$, so $a^m = b$. Then $1 = \log_b b = \log_b(a^m) = m \log_b a$. Upon substitution for m the result follows. A second proof uses Theorem 4.3.2 (iv) with $a = r$.

8. (a) 1.5850; (c) 1.5850; (e) -0.9030; (g) 0.3322; (i) 5.2890.

9. (a) $-1 < x < 7$; (c) $1 < x < 10$; (e) $10 < x < 100$; (g) $e < x < e^2$; (i) $x > 9$.

11. $-\log_b(\sqrt{1 + r^2} - r) = \log_b \dfrac{1}{\sqrt{1 + r^2} - r}$. But

$$\frac{1}{\sqrt{1 + r^2} - r} = \frac{1}{[\sqrt{1 + r^2} - r]} \cdot \frac{[\sqrt{1 + r^2} + r]}{[\sqrt{1 + r^2} + r]} = \frac{\sqrt{1 + r^2} + r}{1 + r^2 - r^2} = \sqrt{1 + r^2} + r.$$

13. (a) Domain: $\{x \mid x < 1\}$, Image: R; (c) Domain: $\{x \mid x \neq 1\}$, Image: R; (e) Domain: $\{x \mid x > 0\}$, Image: $\{y \mid y \geq 0\}$.

14. (a) $L(x) = L(1 \cdot x) = L(1) + L(x)$; hence, $L(1) = 0$.

    (b) $E(x) = E(x + 0) = E(x)E(0)$; hence, $E(0) = 1$ if $E(x) \neq 0$.

section 4.4 (pp. 177–179).

1. (a) 0.0043; (c) 5.4281; (e) 3.8871; (g) $0.3856 - 3$; (i) $0.7716 - 23$.

2. (a) 2.54; (c) 901; (e) 60,000.; (g) 0.000338; (i) 0.0128.

3. (a) $0.1326 - 2$; (c) 1.6276; (e) 2.8862; (g) 2.1411; (i) 0.4343.

4. (a) 545.4; (c) 152,900.; (e) 0.05149; (g) 0.00004053; (i) 0.003333.

5. (a) 269; (c) 122,000.; (e) 4.29; (g) 3.62; (i) 1.22.

6. (a) 84110.; (c) 0.002107; (e) 1.971; (g) 6.04; (i) 1.772.

7. (a) 6.783; (c) 6.606; (e) 31.55; (g) 8.823; (i) 22.45.

9. 1520 cu. in.   11. 35.7 lbs./cu. ft.   13. 3.162×10^{-8} moles/ltr.

answers to selected exercises

15. Each number is placed on the scale, from 0 to 1, at the location of the mantissa of its logarithm; e.g., "2" at 0.3010, "3" at 0.4771, etc. Multiplication involves additon of lengths; i.e., mantissas, on the C and D scales. Similarly, division subtracts these mantissas (lengths).

19. 20 is not 2 times a power of e, as it is two times a power of ten. $\log_e 20 = 2.99573$, $\log_e 2 = 0.69315$.

section 4.5 (pp. 185–186).

1. 32,000.
3. (a) 31,500; (b) 114,880; (c) approx. 1,525,000; (d) approx. 5.6 days.
5. approx. 1,964,260.   7. approx. 7.4 gms.
9. The amount left after 1 hour is calculated to be 94.6%, so the cell has lost some material.
11. approx. 3.11 yrs.   12. (a) $964.20; (c) $975.24; (e) $979.34; (f) $979.39.
13. $138.03.   15. 20.96 lumens.   17. $d = 0.969$.

chapter 5.

section 5.1 (pp. 194–196).

1. (a) $\dfrac{\pi}{4}$; (c) $\dfrac{2\pi}{3}$; (e) $\dfrac{7\pi}{6}$; (g) $\dfrac{11\pi}{6}$; (i) $\dfrac{2\pi}{9}$; (k) $\dfrac{\pi}{12}$; (m) 4π; (o) 2.6π.

2. (a) 30°; (c) 90°; (e) 135°; (g) 210°; (i) 20°; (k) 27° 30′; (m) 131° 46′ 48″; (o) 234° 54′ 49″.

3. (a) π; (c) 27.1 ft.; (e) 34.5 cm.; (g) π in.; (i) 5.86 ft.

4. (a) $\dfrac{\pi}{2}$; (c) $\dfrac{\pi}{2}$; (e) 0.658; (g) 0.217.

5. $\dfrac{2\pi}{3}$.   7. 1.63.   9. 8.75 in.   11. 160π ft/sec $\doteq$ 502.65 ft/sec.

13. 2.2 ft.   15. 4752 rad/min $\doteq$ 756.4 rpm.   17. 1047.2 mph.

section 5.2 (pp. 200–202).

1.

	0	$\dfrac{\pi}{6}$	$-\dfrac{\pi}{6}$	$\dfrac{\pi}{4}$	$-\dfrac{\pi}{4}$	$\dfrac{\pi}{3}$	$-\dfrac{\pi}{3}$	$\dfrac{\pi}{2}$	$-\dfrac{\pi}{2}$
(a) $\sin\theta$	0	$\dfrac{1}{2}$	$-\dfrac{1}{2}$	$\dfrac{\sqrt{2}}{2}$	$-\dfrac{\sqrt{2}}{2}$	$\dfrac{\sqrt{3}}{2}$	$-\dfrac{\sqrt{3}}{2}$	1	-1
(b) $\cos\theta$	1	$\dfrac{\sqrt{3}}{2}$	$\dfrac{\sqrt{3}}{2}$	$\dfrac{\sqrt{2}}{2}$	$\dfrac{\sqrt{2}}{2}$	$\dfrac{1}{2}$	$\dfrac{1}{2}$	0	0
(c) $\tan\theta$	0	$\dfrac{\sqrt{3}}{3}$	$-\dfrac{\sqrt{3}}{3}$	1	-1	$\sqrt{3}$	$-\sqrt{3}$	×	×
(d) θ in degrees	0°	30°	$-30°$	45°	$-45°$	60°	$-60°$	90°	$-90°$

answers to selected exercises

3.

	$\dfrac{2\pi}{3}$	$\dfrac{3\pi}{4}$	$\dfrac{5\pi}{6}$	π	$\dfrac{7\pi}{6}$	$\dfrac{5\pi}{4}$	$\dfrac{4\pi}{3}$	$\dfrac{3\pi}{2}$	$\dfrac{5\pi}{3}$	$\dfrac{7\pi}{4}$	$\dfrac{11\pi}{6}$
(a) $\sin\theta$	$\dfrac{\sqrt{3}}{2}$	$\dfrac{\sqrt{2}}{2}$	$\dfrac{1}{2}$	0	$-\dfrac{1}{2}$	$-\dfrac{\sqrt{2}}{2}$	$-\dfrac{\sqrt{3}}{2}$	-1	$-\dfrac{\sqrt{3}}{2}$	$-\dfrac{\sqrt{2}}{2}$	$-\dfrac{1}{2}$
(b) $\cos\theta$	$-\dfrac{1}{2}$	$-\dfrac{\sqrt{2}}{2}$	$-\dfrac{\sqrt{3}}{2}$	-1	$-\dfrac{\sqrt{3}}{2}$	$-\dfrac{\sqrt{2}}{2}$	$-\dfrac{1}{2}$	0	$\dfrac{1}{2}$	$\dfrac{\sqrt{2}}{2}$	$\dfrac{\sqrt{3}}{2}$
(c) $\tan\theta$	$-\sqrt{3}$	-1	$-\dfrac{\sqrt{3}}{3}$	0	$\dfrac{\sqrt{3}}{3}$	1	$\sqrt{3}$	$\times$	$-\sqrt{3}$	-1	$-\dfrac{\sqrt{3}}{3}$
(d) $\cot\theta$	$-\dfrac{\sqrt{3}}{3}$	-1	$-\sqrt{3}$	$\times$	$\sqrt{3}$	1	$\dfrac{\sqrt{3}}{3}$	0	$-\dfrac{\sqrt{3}}{3}$	-1	$-\sqrt{3}$
(e) $\sec\theta$	-2	$-\sqrt{2}$	$-\dfrac{2\sqrt{3}}{3}$	-1	$-\dfrac{2\sqrt{3}}{3}$	$-\sqrt{2}$	-2	$\times$	2	$\sqrt{2}$	$\dfrac{2\sqrt{3}}{3}$
(g) θ in degrees	$120°$	$135°$	$150°$	$180°$	$210°$	$225°$	$240°$	$270°$	$300°$	$315°$	$330°$

4.

	sin	cos	tan	cot	sec	csc
(a)	$\dfrac{5}{13}$	$\dfrac{12}{13}$	$\dfrac{5}{12}$	$\dfrac{12}{5}$	$\dfrac{13}{12}$	$\dfrac{13}{5}$
(c)	$-\dfrac{8}{17}$	$\dfrac{15}{17}$	$-\dfrac{8}{15}$	$-\dfrac{15}{8}$	$\dfrac{17}{15}$	$-\dfrac{17}{8}$
(e)	$\dfrac{8}{17}$	$\dfrac{15}{17}$	$\dfrac{8}{15}$	$\dfrac{15}{8}$	$\dfrac{17}{15}$	$\dfrac{17}{8}$
(g)	y	x	$\dfrac{y}{x}$	$\dfrac{x}{y}$	$\dfrac{1}{x}$	$\dfrac{1}{y}$

5. (a) 1st; (c) 3rd; (e) 1st. or 4th; (g) 2nd.

6. (a) F; (c) F; (e) F; (g) T; (i) T; (k) T. To verify (a)–(i) use the tables in Exercises 1 and 2. Parts (j), (k), and (l) follow from the definitions.

7. Since θ is in standard position, $\sin\theta = \dfrac{y}{r}$ and $\cos\theta = \dfrac{x}{r}$. Solve these for y and x, respectively, to obtain that the coordinates for $P(x,y)$ are $x = r\cos\theta$, and $y = r\sin\theta$.

section 5.3 (pp. 206–207).

1. (a) .3420; (c) .8693; (e) .6428; (g) .5000; (i) 3.732; (k) .3584;
   (m) 11.43; (o) 3.864.

2. (a) 1.3963; (c) 0.3840; (e) .6690; (g) .4511; (i) 1.1606;
   (k) .5585; (m) 1.49; (o) 1.679.

3. (a) $\beta = 65°$; (c) $b = 16$; (e) $\alpha = 45°$; (g) $a = 18$; (i) $\beta = 60°$.

answers to selected exercises

4. (a) $a = \dfrac{3\sqrt{3}}{2}$, $\beta = 30°$, $b = \dfrac{3}{2}$;   (c) $a = 3$, $b = 3\sqrt{3}$, $\beta = 60°$;

   (e) $a = 4$, $c = 4\sqrt{2}$, $\alpha = 45°$;   (g) $\alpha = 53°$, $b = 9.72$, $c = 16.15$;
   (i) $\alpha = 21°$, $b = 13.05$, $c = 13.98$.

5. 554.42 mi.   7. base 446 ft., altitude 350 ft.   9. 29 ft.

11. 117 ft.   13. 89.9 lbs.

section 5.4 (pp. 212–213).

3. Use Theorem 5.4.2.   (a) Solve Theorem 5.4.2 (i) for $\cos^2 t$.
   (b) Solve Theorem 5.4.2 (i) for $\sin^2 t$.
   (c) $(\cos t + \sin t)^2 = \cos^2 t + 2 \sin t \cos t + \sin^2 t = 1 + 2 \sin t \cos t$.
   (d) Use part (a), $\cos^2 t - \sin^2 t = (1 - \sin^2 t) - \sin^2 t$.
   (e) Use part (b) $\cos^2 t - \sin^2 t = \cos^2 t - [1 - \cos^2 t]$.
   (f) Solve Theorem 5.4.2 (ii) for $\tan^2 t$.

   (g) Note that $\csc t = \dfrac{1}{\sin t}$ and $\sec t = \dfrac{1}{\cos t}$. Divide and simplify.

4. (g) Follows directly from Exercise 1.

5. Both functions are periodic; hence are not one-to-one.

6. Both follow from the fact that $(\cos t, \sin t)$ is a point on the unit circle $x^2 + y^2 = 1$.

7. Yes. $\tan (t + \pi) = \dfrac{\sin (t + \pi)}{\cos (t + \pi)} = \dfrac{-\sin t}{-\cos t} = \tan t$ (see 2a).

8. $(f \circ g)(x) = \sin 2x$, while $(g \circ f)(x) = 2 \sin x$. These are not equal as can be seen
   when $x = \dfrac{\pi}{2}$.

9. (a) $|\sec t| = \left| \dfrac{1}{\cos t} \right| = \dfrac{1}{|\cos t|} = s$, so $1 = s|\cos t| \le s$ from 6(b).

11. See 6 for (i). For (ii) and (iii) see the discussion in Section 52, in particular equations
   (5.2.2) and (5.2.3), respectively.

section 5.5 (pp. 217–219).

1. (a) F; (b) F; (c) F; (d) F; (e) F; (f) T; (g) T; (h) F; (i) T; (j) T.

2. (a) $\tfrac{1}{4}(\sqrt{2} + \sqrt{6})$; (c) $\tfrac{1}{4}(\sqrt{2} + \sqrt{6})$; (e) $2 - \sqrt{3}$; (g) $\sqrt{6} - \sqrt{2}$; (i) $2 - \sqrt{3}$.

3. Write $\tan (\alpha + \beta) = \dfrac{\sin (\alpha + \beta)}{\cos (\alpha + \beta)} = \dfrac{\sin \alpha \cos \beta + \cos \alpha \sin \beta}{\cos \alpha \cos \beta - \sin \alpha \sin \beta}$. Since $\cos \alpha \cos \beta \ne 0$,
   divide numerator and denominator by this.

5. Set $\beta = \alpha$ in (i) and use Theorem 5.5.4.

7. For each appropriate real number t

   $\tan(-t) = \dfrac{\sin(-t)}{\cos(-t)} = \dfrac{-\sin t}{\cos t} = -\tan t$. $\cot(-t) = \dfrac{\cos(-t)}{\sin(-t)} = \dfrac{\cos t}{-\sin t} = -\cot t$.

   $\sec(-t) = \dfrac{1}{\cos(-t)} = \dfrac{1}{\cos t} = \sec t$.

9. (a) $\tfrac{1}{2}\sqrt{2 + \sqrt{3}}$; (b) $\tfrac{1}{2}\sqrt{2 - \sqrt{3}}$; (c) $-\tfrac{1}{2}$; (d) $\tfrac{1}{2}\sqrt{2 - \sqrt{2}}$;
   (e) $\tfrac{1}{2}\sqrt{2 + \sqrt{2}}$; (f) $\tfrac{1}{2}\sqrt{2 - \sqrt{3}}$; (g) $\tfrac{1}{2}\sqrt{2 - \sqrt{2}}$; (h) $\tfrac{1}{2}\sqrt{2 + \sqrt{2}}$;
   (i) Since $\cos \alpha$ and $\sin \alpha$ are both negative, α is in the 3rd quadrant. Therefore,

   $\dfrac{\alpha}{2}$ is in the second quadrant. $\cos \dfrac{\alpha}{2} = -\dfrac{1}{10}\sqrt{10}$.

answers to selected exercises

(j) α is in the 4th quadrant and $\dfrac{\alpha}{2}$ in the second; $\cos\dfrac{\alpha}{2} = \dfrac{2}{13}\sqrt{13}$.

13. (a) T; (b) T; (c) T; (d) F; (e) F; (f) T.

section 5.6 (pp. 222–223).

1. Note that possible examples are (a) $\sin\left(\dfrac{\pi}{4}\right) + \cos\left(\dfrac{\pi}{4}\right) = \sqrt{2}$;

(b) $\sin^2\left(\dfrac{\pi}{4}\right) - \cos^2\left(\dfrac{\pi}{4}\right) = 0 \neq \dfrac{1}{2}$;

(c) $\sin\left(\dfrac{\pi}{4}\right)\cot\left(\dfrac{\pi}{4}\right) = \dfrac{\sqrt{2}}{2}$; (d) $1 + \sin^2\left(\dfrac{\pi}{6}\right) = \dfrac{5}{4}$; (e) $\sin\left(\dfrac{\pi}{2}\right)\sin 1 = .8415$;

(f) $\dfrac{\tan\left(\dfrac{\pi}{3}\right)}{\cot\left(\dfrac{\pi}{3}\right)} = 3$; (g) $\sin\left(\dfrac{\pi}{3} - \dfrac{\pi}{6}\right) \neq \dfrac{\sqrt{3}-1}{2} = \sin\dfrac{\pi}{3} - \sin\dfrac{\pi}{6}$;

(h) $\sin\left(\csc\left(\dfrac{\pi}{2}\right)\right) = .8415$.

3. (f) $2\sin t \cos t - (2\cos^2 t - 1)\tan t = 2\sin t \cos t - 2\cos^2 t \left(\dfrac{\sin t}{\cos t}\right) + \tan t$

$= 2\sin t \cos t - 2\cos t \sin t + \tan t$.

5. (f) $1 - \dfrac{\sin\beta}{\cos\beta}\left[\dfrac{\sin(\alpha-\beta)}{\cos(\alpha-\beta)}\right] = \dfrac{\cos\beta[\cos(\alpha-\beta)] - \sin\beta[\sin(\alpha-\beta)]}{\cos\beta[\cos(\alpha-\beta)]}$

$= \{\cos\beta[\cos\alpha\cos\beta + \sin\alpha\sin\beta] - \sin\beta[\sin\alpha\cos\beta - \cos\alpha\sin\beta]\}\sec\beta\sec(\alpha-\beta)$

$= [\cos\alpha\cos^2\beta + \sin\alpha\sin\beta\cos\beta - \sin\alpha\sin\beta\cos\beta + \cos\alpha\sin^2\beta]\sec\beta\sec(\alpha-\beta)$

$= [\cos\alpha(\cos^2\beta + \sin^2\beta)]\sec\beta\sec(\alpha-\beta)$.

7. (a) F; (b) T; (c) T; (d) F; (e) F; (f) T; (g) F; (h) T. Note on (h)

that $\dfrac{2\tan 76°}{\tan^2 76° - 1} = -\tan 152° = \tan(-152°) = \tan 208°$.

section 5.7 (pp. 228–229).

1. (a) See the appropriate parts of Figures 5.7.3, 5.7.5, 5.7.6, 5.7.7, and 5.7.8.
2. (a) same as $y = \cos x$; (b) same as $y = \sin x$; (c) reflect Fig. 5.7.7 in x axis;
   (d) same as $y = \cos x$; (e) reflect Fig. 5.7.3 in x-axis; (f) reflect Fig. 5.7.5.
7. (a) Fig. 5.7.2 reflected in the x axis; (b) Fig. 5.7.3 reflected in the x axis;
   (c) same as (a); (d) same as $y = \cos x$; (e) Fig. 5.7.5 reflected in the x axis;
   (g) Fig. 5.7.7 reflected in the x axis; (h) Fig. 5.7.8 reflected in the x axis.

section 5.8 (pp. 233–234).

1. (a) Amplitude 3   Period 2π (e) Amplitude 1/2   Period π

   Phase Shift 0   Frequency $\dfrac{1}{2\pi}$ Phase Shift $\dfrac{\pi}{8}$ Frequency $\dfrac{1}{\pi}$

   (c) Amplitude 1   Period 2π 2. (a) Amplitude 2   Period 2π

   Phase Shift $\dfrac{\pi}{4}$ Frequency $\dfrac{1}{2\pi}$ Phase Shift $-\dfrac{\pi}{4}$ Frequency $\dfrac{1}{2\pi}$

answers to selected exercises

2. (c) Amplitude 2   Period 2π (e) Amplitude 1   Period π

     Phase Shift $\dfrac{\pi}{3}$ Frequency $\dfrac{1}{2\pi}$ Phase Shift 0   Frequency $\dfrac{1}{\pi}$

4. (a) Amplitude 5   Period 2π (c) not periodic

     Phase Shift $-.93$ Frequency $\dfrac{1}{2\pi}$ (e) not periodic

5. $i(t) = 20 \sin 120\pi t.$ 7. $p = AE_m \sin^2 \omega t.$

section 5.9 (pp. 239–240).

1. $y = \text{arc sec } x$, iff $x = \sec y$ and $y \in \left[0, \dfrac{\pi}{2}\right) \cup \left(\dfrac{\pi}{2}, \pi\right]$.

     $y = \text{arc cot } x$, iff $x = \cot y$ and $0 < y < \pi$.

     $y = \text{arc csc } x$, iff $x = \csc y$ and $y \in \left[-\dfrac{\pi}{2}, 0\right) \cup \left(0, \dfrac{\pi}{2}\right]$.

2. (a) $\pi/3$; (c) $-\pi/4$; (e) $\pi/6$; (g) $-\pi/3$; (i) $-\pi/4$; (k) $-\pi/6$.

3. (a) 0.2880; (c) 0.4014; (e) 1.0123; (g) 0.4363; (i) -0.1222; (k) 2.2166.

4. (a) u; (c) $\sqrt{1 + u^2}$; (e) $u/\sqrt{1 - u^2}$; (g) $2u/(u^2 - 1)$; (i) $(2 - u^2)/u^2$.

5. (a) $2\sqrt{u^2 - 1}/u^2$; (c) $-1/u$; (e) $uv - \sqrt{(1 - u^2)(1 - v^2)}$;
     (g) $u\sqrt{1 - v^2} + v\sqrt{1 - u^2}$.

6. (a) t; (c) $\pi/2 - t$; (e) $t - \pi/2$; (g) $t - \pi/2$; (i) $t - \pi/2$.

7. (a) $\tan(\tan^{-1} 3 - \tan^{-1} 1) = \dfrac{3 - 1}{1 + 3} = 1/2$, and $0 < \tan^{-1} 1 < \tan^{-1} 3 < \pi/2$.

     Thus, $0 < \tan^{-1} 3 - \tan^{-1} 1 < \pi/2$.

     (c) set $\alpha = \text{arc sin } \dfrac{\sqrt{5}}{5}$ and $\beta = \text{arc sin } \dfrac{2\sqrt{5}}{5}$. We have $0 < \alpha, \beta < \pi/2$ and $\sin \alpha = \dfrac{\sqrt{5}}{5}$,

     $\cos \alpha = \dfrac{2\sqrt{5}}{5}$, $\sin \beta = \dfrac{2\sqrt{5}}{5}$, $\cos \beta = \dfrac{\sqrt{5}}{5}$, $\sin(\alpha + \beta) = 1$, so $\alpha + \beta = \pi/2$.

8. (a) Set $\alpha = \tan^{-1} u$. Then $\tan \alpha = u$ and $\tan(-\alpha) = -u$. With $-\pi/2 < \alpha < \pi/2$,
     we have $-\pi/2 < -\alpha < \pi/2$, so $-\alpha = \tan^{-1}(-u)$. So, $\alpha + (-\alpha) = 0$.
     (c) Set $\alpha = \sin^{-1} u$, then $\sin \alpha = u$ and $-\pi/2 \le \alpha \le \pi/2$. It follows that $\cos(\pi/2 - \alpha) = \sin \alpha = u$ and $0 \le \pi/2 - \alpha \le \pi$, so $\pi/2 - \alpha = \cos^{-1} u$. Result follows.
     (g) Set $\alpha = \sec^{-1} u$. Thus $u = \sec \alpha$ and, since $u \ge \sqrt{2}$, $\alpha \in [\pi/4, \pi/2]$. It follows
     that $\cos \alpha = 1/u$ and $\cos(\pi/2 - \alpha) = \sin \alpha = \sqrt{u^2 - 1}/u$. Hence, $\pi/2 - \alpha \in (0, \pi/4)$
     and $\pi/2 - \alpha = \cos^{-1}(\sqrt{u^2 - 1}/u)$. Thus, $\sin(2\alpha - \pi/2) = -\sin(\pi/2 - 2\alpha) = -\cos(2\alpha) = \sin^2 \alpha - \cos^2 \alpha = (u^2 - 1)/u^2 - 1/u^2 = (u^2 - 2)/u^2$.

section 5.10 (p. 243).

1. (a) $\pi/3, 2\pi/3$; (c) $\pi/3, 4\pi/3$; (e) $\pi/6, 5\pi/6, 7\pi/6, 11\pi/6$;
     (g) $\pi/3, 5\pi/3$; (i) $\pi/2, 3\pi/2, \pi/6, 7\pi/6$.

2. (a) $k\pi, k\pi + (-1)^k \pi/2$, k any integer; (c) $2k\pi \pm \pi/3, 2k\pi \pm 5\pi/6$;
     (e) $2k\pi \pm \pi/3, k\pi - \pi/6$; (g) $k\pi \pm \pi/3$; (i) $k\pi + \pi/6, k\pi - \pi/3$.

3. (a) $\pi/4, 3\pi/4, 5\pi/4, 7\pi/4$; (c) $\pi/8, 5\pi/8, 9\pi/8, 13\pi/8$; (e) $0, 2\pi/3, 4\pi/3, 2\pi$;
     (g) $\pi/6, 5\pi/6, 7\pi/6, 11\pi/6$.

4. (a) $k\pi/2$, k any integer; (c) $k\pi/4$; (e) $k\pi/2$;
     (g) $(2k - 1)\pi/2, (12k + 1)\pi/6, (12k - 7)\pi/6$.

answers to selected exercises

5. (a) $(12k - 1)\pi/6$, $(4n - 3)\pi/2$; (c) $(6k + 1)\pi/6 + (-1)^k \sin^{-1}(3/5)$;
   (e) $1.96 + 2k\pi$, $3.54 + 2k\pi$; (g) $2k\pi - 0.4$, $2k\pi + 1.18$.

6. (a) Let $\alpha = \sin^{-1} t$. Then $\sin \alpha = t$ and $-\pi/2 \le \alpha \le \pi/2$. In this interval $\cos \alpha \ge 0$,
   so $\cos \alpha = \sqrt{1 - \sin^2 \alpha} = \sqrt{1 - t^2}$.

section 5.11 (pp. 250–252).

1. (a) $\alpha = 60°$, $a = b = c = 25$; (c) $\alpha = 120°$, $b = 100$, $c = 100\sqrt{3}$;
   (e) $\gamma = 80°$, $b = 77.6$, $c = 80.3$; (g) $\gamma = 70°$, $a = 39.8$, $c = 48.8$.

2. (b) $\alpha = 25°40'$, $\gamma = 94°20'$, $c = 11.4$; (d) no triangle;
   (e) $\alpha = 70°40'$, $\gamma = 64°20'$, $c = 2.55$, $\alpha' = 109°20'$, $\gamma' = 25°40'$, $c' = 1.22$;
   (g) $\alpha = 43°$, $\gamma = 107°$, $c = 3.83$, $\alpha' = 137°$, $\gamma' = 13°$, $c' = 0.90$.

3. (a) $a = 1$, $\beta = 45°$, $\gamma = 90°$; (c) $c = 1.95$; (e) 7.03; (g) $b = 4.09$.

4. (a) $\alpha = \beta = \gamma = 60°$; (c) $\beta = 90°$, $\gamma = 60°$, $\alpha = 30°$;
   (e) $\gamma = 90°$, $\beta = 22°30'$, $\alpha = 67°30'$;
   (g) no triangle: $\cos \gamma > 1$ is impossible. (Note $a < b + c$.)

5. Sides 14.67 cm and 7.06 cm; angles: $62°$, $118°$.   6. 2.1 min.

9. (a) $\alpha = \beta = 65°$, $c \doteq 0.78$; (c) $\alpha = 47°$, $\beta = 33°40'$; (e) $\gamma = 100°20'$, $\beta = 41°30'$.

10. 160.2 ft.   14. 5.   15. 132.   17. 13.4 mi from 1st, 13.3 mi from 2nd.

19. 24.8 lbs at an angle $82°40'$ from horizontal.   21. 1.1 mi.   23. 2225.

section 5.12 (pp. 258–260).

1. (a) $2(\cos \pi/3 + i \sin \pi/3)$; (c) $2(\cos \pi/6 + i \sin \pi/6)$;
   (e) $\sqrt{2}(\cos \pi/4 + i \sin \pi/4)$; (g) $2(\cos \pi/2 + i \sin \pi/2)$;
   (i) $8(\cos 0 + i \sin 0)$; (k) $7(\cos \pi + i \sin \pi)$.

2. (a) $\sqrt{2}/2 - i\sqrt{2}/2$; (c) $0.8192 + 0.5736i$; (e) $-1/2 - i\sqrt{3}/2$.

3. (a) $2 \operatorname{cis} 2\pi = 2$; (c) $2\sqrt{2} \operatorname{cis} 13\pi/12$; (e) $\operatorname{cis} \pi = -1$;
   (g) $\operatorname{cis} -\pi/12 = \operatorname{cis} 23\pi/12$.

4. (a) -4; (c) $8 \operatorname{cis} \pi/2 = 8i$; (e) $125 \operatorname{cis} 0.36$; (g) $289 \operatorname{cis} 2.16$.

5. (a) $2^{1/3} \operatorname{cis} (12k + 11)\pi/18$, $k = 0,1,2$; (c) $2 \operatorname{cis} (4k + 1)\pi/8$, $k = 0,1,2,3$;
   (e) $\sqrt{2}(1 + i)$, $\sqrt{2}(-1 + i)$, $\sqrt{2}(-1 - i)$, $\sqrt{2}(1 - i)$;
   (g) $2^{1/10} \operatorname{cis} (8k + 5)\pi/20$, $k = 0,1,\dots,4$.

6. (a) 1, $\operatorname{cis} 2\pi/3$, $\operatorname{cis} 4\pi/3$; (c) 1, $\operatorname{cis} 2\pi/5$, $\operatorname{cis} 4\pi/5$, $\operatorname{cis} 6\pi/5$, $\operatorname{cis} 8\pi/5$;
   (e) 1, $\operatorname{cis} 2k\pi/7$, $k = 1, \dots, 6$.

7. (a) ± 2; (c) $2 \operatorname{cis}(2k + 1)\pi/5$, $k = 0,1,2,3,4$; (e) $\operatorname{cis} (4k + 1)\pi/6$, $k = 0, \dots, 5$;
   (g) 3, 1, -1, i, $-i$.

8. If $P(x,y)$ is a point in the plane, $z = (x,y)$ is a complex number. This corresponds
   to the standard form $z = x + iy$ which has polar form $r \operatorname{cis} \theta$. Thus, the pair (r,θ)
   determines P as well. Since $r = |z| = \sqrt{x^2 + y^2}$, it is the length of OP.

chapter 6.

section 6.1 (pp. 268–271).

1. All are equivalent to each other; (a) take 1/5 of second equation;
   (c) first equation of (b) is subtracted from the second;
   (e) first and second equations of (d) are interchanged and the new second equation
   divided by 2.

answers to selected exercises

2. (a) $(1/2, 1/2)$; (c) $(2/5, 8/5)$; (e) $(2, -3)$; (g) $(1, 1/3)$; (i) $(1/6, -1/5)$.

3. $(19/7, -1/7)$.

4. (a) $(6,1)$; (c) $(13/97, 25/97)$; (e) $(21/4, -1)$; (g) $(47/25, 17/25)$.

7. 22 and 42. 9. 67. 11. $[x] = -1, [y] = 2$, so $-1 \le x < 0$, and $2 \le y < 3$.

12. (a) $(1, -1, 0)$; (b) $(1, 1/2, 3/2)$; (c) $(0,1,1)$.

13. Draw three distinct lines with three separate (pairwise) intersections.

14. (a) $(2,1)$; (c) $(1,6), (-1,6), (\sqrt{6}, 1), (-\sqrt{6}, 1)$; (e) $(25,(5 - \log 4)) \doteq (25,4.3979)$.

15. $a = 12/13, b = 9/13$.

section 6.2 (pp. 277–279).

1. (a) $\begin{aligned} x - 2y &= 0 \\ -8y &= 8 \\ (2, -1) \end{aligned}$ (c) $\begin{aligned} x - 5y &= -3 \\ 9y &= 0 \\ (-3, 0) \end{aligned}$ (e) $\begin{aligned} x - \tfrac{1}{4}y &= -1 \\ \tfrac{13}{12}y &= \tfrac{7}{3} \\ (-6/13, 28/13) \end{aligned}$

2. (a) $(2, -3, 4)$; (c) $(-3, 1, -2)$; (e) $(1/2, 1/3, 1/4)$; (g) inconsistent.

3. (a) $(-k, 3k, k), k \in R$; (c) $(-k, 0, k)$; (e) $(-\tfrac{7}{5}k, -\tfrac{3}{5}k, k)$; (g) $(0,0,0)$.

4. (a) $(-7,11,14,-6)$; (c) $(0, 2 - k, -1, k), k \in R$; (e) $(1, -1, -2, 2)$; (g) $(0,0,0,0)$.

5. (a) $(19/34, 19/34)$; (c) $(1, -2)$; (e) $(1,2,2)$; (g) $x = \pi/6 + 2k\pi, k \in Z$.

7. 39 nickles, 13 dimes, 5 quarters. 9. $p(x) = x^3 - 3x^2 + 2x + 4$.

section 6.3 (pp. 286–288).

1. (a), (b), and (c) are row equivalent, (d) is not equivalent to them.

3. (a) $(1,0,1)$; (b) inconsistent; (c) inconsistent; (d) inconsistent; (e) inconsistent; (f) $(0,0,0)$.

4. (a) $(\tfrac{2}{5}k, \tfrac{7}{5}k, k), k \in R$; (c) $(0,0,0,0)$; (e) $(\tfrac{1}{5}k, -\tfrac{8}{5}k, k), k \in R$.

5. (a) inconsistent; (c) $(-\tfrac{1}{5}k, -\tfrac{9}{40}k, -\tfrac{4}{5}k, \tfrac{9}{40}k, k), k \in R$; (e) inconsistent.

6. (a) $\begin{bmatrix} 1 & 0 & -1 \\ 0 & 1 & -1 \\ 0 & 0 & 0 \end{bmatrix}$ (c) $\begin{bmatrix} 1 & 0 & -1 & -2 & -3 \\ 0 & 1 & 2 & 3 & 4 \\ 0 & 0 & 0 & 0 & 0 \\ 0 & 0 & 0 & 0 & 0 \\ 0 & 0 & 0 & 0 & 0 \end{bmatrix}$

7. $y = \tfrac{5}{2}x^2 - \tfrac{5}{2}x - 2$.

9. $x^2 + y^2 + 54x - 35y = 21$.

section 6.4 (pp. 293–295).

1. (a) 3; (c) 3; (e) 1; (g) 12. 2. (a) -6; (c) 0; (e) 120.

3. (a) 1; (c) $\cot^2 \alpha$; (e) $x + y$; (g) 4.

4. (a) $\begin{bmatrix} 1 & -2 \\ 2 & -1 \end{bmatrix}, 3;$ (c) $\begin{bmatrix} 0 & -1 \\ 3 & 5 \end{bmatrix}, 3;$

(e) $\begin{bmatrix} 1 & 0 & -1 \\ -3 & 1 & 1 \\ -1 & 1 & 0 \end{bmatrix}, 1;$ (g) $\begin{bmatrix} 1 & 2 & 1 \\ 3 & 0 & -3 \\ -1 & 2 & 1 \end{bmatrix}, 12.$

answers to selected exercises

5. $\det B = (-1)^2(4) \det \begin{bmatrix} 1 & -1 \\ 0 & -1 \end{bmatrix} + (-1)^3(-2) \det \begin{bmatrix} 3 & -1 \\ 1 & -1 \end{bmatrix} + (-1)^4(0) \det \begin{bmatrix} 3 & 1 \\ 1 & 0 \end{bmatrix} = -12.$

7. Expand the determinant by using minors of the row of zeros. Each term in the expansion will, therefore, have a zero factor.

9. For a 1×1 matrix $[a_{11}]$ we clearly have $\det [ca_{11}] = ca_{11} = c \det[a_{11}]$. Let B be a matrix which is such that each element of the first row of B is c times the corresponding element of the first row of A, $b_{1j} = ca_{1j}$. In this case, the result clearly follows from (6.4.4). If the first row of B is not the row with factor c then each $n-1$ by $n-1$ minor B_{1j} has a row or a column which is c times a row or column of the minor A_{1j} of A. In the latter case, by induction assumption we have $\det B_{1j} = c \det A_{1j}$. Again by (6.4.4) each term has c as a factor.

11. (a) 5; (b) 3; (c) 0, -3; (d) $-3, \dfrac{1}{2} \pm i\dfrac{\sqrt{3}}{2}.$

13. (a), (b), (c), (e), (g), and (h).

15. (a) 1; (b) 1; (c) 1; (d) 0; (e) 1; (f) 0; (g) 1; (h) 1.

section 6.5 (pp. 299–301).

1. (a) $(3/2, 1/2)$; (c) $(-4/3, 3/5)$; (e) $(1, -1)$; (g) $(\sqrt{2}, \sqrt{3})$.
2. (a) $(1,1,1)$; (c) $(1/3, 1/9, 5/9)$; (e) $(-1/2, 15/14, -25/14)$; (g) $(\sqrt{2}, \sqrt{3}, \sqrt{5})$.
3. (a) $\det \neq 0$, so $(0,0)$ only; (c) $\det \neq 0$, so $(0,0,0)$ only; (e) $\det \neq 0$, $(0,0,0)$ only.
4. (a) $(-9/13, 18/13, 1/13)$; (c) $(\csc \theta, -\cot \theta)$; (e) $(1 + 4i, 3 - i, 2 + 3i)$.

5.
$$\det \begin{bmatrix} x & y & 1 \\ x_1 & y_1 & 1 \\ x_2 & y_2 & 1 \end{bmatrix} = \det \begin{bmatrix} x - x_1 & y - y_1 & 0 \\ x_1 & y_1 & 1 \\ x_2 - x_1 & y_2 - y_1 & 0 \end{bmatrix} = (-1)^{2+3} \det \begin{bmatrix} x - x_1 & y - y_1 \\ x_2 - x_1 & y_2 - y_1 \end{bmatrix}$$

7. Use induction on the size of the matrix. For a 1×1 matrix $A = [a_{11}]$, $\det A = a_{11}$. Suppose for any $(n-1) \times (n-1)$ upper-triangular matrix the theorem is true. Let A be an $n \times n$ upper-triangular matrix. Then $\det A$ can be computed using the minors of column 1. Thus, $\det A = a_{11} \det A_{11}$. But, A_{11} is $(n-1) \times (n-1)$ and upper-triangular, so $\det A_{11} = a_{22}a_{33}\dots a_{nn}$. Therefore, $\det A = \prod_{i=1}^{n} a_{ii}$, as desired.

8. (a) -5; (c) -3; (e) -1; (g) 12. 9. -6.

chapter 7.

section 7.1 (pp. 308–310).

1. (a) $8 - 5x - 3y$; (c) $3x^2 - y^2 - xy$; (e) $x^2 + xy + x + y$;
    (g) $x^3 + y^3 - 3xy + 1$; (i) $44x^2 - y^2 + 37x + 1$.
2. (a) $2x^3y - 2x^2y^2$; (c) $6x^2 - 7xy - 5y^2$; (e) $x^2 + y^2 - 9 + 2xy$;
    (g) $9x^2 - y^2 + 4y - 4$;
    (i) $3x^3 - 3x^2y + 3xy^2 + x^4y - x^3y^2 + x^2y^3 - x^2y^2 - xy^3 - y^4$.

3. (a) $(8y)/(3x^3)$; (c) x^3; (e) $\dfrac{y(x-3)(x+3)}{(x+3y)(x-3y)}$; (g) $y/(y+2)$.

4. (a) $\sqrt{13}$; (c) $\sqrt{5}/4$; (e) $\sqrt{6}$; (g) $\sqrt{1006}/6$; (i) 5.
5. (a) the coordinate axes; (c) the line through $(0,0)$ inclined $135°$ with pos. x-axis.
    (e) the single point $(1,2)$;
    (g) two lines intersecting at $(0,0)$, one inclined $45°$ and the other at $135°$ with x-axis.

answers to selected exercises

6. (a) $x = 0$;   (c) $x^2 + y^2 = 1$;   (e) $x^2 + y^2 < 0$.
7. (a) $z^2 = x^2 + y^2$;   (c) $(x + 1)^2 + (y - 4)^2 = 9$;   (e) $y = x$.

section 7.2 (pp. 316–318).

1. (a) 1;   (c) -1;   (e) 5;   (g) 0;   (i) $-1/2$.
2. (a) $x - y = 0$;   (c) $x + y = 0$;   (e) $5x - y + 5 = 0$;
    (g) $y - 1 = 0$;   (i) $x + 2y - 5 = 0$.
5. The slope of the line through $(a,0)$ and $(0,b)$ is $-b/a$. Using the point-slope form, the equation of this line is $y - b = -bx/a$, or $bx/a + y = b$. Then multiplying both sides by $1/b$ gives the desired form.
6. (a) $x + y = 1$;   (c) $x - 3y = 3$;   (e) $2x + y + 2 = 0$;
    (g) $2x - 2y = 1$;   (i) $36x + 15y = 20$.
7. (a) $x = 1$;   (c) $x = 0$;   (e) $x = 5$;   (g) $5x + 3 = 0$;   (i) $x = e$.
8. (a) $x - 2y + 7 = 0$;   (c) $2x + y = 1$;   (e) $3x + y = 2$;   (g) $2x + 3y = 3$;   (i) $x = 4$.
9. (a) $y = 1$;   (c) $x - y = 0$;   (e) $3x - 2y = 1$;   (g) $4x + 3y = 7$;   (i) $2x + 3y = 5$.
10. (a) $(1,0)$;   (c) $(14/3,2/3)$;   (e) $(5,-9)$;   (f) don't intersect.
11. $x + 2y - 10 = 0$.
14. $y = mx + (2 - m)$.   15. $y = \sqrt{3}x + b$.   16. $y = x - 1$.
19. (a) $6/\sqrt{17}$;   (b) $1/\sqrt{17}$;   (c) $11/\sqrt{17}$;   (d) $19/\sqrt{5}$;   (e) $3/\sqrt{20}$;
    (f) $10/\sqrt{17}$;   (g) $7/\sqrt{2}$;   (h) 0.

section 7.3 (pp. 327–329).

1. (a) $(x + 1)^2 + (y - 1)^2 = 1$;   (c) $x^2 + y^2 = 25$;   (e) $(x + 3)^2 + (y - 4)^2 = 9$.
2. (a) $(x - 1)^2 + (y - 3)^2 = 6$;   (c) $(x + 3)^2 + (y + 5)^2 = 25$;
    (e) $(x + 1)^2 + (y + 2)^2 = 17$.
3. (a) center $(-3,4)$, radius 5;   (c) $(-3,2)$, 4;   (e) $(-3/4, -5/4)$, $\sqrt{34}/2$.
6. (a) Vertex $(0,0)$, focus $(0, 1/4)$, directrix $y = -1/4$;
    (c) Vertex $(0,0)$, focus $(-3,0)$, directrix $x = 3$;
    (e) Vertex $(0,0)$, focus $(0,-3)$, directrix $y = 3$.
7. (a) $y^2 = -12x$;   (c) $y^2 = 12x$;   (e) $y^2 = 4x/3$.
8. (a) Vertex $(1,2)$, focus $(1,9/4)$, directrix $y = 7/4$, axis $x = 1$.
    (c) Vertex $(-5,-1)$, focus $(-19/4,-1)$ directrix $x = -21/4$, axis $y = -1$.
    (e) Vertex $(3/2,-11/8)$, focus $(3/2,-15/8)$, directrix $y = -7/8$, axis $x = 3/2$.
9. (a) $y^2 = 8x - 16$;   (c) $x^2 + 8y + 16 = 0$;   (e) $(y - 5)^2 = 12 - 12x$.
10. (a) $x^2 + y^2 = 1$;   (c) $x^2 + y^2 + 5x - 3y = 0$;   (e) $4x^2 + 4y^2 + 7x + 2y - 17 = 0$.
13. (a) $y = x^2 + 1$;   (c) vertical axis is not possible with these points;
    (e) $y = \frac{1}{8}[x^2 + 4x + 21]$.
14. (a) $V: (0,-1)$, $F: (0, -1/2)$, $y = -3/2$;   (c) none;
    (f) $V: (11/12, 85/24)$, $F: (11/12, 81/24)$, $y = 89/24$.
15. (a) horizontal axes not possible;   (c) $x = -y^2/4 + 3y/4$;
    (e) $x = -y^2/3 - y + 28/12$.
16. (a) no parabola;   (c) $V: (9/16,3/2)$, $F: (-7/16,3/2)$, $x = 25/16$.
    (e) $V: (37/12,-3/2)$, $F: (7/3,-3/2)$, $x = 23/6$.

20. If the cross-sectional parabola is placed on a coordinate system so that the vertex is $(0,0)$ and the axis is the x-axis, then the focus is at $(25/24,0)$.

answers to selected exercises

section 7.4 (pp. 338–340).

1. (a) Vertices $(\pm 5,0)$, Foci $(\pm 3,0)$, major axis 10, minor axis 8;
    (c) Vertices $(\pm 13,0)$, Foci $(\pm 12,0)$, major axis 26, minor axis 10;
    (e) Vertices $(0, \pm\sqrt{5})$, Foci $(0, \pm\sqrt{3})$, major axis $2\sqrt{5}$, minor axis $2\sqrt{2}$;
    (g) Vertices $(\pm 4,0)$, Foci $(\pm 2\sqrt{3},0)$, major axis 8, minor axis 4.

2. (a) $\dfrac{x^2}{25} + \dfrac{y^2}{9} = 1$; (c) $\dfrac{x^2}{16} + \dfrac{y^2}{25} = 1$; (e) $\dfrac{x^2}{10} + \dfrac{y^2}{1} = 1$; (g) $16x^2 + 9y^2 = 225$.

3. (a) $\dfrac{(x-2)^2}{4} + \dfrac{(y-3)^2}{9} = 1$; (c) $\dfrac{(x-2)^2}{7} + \dfrac{(y-1)^2}{16} = 1$;

    (e) $32(x+1)^2 + 9(y-2)^2 = 324$; (g) $\dfrac{(x+2)^2}{289} + \dfrac{(y+2)^2}{64} = 1$.

4. (a) Transverse axis 8, conjugate axis 6, vertices, $(\pm 4,0)$, foci $(\pm 5,0)$;
    (c) Transverse axis 24, conjugate axis 10, vertices, $(0, \pm 12)$, foci $(0, \pm 13)$;
    (e) Transverse axis 2, conjugate axis 2, vertices, $(\pm 1,0)$, foci $(\pm\sqrt{2},0)$;
    (g) Transverse axis $2\sqrt{3}/3$, conjugate axis $\sqrt{2}$, vertices, $(0, \pm\sqrt{3}/3)$, foci $(0, \pm\sqrt{5/6})$

5. (a) $y = \pm 3x/4$; (c) $y = \pm 12x/5$; (e) $y = \pm x$; (g) $y = \pm x\sqrt{6}/3$.

6. (a) $\dfrac{x^2}{9} - \dfrac{y^2}{16} = 1$; (c) impossible; (e) $\dfrac{x^2}{16} - \dfrac{y^2}{4/3} = 1$; (g) $\dfrac{y^2}{9} - \dfrac{x^2}{36} = 1$.

7. (a) $\dfrac{y^2}{16} - \dfrac{x^2}{16} = 1$; (c) $\dfrac{(y+1)^2}{5} - \dfrac{(x+1)^2}{4} = 1$; (e) $\dfrac{(x+1)^2}{16} - \dfrac{3(y-3)^2}{4} = 1$.

17. (a) For an ellipse $a > c$, so $0 < e = c/a < 1$;

    (b) Since in an ellipse $a^2 - c^2 = b^2$, we note that $1 - \dfrac{c^2}{a^2} = \dfrac{b^2}{a^2}$. Thus, $1 - e^2 = b^2/a^2$,

    so as e approaches zero, the value of b approaches the value of a;
    (c) For a hyperbola $c > a$, so $e > 1$.

18. (a) $3/5$; (c) $12/13$; (e) $\sqrt{15}/5$; (g) $\sqrt{3}/2$.

19. (a) $5/4$; (c) $13/12$; (e) $\sqrt{2}$; (g) $\sqrt{10}/2$.   20. 25 ft.

section 7.5 (pp. 349–350).

1. (a) $(1,5)$; (c) $(0,1)$; (e) $(6,1)$; (g) $(\sqrt{2} - 1,3)$; (i) $(-1,1)$.
2. (a) $(2 + \sqrt{3}/2, -1/2 + 2\sqrt{3})$; (c) $((1 - \sqrt{3})/2,(1 + \sqrt{3})/2)$; (e) $(5\sqrt{3}/2, -5/2)$;
    (g) $((1 + \sqrt{6})/2, (\sqrt{3} - \sqrt{2})/2)$; (i) $(2 + \sqrt{3}/2, -1/2 + 2\sqrt{3})$.
3. (a) $(1/2 + 2\sqrt{3}, 2 - \sqrt{3}/2)$; (c) $((-1 + \sqrt{3})/2, (1 + \sqrt{3})/2)$; (e) $(5/2, -5\sqrt{3}/2)$;
    (g) $((\sqrt{2} + \sqrt{3})/2,(1 - \sqrt{6})/2)$; (i) $(1/2 + 2\sqrt{3}, 2 - \sqrt{3}/2)$.
4. (a) $x' + y' + 1 = 0$; (c) $x' - y' - 7 = 0$; (e) $2x' - 3y' = 5$;
    (g) $4x'^2 + y'^2 = 4$; (i) $x'^2 - 4y' = 0$.
5. (a) ellipse $9x'^2 + 16y'^2 = 144$; (c) ellipse $4x'^2 + 25y'^2 = 100$;
    (e) parabola $y'^2 = 8x'$; (h) hyperbola $25y'^2 - 9x'^2 = 9$;
    (i) hyperbola $9x'^2 - y'^2 = 36$.
7. (a) $x''^2 - y''^2 = 4$, $\pi/4$; (b) $y''^2 - x''^2 = 8$, $\pi/4$;
    (j) $6x''^2 + 4y''^2 - 12\sqrt{2}x = 6$, $\pi/4$.
11. $A'' + B'' = A\cos^2\Phi + C\sin\Phi\cos\Phi + B\sin^2\Phi + A\sin^2\Phi - C\sin\Phi\cos\Phi +$
    $\qquad B\cos^2\Phi = A(\cos^2\Phi + \sin^2\Phi) + B(\sin^2\Phi + \cos^2\Phi) = A + B.$

section 7.6 (pp. 354–355).

2. (a) $\sqrt{19}$; (c) $\sqrt{118}$; (e) $\sqrt{53}$; (g) $\sqrt{33}$.
3. (a) $(3/2, 7/2, 13/2)$; (c) $(3/2, 1, 5/2)$; (e) $(-1/2, -1, 1)$; (g) $(1/2, 3, -2)$.

answers to selected exercises

4. (a) 19.580; (c) 21.779; (e) 16.218; (g) 14.211 approximately.
5. $|AB|^2 + |BC|^2 = 6 + 3 = 9 = |AC|^2$. Area $3\sqrt{2}/2$.   7. (1, 1, 1).
8. (a) $x^2 + y^2 + z^2 = 4$; (c) $x^2 + (y - 1)^2 + (z + 1)^2 = 4$;
    (e) $(x - 3)^2 + (y + 1)^2 + (z - 2)^2 = 9$; (g) $(x + 5)^2 + y^2 + (z - 1)^2 = 1/4$.
9. $6x - 12y + 4z + 13 = 0$.
11. Each is a plane. (a) is the zy plane; the rest are parallel to the given coordinate plane, the indicated number of units away: (c) xz, 1 unit; (e) yz, 1/2 unit; (g) xy, -3.
12. The coordinate planes.

section 7.7 (pp. 364–366).

1. (a) $-1, 0, 0$; (c) $\sqrt{3}/3, \sqrt{3}/3, \sqrt{3}/3$; (e) $-4/\sqrt{41}, 3/\sqrt{41}, -4/\sqrt{41}$;
    (g) $14/\sqrt{365}, 12/\sqrt{365}, -5/\sqrt{365}$; (i) $3/\sqrt{91}, 9/\sqrt{91}, 1/\sqrt{91}$.

2. (a) $y = z = 0$; (c) $x = y = z$; (e) $\dfrac{x - 3}{-4} = \dfrac{y + 1}{3} = \dfrac{z - 9}{-4}$;

    (g) $\dfrac{x + 13}{14} = \dfrac{y}{12} = \dfrac{z - 5}{-5}$; (i) $\dfrac{x}{6} = \dfrac{y + 3}{18} = \dfrac{z + 2}{2}$.

3. (a) $y = z = 0$; (c) $\sqrt{3}x = \sqrt{3}y = \sqrt{3}z$;

    (e) $\dfrac{-\sqrt{41}(x - 3)}{4} = \dfrac{\sqrt{41}(y + 1)}{3} = \dfrac{-\sqrt{41}(z - 9)}{4}$;

    (g) $\dfrac{\sqrt{365}(x + 13)}{14} = \dfrac{\sqrt{365}y}{12} = \dfrac{-\sqrt{365}(z - 5)}{5}$;

    (i) $\dfrac{\sqrt{91}x}{3} = \dfrac{\sqrt{91}(y + 3)}{9} = \dfrac{\sqrt{91}(z + 2)}{1}$.

4. (a) $x = 1 - t$ (c) $x = t$ (e) $x = 3 - 4t$ (g) $x = -13 + 14t$
        $y = 0$ $y = t$ $y = -1 + 3t$ $y = 12t$
        $z = 0$ $z = t$ $z = 9 - 4t$ $z = 5 - 5t$
    (i) $x = 3t, y = -3 + 9t, z = -2 + t$.
5. (a) $\cos\theta = 0$; (c) $\cos\theta = 0$; (e) $\cos\theta = -11\sqrt{26}/78$; (g) $\cos\theta = 97/2\sqrt{2701}$.
6. $45°, 45°, 90°$.   7. $107°, 26°, 47°$.
8. $x = 1 + 2t$ 9. $x = 8 + 2t$
    $y = 1 - 5t$ $y = 3 - 3t$
    $z = 1 + 3t$ $z = 2 + t$
13. (a) $z = 0$; (c) $x + y + z = 1$; (e) $x + y + z = 2$;
    (g) $3x - 7y + 3z + 11 = 0$; (i) $6x - 4y + z = 31$; (k) $x - 6y - 4z + 18 = 0$.
18. The intersection is the line whose parametric equations are
    $x = -3t, y = 1 + 2t, z = t$.
19. $x + 3z = 0$.   20. Solve the three equations simultaneously; the point is $(\frac{11}{5}, \frac{-7}{5}, \frac{11}{5})$.
23. $3x - 2y + 9z + 85 = 0$.
25. (a) 6; (c) 2; (e) 2; (g) 2; (i) 85.

section 7.8 (pp. 372–373).

1. (a) $r = 4$; (c) $\theta = \pi/4$; (e) $r = 4\sin\theta$; (g) $r = -4\cos\theta$.

3. (a) $r = 3\sec\theta$; (c) $\theta = \pi/4$; (e) $r = 5$; (g) $r = \dfrac{2\cos\theta}{1 - \cos^2\theta}$;

    (i) $r^2 = \dfrac{36}{4 + 5\cos^2\theta}$.

answers to selected exercises

4. (a) $x^2 + y^2 = 36$; (c) $y^2 = 2x - x^2$; (e) $x = 6$;
   (g) $x^3 + y^2 = 2x + 2\sqrt{x^2 + y^2}$; (i) $(x^2 + y^2)^2 = x^2 - y^2$.

5. $d = [r_1^2 + r_2^2 - 2r_1 r_2 \cos(\theta_1 - \theta_2)]^{1/2}$.

6. $r^2 - 10r \cos(\theta - \pi/3) + 21 = 0$.

index

index

index

index

index